Mites (Acari)
for Pest Control

Mites (Acari) for Pest Control

Uri Gerson[1], Robert L. Smiley[2] and Ronald Ochoa[2]

[1] *Department of Entomology, Faculty of Agricultural, Food and Environmental Sciences, Hebrew University, Rehovot, Israel*

[2] *Systematic Entomology Laboratory, US Department of Agriculture, Agricultural Research Service, Beltsville, MD, USA.*

Blackwell
Science

© 2003 Blackwell Science Ltd

Blackwell Science Ltd, a Blackwell
Publishing Company
Editorial Offices:
Osney Mead, Oxford OX2 0EL, UK
　Tel: + 44 (0) 1865 206206
Blackwell Science, Inc., 350 Main Street,
Malden, MA 02148-5018, USA
　Tel: + 1 781 388 8250
Iowa State Press, a Blackwell Publishing
Company, 2121 State Avenue, Ames,
Iowa 50014-8300, USA
　Tel: + 1 515 292 0140
Blackwell Publishing Asia Pty, 550 Swanston
Street, Carlton, Victoria 3053, Australia
　Tel: + 61 (0)3 9347 0300
Blackwell Wissenschafts Verlag,
Kurfürstendamm 57, 10707 Berlin, Germany
　Tel: + 49 (0)30 32 79 060

First published 2003 by
Blackwell Science Ltd

Library of Congress
Cataloging-in-Publication Data
has been applied for

ISBN 0-632-05658-4

A catalogue record for this title is available
from the British Library

Typeset and produced by Gray Publishing,
Tunbridge Wells, Kent

For further information on Blackwell
Science, visit our website:
www.blackwellpublishing.com

Contents

Acknowledgements

The following colleagues kindly read and commented critically on various chapters: Dr Val Behan-Pelletier (Agriculture and Agri-Food Canada, Ottawa, Ontario, Canada), Dr Heather Proctor (Australian School of Environmental Studies, Griffith University, Queensland, Australia), Prof. Zhi-Qiang Zhang (Landcare Research, Auckland, New Zealand), Dr Yael Argov (The Israel Cohen Institute for Biological Control, Bet Dagan, Israel), Dr Eric Palevsky (Department of Entomology, Volcani Center, Bet Dagan, Israel), Prof. Baruch Rubin and Dr Moshe Coll (Faculty of Agricultural, Food and Environmental Sciences, Rehovot, Israel). The senior author wishes to extend special thanks to Dr R. Mozes-Koch (Faculty of Agricultural, Food and Environmental Sciences, Rehovot, Israel) for help with esoteric terms and methodology. The special assistance of Prof. Zhang and Drs Behan-Pelletier and Ian M. Smith (Agriculture and Agri-Food Canada, Ottawa, Canada) and Dr Calvin C. Welbourn (Florida Department of Agriculture and Consumer Services, Gainesville, Florida, USA) is also appreciated.

For the authorisation to use figures in this book we thank Drs G.W. Krantz (Oregon State University, Oregon, USA), Dr David R. Cook (Paradise Valley, Arizona, USA), Dr Vikram Prasad (Indira Publishing House, Michigan, USA) and Prof. Harold A. Denmark (Florida Department of Agriculture and Consumer Services, Gainesville, Florida, USA). Thanks are also due to Dr M. Bertrand (*Acarologia*, Paris, France), Ms J. Byron (*Hilgardia*, California, USA), Mrs I. de Boer (Kluwer Academic Publishers, Dordrecht, The Netherlands) and Prof. A. Fain (Institut Royal des Sciences Naturelles de Belgique, Bruxelles, Belgium) for permission to use certain figures. In addition, we express our thanks to Christopher J. Fall, BioQuip Products, Inc. (Gardena, California, USA, e-mail: bioquip@aol.com), for permission to use their figures.

Finally, we extend our appreciation to Dr Michael Schauff (Systematic Entomology Laboratory, USDA, ARS, Maryland, USA), for support during the preparation of this publication, and to Ethan C. Kane (University of Maryland, Beltsville, USA) and Eric F. Erbe and Christopher Pooley (Electron Microscopy Unit-SGIL, USDA ARS), for help with figures and SEM photo digital preparations. Special thanks are due to Dr Jack R. Coulson (Director, Biological Control Documentation Center, USDA, ARS, Maryland, USA), for preparing the Foreword.

Abbreviations

ABA	Acarine biocontrol agent
AChE	Acetylcholinesterase
AI	Active ingredient (of a pesticide)
AIRS	Airborne Insect Release System
Bt	*Bacillus thuringiensis*
CGM	Cassava green mite, *Mononychellus tanajoa*
COI	Cytochrome oxidase subunit I
CPB	Colorado potato beetle, *Leptinotarsa decemlineata*
CRB	Carbamates (group of insecticides)
DDT	Dichlorodiphenyltrichloroethane, an OCL pesticide
DNA	Deoxyribonucleic acid
EI	Ecoclimatic index
ELISA	Enzyme-linked immunosorbent assay
ERM	European red mite, *Panonychus ulmi*
FA	Fluctuating asymmetry
F/M	Female to male ratio
GM	Genetically modified (plants)
GST	Glutathione *S*-transferase
HQ	Host quality (model)
IGP	Intraguild predation
IGR	Insect growth regulator
IOBC	International Organisation for Biological Control
IPM	Integrated pest management
km	Kilometre
L/D	Light/dark (hours)
LMC	Local mate competition (model)
Lx	Lux (unit of illumination)
MFO	Mixed function oxidase
MRF	Mass-rearing facility
OCL	Organochlorines (group of insecticides)
OP	Organophosphates (group of insecticides)
OSR	Operational sex ratio (model)
PAGE	Polyacrylamide gel electrophoresis
PCR	Polymerase chain reaction
PE	Postcolonisation evolution
PYR	Pyrethroids (group of insecticides)
RAPD	Random amplified polymorphic DNA
RFLP	Restriction fragment length polymorphism
RH	Relative humidity

RLEM	Red-legged earth mite, *Penthaleus major*
r_m	Intrinsic rate of increase
R_0	Net reproductive rate
RRD	Rose rosette disease
SCR	Southern corn rootworm, *Diabrotica undecimpunctata howardi*
SEM	Scanning electron microscope
SJS	San José scale, *Quadraspidiotus perniciosus*
SPB	Southern pine beetle, *Dendroctonus frontalis*
SR	Sex ratio (proportion of females from total population; females/males + females)
TRM	Tomato russet mite, *Aculops lycopersici*
TSSM	Two-spotted spider mite, *Tetranychus urticae*
TSWV	Tomato spotted wilt virus
VAM	Vesicular–arbuscular mycorrhizal fungi
WA	Western Australia
WFT	Western flower thrips, *Frankliniella occidentalis*

Foreword

The initial publication of *Acarine Biocontrol Agents*, published in 1990, was a most welcomed reference work for practitioners of biological control. A Chinese translation of the book was published in 1996. This new book provides important additional and updated information. As in the first, this new version discusses, in detail, the 34 acarine families (five more than in the first book) that contain mites useful for the control of insects, mites and nematodes, as well as of weeds (a subject new to this edition), and provides an illustrated taxonomic key for their identification. This information and the list of relevant publications are brought up to date.

Mites have been used in various ways for biological control, and a number of species, particularly phytoseiids, are sold commercially throughout the world for biological control. My specific interest in the subject matter relates to the use of mites in 'classical biological control', i.e. the importation of exotic species for the control of introduced pests. The authors have all been involved in taxonomic and other research on mites for many years, with many publications to their credit. Their authorship of such a comprehensive work relating to biological control as this book is most fitting. In regard to my specific interest: Uri Gerson long ago reminded me that the first biological control agent involved in classical biological control was a mite; Robert Smiley is the describer of an acarine parasite introduced by the US Department of Agriculture (USDA) (but not established) in the USA as a biological control agent of the Mexican bean beetle; and Ronald Ochoa has provided comments concerning exotic mites proposed for introduction in North America to me in my role as a member of USDA's Technical Advisory Group for Biological Control of Weeds. Several species of mites have recently been introduced throughout the world for the biological control of weeds, and I am pleased to see this subject covered in the new version.

To conclude, I can only repeat the words of the author of the Foreword to the first edition, Lloyd Knutson, that this updated book 'will be of great value to the biological control worker and to acarologists, the specialist and non-specialist, the fieldperson and the theoretician'. It will certainly be a welcomed addition to the ARS Biological Control Documentation Center library.

Jack R. Coulson
Director, Biological Control Documentation Center
US Department of Agriculture, Agricultural Research Service

Introduction

Biological control provides an environmentally safe, cost-effective and energy-efficient means of pest control, either alone or as a component of integrated pest management. The predatory mites in the family Phytoseiidae are examples of biological control agents that have been recognised only recently as effective components of agricultural systems. The relative slowness with which this fact was recognised suggested to us that other mite groups may be overlooked at present but be capable of serving as effective biological control agents as soon as we gain additional knowledge.

This paragraph, which opened the introduction to the first ever conference on the biological control of pests by mites (Hoy et al., 1983a), was also the starting point of our first book (Gerson & Smiley, 1990) and serves the same purpose in the present, much enlarged version.

In the past there was little communication between acarologists and biological control practitioners; the conference whose proceedings were edited by Hoy et al. (1983a) was convened to serve as a meeting ground for these two groups of scientists. The wide choice of topics discussed attested to advances that had been made as well as to the very many goals that were yet to be attained. One of the major problems recognised was the need to educate pest management practitioners that mites (Acari) could serve as biocontrol agents. Professionals engaged in economic entomology (including medical and veterinary applications) often recognise mites and record their presence in association with pests. However, being untrained in acarology, these practitioners seldom identify the Acari and often ignore them or, at best, refer their specimens to overworked mite specialists. Answers are long in coming and interest raised in the observer could be dissipated by that time, or the situation in the field might have changed. The present volume, as was our first version, is intended to serve as an introduction to the use of mites in the biological control of pests and includes data published up to the end of the year 2000; a few later publications are also included.

This book does not discuss all mite families that have ever been recorded as being associated with pests. Only families with one or more members that are known or are postulated to have an adverse effect on pest numbers and/or to reduce their damage are included, comprising a total of 34 families of acarine biological agents, abbreviated throughout to ABAs. Members of many additional families are known to prey on or to parasitise pests (Smiley & Knutson, 1983), but the lack of quantitative data about their effects (if any) on the prey/host populations precluded their inclusion. Cases in point are some terrestrial and aquatic families of the prostigmatic Parasitengona, and the terrestrial Caligonellidae. The former include several families that could have potential use in pest control (Welbourn, 1983), as well as many families of predatory and parasitic water mites. An example is the Sperchontidae; one of its members was believed by Davies (1959) to have some controlling effect on a group of black flies (Diptera: Simuliidae). As to the Caligonellidae, some species were assumed to feed on the eggs of pestiferous spider mites (Tetranychidae) (Summers & Schlinger, 1955; Dosse, 1967), but no quantitative data are available. Other potential species are noted in Chapter 47.

Also excluded are common but mostly harmless phoretic mites, found on the bodies of numerous pests, including beetles, flies, grasshoppers and termites. When occurring in very large numbers on certain pests (e.g. house flies) they may cause their death and/or a decline in fecundity (e.g. Elzinga & Broce, 1988), but these seem to be isolated cases whose potential for exploitation is not clear. Nor did we include families whose members are natural enemies of non-pest insects and mites and of non-weed plants; when these mites become better known, some could be found to affect pest populations. Insect–mite associations are listed in many general reviews (Costa, 1969; MacNulty, 1971; Lindquist, 1975; Karg & Mack, 1986; Hunter & Rosario, 1988; Houck & OConnor, 1991; Kaliszewski *et al.*, 1995; and others). Thompson and Simmonds (1965) compiled a catalogue of the Acari associated with pests.

The 34 families discussed herein probably represent only a portion of the acarine families that could and should play a role in biological pest control. As noted, biological control is concerned with the manipulation of natural enemies of pests; the ability to recognise these beneficial organisms is an essential first step in this endeavour. We feel that most work is only at the initial, exploratory stage, where the actual recognition of mites active in pest control and of their capabilities is still deficient. We hope that this book will promote further research on exploring, developing and realising the potential of mites in pest control.

Following this Introduction we present a brief historical review of the use of ABAs, provide basic definitions and overview current work. This is followed (in Chapters 2–37) by a discussion of acarine form and structure, basic information on collecting, examining and identifying mites, some notes on rearing and then keys to each of the 34 families (presented alphabetically). Each family is briefly diagnosed, the biology of its members summed up and their use in biocontrol elaborated. Following this 'vertical' approach, we address topics of general, 'horizontal' interest in Chapters 38–47. These include:

- the search for, introduction, mass-rearing and establishment of ABAs
- the effect of sex ratios on the performance of ABAs
- the influence of host plants
- mutual interference between ABAs and intraguild predation
- the effect of pesticides on ABAs
- the desired attributes of ABAs and how to demonstrate their efficacy
- environmental constraints on the use of ABAs and risk assessment
- economic and commercial aspects.

In addition, there are short chapters on ABAs as predators of:

- soil pathogens
- nematodes
- weeds
- scale insects
- stored product pests.

Finally, we propose a few avenues for future work and append a Glossary.

Two chapters from the first version were deleted: Chapter 33, about ABAs that attack locusts and grasshoppers, and Chapter 35, on mites affecting aquatic Diptera of medical importance. The former was removed because extensive exploration for the natural enemies of locusts and grasshoppers failed to show that mites have any important effect on their populations (Lomer & Prior, 1992). In addition, a recent evaluation of the European grasshopper mite, *Eutrombidium trigonum* (Hermann), suggests that this natural enemy has little effect on its hosts (Wohltmann *et al.*, 1996) (for a dissenting view, see Belovsky *et al.*, 2001). Species affecting grasshoppers are discussed in the chapters on the Podapolipidae and Trombidiidae. Chapter 35 was deleted because water mite communities are considered to be affected mainly by abiotic factors, such as geomorphological components, temperature and type of water body (standing or flowing) (Di Sabatino *et al.*, 2000). It follows that reciprocal interactions with pest or host prey are, at best, of minor importance. In addition, little new information on the ability of ABAs to limit the numbers of these pests has been published since 1990; the pertinent data were incorporated into the chapters on water mites (Chapters 7, 20, 22 and 27, as well as 47). Limitations on the use of water mites for pest control are discussed in Chapter 27.

Some replication was unavoidable in this 'vertical' and 'horizontal' presentation. In order to minimise redundancies interested readers are referred to other relevant chapters, as appropriate.

Chapter 1
History, definitions and current work

A brief history

The first mite to be recognised for its ability to reduce pest populations appears to have been *Hemisarcoptes malus* (Shimer), feeding on the oystershell scale, *Lepidosaphes ulmi* (L.) (Hemiptera: Diaspididae). The discoverer (Shimer, 1868a) believed that '... it can be seen that, although it [*H. malus*] may not entirely exterminate the Apple Bark-louse [*L. ulmi*], yet it is exerting a wholesome, restraining influence, doing much more than man could do'. After another year's observations he became more enthusiastic:

> ... the young [mites] appear abundantly in the early spring, and destroy many young bark lice ... This acarian is the most formidable enemy of the apple bark louse extant, and is far more effectual than all other combined restraints in nature, the most zealous efforts of man included ... Last year when I discovered it, I considered it of great importance, and now after fully completing its natural history, I know that its importance is much greater that I then anticipated (Shimer, 1868b).

The recognition of the beneficial effect of *H. malus* prompted its transfer from eastern to western Canada in 1917 in order to control *L. ulmi* (Tothill, 1918b). The project was rated a successful biological control project (Turnbull & Chant, 1961). Further efforts at transferring mites of this genus to various parts of the world are detailed in Chapter 19 (Hemisarcoptidae).

The economic importance of *Hemisarcoptes* was reiterated by Riley (1873), and could have been in the background of the first international shipment of natural enemies, which took place that year. The grape phylloxera (*Daktulosphaira vitifolii* Fitch) had become a major pest of grapes in France at that time. Riley (1874a), who had just discovered a member of the family Acaridae (identified as *Tyroglyphus phylloxerae* Riley & Planchon) feeding on this root-infesting pest in the USA, believed that the predator 'renders efficient aid in keeping it [the grape phylloxera] in check in this country' (Riley, 1874b). The mites were shipped to France, arrived safely, were released, became established, but failed to have any effect on the pest (Howard, 1930). About 100 years later, however, Rack & Rilling (1978) found *Tyrophagus putrescentiae* (Schrank) (Chapter 4, Acaridae) that invaded the leaf galls of the grape phylloxera and fed on its dead and living adults and eggs. As *T. putrescentiae* is a common soil-borne species, there is no way of clearly associating it with the Riley introduction.

Pyemotes (= *Pediculoides*) was recognised in the 1880s as a natural enemy of many insect pests (Webster, 1910). A *Pyemotes* sp. was mass reared and tried against the cotton boll weevil (*Anthonomus grandis* Boheman) in Mexico (Rangel, 1901). That could have been the first report of mass-producing an acarine biocontrol agent

(ABA), as well as suggesting alternate, easier-to-rear hosts for the mite. The *Pyemotes* sp. was introduced into Texas and during the 1903 season released to control the weevil in the field, but it failed to penetrate the cotton squares wherein the weevil feeds and the effort was abandoned (Hunter & Hinds, 1904).

The role of cheyletids in reducing populations of pest mites infesting stored products was recognised by Ewing (1912). He reported that 'in a short period of only a few days this predatory species [*Cheyletus*] had multiplied and destroyed about 95% of the pernicious Tyroglyphids'. *Cheyletus eruditus* (Schrank) was used in early theoretical work on predator–prey interactions (Gause *et al.*, 1936). Nowadays that species is mass-reared and used to control stored food mites in the Czech Republic (Zdárková & Horák, 1990).

Interest in members of the family Phytoseiidae (Chapter 26) as biocontrol agents arose from observations on pestiferous mites. An early record (Parrott *et al.*, 1906) mentioned that *Seius pomi* Parrott (currently placed in the genus *Metaseiulus*) 'was very abundant this season upon infested trees and undoubtedly materially assisted in reducing the numbers of blister mites' [*Eriophyes pyri* (Pagenstecher)]. Students of spider mites recognised the beneficial effects of phytoseiids early on. A phytoseiid (then called '*S. pomi*') was observed by Ewing (1914) to be 'perhaps the most efficient of any of the natural enemies of the red spiders', and, when occurring on infested potted violets in a greenhouse, 'caused the spider mites rapidly to diminish in numbers. The attacks of following [pest] generations caused the red spiders to be practically overcome by their destroyers'. McGregor & McDonough (1917) later noted that *Laelaps macropilis* Banks (now placed in *Phytoseiulus*) 'was responsible for the marked decimation of the red spider which occurred in late summer'. Notwithstanding such early recognition, it was not until 30 years later, with the emergence of spider mites as major crop pests, that interest in the Phytoseiidae was rekindled. Members of this family are currently the main ABAs used in the biological control of mite pests.

Weed control with Acari was first considered in 1924, when the spider mite *Tetranychus desertorum* Banks (then called *Tetranychus opuntiae* Banks), which had accidentally been introduced into Australia, was found to be an important feeder on prickly-pear cactus (*Opuntia inermis*). The mite, initially considered for controlling the cactus because its heavy infestations restricted weed fruiting, was later superseded by more efficient natural enemies (Hill & Stone, 1985). Another spider mite was more recently introduced into New Zealand in efforts to control another weed (Chapter 35). A further Australian discovery of an ABA was *Bdellodes lapidaria* (Kramer) (initially placed in *Biscirus*), recognised as an important predator of the lucerne flea, the collembolan *Sminthurus viridis* (L.), a major pest of pastures (Womersley, 1933). The mite was later introduced into South Africa, where it also controlled collembolan pests (Chapter 9).

Early observations on mosquitoes parasitised by water mites (Howard *et al.*, 1912) led to the consensus that these parasites were not very harmful to their hosts. Uchida & Miyazaki (1935) were apparently the first to consider such mites as important enemies of mosquitoes, as insects bearing more than four mites 'were inactive and did not attack human beings'.

'The economic value of mites as predatory agents in the control of bark beetles has never been realized', wrote Rust (1933), and despite much research (reviewed by Kinn, 1983a), such realisation is still to be achieved.

This brief overview of early biocontrol efforts with mites emphasises the fact that Acari were recognised as natural enemies of pests (including weeds), and employed against them, from the beginning of the emergence of biocontrol as a separate discipline. However, notwithstanding some successes, it was only from the 1950s that Phytoseiidae and other mite families were recognised for their potential as biocontrol agents and began to draw the attention that they deserve.

One measure of the expanding interest in the topic is the increasing number of papers published on ABAs in the acarological literature. These include the periodicals *Acarologia* and *International Journal of Acarology* (both of which publish mainly systematic papers), *Experimental and Applied Acarology* (which does not include systematic contributions) and *Systematic and Applied Acarology*, and the Proceedings of the various International Congresses of Acarology. It should be noted that in many cases no clear distinction was made between biological control *of* mites and biological control *by* mites; students may thus be somewhat confused, like the narrator in Richard Hull's *The Murder of my Aunt*.

Gerson & van de Vrie (1979) presented the first critical overview on the role of mites in the biological control of pests. Xin (1985) later published a short review in Chinese. The topic was elaborated at the aforementioned first-ever conference on the biological control of pests by mites (Hoy *et al.*, 1983a). The first version of this book was published in 1990, and translated into Chinese and published (with a specially prepared update) in 1996 by Fudan University Press (ISBN7-309-01617-3). Other recent summations include reviews by Grout (1998), Lo (1996) and Petanović (1993), and a new book in Chinese by Xin *et al.* (1998). Karg (1994) published a major work on predatory mites in the suborder Mesostigmata.

Definitions

Pests

In the present context pests are invertebrates (insects, mites, nematodes, slugs) that harm their plant or animal hosts to some economic extent (including disease transmission), impair the health of humans or domestic animals and/or are nuisances (e.g. house flies). Also included are weeds, plants that repress and/or successfully compete with agricultural plants and, unless controlled, cause yield reductions or clog up waterways. The present interpretation of pests further incorporates soil-borne pathogenic fungi, as discussed below.

Biological control of pests by mites

Owing to the different terms that are used to describe the same process (e.g. attaining pest control by biological means) by practitioners of the various biocontrol disciplines, the present definition (actually definitions) of biological control requires greater elaboration. In regard to the biological control of invertebrate pests, the current definitions deviate only slightly from the classical approach (e.g. DeBach, 1964; DeBach & Rosen, 1991). Thus, the definition of biological control of pests by mites indicates acarine actions (e.g. predation, parasitism, parasitoidism, phytophagy, competition, disease transmission, other activities and any combination thereof) that reduce pest numbers and/or the extent of their damage to below accepted economic (and medical or veterinary) injury levels. This comprises natural as well as applied (brought about by human intervention) reductions. Such biological control usually comprises three basic strategies: importation, preservation and augmentation.

Importation

This consists of introducing and releasing exotic ABAs in order to control (mostly) introduced pests, also called inoculative releases or 'seeding'. The aim is to establish a stable predator–prey association between the exotic natural enemy and the pest, the numbers of which will be kept below a certain economic injury level. Long-term stability should be brought about by predator-generated reductions in prey numbers that would in turn generate declines in ABA numbers, thus avoiding overexploitation. Pest numbers would then increase again, engendering the growth of predator populations that again reduce pest numbers, and so on. This implies that the ABA should be able to subsist and multiply on the prey (or on available alternate food), and to occupy a similar range of habitats. The importation of ABAs (as well as the first-ever overseas transport of a natural enemy) began with the above-mentioned transfer of '*Tyroglyphus phylloxerae*' from the USA to France. Modern introductions include many phytoseiids from and to different parts of the world (McMurtry, 1982), importing Bdellidae from the Mediterranean region to control pasture pests in Australia (Wallace, 1974) and introducing European Eriophyidae (Chapter 15) into the USA for weed control (McClay *et al.*, 1999).

Conservation

This strategy refers to promoting biocontrol by manipulating the environment. It consists of alleviating or eliminating any factors that may be adverse to ABAs, or adding essential components, or both. Pesticides are the dominant adverse environmental factor, as it is now recognised that the lethal effect of agricultural toxicants on predators was probably the main reason for the great surge in spider mite damage after World War II (Huffaker *et al.*, 1970). This realisation promoted intensive efforts that were intended to reduce the usual mutually exclusiveness between ABAs and pesticides, efforts that are moving along several avenues. The first has several facets. One comprises assaying all agricultural chemicals (insecticides, fungicides and herbicides), those in use as well as those about to be put in use, for their effects on natural enemies (Anonymous, 1998; Sterk *et al.*, 1999). Another facet is using, as far as possible, only pesticides that have little or no effect on ABAs (usually termed 'selective pesticides'). The oystershell scale in eastern Canada (Nova Scotia) is naturally controlled by a hymenopterous parasitoid, *Aphytis*, and by *Hemisarcoptes*. Sulfur treatments that were applied against plant diseases eliminated both enemies, with subsequent upsurges in pest populations and much injury to apple trees. The implementation of a harmonised pest control programme, in which only selective chemicals were used, resulted in the return of both parasitoid and mite, followed by a decline in the pest's populations (Pickett, 1965). Another approach to promote conservation of ABAs is the use of phytoseiids that are resistant to pesticides (Hoy, 1992; Chapter 26). Most phytoseiids that are nowadays available in commerce are resistant to many pesticides. The use of cultural practices to create conditions under which ABA development will be undisturbed is another means of conservation. Populations of the house fly predator, *Macrocheles muscaedomesticae* (Scopoli), which develop in animal manure, often decrease after cleaning out the animal-rearing facility, because they need three to four weeks to become fully re-established in disturbed manure. The harm due to total cleaning could be lessened by gradual or alternate manure removal, leaving refugia of natural enemies within piles or entire rows of droppings (De Jesus & Rueda, 1992; Chapter 23).

ABAs often need shelter and alternate food in order to survive in the field. The weed *Ageratum conyzoides*, planted as cover crop among citrus trees in China, encouraged increases in the number of phytoseiids inhabiting the grove and in their abundance, probably due to ameliorating the climate within the orchard (Liang & Huang, 1994). Many phytoseiids use pollen as an alternate diet. The pollen of *Casuarina* trees provided all required nutrients for *Euseius addoensis* (Van de Merwe & Ryke), an important predator of citrus mites, leading Grout & Richards (1992b) to recommend that these trees should be the first choice for windbreaks in the eastern Cape Province of South Africa (Chapter 41). Conservation as well as the next strategy, augmentation, may encourage exotic as well as indigenous ABAs.

Augmentation

This strategy refers to manipulating the populations ABAs, mostly by increases due to repeated releases. Implicit in using this strategy is the certainty that the ABAs are capable of controlling the pest, but were prevented from exerting their full beneficial effect owing to various factors, such as unsynchronised phenology, prey overexploitation or chemical applications. To reduce the pest increases that may occur in the absence of natural enemies, ABAs are introduced into the system either before the target prey arrives or while its numbers are still low. The predator initially feeds on other diets (e.g. pollen) or, as in the pest-in-first method, on the actual pest (Markkula & Tiittanen, 1976). Overexploitation may occur with predators such as *Phytoseiulus persimilis* Athias-Henriot, which act like 'biotic acaricides', consume all prey and move away. With the predator gone, pest populations rise again and it is necessary to augment predator populations with another release. Disruptions in prey–predator systems after a non-selective chemical application may be remedied by additional releases, or by using ABAs that had developed resistance to toxicants in the field or were selected for this trait by genetic improvement (Hoy, 1992).

Infochemicals, information-bearing chemicals, produced by spider mite-infested leaves, which entice predators to damaged sites, could possibly be incorporated into the plant genome and serve to attract and possibly hold the natural enemies there before any injury occurs (Dicke, 1995). Many other ABAs locate their prey or hosts by chemical means, and might be used to augment their populations in the future.

These are the main components of the biological control of arthropod pests, and all entail direct aggression of natural enemies against the prey. In some cases the 'prey' could consist of soil-borne, or plant-infecting, pathogenic fungi (Enami & Nakamura, 1996; English-Loeb *et al.*, 1999) or of fungi that contaminate fruit (Mendel & Gerson, 1983). ABAs can also transmit diseases of pests (Schabel, 1982), cause pest eggs to fall off branches (McClure, 1995), compete with other herbivores without causing plant damage (Karban *et al.*, 1997) and promote the presence of other predators by serving as their alternate food.

Miscellaneous

Definitions of the biological control of soil-borne pathogens and of weeds will be presented in Chapter 38.

Other methods of control that are based on pest biology or host (whether plant or animal) biology were excluded from these definitions. These methods comprise plant resistance to pests (including that expressed by transgenic plants), sterile-male techniques or mass-trapping of males by pheromones. Nor do we include the non-reproducing pesticidal *Bacillus thuringiensis* (Bt) compounds, even though they are derived from a bacterium.

Integrated pest management

This strategy (usually abbreviated to IPM) is understood to be a pest population management system that uses any suitable and available techniques in a compatible manner, in order to reduce and maintain pest numbers and/or their injury levels below the economic damage threshold levels (see also Kogan, 1998). Implicit in this definition are two important clauses. First, pesticides can be used, but all efforts should be taken to limit their use in order to avoid environmental disruptions. Second, the aim of IPM is not to kill as many pests as possible, but to reduce their numbers and/or to decrease their adverse effects to an acceptable level (as above). Although perceived to be the only rational approach to solving long-term pest problems without undue environmental injury, IPM has not been universally accepted. Lack of suitability to the specific needs of different growers, along with insufficient emphasis on and demonstration of economic benefits, are major obstacles to the much greater adoption of IPM (Wearing, 1988; Trumble, 1998).

Shortcomings of biological control

Having presented these definitions, and despite the authors' being firm proponents of biocontrol, a seldom-discussed caveat is now added. Biological control, although at times claimed to be the only environmentally safe answer to all pest-control problems, is not a panacea. The use of natural enemies (by itself or in the context of IPM) has several drawbacks of which beginning practitioners should be aware; grower education could go a long way to assuage most problems.

First, there are environmental objections to 'classical' biocontrol, especially that exotic enemies may attack indigenous non-target organisms (including other natural enemies and flora), threaten the integrity of the endemic biodiversity or become pests themselves. The introduction of genetically modified ABAs is another hazard. These topics are discussed in Chapter 45.

Not all pests have natural enemies. Arthropods affecting humans or domestic animals, for instance, seem to be almost devoid of specific natural enemies, especially while residing on their hosts. In addition, some agricultural pests have many predators and parasites, but none that exert sufficient control.

If an efficient, commercially available natural enemy (a 'biopesticide', Chapter 46) is at hand, its use may be an attractive, environmentally friendly option. Although rapid pest control can be expected from the 'biopesticide', the practitioner (and grower) should bear in mind, as noted, that with the control of the pest, the natural enemy may also depart. Should the pest then reappear, another release of the 'biotic acaricide' would be needed, just like another chemical application.

When striving for long-term control, the pest- and damage-reducing outcome of applying a natural enemy is seldom immediately evident, and any reductions, when they occur, are less dramatic than the effect of pesticides. When a grower decides to use natural enemies, either the crop should be somewhat tolerant to the pest or an IPM programme has to be initiated (see below). In the first case little or no damage to crop yield is incurred while the natural enemy controls the pest. In the latter case pest populations could first be reduced to below their economic injury levels with pesticides that have a minimal effect on the natural enemy, which is then being released to maintain low pest levels (e.g. Easterbrook, 1992). In all cases, an interval takes place between the introduction of the natural enemy and the resulting control.

When a crop is concurrently attacked by several pests (which is common), natural enemies alone may not suffice, because some pests do not have efficient natural enemies and/or their use is incompatible with other control strategies. Pest control must then be synchronised within IPM programmes. This, in turn, requires more basic as well as practical 'know-how' (e.g. the tolerance of natural enemies to chemicals and economic injury levels) from growers than is needed when using pesticides. The application of IPM also demands more reliance on others, which detracts from the growers' independence.

Another often overlooked issue is that pests are not meant to be eradicated in long-term biocontrol projects; as noted, the interest of the grower may actually lie in keeping small, subinjurious pest populations in the field, to provide prey or hosts for the natural enemies.

Biological control is not always predictable, and its outcome often depends on random and environmental factors. An unexpected spray, a heat wave, a cold spell or strong winds could adversely affect the natural enemies and disrupt ABA activities.

Last, but not least, there may be no economic advantage to using biological control. A large-scale release of the very efficient *P. persimilis* in strawberry beds resulted in significant reductions in spider mite numbers, but there was no difference in yield value between release and control plots (Oatman *et al.*, 1976).

Current studies on ABAs

Research on and application of ABAs is proceeding with many mite groups, targeting diverse pests. A portrait of the ongoing research conducted on ABAs, world-wide, was obtained by accessing computerised databanks (AGRICOLA and CABI Abstracts) and collating publication data on the relevant mite families (Gerson, 2001). AGRICOLA data were available from 1970 to 1997, but those of CABI Abstracts only from 1984. Owing to the structure of the databanks the entire period had to be divided into three parts: 1970–1984, 1984–1991 and 1991–1997. The criteria used for family inclusion were that at least one member was a proven ABA, that at least 40 references were available for that family, and that the bulk of the published work dealt with species that are natural enemies of pests. The latter proviso excluded major groups such as the pestiferous gall and rust mites (Chapter 15) and spider mites (Chapter 34), a few of which had been released for weed control, or the Acaridae, pests of stored products. The Podapolipidae (Chapter 28) were also excluded, because about half of the references were by a single author (R.W. Husband). Data on 30 families were thus collated, but only 11 remained for the next analysis.

In total, 3904 references relating to the 11 families were obtained, but some replication in the resultant data could have occurred, because two databanks were used and the periods overlapped to a certain degree. Then, some papers were tallied more than once because they mentioned members of more than a single family. Consequently, the analysis should be taken only as a comparative representation of research conducted on ABAs during the surveyed period.

The Hemisarcoptidae had the fewest references, whereas the Phytoseiidae the most. The rate of publication pertaining to all families increased steadily during the three periods, because approximately 650 papers were added from period to period, and half were published during 1991–1997. Despite shifts in the emphasis given to various families during the three periods, the Phytoseiidae dominated the literature, as papers on this family accounted for more than half of all publications, maintaining a majority of more than 50% throughout all intervals. The Laelapidae (Chapter 21) came a distant second, and many of the citations probably refer to the bee parasites *Varroa* and *Tropilaelaps*.

Another period-based analysis (Gerson, 2001) was limited to eight families about which over 99 papers had been published. The Cheyletidae, Laelapidae, Macrochelidae, Phytoseiidae (together comprising *c.* 83% of all papers) had a steady growth in numbers of publications from one period to the next. Papers on the Trombidiidae increased greatly during the 1991–1997 period, whereas research on the Cheyletidae was initially steady and then increased. Interest in the Macrochelidae levelled off and studies on the Pyemotidae declined (from 4.1% in the first to 1.8% in the last period). A later analysis of papers that were published during the years 1998–2001 provided similar results, except for great decreases in interest on the Bdellidae and Macrochelidae, and an increase in the Ascidae. As a result, the latter, along with the Cheyletidae, Laelapidae (with the above proviso) and Phytoseiidae, accounted together for around 83% of all publications. Interest in the Phytoseiidae remained steady, 57.9% compared with 56.6% for the period 1970-1997.

The changes in the rate of publication on the various families during these periods reflect biocontrol efforts and successes, and thus also indicate recent advances. The continued dominance of the Phytoseiidae is due to the growing commercial success of its members and to their broadening spectrum of targets. Almost a dozen phytoseiid species, predators of spider mites and thrips, are now in commerce (Anonymous, 2000). In Africa the outstanding, continent-wide, success with phytoseiids was the control of the cassava green mite, *Mononychellus tanajoa* (Bondar) with *Typhlodromalus aripo* De Leon. This success prompted several wildly enthusiastic headlines in the international press. *The Financial Times* proclaimed 'Mighty Mite Cuts Cassava Losses' (14, May, 1997) and *The Washington Post* of 19 May 1997, likewise succumbed to the tempting play on words, entitling its article 'A Mite Strikes Mighty Blow Against World Hunger'. At that time the mite was credited with a 60 million dollar annual increase in cassava profits. Other phytoseiids are being tried to control citrus thrips (Grout, 1994) and whiteflies (Hemiptera: Aleyrodidae) (Chapter 26).

Moving on to other families, interest in the Ascidae was driven by success in controlling mushroom insects (e.g. Rudzińska, 1998). The Laelapidae were also used against mushroom pests, as well as against flower bulb pests (Lesna *et al.*, 1995; Ali *et al.*, 1997; Enkegaard *et al.*, 1997). The predatory Stigmaeidae were more recently recognised as being capable of regulating spider mite numbers at very low levels (MacRae & Croft, 1996). Interest in the Trombidiidae is mainly due to demonstrating their potential to control aphids (Hemiptera: Aphidoidea) (Zhang, 1998). The Cheyletidae are recognised as natural enemies of stored food mites (Zdárková & Horák, 1990) and perhaps other storage pests (Sinha, 1988). The reduced interest in the Macrochelidae could reflect a re-evaluation of their role in filth fly control (Axtell, 1991). The drop in papers on the Pyemotidae may be due to concerns of their causing human dermatitis and to a failure to control the imported red fire ant in the USA (Thorvilson *et al.*, 1987).

Other new research includes studies showing reductions of pest slug populations due to mites (Raut & Panigrahi, 1991) and exploring mites as a factor in lessening damage of plant-pathogenic fungi (English-Loeb *et al.*, 1999). Another advance was showing that pestiferous spider mites were displaced by competing but more benevolent phytophagous members of the same family (Karban *et al.*, 1997). Outstanding recent basic contributions to the study of pest control by mites include the exploration of transtrophic interactions between plants, herbivorous mites and the predators that prey on them, which are mediated by infochemicals (Dicke *et al.*, 1990). An exploration of the role of plant domatia in maintaining predatory mites on leaves (Walter & O'Dowd, 1992) was another discovery, as was an examination of the transformation of a mite that is parasitic on one group of insects into becoming a parasite of another group (Houck, 1994). The ongoing stabilisation of the systematics of the Phytoseiidae (Chant & McMurtry, 1994) should also be noted.

The last decade of the twentieth century witnessed a substantial increase in research on the biological control of pests by mites. In particular, the addition of five families to the roster attests not only to the broadening familial basis for pest control by mites, but also to the widening recognition of their potential in this area. Sadly, however, it is a fact of current scientific life that even as more and more mites are recognised as being capable of reducing pest populations, fewer and fewer practitioners of mite taxonomy and biology are available to provide essential assistance to such efforts. The continuing decline in financial support for systematic biology is bound to impact very negatively on biocontrol efforts and will thus contribute to environmental degradation, in the short as well as in the long run.

Chapter 2
The Acari

Form and function

Introduction

The mites or Acari (including the ticks) are placed in the class Arachnida, which differs from other arthropods by a body composed of two main divisions [cephalothorax (the fused head and thorax) and abdomen], four pairs of legs, and lack of wings and antennae. Mites differ from other arachnids by the total loss of body segmentation, resulting in the complete fusion of their bodies; the apparent segmentation in some mites (e.g. Pyemotidae, see Fig. 3.29) is probably secondary (Baker & Wharton, 1952). The reader is referred to Baker (1999a) and to Smiley (1991) in order to supplement the following concise synopsis of acarine form and function. This brief discourse notes only structures of help to students and biocontrol practitioners in the identification of acarine biocontrol agents (ABAs). Specific details are provided in the key to families, within the boxes added after each family identification (see Chapter 3).

Integument

The acarine integument consists of an epidermis that secretes the overlying cuticle. It is made up of the thin external epicuticle, consisting of cement and wax layers, and the thicker internal procuticle. The integument is usually pale after each moult, but it soon attains the final coloration (which may remain white, or become brown, red, green and combinations thereof). Many sensory organs are cuticular (see below). Most Acari are sclerotised to various degrees, the Cryptostigmata usually being quite dark, whereas the Astigmata, owing to breathing through the integument, are usually milk coloured (but their deutonymphs, or hypopodes, are sclerotised). The integument has been studied and reviewed extensively (Alberti & Coons, 1999); for brevity Krantz (1978) is followed here.

The gnathosoma

The anterior part of the acarine body, the gnathosoma or capitulum, consists of the mouthparts, the external pedipalpi (usually called palpi) and the internal chelicerae, flanking and lying above the oral opening (see Figs 3.1, 3.2). The gnathosoma is separated from the main body (the idiosoma) by a suture. The palpi usually consist of several segments, named like the leg segments (coxa, trochanter, femur, genu, tibia and tarsus), but with the prefix 'palp' or 'palpal'. The apical segments (palpal tibia or palptibia and palpal tarsus or palptarsus) are often armed by spines or claws (Fig. 3.2, apotele; Fig. 3.24). In other cases the palpus is reduced (Figs 3.3, 3.10, 3.34). The chelicerae, which are used for wounding, sucking or chewing, may be dentate (Fig. 3.16), smooth (Fig. 3.20) or needle-like (Figs 3.37, right; 3.44). In some groups the male chelicerae are modified to function as sex organs (spermatactyls), transferring spermatophores to the female (see below). The lower part of the gnathosoma, the hypostome, bears retrose teeth in the Metastigmata (ticks) (Fig. 3.64). In some groups the anterior part of the idiosoma overhangs the gnathosoma, enclosing it in a cavity or camersotome, into which the mouth parts might be retracted (e.g. Camerobiidae). The gnathosoma bears the breathing organs (stigmata) and breathing tubes (peritremes) of many Prostigmata (Figs 3.38, 3.42, right).

Respiration in most Acari is through external openings, called stigmata, located at various positions on the body. They may be dorsal (Prostigmata, Fig. 3.38), ventral (Mesostigmata, Fig. 3.2, and Metastigmata), hidden (Cryptostigmata) or non-existent (Astigmata); their location serves to name the suborders. The larvae of the Cryptostigmata and the parasitengone Prostigmata bear a small ventral breathing pore, or urstigmata, between coxae I and II. Air passes from the stigmata into the body via dorsal or ventral tubes called peritremes; in the Astigmata gas exchange is through the integument. In some Prostigmata the peritremes are located beyond the gnathosoma (Fig. 3.39) or on the prodorsum. The respiratory system of the Acari has been thoroughly reviewed by Alberti & Coons (1999).

The idiosoma

The idiosoma makes up most of the acarine body, and is conventionally subdivided into various regions (see Fig. 3.1). For brevity the dorsal parts of the propodosoma and hysterosoma are herein termed 'prodorsum' and 'hysteronotum', respectively; they may be separated by a sejugal furrow (Figs 3.18, 3.27). The dorsum and venter of the idiosoma can be nude or bear plates (or shields). When both sides of the idiosoma are plated, those on the dorsum are stronger and more variable, and cover part or all the idiosoma. The dorsal plate may be whole or subdivided, its surface smooth (Fig. 3.69) or ornamented (Figs 3.1, 3.11, 3.31, 3.43). The simple eyes, when present, usually consist of one or two pairs (Figs 44 to 3.46), located on the prodorsum. Dorsally the idiosoma bears a variety of sensory receptors, most in the form of setae. The most common are the sensilli, comprising a cuticular structure in the form of a spine, a seta or a pore, which receives external stimuli and conveys them to the nervous system. They sense tactile stimuli (then called mechanoreceptors), chemical stimuli (chemoreceptors) or both; other sensilli are sensitive to humidity. The setae occur in an endless variety of forms and shapes, often being hair-like (Figs 3.9, 3.14), expanded (Figs 3.1, 3.42) or pectinate (Fig. 3.18). Many of these setae also occur on the venter (where they are usually weaker) and on the legs, especially on the anterior appendages.

In addition, specialised sensory setae, or trichobothria, are borne on the pronotum of the Cryptostigmata (Fig. 3.13) and of many Prostigmata [Tarsonemidae, where they are called 'pseudostigmatic organs' (Fig. 3.30), Erythraeidae (Fig. 3.45), Bdellidae, Cunaxidae (Fig. 3.28) and Tydeidae (Fig. 3.31)]. The Eryenetidae have a second pair on the hysteronotum (Fig. 3.32). The trichobothria differ from other setae, being usually longer or club-shaped (Fig. 3.29). As noted, the prodorsum of some Prostigmata bears the breathing organs (Figs 3.29, 3.30).

The acarine venter often carries the setae, stigmata and peritremes (Figs 3.2, 3.4, 3.5) on plates, as well as the genital and anal apertures (Figs 3.3, 3.7, 3.52). These orifices may be conjoined (Fig. 3.3) or separated (Figs 3.2, 3.7, 3.8), usually bearing a constant number of setae.

Some mites reproduce asexually, but most are sexual, sperm transfer being by a variety of direct and indirect ways. Direct transfer is via an aedeagus (many Prostigmata, Astigmata) or by conveying sperm packets (spermatophores) into the female's genital or extragenital apertures (many Mesostigmata, ticks). Indirect sperm transfer occurs when the male deposits spermathophores onto substrates from which they are taken up by the female (e.g. Eriophyidae). Sperm transfer in the Acari was reviewed by Evans (1992).

The legs

Acarine larvae usually emerge with three pairs of legs, the fourth being added before becoming protonymphs (except for the *Eriophyoidea*, Fig. 3.22, with two pairs of legs, and the Podapolipidae, Fig. 3.23b, with three pairs or fewer). Each leg consists of six segments, namely the basal coxa, then trochanter, femur, genu, tibia and the apical tarsus (Fig. 3.2), but segments may be fused (Fig. 3.35) or subdivided (Fig. 3.59). A seventh, terminal segment (the apotele) is usually modified to become the claws and empodium (Figs 3.2, 3.32, left) or only claws (Fig. 3.15), an empodium alone, an empodium in the form of a feather (Fig. 3.22) or as a sucker-like organ (Fig. 3.19). The two anterior pairs of legs carry more sensilli than other legs. Leg I of some mites is longer than the others and devoid of claws (Fig. 3.67). Ticks bear a cluster of sensory organs, called Haller's organ, on their anterior legs (Fig. 3.65).

Collecting ABAs

Many acarologists have developed techniques for collecting, examining and preserving mites; in consequence, the literature abounds with such methods, collated by Baker (1999a), Evans (1992), Hughes (1976) and Woolley (1988). Herein, methods that are more pertinent to ABAs, although applicable to other mites also, are emphasised. Some of this information is derived from the personal experiences and publications of Edward W. Baker and Robert L. Smiley (Retired Acarologists, Systematic Entomology Laboratory, Agricultural Research Service, United States Department of Agriculture at Beltsville, Maryland, USA). Unless otherwise noted, the figures in Chapter 3 were drawn by these scientists.

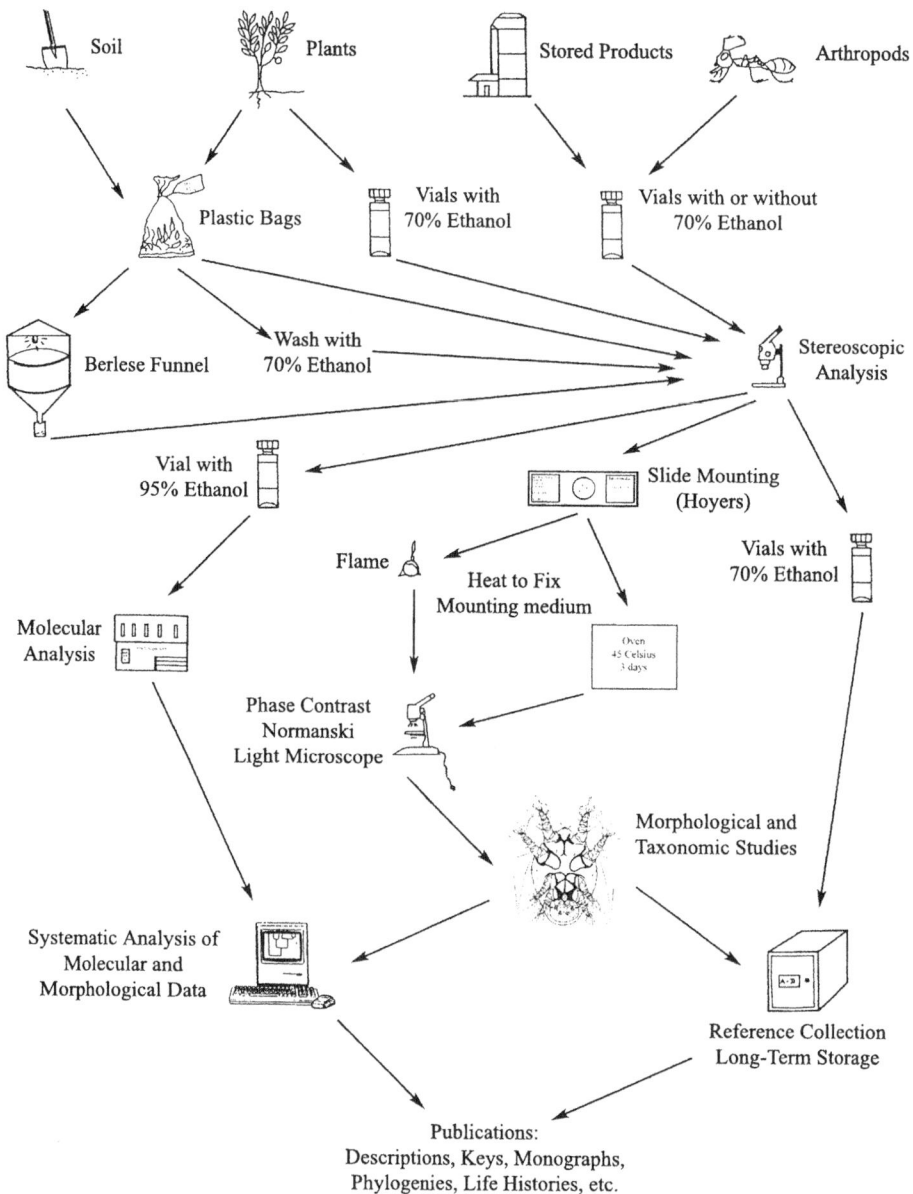

Fig. 2.1   Steps in the collection, processing and identification of mite samples.

Figure 2.1 shows the basics of mite collecting, examining and preservation. Samples may be obtained from the soil, plants, food stores or arthropods, brought to the laboratory and processed by various means, mounted on microscopic slides or kept in alcohol, and examined. Other specimens can be used for chemical or molecular studies (below). The obtained data are recorded and further analysed, as neces-

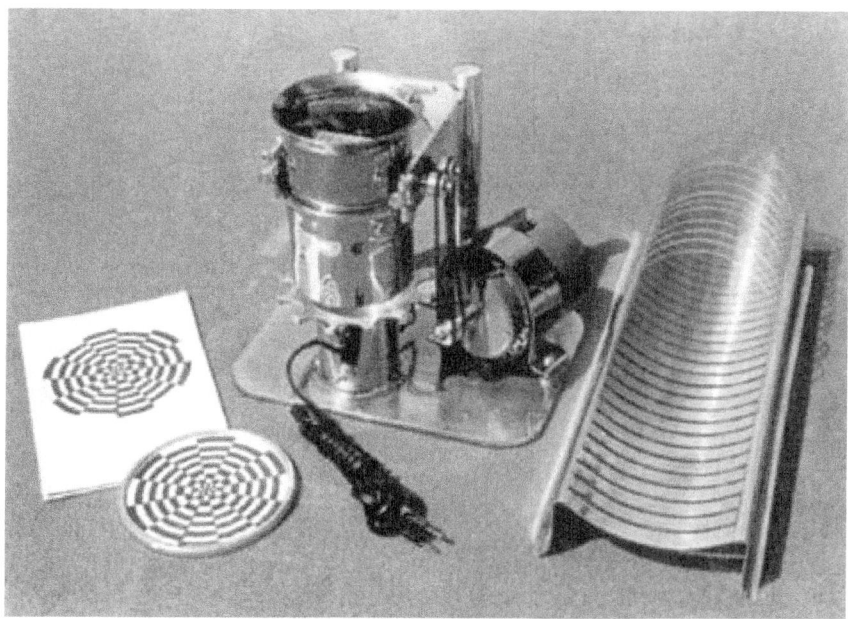

Fig. 2.2   A mite brushing machine. (Courtesy of BioQuip Products, Inc.)

sary. A device that is useful for the qualitative and quantitative collection of mites from plants is the leaf-brushing machine (Henderson & McBurnie, 1943), operated by inserting a leaf between the whirling brushes, first from one end, then from the other. In the BioQuip machine (Fig. 2.2) the sampled leaf is run through a pair of rotating goat hair brushes that deposit the mites onto a plate placed on a rotating table below a basal aperture. The table's rotations ensure the mites' even distribution on the plate, facilitating their fast and accurate counting. This machine was the most rapid and satisfactory method of estimating phytophagous and predatory mite numbers on apple leaves (Morgan *et al.*, 1955). Hugo Aguilar (University of Florida, Lake Alfred, FL, USA, personal communication) obtained similar results when processing strawberry, rose and African oil palm leaves in Costa Rica. The main disadvantages of this method are that it cannot distinguish between live and dead mites, or between the leaf sides on which they had been, and that it obscures variation between leaves within a sample.

Sieves (Fig. 2.3), commonly used to collect arthropods from ground litter, bird nests and similar materials, are also used; screens with 2.5–3.0 mesh/cm suffice for general purposes. The sifted material is placed into the container and shaken over a white pan or white cloth, from which any fallen mites can be collected with an aspirator (below), a brush or forceps. Mites are collected from leaves by beating them onto the sieve that is connected to a funnel, which in turn leads to a container. The latter may hold alcohol (to obtain dead specimens) or wetted plaster-of-Paris (for live mites).

Fig. 2.3   Standard brass and steel test sieves. (Courtesy of BioQuip Products, Inc.)

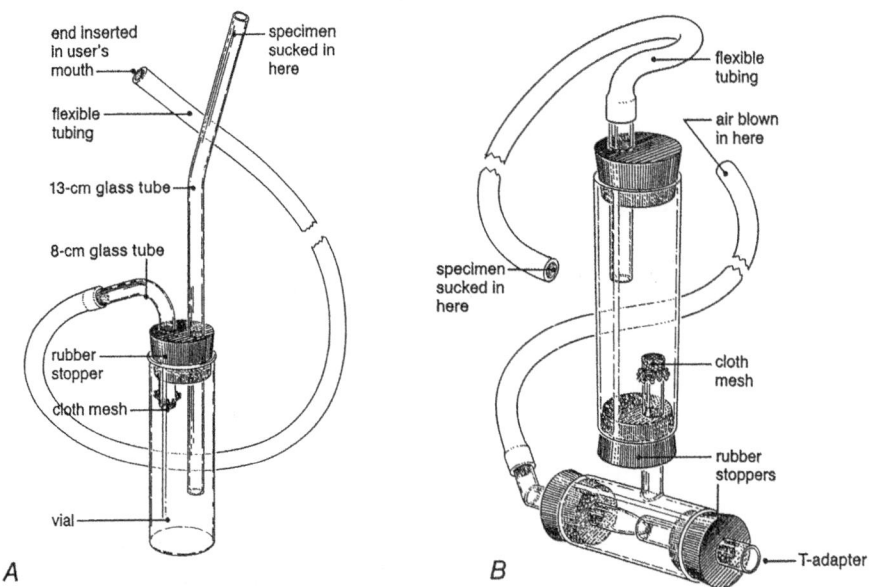

Fig. 2.4   Aspirators. (A) Assembled vial aspirator; (B) tube aspirator with blow-type T-adaptor.

Fine brushes or aspirators (Fig. 2.4A, B) are used to pick mites off surfaces. For full details of their construction and uses, see Evans (1992) and Steyskal *et al.* (1986). The vial aspirator (Fig. 2.4A, known in the UK as 'Pooter') is a convenient device for collecting mites. When using aspirators it is advisable to keep small pieces of absorbent tissue therein, to prevent the accumulation of moisture, and care should be taken not to inhale harmful substances or organisms. To avoid such health risks, as

Fig. 2.5   Backpack aspirator. (Courtesy of BioQuip Products, Inc.)

well as to sample larger surfaces, the faster and more effective backpack vacuum aspirator (Fig. 2.5) is recommended.

The Berlese funnel (Figs 2.6–2.8) is a useful and effective device for collecting mites from ground litter, nests and plant material. Many funnel variations, stationary or portable, have been designed and manufactured; they are cleaner and more efficient than manual sifting. Samples to be processed are placed onto a screen and mites are driven therefrom (by a light bulb) through the body of the funnel, towards collecting containers (as above). Low-wattage bulbs are preferred in portable funnels because their weak heat prevents slow-moving mites from dying before they can fall into the container. BioQuip Products Inc. manufactures several portable funnels; one is fabricated around a 13 litre (3.5 gallon) sturdy bucket (Fig. 2.7) and another, which is collapsible, is made of unbleached muslin, to allow moisture evaporation. Where no power sources are at hand, biodegradable heat packs, which maintain a temperature of around 38–65°C, can be used.

Mites on plants can also be obtained by beating over white or coloured water pans or by washing samples in alcohol. Leaves, parts of branches, fruits or chips of bark are immersed in 70% alcohol for 2–3 days, and the alcohol is then passed through a sieve padded with white or black filter paper, on which the mites may be seen. Acari parasitic on insects can be obtained by inspecting the hosts; individual host examinations also provide quantitative data on parasitisation rates. Free-living water mites (Chapter 7) may be collected with a dip net or the Birge net. The former consists of a narrow mesh bag to scoop up mites that are visible in the water. The Birge net

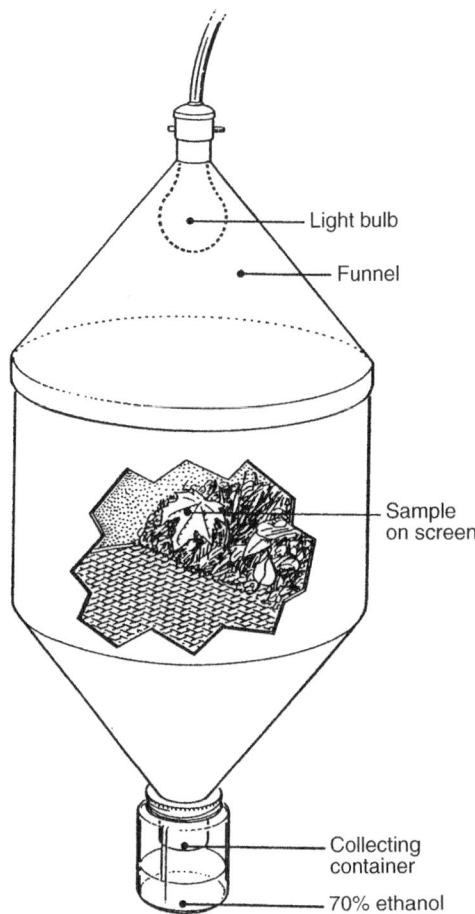

Light bulb

Funnel

Sample
on screen

Collecting
container

70% ethanol

Fig. 2.6   Berlese funnel. (Courtesy of BioQuip Products, Inc.)

(Fig. 2.9) is designed to collect from weedy areas along shores where water mites abound (Baker & Wharton, 1952). The usually reddish Acari are then collected from a white porcelain tray. Barr (1973) summed up methods for collecting and process-ing water mites.

Sticky strips are used to study the insect-mediated transport of ABAs and their wind-borne dispersal (e.g. Phytoseiidae, Chapter 26). Quantitative records of preda-tor presence on plant parts can be kept by photographing leaves with prey and ABAs (Sircom, 2000). The advantage of this method lies in records being available long after they were made, although it may be difficult to distinguish between live and dead ABAs.

Collected mites are often preserved in 70% ethanol (with a drop of glycerine to preserve them in case of evaporation) or in an AGA solution (one part glycerine, one part glacial acetic acid, five parts distilled water and eight parts 95% ethyl alcohol), which prevents the specimens from hardening. To preserve eriophyids

Fig. 2.7   Berlese funnel. (Courtesy of BioQuip Products, Inc.)

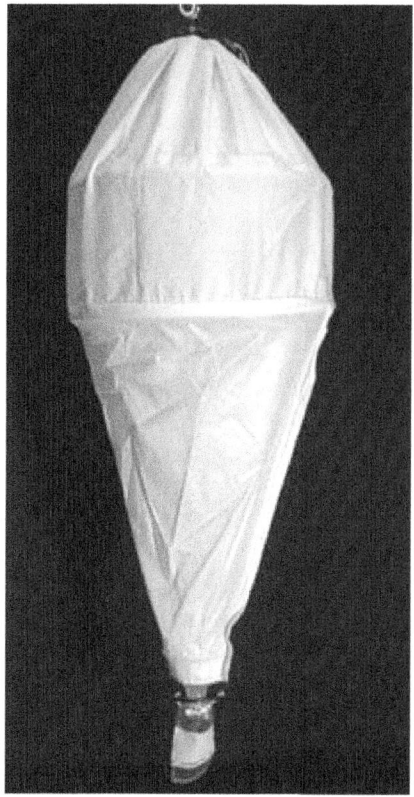

Fig. 2.8   Collapsible Berlese funnel. (Courtesy of BioQuip Products, Inc.)

Fig. 2.9   Birge net for collecting water mites.

(Chapter 15) Jeppson *et al.* (1975) recommended a thin soup of sorbitol in 25% iso-propyl alcohol, to which a small amount of iodine crystals can be added, to aid con-servation and prevent moulding.

Mounting and preserving mites

Mites can be cleared in warm lactic acid or placed directly in mounting solutions and covered by a cover slip. A universal mounting medium for microscopic examinations is Hoyer's solution (20 g glycerine, 30 g gum arabic; 40 g distilled water, 200 g chlo-ral hydrate, and modifications thereof). This fixative is often used for building up ref-erence collections of permanent slides. Hoyer's solution has some advantages over other media, including less discoloration over time and, owing to the solubility of chloral hydrate, the option of remounting. Slides in the authors' collections, fixed in Hoyer's solution and prepared over a period of over 50 years, remain in good condi-tion. Prepared slides are placed in an oven (*c.* 40°C) for 1–2 weeks, to complete clear-ing and harden the fixative. Since Hoyer's solution absorbs moisture, cover slips should be ringed after drying with neutral substances (e.g. Glyptal or other insulating varnishes). The date and site of collection, as well as the collector's name, are then written on a label and stuck on the slide. Damaged or deteriorated slides can be remounted by soaking in warm water within a small dish; the cover slip floats off with-out damaging the specimen, which can be remounted. Fain (1980) developed a method for remounting damaged slides without touching their cover slips.

Examining mites

Improvements made since the 1950s in the microscopic examination of mites include the wide application of phase contrast and Nomarski interference microscopy as well as scanning electron microscopy (SEM). The latter has much improved our ability to study and understand fine details of the acarine external morphology (e.g. Walter & Proctor, 1999), supplementing descriptions based on light microscopy. The best SEM results were formerly obtained when scanning mites freshly killed in liquid air, nitrogen, 5% potassium hydroxide, boiling water or by freeze-drying (Krantz, 1978).

Fig. 2.10 An eriophyid mite (*Aceria anthocoptes*) infesting its weed host, Canada thistle. This mite is a potential control agent of the weed (magnified about 700x).

Wergin *et al.* (2000) developed a new technique, low-temperature scanning electron microscopy (LT-SEM) or cryofixation. A mite is frozen in its natural state on the host, without moving or distortion, allowing the examination of eriophyid mites on their feeding sites at magnifications of ×700 (Fig. 2.10).

Some students prefer to examine mites as wet mounts. Cleared specimens are kept in 70% alcohol or the AGA solution in small vials, which are placed within a jar with the same preservative. Before examination specimens are transferred to a drop of glycerine or lactic acid on a concavity slide and manipulated under the microscope with thin needles.

Identifying mites

Mites are identified to orders, families, genera and species by morphological characters (see Chapter 3 and the above-noted textbooks), which usually suffice for separating the higher categories. The identification of species (and subspecific categories) may, however, require additional methods. Congdon & McMurtry (1985) used mating compatibility for separating species of the phytoseiid *Euseius*, whereas Leal *et al.* (1989) distinguished between four species of the acarid *Tyrophagus* by their unique hydrocarbons. Electrophoretic, immunological and molecular methods have more recently begun to be widely used, as noted below.

Electrophoretic analysis of isozyme patterns and immunological techniques

Hemisarcoptes spp. are natural enemies of armoured scale insect (Hemiptera: Coccoidea), whose deutonymphs (hypopodes) are disseminated by *Chilocorus* spp. (Coleoptera: Coccinellidae) (Chapter 19). Deutonymphs of more than one species may occur on the same beetle, but their separation is difficult. Houck (1989) discerned between two species by using isozymes assayed by horizontal starch gel electrophoresis. The isozyme patterns of adults and deutonymphs of the same species were consistent, despite their vastly different appearance. Electrophoresis even enables the recognition of strains or populations of ABAs. The presence of a pesticide-resistant strain of the phytoseiid *Neoseiulus fallacis* (Garman) was determined from field samples by markers obtained with isoenzyme focusing electrophoresis (Navajas *et al.*, 2001). Solomon *et al.* (1996) discussed the application of electrophoretic methods to detect and quantify the presence of prey remnants in the gut of predators. Immunological techniques are used when the student needs to identify an ABA's range of prey, and are discussed in Chapter 44.

Molecular methods and the polymerase chain reaction

The few examples presented below show that molecular methodology is an exciting development that may have a major impact on acarine systematics, although it is not without pitfalls (Mozes-Koch & Gerson, 1998). The application of the polymerase chain reaction (PCR) to assist in studies of mite phylogeny was demonstrated by Kaliszewski *et al*. (1992), who sequenced DNA from 17 mite species referable to all major taxa (except for Astigmata).

Random amplification of polymorphic DNA

Random amplification of polymorphic DNA (RAPD) has the advantage that no prior knowledge of DNA sequences is needed. It was used to distinguish between three species of the phytoseiid *Typhlodromalus*, based on DNA pooled from five females of each species. Five of the eight tested primers could individually be used to identify the species. Edwards *et al*. (1998) believed that RAPD-PCR, which is quicker than other techniques, would be helpful in separating between cryptic species, provided it was not the only basis for identification.

Genetic markers

The minute size and frequent hidden locations of many eriophyids (Chapter 15) present difficulties in their identification, especially if more than one species occurs in a given habitat. Using PCR of restriction fragment length polymorphism (RFLP) analysis from 10–20 mites, Fenton *et al*. (1995) prepared profiles of the DNA banding patterns of various *Cecidophyopsis* spp. (Eriophyidae) infesting *Ribes*. Navajas *et al*. (1997) used analyses of nucleotide sequences of a fragment of cytochrome oxidase subunit I (COI) to demonstrate phylogenetic relationships among 20 species of spider mites (Chapter 34). The resulting phylogeny was compatible with conventional systematics that are based on morphological characters, and indicated that two species in *Tetranychus* should be placed in *Amphitetranychus*.

Rearing in the laboratory

Methods for mass rearing ABAs are considered in Chapter 39, whereas the present discussion (which omits phytophagous mites) is limited to culturing individual mites in the laboratory. Diverse methods were developed for this purpose, usually dependent on the ABAs' diets and idiosyncrasies (e.g. leaves or waxed paper for walking, soil for hiding, threads for ovipositing). As mites are to be individually observed, rearing units are usually small, often consisting of glass slides or Perspex plates on which rings of a plastic material (or rubber or candle wax) are glued, or Petri dishes, boxes of cover slips and the like. Moisture-holding substrates, such as filter paper, plaster of Paris or peat, are usually placed at the bottom of the containers. Barker (1991) reared stored-product cheyletids (Chapter 12) in microconcavity tissue-culture cells to which seeds (providing hiding sites) and prey were added, whereas Avidov *et al*. (1968) cultured a plant-inhabiting cheyletid within depressions that were drilled into blocks of a plaster-of-Paris and charcoal mixture. The cells were closed with glass slides and wetted as needed. Stigmaeidae (Chapter 32) were reared on leaves placed on a plastic foam pad, saturated with water and confined by absorbent cotton strips (Yue & Tsai, 1995).

ABAs and their prey (or hosts) can be reared on the host plant itself. Abou-Awad & Reda (1992) used leaf discs placed on saturated cotton in Petri dishes as rearing arenas for rust mite prey and their predators. This method was also used for cheyletids that feed on spider mites (Kanavel & Selhime, 1967). An arena with entire citrus leaves and a wick to supply water was devised by Abou-Setta & Childers (1987) to observe phytoseiids. These authors, as well as Overmeer (1985), discussed other mite-culturing methods. Stigmaeidae and Phytoseiidae were studied with their prey on tea leaves placed in vials, the bases of which were open to water in a tray (Oomen, 1982). Gerson & Blumberg (1969) observed Eupalopsellidae (Chapter 17) on lemon fruit within glued-on rubber rings, the tops of which were smeared with machine oil to deter escape. *Hemisarcoptes* spp. were cultured on scale insect hosts grown on hardy fruit (e.g. green lemons), potatoes tubers or cactus pods (Izraylevich & Gerson, 1995d). Brickhill (1958) placed two tydeids (Chapter 36) on leaves or darkened wax paper floating on water in a small, open dish, adding nutrients as needed. Others used leaf discs floating on water or placed on water agar. Sorensen *et al.* (1983) reared Bdellidae (Chapter 9) on discs punched out of vine bark and bordered with petroleum jelly. The discs were fastened with a pin and inserted into jars that maintained a controlled humidity. Erythraeidae (Chapter 16) and Trombidiidae (Chapter 35), which have a long life cycle, were individually confined in glass vials with moist soil or plaster-of-Paris with activated charcoal as substrate, and prey or water added as needed (Young & Welbourn, 1987; Zhang, 1992). Similar methods were used to rear Anystidae (Chapter 6) (Otto & Halliday, 1991), Laelapidae (Chapter 21) (Ydergaard *et al.*, 1997) and Parasitidae (Chapter 25) (Berry, 1973). Macrochelidae and Uropodidae (Chapters 23 and 37, respectively), which feed on house fly juveniles and nematodes, were cultured in small containers with wetted cow manure and prey (O'Donnell & Axtell, 1965).

Parasitic species such as *Pyemotes* (Chapter 30) are reared on their hosts (e.g. beetle pupae) or on an artificial diet (Bruce, 1983). Parasitic water mite larvae (e.g. Arrenuridae, Chapter 7) are collected with their hosts (e.g. mosquitoes) and held until the larvae drop off (Rajendran & Prasad, 1994), and then placed in small containers with suitable prey. Another option is to collect the free-living adults (e.g. Pionidae, Chapter 27), place them, along with debris from their habitat, in water within small vessels, and observe feeding and oviposition (Ellis-Adam & Davids, 1970).

Observing live ABAs

Observations are usually conducted through a stereomicroscope with a 'cold' light source (e.g. fibre optics). When observing several mites in the same cage it is convenient to tag individuals with plastic paints (Hunter, 1960) or non-toxic coloured enamel (Moser & Roton, 1970). When concurrently manipulating more than a single species within a cell, separate brushes or needles should be used.

Behaviour can be observed directly, with closed circuit television or video equipment, which also enables recording and further analysis. Wharton & Arlian (1972) followed the predation process of a cheyletid by closed circuit television. A time series of the behaviour of phytoseiid juveniles and their interspecific interactions was obtained by setting the shutter of a video camera to film the predators at uniform, preselected periods (Palevsky *et al.*, 1999, and Chapter 44). Bowie *et al.* (1999) used a similar system, directly connected to an image-analysis program, which immediately provided a quantitative analysis of an ABA's behaviour in response to pesticide stress.

Chapter 3
An illustrated key to the relevant acarine families

Only 34 out of the several hundred families of Acari that have been named are postulated at this time to play a role in biological pest control. This could become a problem for students trying to use the following key, because it includes only those 34 families. With that in mind, 'boxes' that contain further characters to help in the unequivocal recognition of specific families are placed along with the keys. In addition, a diagnosis precedes each family chapter.

The Acari may be arranged in several orders and suborders that have been given different names by various authors. As these names are still being used in many publications, the systems are listed and compared in Table 3.1.

Table 3.1 Comparative listing of names proposed for the higher acarine taxa (orders and suborders)

System I		System II	System III	System IV
Order	**Parasitiformes**			
	Onychopalpida			
	Holothyroidea	Tetrastigmata	Holothyrida	Holothyrina*
	Notostigmata	Notostigmata	Opilioacarida	Opilioacarida*
Parasitiformes				
	Ixodides	Metastigmata	Ixodida	Ixodida†
	Mesostigmata	Mesostigmata	Gamasida	Mesostigmata
Order	Acariformes			
	Trombidiformes	Prostigmata	Actinedida	Prostigmata
	Tetrapodili			Eriophyoidea
	Sarcoptiformes			
	Acaridiae	Astigmata	Acaridida	Astigmata
	Oribatei	Cryptostigmata	Oribatida	Cryptostigmata

System I: Baker & Wharton (1952); system II: Evans *et al.* (1961); system III: Krantz (1978); system IV: Smiley (1991), which is adopted in the present book.
* It is unlikely that users of this book will encounter mites belonging to these very rare suborders.
† The Ixodida or ticks are included in the key in order to show their position in acarine classification.

Key to orders, suborders and families

1. Body with one to four lateroventral hysterosomal stigmata (Figs 3.2, 3.5)
                                                            29 (Order Parasitiformes)

   Body either devoid of stigmata, or inconspicuous, or stigmata located on or near gnathosoma (Figs 3.1, 3.3, 3.39)   2 (Order Acariformes)

2. Stigmata absent or inconspicuous; in latter case, never located on or near the gnathosoma (Table 3.2, Figs 3.3, 3.7)
                                      3 (Suborders Astigmata and Cryptostigmata)

   Stigmata present, opening either on gnathosoma or on the anterior part of the prodorsum (Table 3.2, Figs 3.1, 3.39)   7 (Suborder Prostigmata)

3. Adult integument strongly sclerotised, body brown to black. Sensilli located on prodorsum (Figs 3.6, 3.13). Genital and anal orifices longitudinal, similar, covered by trapdoor-like valves; genital orifice usually flanked by three pairs of discs (Fig. 3.7). Sexes usually similar. Empodium, when present, claw-like, not borne on pretarsus   4 (Suborder Cryptostigmata)

   Adult integument weakly sclerotised or lacking, body whitish (except hypopus, see box and Figs 3.9–3.12). Genital orifice transverse or U-, V- or Y-shaped, usually flanked by two pairs of discs; anal orifice dissimilar (Figs 3.3, 3.8). Sexes often dissimilar. Empodium claw-like or sucker-like (Figs 3.15, 3.19). No sensilli on prodorsum   6 (Suborder Astigmata)

> The deutonymph or hypopus (Figs 3.9–3.12) of the Astigmata is usually dark brown. It is distinguishable from the Cryptostigmata by lacking mouth parts and prodorsal sensilli and by bearing anal suckers or claspers.

Table 3.2 Additional differences among the three suborders of the Acariformes, intended to separate mites whose stigmata are difficult to see or non-existent

Character	Cryptostigmata and Astigmata	Prostigmata
Tenent hairs on empodium	Absent	May be present
Empodium	Claw-like or sucker-like	Claw-like or pad-like
Chelicerae	Chelate and dendate	Stylettiform
Thumb–claw on palpus	Absent	Often present
Venter of opisthosoma	May have anal claspers or suckers	Never with anal claspers or suckers
Heteromorphic deutonymph (hypopus)	Sometimes present (only in the Astigmata)	Never present

4. Pteromorphs not auriculate (ear-like), or absent   5

   Pteromorphs present, auriculate (Fig. 3.6)   Galumnidae

   > Pteromorphs movable, anteriorly rounded; lamellae weak or lacking; cuti-
   > cle smooth with four porose areas and sclerites around the posterior setae
   > (Fig. 3.7); each genital plate with six pairs of setae.

5. Pteromorphs developed, directed downwards; inner margins of lamellae fastened
   to each other (Fig. 3.13)   Ceratozetidae

   > Genital plate with six pairs of setae; cuspis usually present.

   Pteromorphs little developed or absent; inner margins of lamellae not fastened
   to each other (Fig. 3.14)   Scheloribatidae

   > Genital plate with two to four pairs of setae; cuspis usually absent.

6. Empodial claws large and distinct (except in hypopodes) (Fig. 3.15); chelicerae
   chelate–dendate (Fig. 3.16); setae *sce* and *sci*, when present, inserted on a horizontal plane (Fig. 3.17); genital and anal plates separated, genital plate at level
   of coxae IV (Fig. 3.8)   Acaridae

   > Prodorsum often bears a plate-like sclerite (Fig. 3.17), dorsum divided by a
   > sejugal furrow, hysteronotum with long, slightly barbed setae (Fig. 3.18).
   > Hypopodes without eyes (Fig. 3.9) and without a median ventral sucker
   > anterior to the anal plate (Fig. 3.10).

   Empodial claw absent, replaced by a sucker-like organ (Fig. 3.19); chelicerae
   chelate but not dendate (Fig. 3.20); setae *sce* and *sci* inserted on a vertical plane
   (Fig. 3.21); genital and anal plates confluent (Fig. 3.3)   Hemisarcoptidae

   > Hypopodes with eyes (Fig. 3.11) and a median ventral sucker anterior to the
   > anal plate (Fig 3.12).

7. Found in aquatic habitats, often capable of swimming   25

   Rarely found in aquatic habitats; terrestrial, free-living mites   8

8. Adults with four pairs of legs (Figs 3.1, 3.2)   10

   Adults with one to three pairs of legs (Figs 3.22, 3.23)   9

9. Body worm-like, with two pairs of legs (Fig. 3.22)   Eriophyidae

> Minute mites (0.15–0.35 mm long). Chelicerae stylettiform; genital aperture transverse, located behind legs; tarsi without claws but with a feather-like empodium ('featherclaw').

Body of female sac-like or elongate, but not worm-like; with one to three pairs of legs (except for *Chrysomelobia*) (Fig. 3.23A)   Podapolipidae

> Body of female with very few dorsal setae. Chelicerae needle-like, palpi rudimentary. Larviform female with three pairs of developed legs (Fig. 3.23B).

10. With a distinct palpal claw or palpal thumb–claw process (Figs 3.24, 3.25)   17

    Without a palpal thumb–claw process (Figs 3.26–3.28)   11

11. Cheliceral bases fused or, if not fused, incapable of scissor-like motion over gnathosoma; prodorsum with one pair of sensilli (Fig 3.29–3.31)   13

    Cheliceral bases not fused; chelicerae hinged at base, moving scissor-like over gnathosoma; prodorsum with two pairs of long sensilli (Figs 3.27, 3.28)   12

12 Gnathosoma snout-like, palpi terminating with strong, unbranched setae, without spines, spurs or apophyses (Fig. 3.27)   Bdellidae

> Large (up to 4 mm), red–brown or green mites.

Gnathosoma cone-like, palpi terminating with a claw (except in *Parabonzia*), with or without multibranched setae and with spines, spurs or apophyses (Fig. 3.28)
                    Cunaxidae

> Palpi with three to five segments, tibia IV with one sensillus.

13 Prodorsal sensilli of female usually globular; tarsal claws present or absent on legs II–III, but always with a pulvillus (Figs 3.29, 3.30)   15

    Prodorsal sensilli of female elongate, setaceous or serrate; tarsal claws on legs II–IV (and I when present) with a pad-like empodium (Figs 3.31, 3.32)   14

14 Female prodorsum and legs without net-like ornamentation, hysteronotum without sensilli (Fig. 3.31, left), genital area without discs. Tibia I devoid of an ereynetal organ (Fig. 3.31, right)   Tydeidae

> Chelicerae fused or contiguous, needle-like; idiosoma striated, may be partly or completely reticulated.

Female prodorsum and legs with net-like ornamentation, hysteronotum with sensilli (Fig. 3.32, right), genital area with discs. Tibia I with ereynetal organ (Fig. 3.32, left)   Ereynetidae

> Free-living or parasitic soft-bodied mites.

15   Prodorsum of female with globular or seta-like sensilli   16

Prodorsum of female without sensilli (Fig. 3.33)   Acarophenacidae

> Gnathosoma partly or completely fused into propodosoma, palpi indistinct; dorsum secondarily segmented; anterior ventral plates with three or fewer pairs of setae (Fig. 3.34). Obligate parasites.

16. Leg IV of female with a terminal whip-like seta, no pretarsus, claw and pulvillus (Fig. 3.30); gravid female usually not physogastric   Tarsonemidae

> Dorsum secondarily segmented; leg IV of male usually terminates with a large claw (Fig. 3.35).

Leg IV of female with pretarsus, claw and pulvillus, but without a terminal whip-like seta; gravid female physogastric (Fig. 3.36)   Pyemotidae

> Dorsum secondarily segmented; chelicerae stylettiform; leg I with a claw but usually without empodium; legs II–IV of female bear claws and empodia (Fig. 3.29); trochanter IV subtriangular. Obligate parasites.

17. Chelicerae not whip-like and do not emerge from an eversible stylophore   18

Chelicerae whip-like, emerge from an eversible stylophore (Fig. 3.37)
                                                              Tetranychidae

> Tarsus I often with two pairs of duplex setae; phytophagous mites.

18. Chelicerae and rostrum not fused into a cone; peritremes neither arched nor in an M- or W-shaped configuration   19

Chelicerae and rostrum fused into a cone; peritremes arched or form an M- or W-shaped configuration (Figs 3.1; 3.38)   Cheyletidae

> Gnathosoma prominent with a strong thumb–claw complex; palpal tarsus often bears sickle-like and/or comb-like setae (Fig. 3.38). Dorsal setae in many shapes: hair-like, club-like or shell-like.

19   Peritremes not emergent; a suture usually present between prodorsum and hysteronotum   20

Peritremes emergent above chelicerae; no suture between prodorsum and hysteronotum (Fig. 3.39)   21

20   Palpal tibiae with claw; chelicerae strongly hooked distally (Fig. 3.40); all tarsi with two claws that may be combed, toothed or pilose, and with claw-like, brushlike or bell-like empodia   Anystidae

> Large, fast-running, usually reddish mites; palpal tarsus longer than palpal claw (Fig. 3.41).

Palpal tibiae without claw; chelicerae chelate, weakly developed; all tarsi with two claws and tenent hairs, but without empodia; peritremes emerge sideways above chelicerae (Fig. 3.39)   Pterygosomatidae

> Species that parasitise lizards have a very broad body; others attack scorpions and various insects.

21   Prodorsum with crista metopica and sensilli (Figs 3.45, 3.46)   24

Prodorsum without crista metopica and sensilli   22

22   Peritremes embedded in dorsal surface of stylophore (Fig. 3.42, above); tibiae at least twice as long as respective tarsi (Fig. 3.42, below)   Camerobiidae

> Legs very long ('stilt-like'); weak palpi borne on a ventrally directed gnathosoma.

Peritremes not located on dorsal surface of stylophore; tibiae subequal to tarsi in length   23

23   Cheliceral bases partially or completely separated (Fig. 3.43); empodia (rarely absent) consist of three pairs of tenent hairs as long as claws   Stigmaeidae

> Palpal tarsus may terminate in a three-pronged sensillus.

Cheliceral bases completely fused (Fig. 3.44); empodia consist of one or two pairs of long raylets that are much longer than claws   Eupalopsellidae

> Palpi and chelicerae very long, bearing a reduced palpal thumb–claw complex.

24. Chelicerae long, retractable; crista metopica on prodorsum with two pairs of sensilli (Fig. 3.45). Larvae without urstigmata   Erythraeidae

> Large red mites with long and straight chelicerae.

Chelicerae short, non-retractable; crista metopica on prodorsum with a single pair of sensilli (Fig. 3.46). Larvae with urstigmata   Trombidiidae

> Large, usually red mites covered by a dense coat of setae.

25. Lateral eyes not placed in distinct capsules, or capsules not incorporated into the heavily sclerotised dorsum; palpus not chelate   26

Lateral eyes placed in distinct capsules that lie on the soft, papillate or lined integument (Fig. 3.47); palpus chelate (e.g. distal end of tibia extends beyond insertion of tarsus)   Hydryphantidae

> Palpal tibia bears a dorsal distal process (Fig. 3.48, upper part). Coxal plates in four groups; genital field with acetabula and movable genital flaps (Fig. 3.48, lower part)

26. Palpal telofemur without a ventral seta; leg IV with claws   27

Palpal telofemur with a ventral seta (Fig. 3.49); no claws on leg IV (Fig. 3.50)
                    Limnesiidae

> Coxal plates arranged in four groups of two each, seldom forming a single group. Coxal plates I–III fused, suture lines present; coxal plate IV large, triangular, its acetabula beyond posterior lateral margin (Fig. 3.51).

27. Body heavily sclerotised with closely fitting dorsal and ventral plates   28

  Body usually soft, males may have large dorsal and ventral sclerites that do not form closely fitting plates   Pionidae

> Capitulum separated from coxae, which may be apart or fused into two or three groups. Posterior margin of coxa IV with projections or apodemes (Fig. 3.52). Chelicerae separated medially. Palpal tibia with small setae or protuberances on inner margin and a sclerotised peg at distal end (Fig. 3.53).

28 Genital acetabula on wing-like plates that are incorporated into ventral plate (Fig. 3.54); venter of palpal tibia bulged (Fig. 3.55)   Arrenuridae

> Coxal and ventral plates fused. Genital aperture located beyond coxa IV (Fig. 3.54).

  Genital acetabula lying free in the gonophore (Fig. 3.56); venter of palpal tibia not bulged (Fig. 3.57)   Mideopsidae

> The family Mideopsidae does not include ABAs, but its characters and the accompanying figures are provided to show similar morphological structures among the genera.

> The last four families are water mites, whose parasitic larvae are the stage more likely to be encountered in biocontrol studies. As these larvae are heteromorphic, a key to separate them (without the Mideopsidae) (adapted from Cook, 1974, and from Mullen, 1974) is appended.
>
> 28a. 28a. Basifemur undivided (e.g. leg, excluding coxa, with five segments) (Fig. 3.58)   28b
>
>   Basifemur divided (e.g. leg, excluding coxa, with six segments) (Fig. 3.59)   Hydryphantidae
>
> 28b. Coxal plates fused, but suture lines evident between coxae   28c
>
>   Coxal plates separated from each other by a medial membranous area   Arrenuridae
>
> 28c. Coxal plate I clearly separated from fused coxal plates II–III (Fig. 3.60)   Pionidae
>
>   Coxal plates I–III fused, but suture lines evident (Fig. 3.61)   Limnesiidae

29. Non-leathery mites whose adults are usually <1 mm long; hypostome without retrose teeth; palpus with a tined apotele (Fig. 3.62); tarsus I without Haller's organ (Fig. 3.63); stigmata located either between coxae II and III or between coxae III and IV (Figs 3.2, 3.4)   30 (Suborder Mesostigmata)

    Large, leathery mites (adults usually >1 mm long); hypostome armed with retrose teeth (Fig. 3.64); stigmata located either between coxae III and IV or beyond coxae IV; palpus without a tined apotele; Haller's organ located on tarsus I (Fig. 3.65)   Suborder Metastigmata or Ixodides, ticks

    > Ticks, external parasites of vertebrates, are not known to affect populations of agricultural or veterinary pests, although they may parasitise pests (e.g. horse flies, Tabanidae; Boshko & Skylar, 1981). Being vectors of many diseases, ticks were listed (among arthropods) as having potential for biological warfare (Lockwood, 1987). Their use against especially noxious vertebrate pests might thus be considered, subject to strict environmental considerations.

30. Coxae III and IV without leg grooves; stigmata located between coxae III and IV
                                                                    31

    Coxae III and IV with leg grooves; stigmata located between coxae II and III (Fig. 3.4)   Uropodidae

    > Dark brown soil mites whose deutonymphs are often dispersed by beetles.

31. Epigynial or genital plate not triangular (Fig. 3.67)   32

    Epigynial or genital plate triangular (Fig. 3.66)   Parasitidae

    > Venter with two large metasternal shields, located in front of the triangular epigynial plate; peritremes straight.

32. Leg I with claws and empodia; peritremes not looped around stigmata   33

    Leg I without claws and empodia; peritremes looped around stigmata (Fig. 3.67)
                                                                    Macrochelidae

    > Legs I longer and thinner than other legs; females bear small metasternal plates.

33. Ventrianal plate present, surrounding the subterminal anus   4

    Ventrianal plate absent (Fig. 3.68)   Otopheidomenidae

> Weakly sclerotised parasites of insects with a reduced number of dorsal setae and very short peritremes.

34. Dorsum with more than 23 pairs of setae   35

    Dorsum with fewer than 23 pairs of setae (Fig. 3.69)   Phytoseiidae

> Dorsal plate entire; legs I with ambulacra; metasternal plates small (Fig. 3.70).

35. Epigynial plate with four pairs of setae; ventrianal plate with only three setae (Fig. 3.2)   Laelapidae

> Female genital plate flask-shaped (Fig. 3.2); metasternal shields small; peritremes straight; legs I with ambulacra.

    Epigynial plate with one pair of setae; ventrianal plate with more than three setae (Fig. 3.5)   Ascidae

> Female genital shield not flask-shaped; peritremes straight; metasternal shields small; legs I with ambulacra.

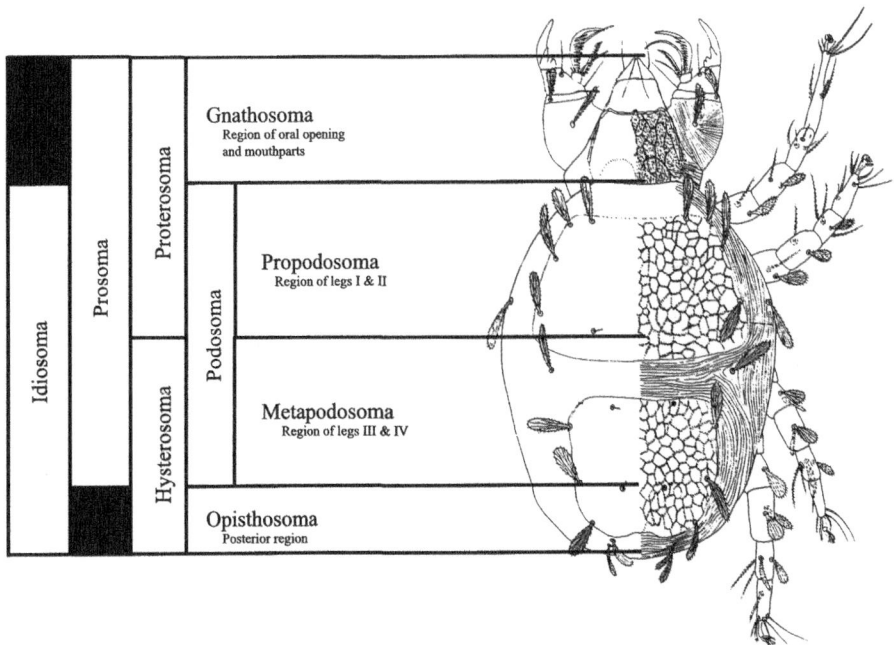

Fig. 3.1   Prostigmata, Cheyletidae, *Cheyletus misonnei*, female, dorsal view, indicating the sub-divisions of the acarine body. (From Gerson *et al.*, 1999, with kind permission from Prof. Alex Fain.)

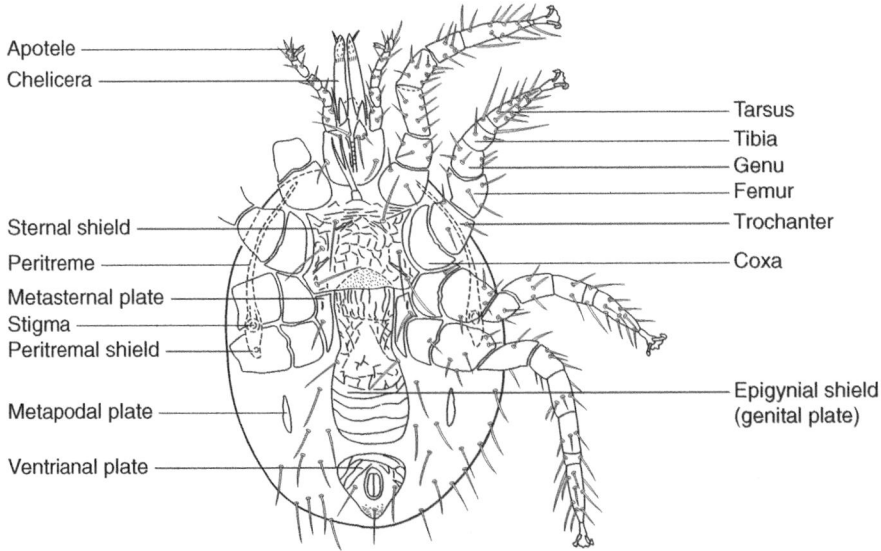

Fig. 3.2   Mesostigmata, Laelapidae, *Androlaelaps casalis*, female, ventral view.

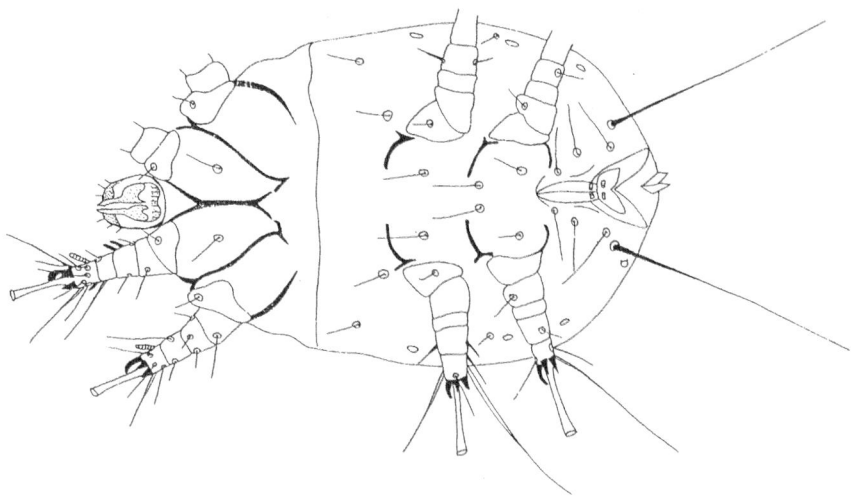

Fig. 3.3 Astigmata, Hemisarcoptidae, *Hemisarcoptes coccophagus*, female, ventral view. (From Gerson & Schneider, 1981, with kind permission from *Acarologia*.)

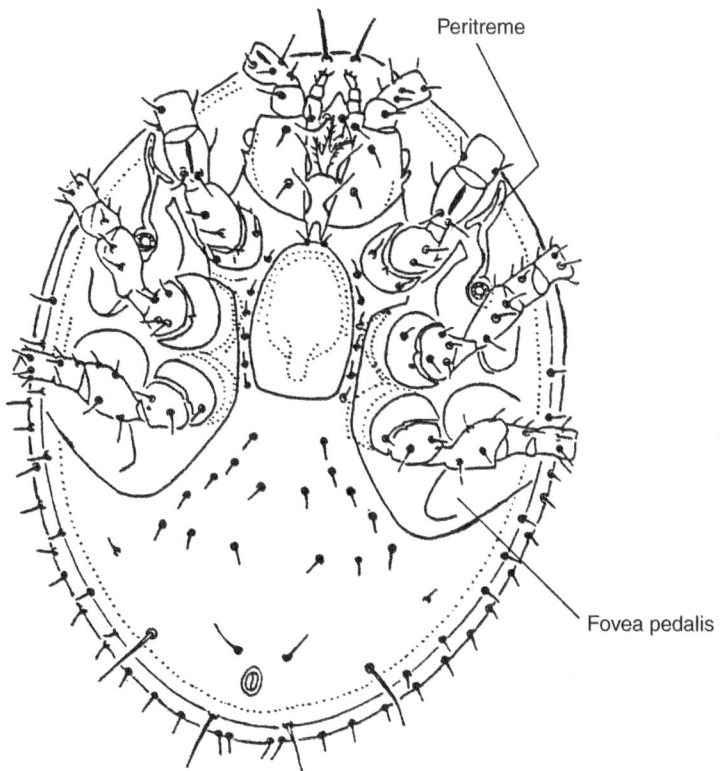

Fig. 3.4 Mesostigmata, Uropodidae, *Uroobovella marginata*, female, ventral view.

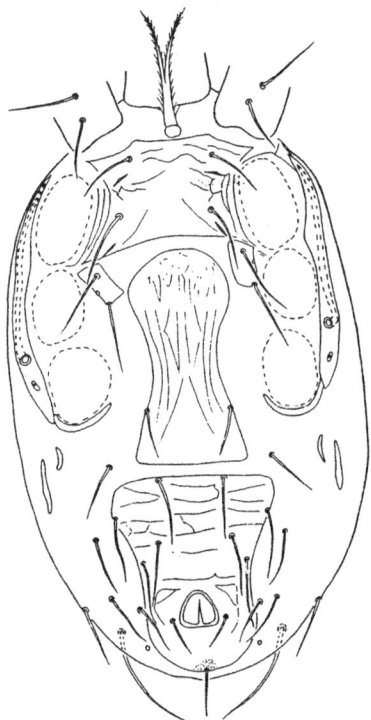

Fig. 3.5   Mesostigmata, Ascidae, *Blattisocius* sp., female, ventral view.

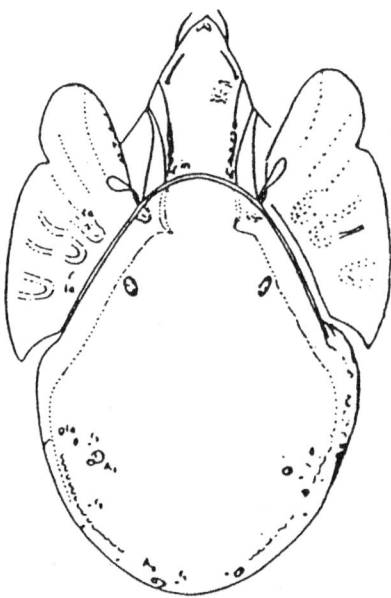

Fig. 3.6   Cryptostigmata, Galumnidae, *Orthogalumna terebrantis*, female, dorsal view.

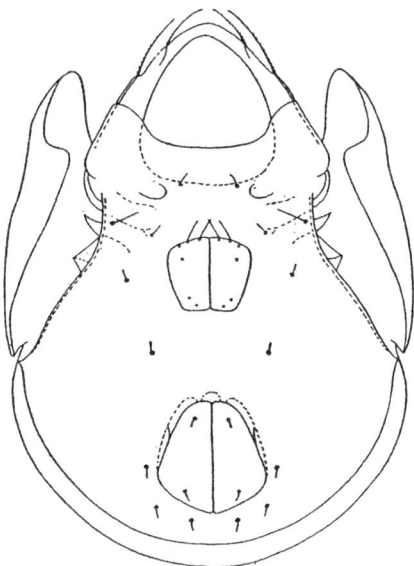

Fig. 3.7 Cryptostigmata, Galumnidae, *Galumna virginiensis*, female, ventral view.

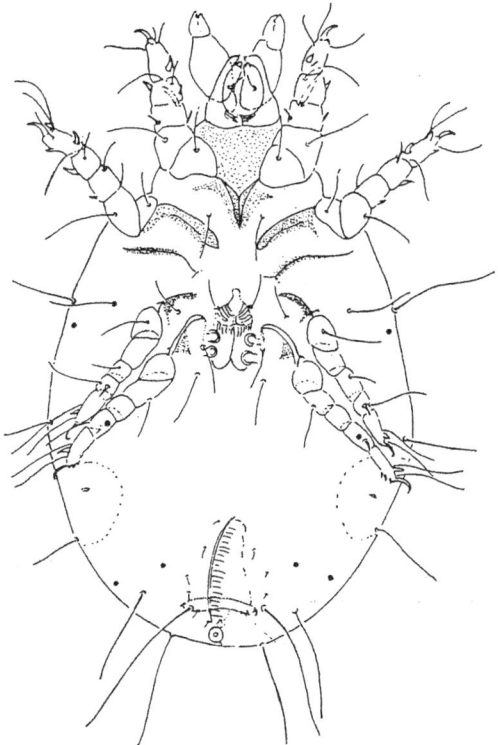

Fig. 3.8 Astigmata, Acaridae, *Rhizoglyphus echinopus*, female, ventral view. (From Manson, 1972, with kind permission from *Acarologia*.)

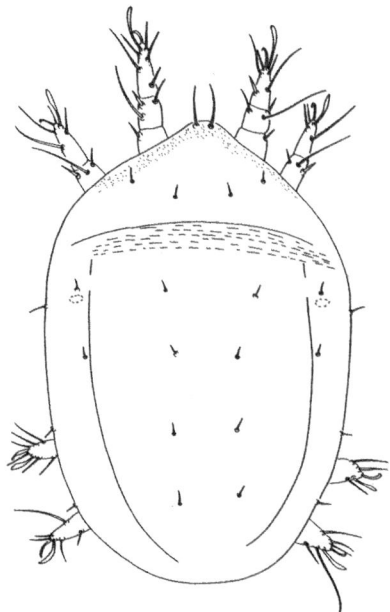

Fig. 3.9   Astigmata, Acaridae, *Rhizoglyphus robini*, hypopus, dorsal view.

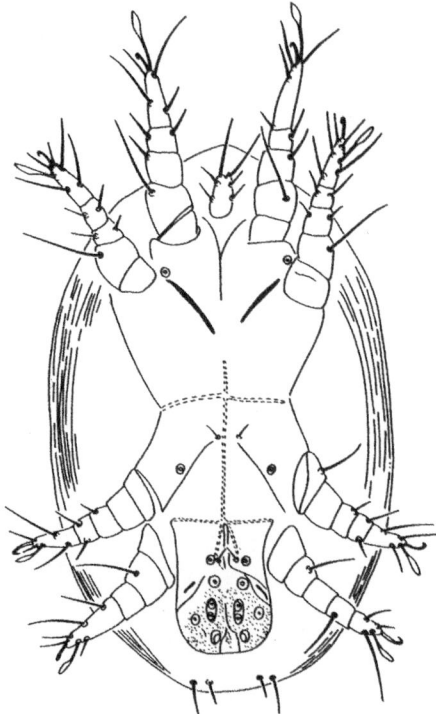

Fig. 3.10   Astigmata, Acaridae, *Rhizoglyphus robini*, hypopus, ventral view.

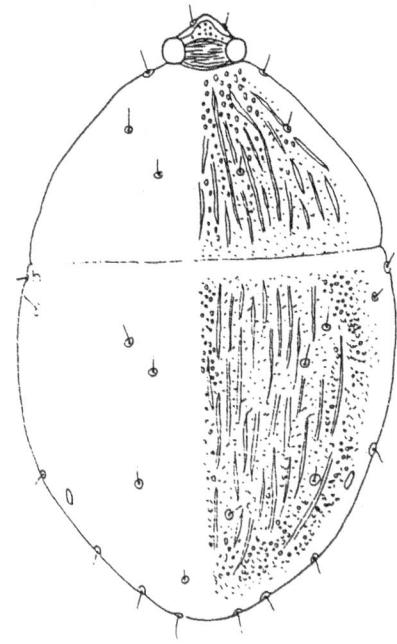

Fig. 3.11 Astigmata, Hemisarcoptidae, *Hemisarcoptes coccophagus*, hypopus, dorsal view. (From Gerson & Schneider, 1982, with kind permission from *Acarologia*.)

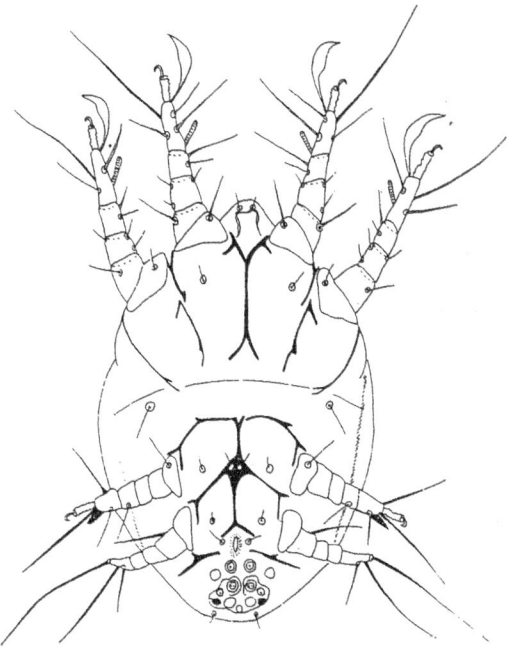

Fig. 3.12 Astigmata, Hemisarcoptidae, *Hemisarcoptes coccophagus*, hypopus, ventral view. (From Gerson & Schneider, 1982, with kind permission from *Acarologia*.)

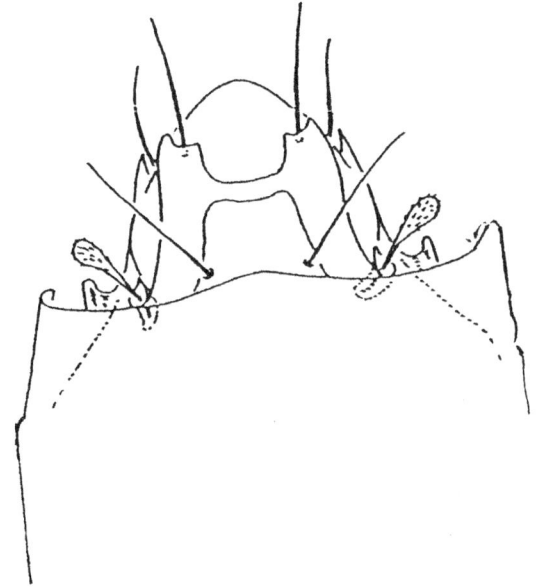

Fig. 3.13 Ceratozetidae, dorsal view.

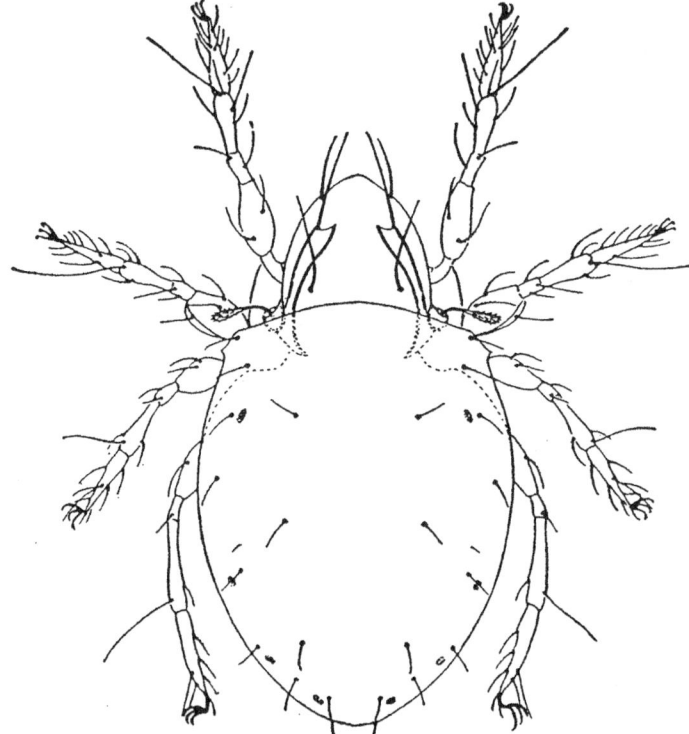

Fig. 3.14 Scheloribatidae, dorsal view.

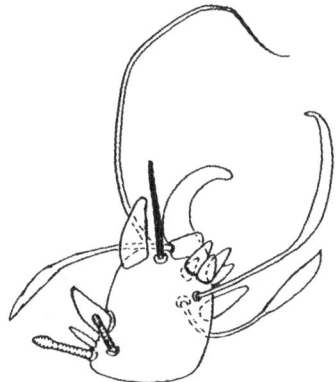

Fig. 3.15   Acaridae, *Rhizoglyphus robini*, tarsus I.

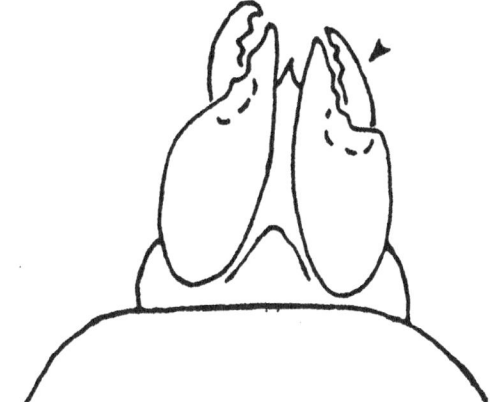

Fig. 3.16   Acaridae, chelicerae.

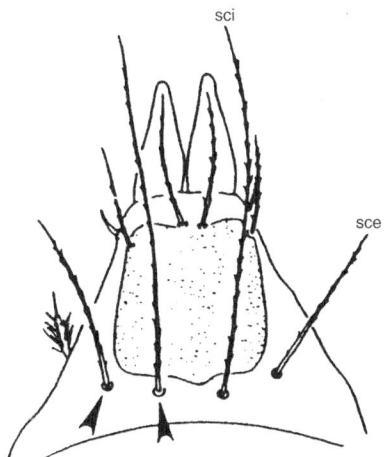

Fig. 3.17   Acaridae, *Tyrophagus putrescentiae*, prodorsum.

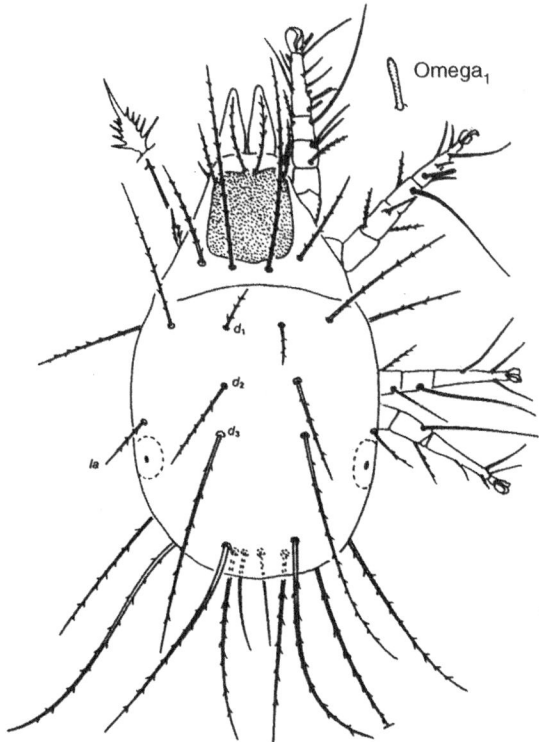

Fig. 3.18 Acaridae, *Tyrophagus putrescentiae*, female, dorsal view.

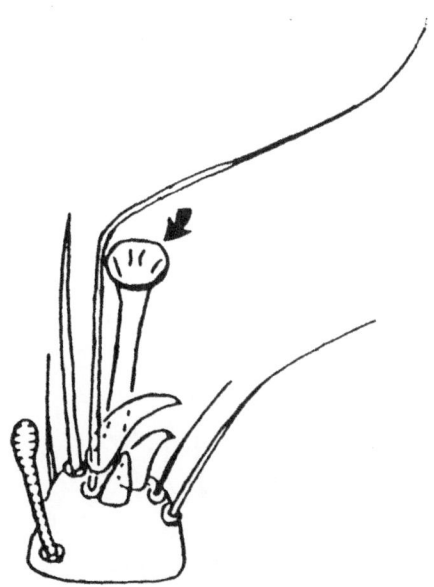

Fig. 3.19 Hemisarcoptidae, *Hemisarcoptes malus*, tip of tarsus I.

Fig. 3.20 Hemisarcoptidae, *Hemisarcoptes malus*, a chelicera.

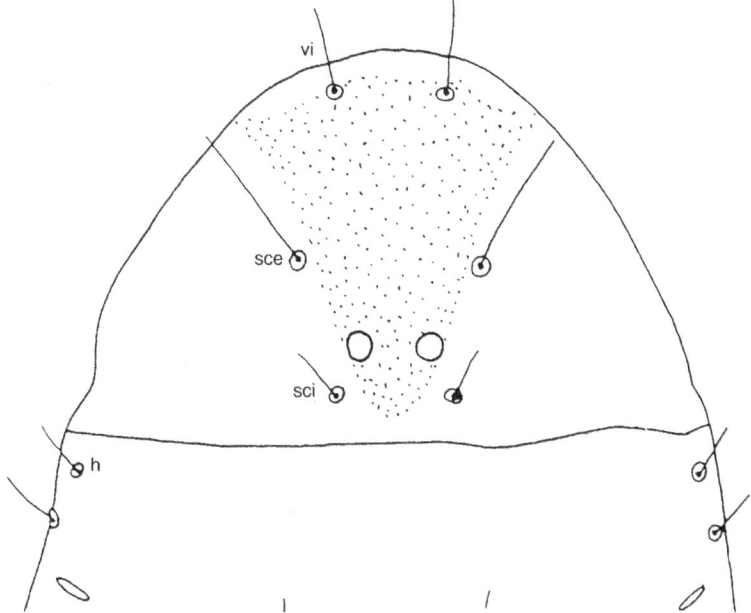

Fig. 3.21 Hemisarcoptidae, *Hemisarcoptes coccophagus*, prodorsum. (From Gerson & Schneider, 1981, with kind permission from *Acarologia*.)

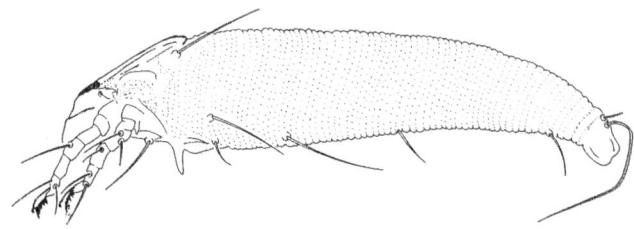

Fig. 3.22 Eriophyidae, *Eriophyes ficus*; female, lateral view.

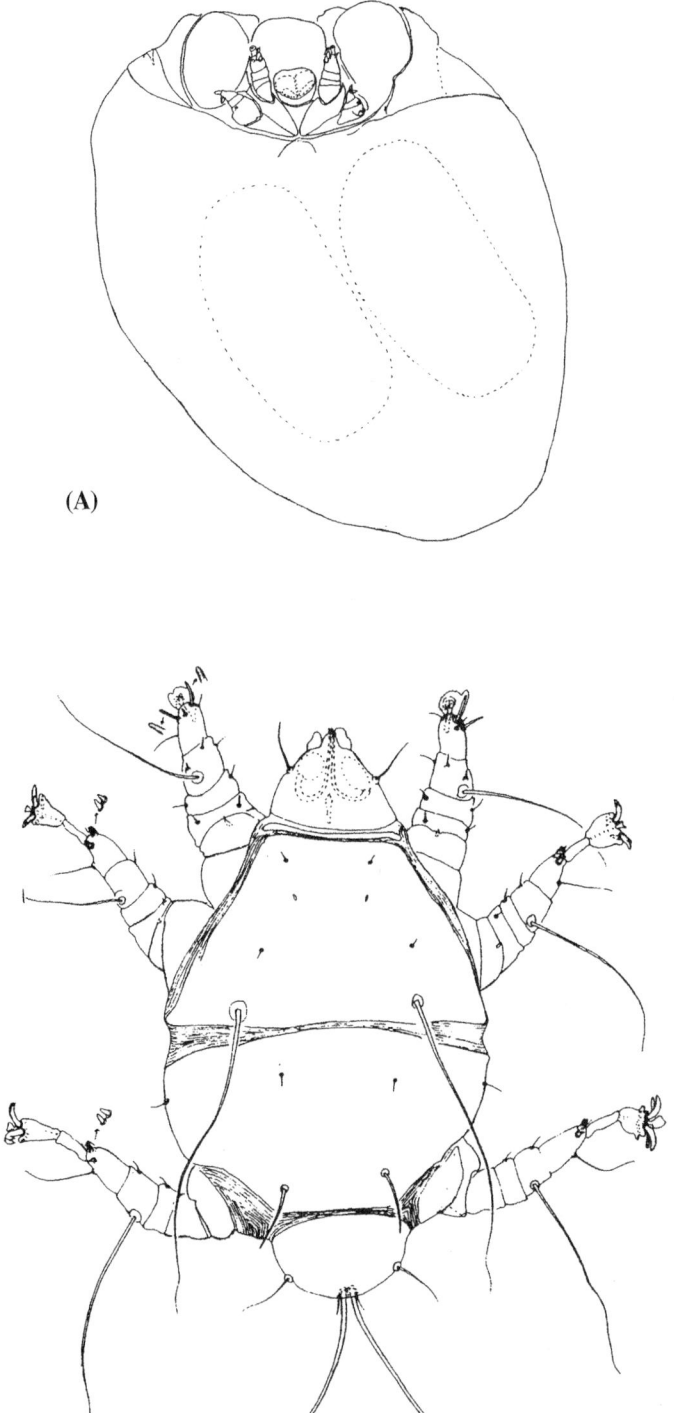

Fig. 3.23   Podapolipidae, *Coccipolipus epilachnae*. (A) female; (B) larviform female.

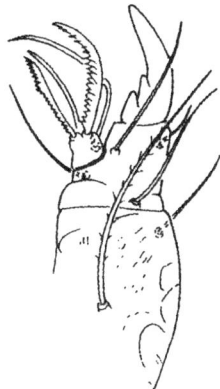

Fig. 3.24 Cheyletidae, palpus, dorsal view.

Fig. 3.25 Tetranychidae, palpus, dorsal view.

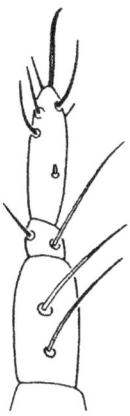

Fig. 3.26 Tydeidae, *Tydeus*, palpus.

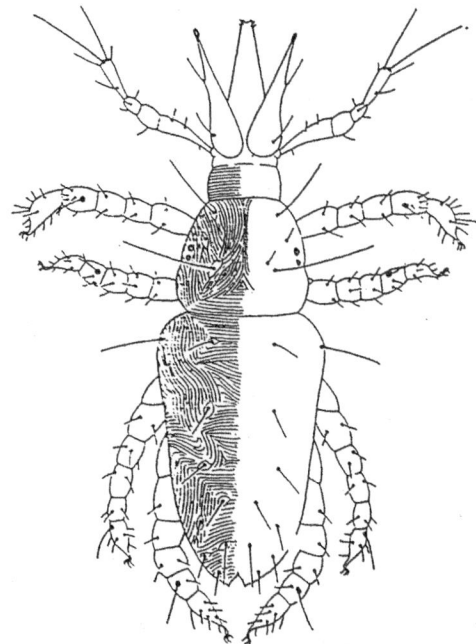

Fig. 3.27   Bdellidae, *Bdella willisi*, female, dorsal view.

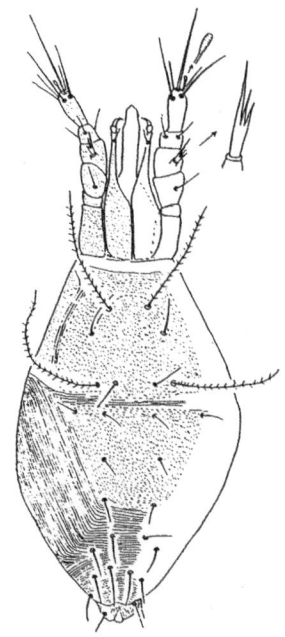

Fig. 3.28   Cunaxidae, *Parabonzia bdelliformis*, dorsal view of gnathosoma and propodosoma.

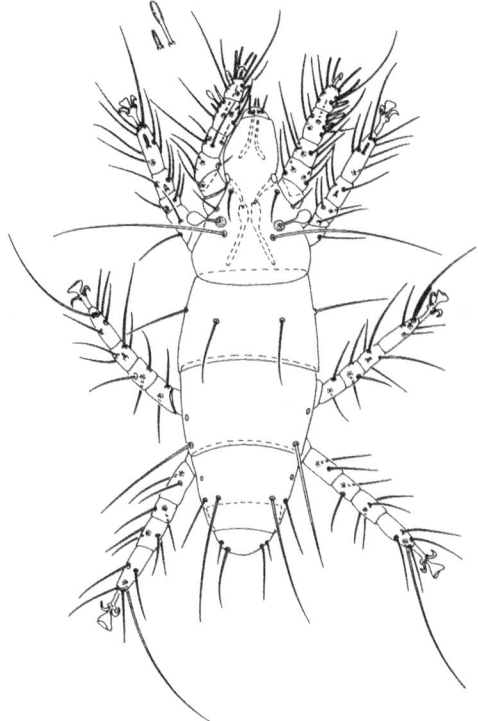

Fig. 3.29   Pyemotidae, young female, dorsal view.

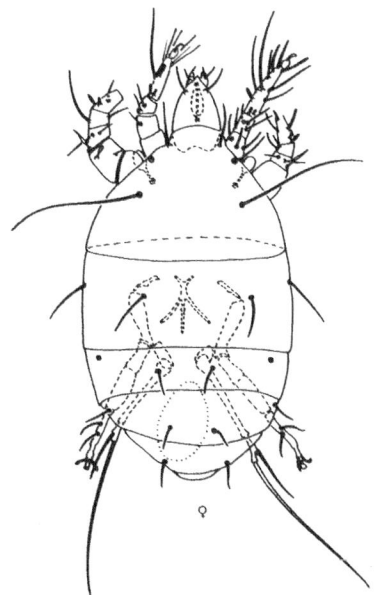

Fig. 3.30   Tarsonemidae, female, dorsal view.

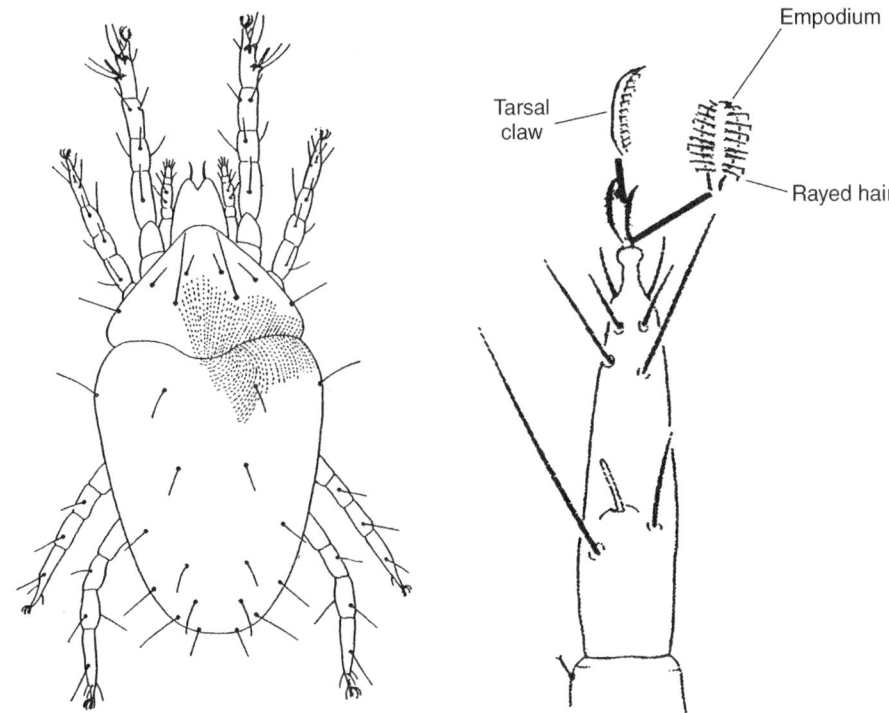

Fig. 3.31 Tydeidae, *Tydeus starri*; left: female, dorsal view; right: tarsus I.

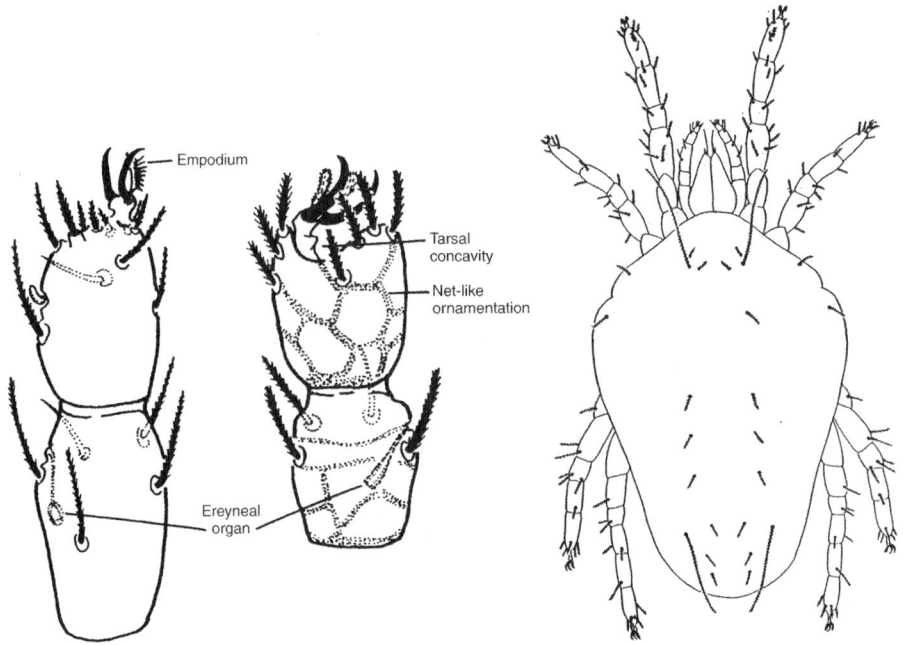

Fig. 3.32 Ereynetidae, *Opsereynetes robustus*; right: female, dorsal view; left: tibia and tarsus of leg I, lateral views. (From Krantz, 1978, with kind permission from Prof. G.W. Krantz.)

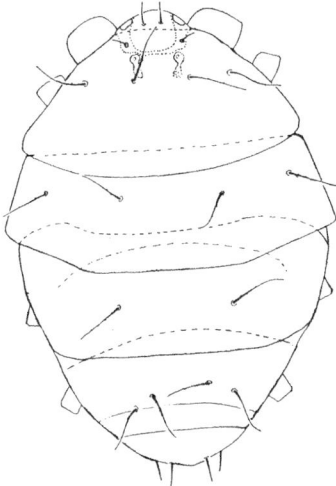

Fig. 3.33   Acarophenacidae, *Acarophenax nidicolus*, female, dorsal view. (From Cross & Krantz, 1964, with kind permission from *Acarologia*.)

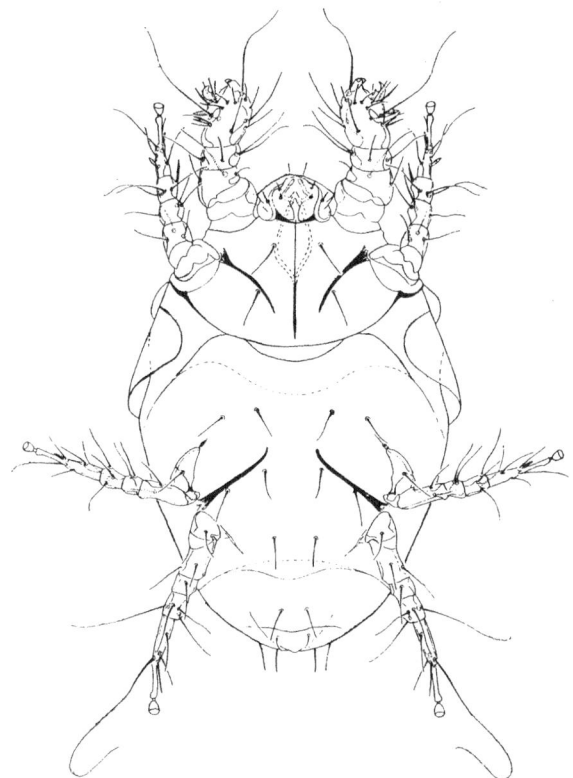

Fig. 3.34   Acarophenacidae, *Acarophenax nidicolus*, female, ventral view. (From Cross & Krantz, 1964, with kind permission from *Acarologia*.)

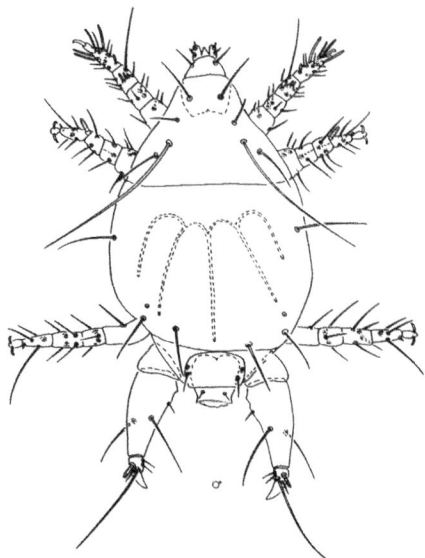

Fig. 3.35 Tarsonemidae, male, dorsal view.

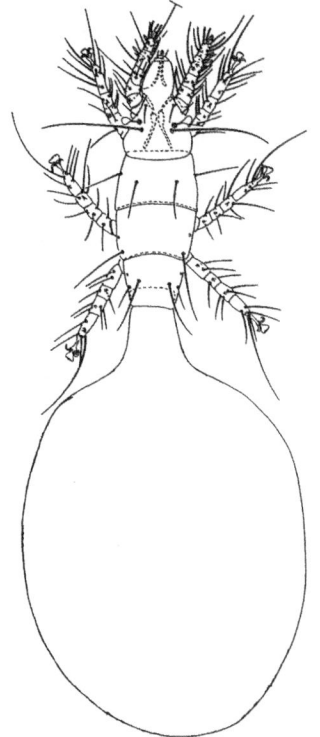

Fig. 3.36 Pyemotidae, physogastric (gravid) female, dorsal view.

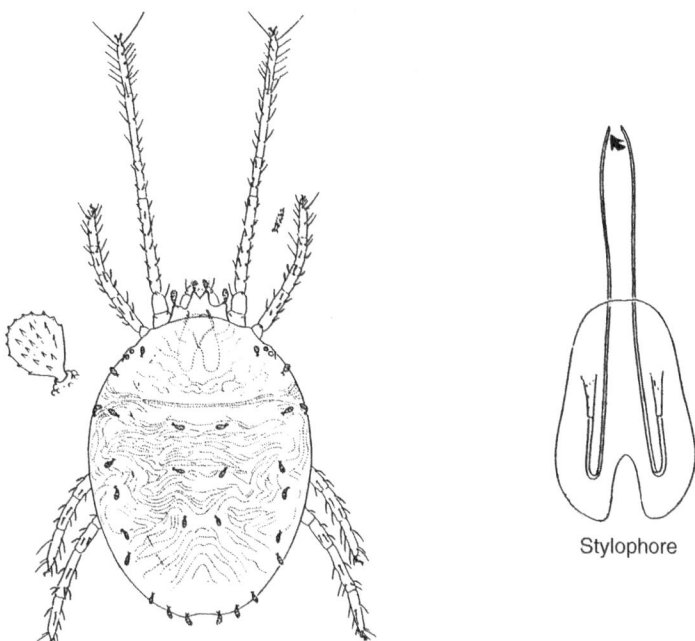

Fig. 3.37 Tetranychidae, *Bryobia praetiosa*; left: female, dorsal view; right: stylophore and chelicerae.

Fig. 3.38 Cheyletidae, *Cheyletus eruditus*, female, dorsal view of gnathosoma.

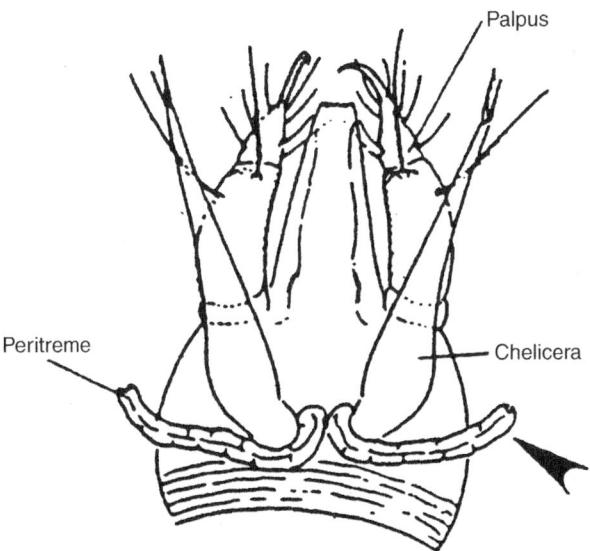

Fig. 3.39   Pterygosomatidae, *Pimeliaphilus cunliffei*, dorsal view of gnathosoma.

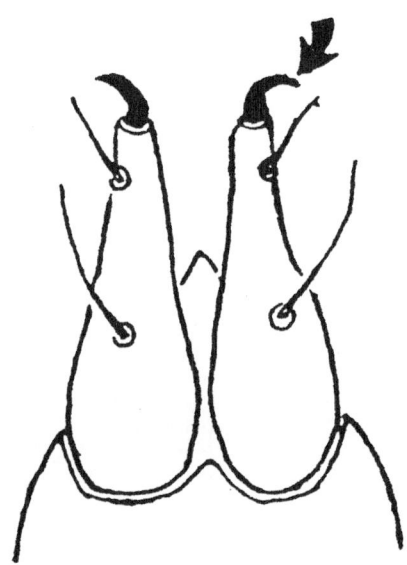

Fig. 3.40   Anystidae, *Anystis sp.*, chelicerae with terminal hook.

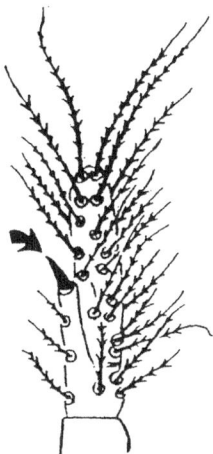

Fig. 3.41 Anystidae, *Bechsteinia* sp., palpal tibia and tarsus.

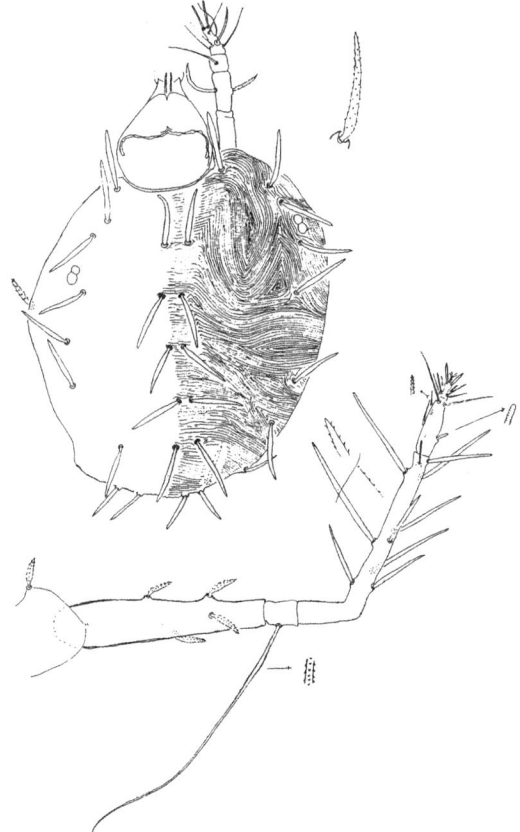

Fig. 3.42 Camerobiidae, *Neophyllobius lorioi*; above: female, dorsal view; below: leg.

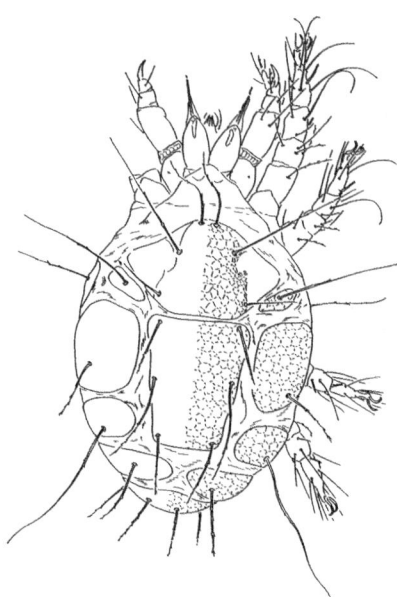

Fig. 3.43 Stigmaeidae, *Stigmaeus scaber*, female, dorsal view. (From Summers, 1962, with kind permission from *Hilgardia*.)

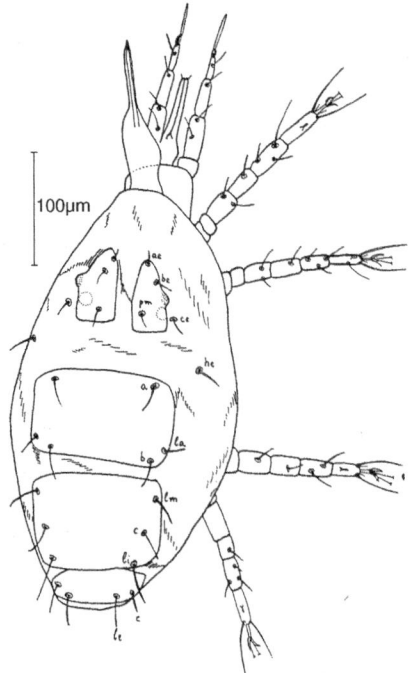

Fig. 3.44 Eupalopsellidae, *Eupalopsis maseriensis*, female, dorsal view.

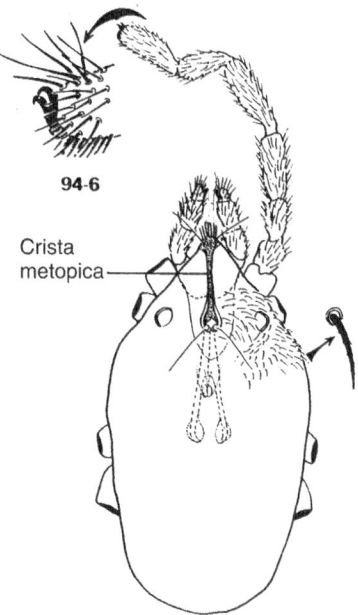

Fig. 3.45 Erythraeidae, *Balaustium*, female, dorsal view. (From Krantz, 1978, with kind permission from Prof. G.W. Krantz.)

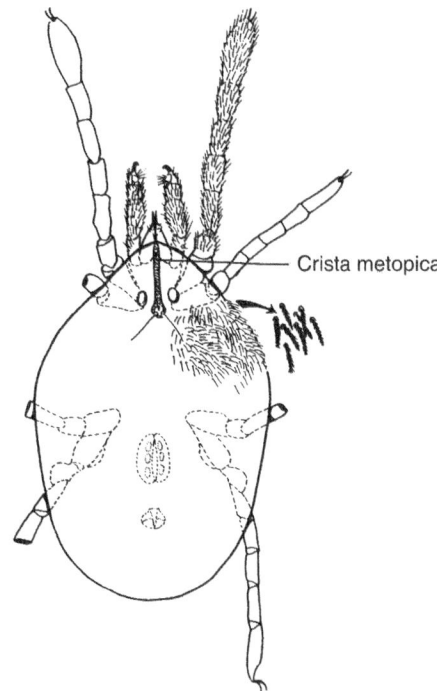

Fig. 3.46 Trombidiidae, *Microtrombidium*, female, dorsal view. (From Krantz, 1978, with kind permission from Prof. G.W. Krantz.)

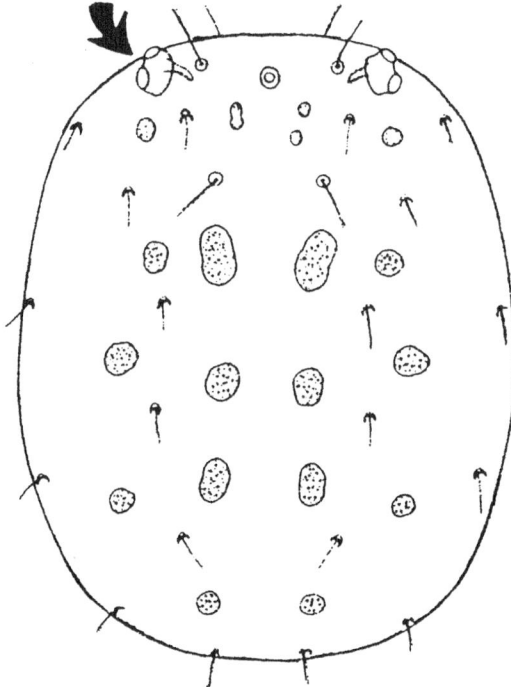

Fig. 3.47   Hydryphantide, *Thyas stolli*, dorsal view.

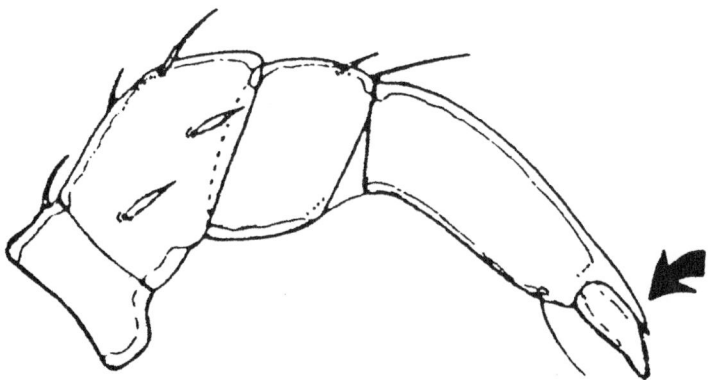

Fig. 3.48   Hydryphantide, palpus.

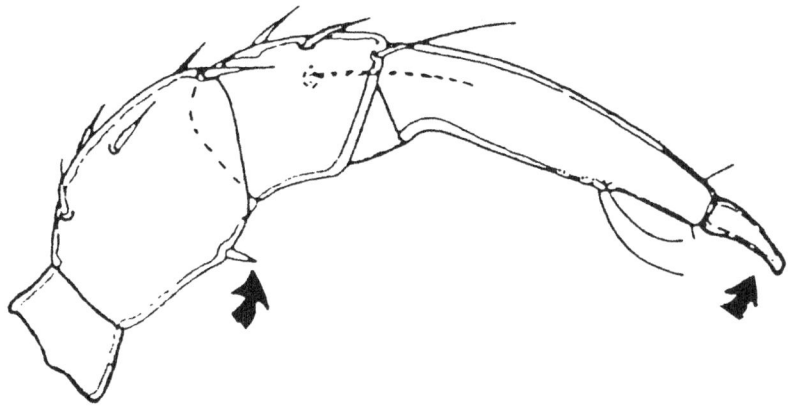

Fig. 3.49   Limnesiidae, *Limnesia lembangensis*, palpus.

Fig. 3.50   Limnesiidae, *Limnesia pinguipalpis*, leg IV.

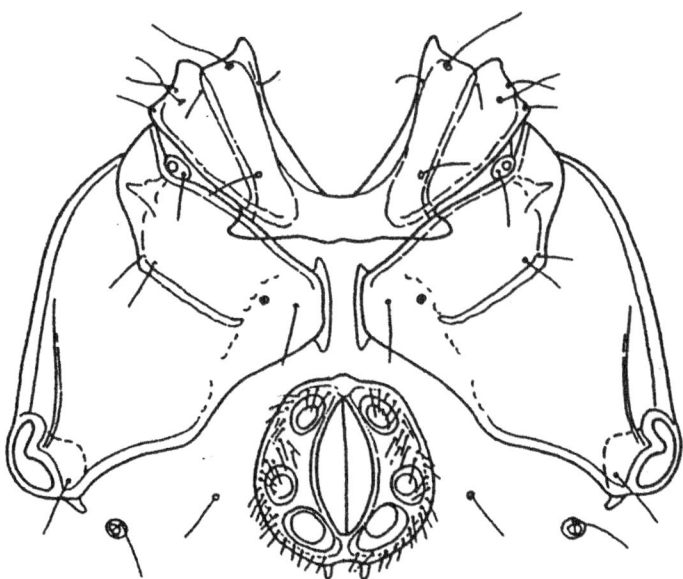

Fig. 3.51 Limnesiidae, *Limnesia lembangensis*, female, ventral view. (From Cook, 1974, with kind permission from Prof. D.R. Cook.)

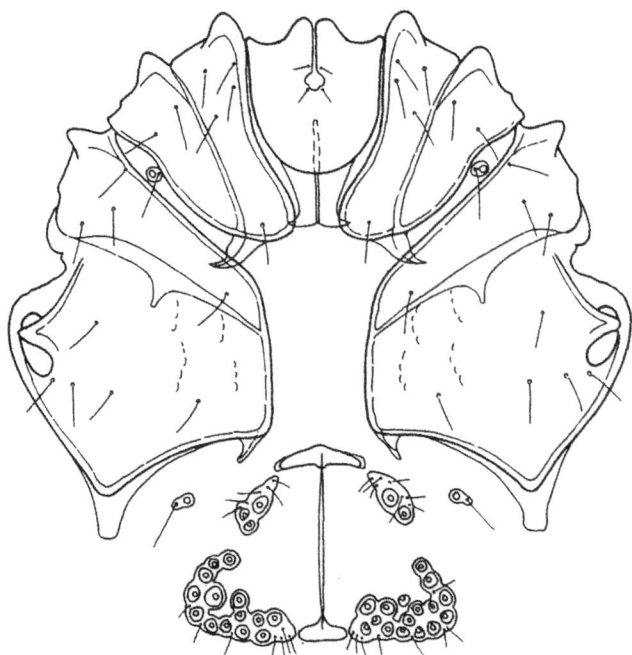

Fig. 3.52 Pionidae, *Piona catatama*, ventral view. (From Cook, 1974, with kind permission from Prof. D.R. Cook.)

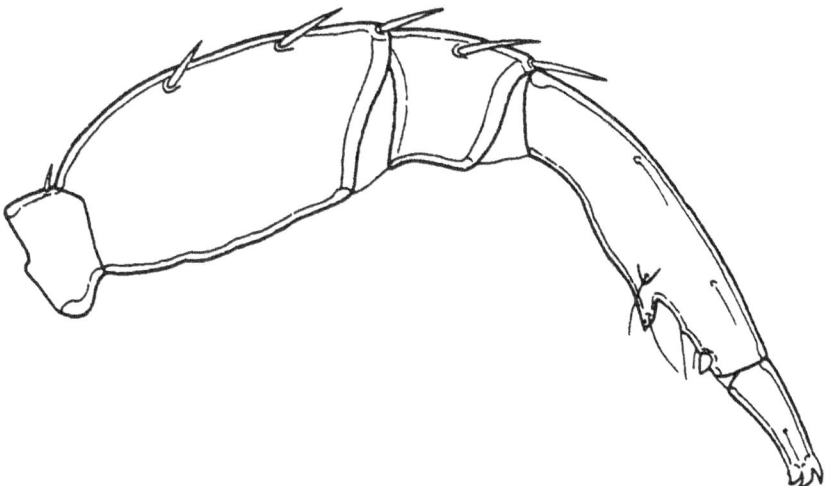

Fig. 3.53 Pionidae, *Piona mahisa*, male palpus. (From Cook, 1974, with kind permission from Prof. D.R. Cook.)

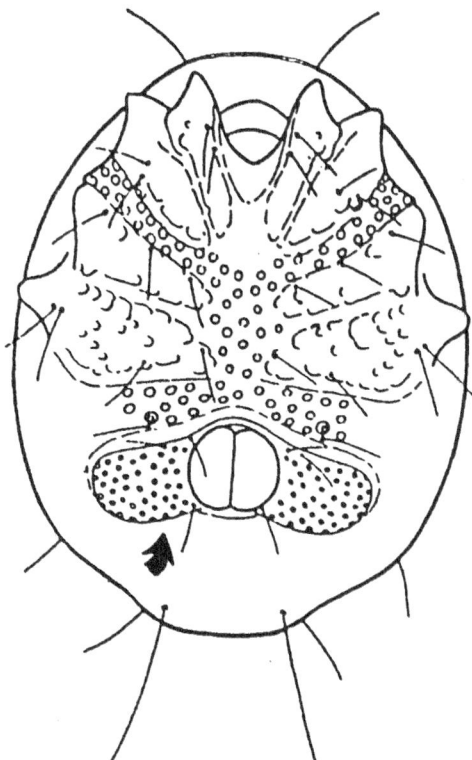

Fig. 3.54 Arrenuridae, *Arrenurus liberiensis*, female, ventral view. (From Cook, 1974, with kind permission from Prof. D.R. Cook.)

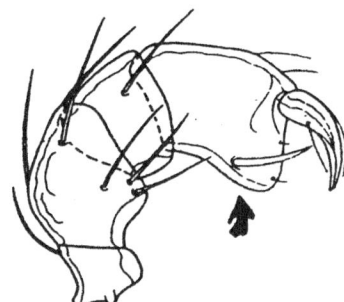

Fig. 3.55   Arrenuridae, *Arrenurus pseudoaffinis*, palpus. (From Cook, 1974, with kind permission from Prof. D.R. Cook.)

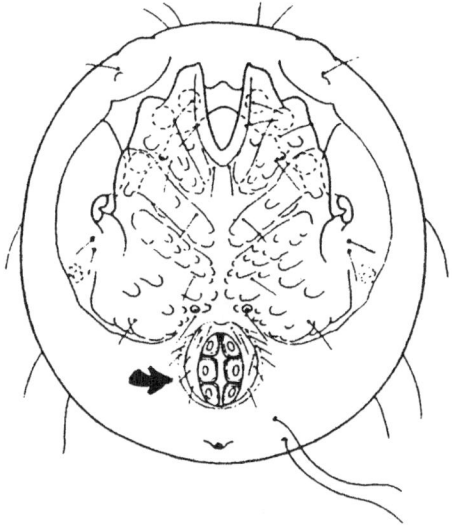

Fig. 3.56   Mideopsidae, female, ventral view.

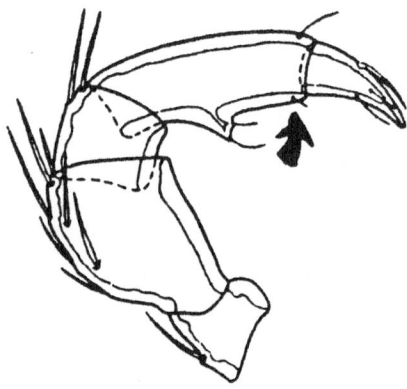

Fig. 3.57   Mideopsidae, *Mideopsis fibrosa*, palpus.

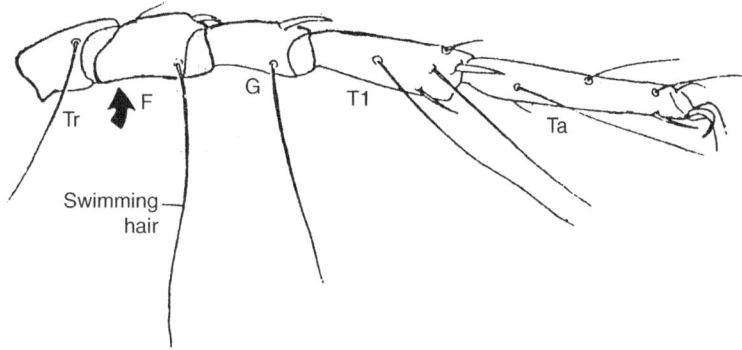

Fig. 3.58 Arrenuridae, larva, leg I.

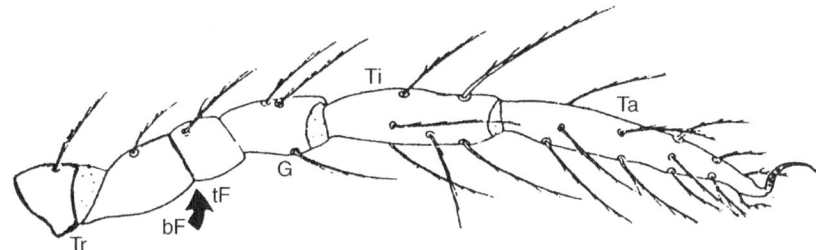

Fig. 3.59 Hydryphantidae, larva, leg I.

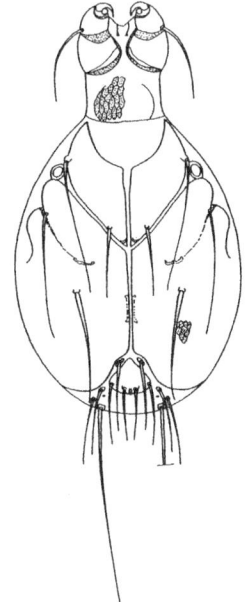

Fig. 3.60 Pionidae, *Piona interrupta*, larva, ventral view.

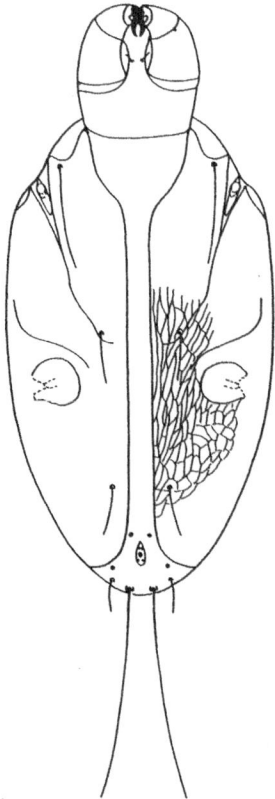

Fig. 3.61 Limnesiidae, *Limnesia marshalliana*, larva, ventral view. (From Cook, 1974, with kind permission from Prof. D.R. Cook.)

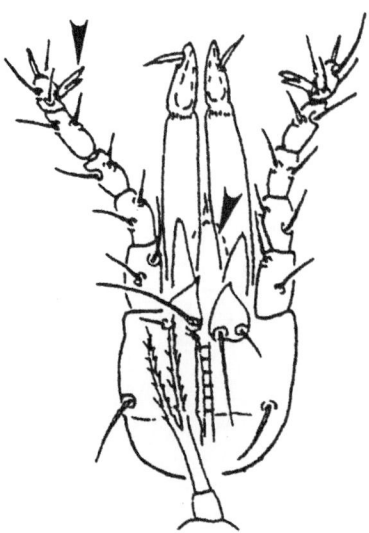

Fig. 3.62 Mesostigmata, venter of gnathosoma.

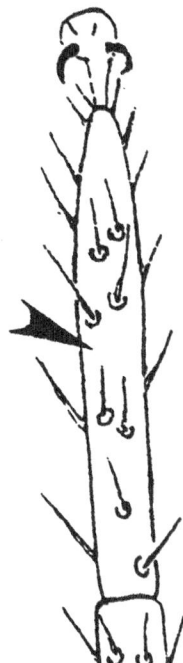

Fig. 3.63   Mesostigmata, leg I.

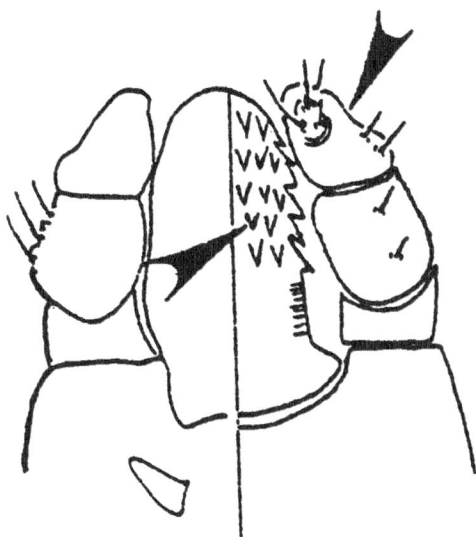

Fig. 3.64   Metastigmata, venter of gnathosoma.

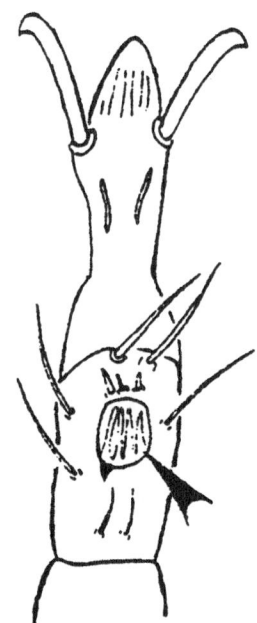

Fig. 3.65   Metastigmata, leg I.

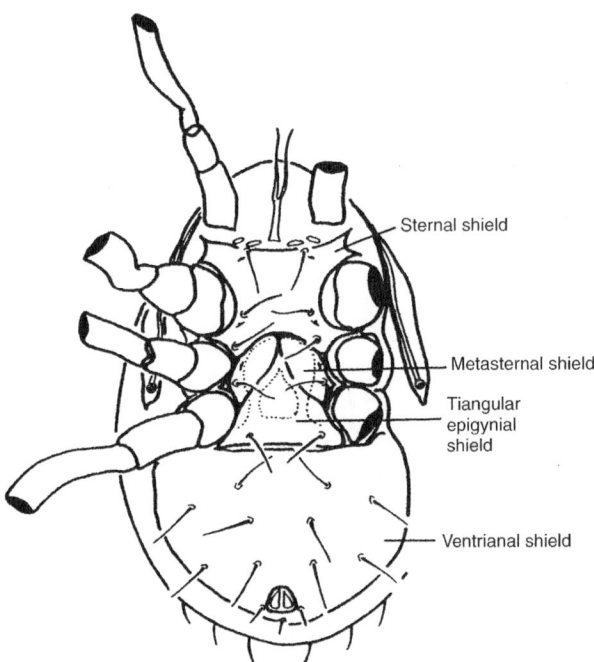

Fig. 3.66   Parasitidae, female, ventral view. (From Krantz, 1978, with kind permission from Prof. G.W. Krantz.)

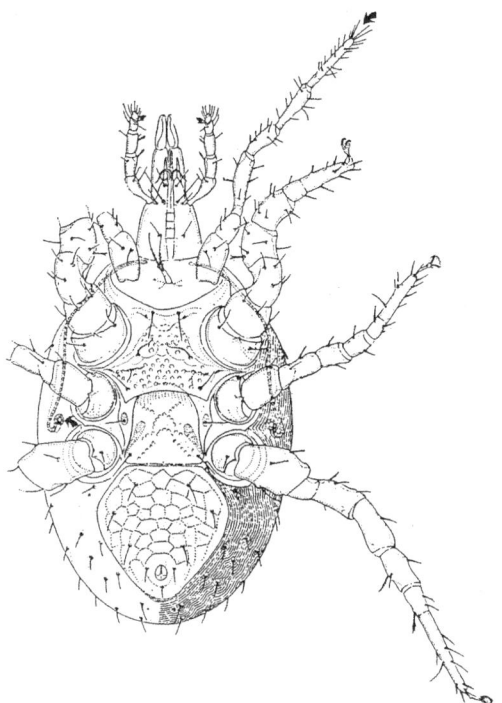

Fig. 3.67   Macrochelidae, *Macrocheles muscaedomesticae*, female, ventral view.

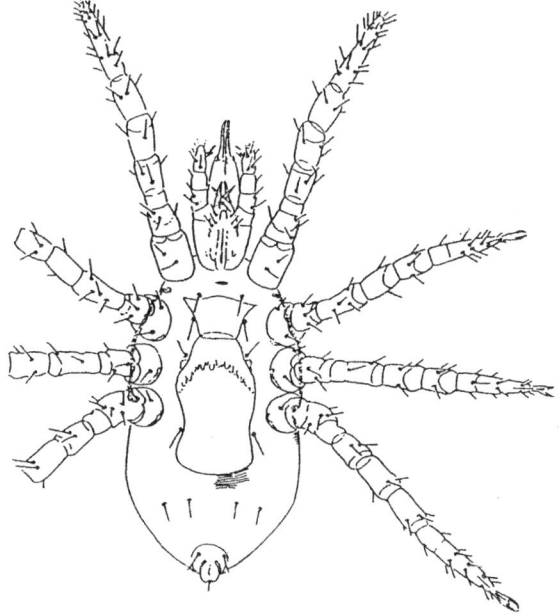

Fig. 3.68   Othopheidomenidae, female, ventral view.

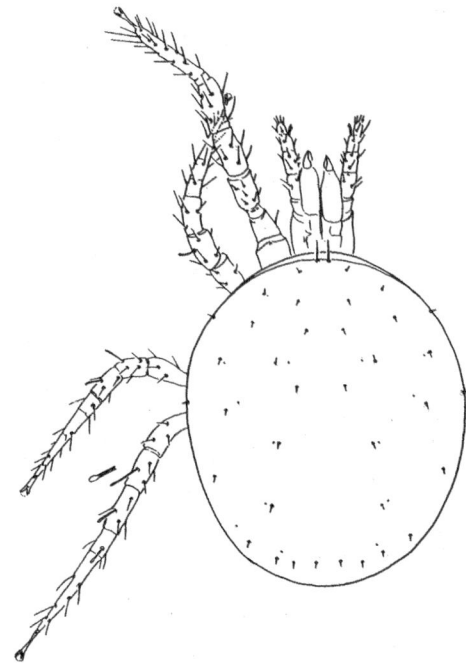

Fig. 3.69 Phytoseiidae, *Iphiseius degenerans*, female, dorsal view.

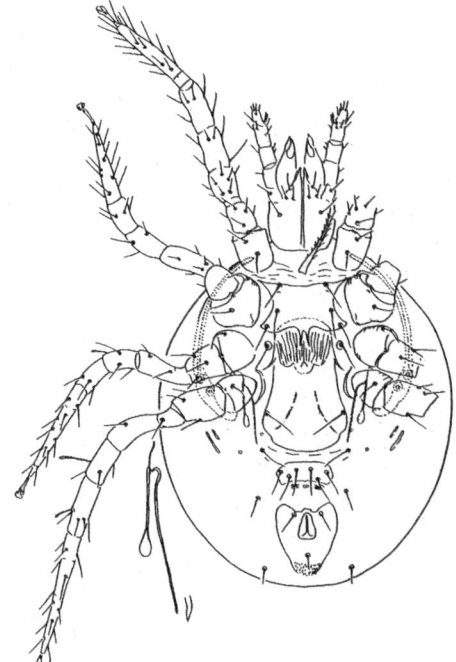

Fig. 3.70 Phytoseiidae, *Iphiseius degenerans*, female, ventral view.

Chapter 4
Acaridae

Diagnosis

Acaridae are whitish, slow-moving mites whose prodorsum is usually covered by a shield-like sclerite. They have a sejugal furrow, carry long, slightly barbed setae on the hysteronotum, and their tarsi have well-developed empodial claws. Owing to their dentate mouthparts acarids can consume particulate foods. The hypopodes (hetero-morphic deutonymphs) possess dissimilar morphological aspects and require sepa-rate keys for their determination; keys to the then-known hypopodes were presented by Turk & Turk (1957) and Zakhvatkin (1959). The family, defined by OConnor (1982), is fairly large, with close to 100 genera, of which at least one-third are known only as hypopodes.

General biology

The Acaridae abound in leaf litter and in the upper strata of soils rich in organic mat-ter, as well as in decomposing animal droppings. Many species are associated with the nests or bodies of invertebrates and vertebrates (including parasites). A few occur on plants or in standing waters and they may contaminate laboratory cultures. Acaridae are especially abundant in stored foods, of which they can be major pests (e.g. *Acarus siro* L.). Some common species consume vegetative as well as animal foods (Gerson *et al.*, 1991). Reproduction is usually sexual, the sex ratio (females/females+males) being 0.5. Acarids emit a diversity of aggregation, alarm and sex pheromones (Kuwahara, 1991, and former papers; Mori *et al.*, 1998), some of which attract predators or have an adverse effect on fungi (see below). When the younger stages of most non-parasitic Acaridae are exposed to adverse conditions [e.g. low relative humidity (RH), extreme temperature, lack of adequate food] they moult to become hypopodes, whose development may be continued when suitable conditions return.

Tyrophagus putrescentiae (Schrank)

This species, known as the copra or mould mite, is a pest of stored grains, cheese, commercial mushroom beds and even fungal cultures in the laboratory (Hughes, 1976). When kept on yeasts, various fungi or wheat germ, the mite completed a generation in 2–3 weeks at 20–25°C under near-saturation conditions, producing around 500 eggs/female (Rivard, 1961). Offered animal food (live and dead aphids;

see below), it layed almost 450 eggs/female (Eraky, 1995). Although high humidities are optimal, *T. putrescentiae* can also develop and reproduce at 70% RH. In the soil the mite may attack living immobile insect stages (especially eggs) or nematodes. Feeding on eggs of the southern corn rootworm, *Diabrotica undecimpunctata howardi* Barber (Coleoptera: Scarabaeidae) (SCR), a major peanut and corn pest in North Carolina, the mite was seen as an important mortality factor for this insect (Brust & House, 1988). *Tyrophagus putrescentiae* was strongly attracted to SCR eggs, all mite stages aggregating around them within a few hours. It completed several generations when offered only these eggs, onto which the mite placed its own eggs. The predator initially removed small areas of the chorion and, after a suitable hole was created, moved in and devoured the contents from the inside; a single SCR egg was consumed in about 5 h. Juvenile mites were unable to penetrate the intact chorion, feeding on previously-damaged eggs from the inside. When offered a choice between moist filter paper and SCR eggs, significantly more mites moved towards the eggs from a distance of 10 cm. *Tyrophagus putrescentiae* preferred these eggs over fungi, other organic debris or dead arthropods. SCR larvae were not attacked.

The presence of even a single mite with 40 SCR eggs in Petri dishes reduced their hatch by 20% (Fig. 4.1), and with ten mites/dish, reduction came to 90% (Brust & House, 1988). In a greenhouse experiment, conducted in pots planted to

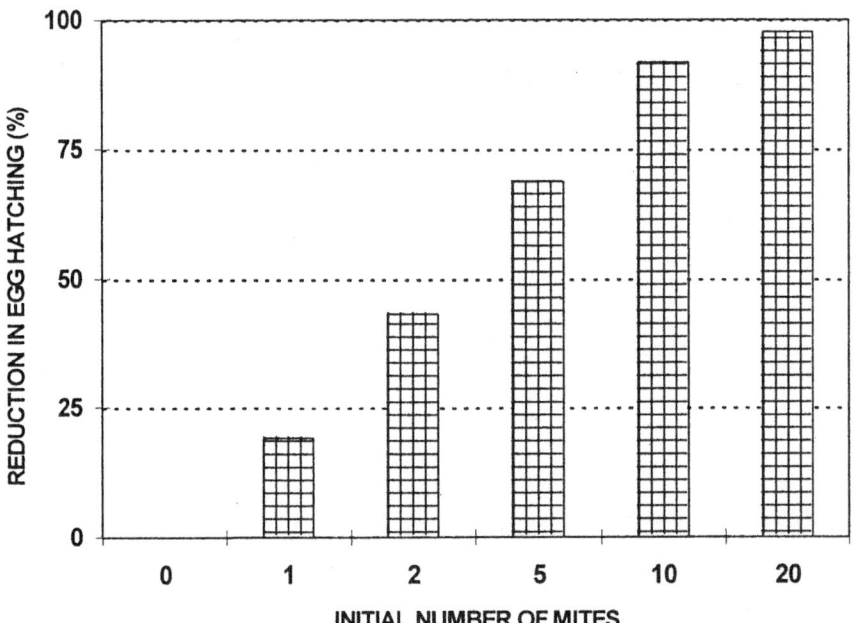

Fig. 4.1 Reductions in the hatch of 40 eggs of the southern corn rootworm, 10 days after being placed in a Petri dish with various numbers of *Tyrophagus putrescentiae*. (Data from Table 2 in Brust & House, 1988.)

peanuts, mites (in batches of 25) and SCR eggs (100/pot) were placed together; the controls consisted of pots with insect eggs but no mites. Mite presence reduced the emergence of adult insects by approximately 50%. In the field *T. putrescentiae* located SCR egg masses placed randomly in the soil, being four times more common in no-tillage systems than in tilled soils. Brust & House (1988) advocated increasing mite numbers in infested soils and manipulating the agroecosystem by no-tillage and by adding organic matter to promote predator activities.

The mite invaded leaf galls produced by the grape phylloxera, *Daktulosphaera vitifolii* (Fitch) (Hemiptera: Phylloxeridae), a major pest of grapes, and fed on its dead and living adults and eggs. Offered only live phylloxera in the laboratory, the mite required about the same period (2–3 weeks) for its development at 23°C and 85% RH as on yeasts, indicating the suitability of the insect diet. Rack & Rilling (1978) believed that the mite's sensitivity to low temperatures and humidities precluded its use for pest control in the field. Live and dead individuals of the aphid *Rhopalosiphum padi* (L.) (Hemiptera: Aphididae), a pest of cereal roots, also sufficed for the predator's reproduction in the laboratory (Eraky, 1995).

An array of detrivorous, plant-parasitising and predatory nematodes was offered to *T. putrescentiae* placed on water agar in the laboratory (Bilgrami, 1994). Maximal feeding (after 24 h at 28°C) was on the migratory juveniles of plant-parasitic, sedentary endoparasitic nematodes, although some of these prey were little eaten. Minimal feeding (but still around 40%) occurred on the predatory forms, which may have shown some resistance to *T. putrescentiae*. When the mite came upon nematodes, the latter were grasped anywhere along their length and consumed within 1–2 min. All mite stages fed on nematodes, in the laboratory devouring more than 700 juveniles of the root-knot nematode, *Meloidogyne javanica* (Treub) during 24 h, even penetrating the pests' gelatinous egg masses and feeding therein (Walia & Mathur, 1995). The preference of *Tyrophagus similis* Volgin for nematodes appeared to be more pronounced than that of *T. putrescentiae*, (Walter *et al.*, 1986), indicating a gradation of nematophagy within members of this genus. Some *Tyrophagus* devour anhydrobiotic nematodes, showing that feeding on these pests is not restricted to near-saturation soil conditions.

Opinions differ as to the nematode-finding mode of *T. putrescentiae*. Bilgrami (1994) believed that predation depended on chance encounters between mite and nematodes. As such encounters are more frequent when more prey is available, the rate of predation also increases, apparently leading to a density-dependent relationship. Walter *et al.* (1986) postulated that the mite could be attracted by chemical cues that the nematodes produce. Observations that injured nematodes attract *T. putrescentiae* to come and feed together suggests that the mites respond to prey secretions (Bilgrami, 1994). The rapid aggregation of all mite stages on SCR eggs, and the results of the attraction experiments noted above, further support the hypothesis that *T. putrescentiae* locates its prey mostly by olfactory means. Indirect confirmation for this premise comes from the attraction of acarid mites to fungal metabolites (Okabe & Amano, 1990; see Chapter 38, ABAs as enemies of soil-borne pathogenic fungi).

Čatská & Smrž (1989) postulated that the selective feeding of *T. putrescentiae* on plant-pathogenic soil fungi could prevent the injury that these fungi cause to apple seedlings (see Chapter 38). Volatiles secreted by *T. putrescentiae*, living in the faeces of the granary weevil, *Sitophilus granarius* (L.), attracted both sexes of a parasitoid of this pest. The mite thus served as 'matchmaker' in addition to mediating host finding by the parasitoid (Ruther & Steidle, 2000).

Rhizoglyphus spp.

Plant-parasitic nematodes, including *Ditylenchus*, *Heterodera* and *Longidorus*, were consumed by the bulb mite, *Rhizoglyphus echinopus* (Fumouze and Robin). Small nematodes were completely devoured, whereas larger ones were cut into pieces and sucked out; nematode cysts were attacked only after some initial hesitation. These observations led Sturhan & Hampel (1977) to attribute an important role to the mite in regulating soil nematodes. *Rhizoglyphus echinopus* is a polyphagus pest of bulbs, corms and tubers (Hughes, 1976), but could, like *T. putrescentiae*, be beneficial under special conditions. Another acarid, *Sancassania ultima* Samšiňák, is a rapacious feeder on all stages of gall-making *Meloidogyne* spp. (Sell, 1988). The mite was strongly attracted to undamaged roots, where it searched for nematodes, an attribute that adds to its biological control potential. *Sancassania* spp., as well as *Rhizoglyphus* spp., are voracious consumers of eggs of *Ascaris* (round-worms parasitic in humans, primates and domestic animals). Five adult mites placed on 0.5 g roundworm eggs consumed about 75% of the eggs in 32 days (Lysek, 1963).

Rhizoglyphus robini (Claparède) produces a cuticular anti-fungal compound that inhibited the growth of the mycelia of several plant-pathogenic fungi (Leal *et al.*, 1990). For more details, see Chapter 38, ABAs as enemies of soil-borne pathogenic fungi.

Miscellaneous

Acarids may indirectly reduce pest numbers by transmitting entomopathogenic fungi. *Sancassania phyllognathi* Samšiňák dispersed the spores of *Beauveria bassiana*, applied against a soil pest in a pot, horizontally (*c.* five cm) as well as vertically (*c.* 0.5 cm) (Samšiňáková & Samšiňák, 1970). The mites did not affect the fungus in any way. In addition, the mycoparasite *Coniothyrium minitans*, proven to have biocontrol activity against the plant pathogen *Sclerotina sclerotiorum*, was dispersed to a minimal distance of 55 mm in the soil by *A. siro*. Transmission was external as well as internal, as around 90% of the ingested and excreted spores were viable (Williams *et al.*, 1998; see Chapter 38).

Acarids (e.g. *T. putrescentiae*, *Acarus* spp.) cultured on wheat bran serve as alter-nate diets in mass-rearings and field releases of phytoseiid mites (Chapters 26 and 39). Acarids may thus be beneficial as predators of arthropod, nematode and fungal pests, as producers of attractants for natural enemies of other pests, and as vectors of insect and plant-pathogenic diseases, in addition to serving as alternate diets for various predators. However, acarids may feed on biocontrol agents such as beneficial nematodes or mycopathogenic fungi used for pest control. Methods for their mass-rearing are available (Rivard, 1961; Walter *et al.*, 1986; Gerson *et al.*, 1991). A concern to be kept in mind is the propensity of *T. putrescentiae* to cause allergy (Green & Woolcock, 1978).

Chapter 5
Acarophenacidae

Diagnosis

These heterostigmatid mites can be recognised by the gnathosoma that is partly or completely fused into the propodosoma and by the indistinct palps. The anterior ventral plate carries three or fewer pairs of setae and the females do not bear prodorsal sensilli. Leg I is thicker than the other legs, and may have a strong, sessile claw. These are small (<200 μm in length), elongate to oval mites that parasitise the eggs of beetles or thrips (Thysanoptera).The Acarophenacidae is a small taxon, with six described genera and about 30 named species. The family was defined by Mahunka (1970) and a key to the genera and a list of species were provided by Magowski (1994); Goldarazena *et al.* (2001) recently added ten new species. The antiquity of the family was established when a Mesozoic Siberian species, associated with the male of a primitive scale insect (Coccoidea), was discovered in amber (Magowski, 1994).

General biology

The acarophenacids, like other Heterostigmata, undergo a reduced life cycle. The juveniles complete their entire development inside the body of the female, whose hysterosoma expands to accommodate them (it becomes physogastric). Mated females of *Acarophenax* are dispersed by the insects on whose eggs they feed, apparently without harming the vectors. As the insects begin to oviposit, the mites disembark and attack the eggs, becoming physogastric within a few hours. Feeding lasts for 1–2 days, causing the egg to shrivel. The entire life cycle requires about 4–5 days. The mites seem to reproduce by strict sib-mating, as the (usually single) male, which develops slightly more quickly than its sisters, inseminates them inside the mother's body before they leave. The emerging females (up to a dozen for *Acarophenax tribolii* Newstead & Duval, about two dozen for *Acarophenax dermestidarum* Rack) then begin to seek suitable hosts. Some species seem to require specific (but unknown) conditioning to initiate their life cycles (Newstead & Duvall, 1918). Owing to the facts that a single host individual suffices for their development, that only a single mite stage feeds on the host, and that the latter does not survive the attack, Lindquist (1983) considered the Acarophenacidae to be parasitoids. However, as *Adactylidium* spp. suck the body fluids of their adult hosts' bodies, they can also be considered to be ectoparasites.

Parasites of beetles

Acarophenax mahunkai Steinkraus & Cross is an egg parasitoid of the sentinel dark-ling beetle (or lesser mealworm), *Alphitobius diaperinus* (Panzer), a cosmopolitan pest of poultry. More than half of the beetle egg masses examined in Arkansas were parasitised by one or more mites (Steinkraus & Cross, 1993). The mites greatly reduced pest numbers, because only 24% of the attacked egg masses hatched, compared with 73% of the unparasitised masses (the fate of the remaining eggs was not stated). This parasitoidism thus caused about two-thirds of natural egg mortality. An average of 27 (range 7–54) females was produced by each physogastric mite, along with about 1.7 males; this resulted in a 0.94% sex ratio. Mite development was faster than that of the host eggs, resulting in some attachment and feeding of the second generation parasitoid progeny on the remaining healthy mealworm eggs. Such mites did not become as large as those that had fed on younger eggs, probably because of reduced host suitability. Young mite females attacked host eggs soon after emergence, indicating that they did not need any stimuli to feed. The mite appears to be a specific enemy of *A. diaperinus*, because it did not attack young eggs of several beetle and moth species offered, even in the absence of mealworm eggs. Sexually mature mite females disperse while adhering to the ventral regions of the beetles' bodies, apparently without causing any harm. The high host specificity of *A. mahunkai*, its rapid development, its phoretic dispersal and apparent harmlessness to animals and humans, suggested to Steinkraus & Cross (1993) that this mite could be a useful biological control agent in closed poultry houses or in stored product situations.

Acarophenax lacunatus (Cross & Krantz) is an egg parasitoid of the lesser grain borer, *Rhyzopertha dominica* (Fabricius) (Coleoptera: Bostrichidae), a major pest of many cereals in storage. The optimal temperature and relative humidity (RH) for mite development were 30°C and 60% RH, under which conditions *A. lacunatus* produced 17 progeny. Placing a few physogastric mites into jars containing 420 g whole wheat and 50 beetles reduced pest numbers as well as damage. The addition of two mites/jar resulted in complete suppression of beetle eggs and first instar larvae in 45 days, whereas the total elimination of adults required six mites and 60 days. Increasing the number of mites/jar decreased the losses caused by the pest. Ten mites/jar reduced beetle damage by 70–80% (Fig. 5.1), and the presence of even two mites/jar decreased injury by more than 50%. Faroni *et al.* (2000) concluded that *A. lacunatus* could be a useful biocontrol agent of the lesser grain worm. The mites performed better at high temperatures, suggesting that they would be more suited for tropical regions.

Acarophenax tribolii Newstead & Duvall feeds on the eggs of various tenebrionid beetles (mostly *Tribolium* spp.), which infest stored products throughout the world; the mite is also dispersed by these beetles. This mite, a nuisance species in laboratory beetle cultures, had up to 15 progeny, only one of which was a male (Newstead & Duvall, 1918). Rack (1959) reported that *A. dermestidarum* attacked the eggs of several *Dermestes* spp., common stored product beetles.

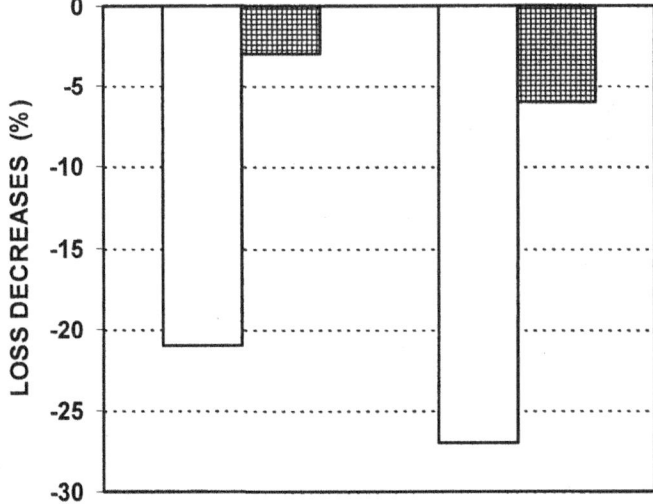

Fig. 5.1 Effect of ten females of *Acarophenax lacunatus* on reductions in the weight of whole wheat, caused by 50 *Rhyzopertha dominica*, 45 (left) and 60 (right) days after being infested by the pest. White bars: mites added; cross-hatched bars: controls, no mites. (From data in Faroni *et al.*, 2000.)

Parasites of thrips

All known species of *Adactylidium* are associates of tubuliferous thrips (Thysanoptera: Tubulifera) (Goldarazena *et al.*, 2001). The laurel thrips, *Gynaikothrips ficorum* (Marchal), a leaf-rolling pest of *Ficus* trees in warmer climates, is attacked by *Adactylidium ficorum* Goldarazena & Ochoa in Egypt . The mite raised a generation in 4 days and produced six to nine progeny (including a single male) at 30°C, while feeding on a single thrips egg. Elbadry & Tawfik (1970) believed that its effectiveness as a biological control agent was enhanced by its short life cycle. This, or a similar species, was the most abundant natural enemy of *G. ficorum* in Brazil, present in 11 out of 16 leaf samples obtained by Bennett (1965) (who called the mite ?*Pyemotes* sp.) from various parts of that country. At some locations (Belem; Manaus) most eggs were attacked, contributing to the pest's control. Eggs as well as young thrips nymphs were attacked, and the mite also occurred on large nymphs and adults. Some *Adactylidium* may suck the haemolymph of the latter; however, as adult thrips serve to transport the mite between patches, the effect of this damage is not known. The host-parasitoid data collated by Goldarazena *et al.* (2001) suggest that *Adactylidium* spp. are not host specific, as some species were obtained from several different host thrips genera.

The small size of the Acarophenacidae allows them to attack hosts hiding in sites (such as bark crevices, gall or rolled leaves) that are too small for other natural enemies. It is possible, because of such cryptic habitats, along with their minuteness, that these natural enemies have not sufficiently been noticed in the past, and that the discovery of more species will enhance their potential as ABAs.

Chapter 6
Anystidae

Diagnosis

Anystidae are large (0.5–1.5 mm), reddish, soft-bodied mites with a palpal thumb–claw complex, whose tarsus overhangs the claw. The family is being revised by Otto (2000 and earlier papers); Meyer & Ueckermann (1987) studied the South African species. The Anystidae has about 20 genera and 100 described species.

General biology

The Anystidae, and especially members of the common and best known genus, *Anystis*, abound in soils and on plants. They usually have a long development time and deposit their eggs in masses. Anystids are fast runners and appear to depend on chance encounters for finding prey, being apparently generalist predators that feed on any small arthropods that they can catch (live or dead). The prey include many mites as well as psocids (Psocoptera), thrips (Thysanoptera), springtails (Collembola), various Hemiptera, such as bugs (Miridae), leafhoppers (Cicadellidae), psyllids (Psyllidae), aphids (Aphidoidea), bast scales (Margarodidae), armoured scales (Diaspididae), small butterfly caterpillars (Lepidoptera) and eggs of pestiferous flies (Diptera) (Baker, 1967). Anystids also take non-living food, such as exudates of extrafloral nectaries (Pemberton, 1993) and even sugar solutions (Otto & Halliday, 1991). All stages of *Anystis* are cannibalistic, the bigger and stronger individuals attacking the smaller stages, including those in moult. *Anystis* spp. are considered to be components of groups of natural enemies that can affect the populations of diverse pests, and may even be controlling them (Grobler, 1962; Cheng *et al.*, 1983; James, 1995). One species has been introduced into Australia and South Africa, where it successfully controls several pasture pests.

Anystis as a predator of fruit tree pests

In North America *Anystis agilis* (Banks) feeds on pests infesting apples (MacPhee & Sanford, 1961), citrus (Mostafa *et al.*, 1975), vineyards (Sorensen *et al.*, 1976) and forest trees (Neilson, 1963). The mite is readily recognisable by its bright red colour and characteristic figure-of-eight, whirling running pattern. Mites overwinter under tree bark or in the soil, raising two annual generations (May–June and September–October). A female deposits one to three egg clusters in the ground, each consisting of about 23 ova. The mites move onto new growth as soon as it appears, resting in the inner crowns of vines or around and under the citrus fruit button (calyx). They avoid direct sunlight and search for prey mainly on the leaf underside, along the veins and edges. *Anystis agilis* is a fast runner: ten undisturbed, fed females covered an average distance of 84 cm/min, attaining a speed of 180 cm/min when disturbed (Sorensen *et al.*, 1976).

Several predatory mites ('red mites') were reported in the 1950s to feed on eggs of the spruce budworm, *Choristoneura fumiferana* (Clemens), during this moth's outbreak in eastern North America. However, only *A. agilis*, the most common species, was identified. Its populations had a numerical response to pest densities, up to the level of 100 eggs/10 sq. feet (approx, 100 eggs/m^2) (Neilson, 1963).

In the laboratory *A. agilis* preferred the grape leafhopper (*Erythroneura elegantula* Osborn) to the two-spotted spider mite (TSSM, *Tetranychus urticae* Koch). Female predators readily abandoned partially consumed TSSM to capture nearby leafhoppers. More significantly, *A. agilis* was tenacious when feeding on the latter, ignoring disturbances by a hand-held brush or by encounters with spider mites (Sorensen *et al.*, 1976). Indirect evidence for the greater suitability of leafhoppers as diet was obtained by observing the predator's different colour and behaviour when reared on either of the two prey species. On TSSM *A. agilis* became dark red, sluggish and rather un-coordinated, whereas when offered leafhoppers, it retained its bright orange hue and moved normally. Fecundity was only marginally higher when feeding on leafhoppers, compared with spider mites (33 vs 29 ova/female). Female predators consumed a daily average of 5.6 leafhopper nymphs. When offered only spider mites, the trito-nymphs and females of *A. agilis* ingested 23–40 prey/day. Smaller spider mites were taken to a much lesser extent by most predator stages (Fig. 6.1).

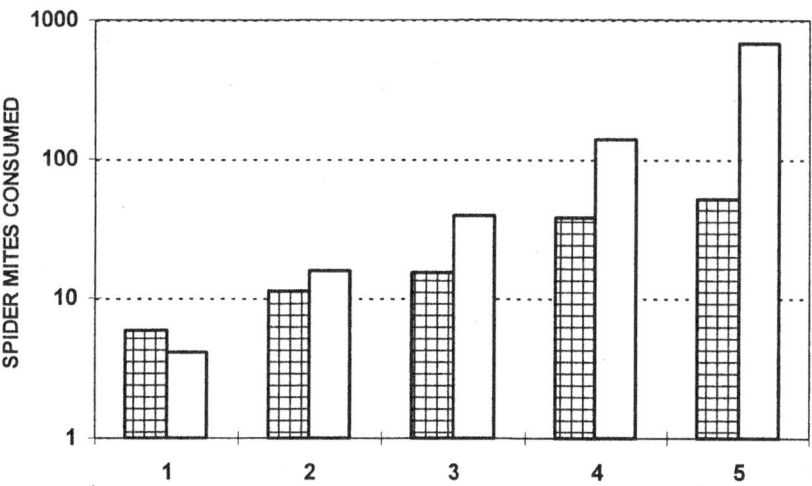

Fig. 6.1 Consumption of juveniles (cross-hatched bars) and adults (white bars) of the two-spotted spider mite, *Tetranychus urticae*, by all active stages of *Anystis agilis* (1: larvae; 2: protonymphs; 3: deutonymphs; 4: tritonymphs; 5: adults). (Based on Table 2 in Sorensen *et al.*, 1976.)

Adult leafhoppers were paralysed 30 s after being attacked, suggesting to Sorensen *et al.* (1976) that *A. agilis* injected a toxin into its prey. This attribute allows the predator to attack prey larger than itself. Evasive tactics such as springtail jumping, physical barriers such as spider mite webbing and a thick integument were more effective than size alone in protecting potential prey. Starved *A. agilis* females lived for 4 days; with water they survived for 2 days more, and those offered vine leaf exudates lived for another 2 days, indicating the importance of this natural nutrient.

Long-term studies of the effect of spray programmes on the arthropod fauna of apple orchards in Nova Scotia revealed an inverse ratio between *A. agilis* numbers and those of the pestiferous clover mite, *Bryobia praetiosa* Koch. The latter developed large populations in orchards that had been sprayed with an arsenical pesticide, known to be toxic to the predator. This indirect evidence lent weight to the postulated role of *A. agilis* in controlling the pest (Lord, 1949).

The predator attacked the citrus thrips [*Scirtothrips citri* (Moulton)] more often than the citrus red mite (*Panonychus citri* McGregor) or aphids on California citrus, and deposited only when feeding on the thrips (Mostafa *et al.*, 1975). In Russia, near Moscow, *Anystis baccarum* (L.) cleaned up spider mite infestations on blackberries within a few days when the predator/prey ratio was 1/30 to 1/50 (Lange *et al.*, 1974a). The predator was released on pest-infested plants immediately after being collected in large numbers from oak and pine litter (Lange *et al.*, 1974b).

Several species of *Anystis* partake in the control of diverse Hemipteran pests. In British Columbia the predator fed on all instars and morphs of aphids that infested alfalfa (Frazer & Nelson, 1981). *Anystis agilis* was an important predator of the woolly pine needle aphid, *Schizolachnus pini-radiatae* (Davidson) in Canada. Mortality of overwintering aphid eggs, tentatively attributed to the mite, was 30% in early October, rose to 53% by mid-November and then declined, probably owing to seasonal cold (Grobler, 1962). The long generation time and cannibalistic habits of *A. agilis* precluded its being considered for mass-rearing in the laboratory, but conservation efforts in South African agricultural systems were advocated. *Anystis baccarum* preyed on nymphs of the citrus psylla, *Trioza erytreae* (Del Guercio) infesting citrus, and on the yellow pecan aphid, *Monelliopsis pecanis* Bissell (Berg *et al.*, 1995). An *Anystis* sp. was one of several predators reducing populations of the pine bast scale, *Matsucoccus matsumurae* Kuwana (Margarodidae) in China (Cheng *et al.*, 1983). A single *A. baccarum* devoured about 100 newly hatched larvae (crawlers) of the pine pest *Hemiberlesia pitysophila* Takagi (Diaspididae) in the laboratory. One to four weeks the predator was released in a pine plantation, it reduced young scale populations by more than 30% (Wu & Shen, 1990).

Anystis as a predator of pasture pests

Efforts to find natural enemies of the red-legged earth mite, *Halotydeus destructor* (Tucker) (RLEM) (Penthaleidae) and of the blue oat mite, *Penthaleus major* (Dugés) (Penthaleidae), have been continuing in Australia since the initial outbreaks of these pasture pests in 1917 and 1921, respectively (Otto & Halliday, 1991). Pre-World War

II surveys to collect natural enemies in Europe resulted in finding an *Anystis* sp. that was feeding on *P. major* and various springtails in southern France. The predator was initially identified as *A. baccarum*, a species that had already been found in Australia, leading to the cancellation of the proposed importation. An Australia-wide post-World War II survey of anystid and bdellid (Chapter 9) mites revealed that *A. baccarum* occurred mostly in horticultural habitats but very rarely in pastures. This suggested that the European species was not conspecific with *A. baccarum*, paving the way for an additional European survey. The resultant data confirmed that in Europe, as in Australia, *A. baccarum* colonised mainly horticultural habitats. The species formerly named *A. baccarum*, at that time termed *Anystis* sp. A, was rarely collected anywhere except in pastures and meadows, where it co-occurred with the pests that it was intended to control. This predator appeared to be adapted to Mediterranean climates (Wallace, 1981). *Anystis* sp. A, later called *A. salicinus* (L.), is now known as *Anystis wallacei* Otto (Otto, 1992).

Data on the sites where *A. wallacei* had, or had not, been found in North Africa and in western Europe were then used to draw a bioclimatic map of the species, applying the climate modelling program CLIMEX (Chapter 39). The map showed the suitability of various Mediterranean countries for the predator, predicting that it would occur along most of the region's coastal areas. Another map, based on the same data but incorporating climatic measurements from Australia, was then prepared. The new map showed the many suitable coastal regions that lie above the 35° parallel where *A. wallacei* could be expected to establish, indicating its optimal release sites (Otto & Halliday, 1991).

About 18,000 mites were collected in 1965 by vacuum suctioning in southern France and sent to Western Australia (WA), where they were released at four locations. Three years later *A. wallacei* was established at two of these sites and during the following years it reduced RLEM numbers by 25–80% (Wallace, 1981). Despite these promising results, it became apparent that the natural spread of *A. wallacei* was too slow (about 50 m/year). A programme intended to increase its distribution by further releases was undertaken in WA during 1988–1990. Predators were mass-collected by suctioning in fields where they were established, maintained in aerated plastic containers padded with moistened paper towels and kept cool until used, within 1–2 days (Michael *et al.*, 1991). Releases usually consisted of batches of about 100 mites/m^2, placed at several marked sites in the same field. Over 572,000 *A. wallacei* were thus released. The minimum number of released mites that enabled establishment was 1104. A major factor that determined the predator's establishment was grazing management. Establishment failed in heavily grazed plots, whereas it was usually successful, even with low mite numbers, in fields with satisfactory pasture conditions. The predator was introduced into Tasmania in 1993 and became established at a few sites, but does not seem to be doing well in that island's prevailing temperate climate (Ireson *et al.*, 2001).

Sampling in 1990 showed that RELM populations were reduced by over 50% wherever predator density exceeded 40/m^2 early in the season. Pest numbers in fields with *A. wallacei* reached only one-fifth of their levels at predator-free sites. Part of the success of *A. wallacei* was attributed to its early autumnal emergence, which occurs before that of the prey (Michael *et al.*, 1991). The predator is consequently present and active in the field at the time the juveniles of RELM hatch, and it immediately attacks them. Numbers of prey taken by *A. wallacei* were observed in boxes with soil and growing pasture plants. Discrete numbers of pests and predators were added, left for 5 days and counted. With 100 *A. wallacei*/m², each predator would kill about 4.4 RELM/day. With 400 predators, daily kill came only to 2.2/predator (Michael, 1995), suggesting that intraspecific inhibition and/or cannibalism would operate when too many *A. wallacei* were present in the same area. Controlled trials intended to gauge the efficacy of the predator (Chapter 44) showed that in the field *A. wallacei* reduced RELM numbers by about two-thirds (from 35,000 to 12,100/m²). Michael (1995) further calculated that 100 *A. wallacei*/m² would kill 16,000 RLEMs within one pest generation. Although *A. wallacei* also feeds on the lucerne flea, the collembolan *Sminthurus viridis* (L.), it does not control this pest.

The long life cycle of *A. wallacei* was considered a handicap when assessing its efficacy, as such a predator would not be expected to respond quickly to rapid increases in prey numbers (Michael, 1995). However, this referred only to the predator's numerical response, not to its functional response, which remains to be investigated. In contrast, the propensity of the predator's eggs to hatch after dry spells of different duration was considered an advantage. The benefit was two-fold: first, *A. wallacei* would already be in the field when the pests began to emerge, and second, small individuals of *A. wallacei*, attacking small RLEMs, provided the highest rate of pest kill (Michael, 1995). Otto & Halliday (1991) voiced other doubts concerning the efficacy of *A. wallacei* as a predator of RLEM. When given a choice of springtails as well as RLEM adults and eggs, the predator attacked springtails and most RLEMs survived. This suggests that in field situations, wherever Collembola, the preferred prey, were present, *A. wallacei* would not kill many RLEMs. Otto & Halliday (1991) strove to resolve the apparent contradiction between their data and those of others about the efficacy of *A. wallacei*, by invoking the movement patterns of prey and predator. RLEM tends to leave its feeding sites on plants in response to slight disturbances that may be caused by *A. wallacei*. Once off its food plant, the pest would be more susceptible to indirect mortality factors.

The predator was introduced into the Western Cape area of South Africa, where it serves as an effective predator of the pest (called the black sand mite there, Meyer; 1996).

For mass-rearing, *A. wallacei* were kept within small containers in the dark, on a substrate of moistened plaster of Paris and organic material, with springtails (reared under the same conditions) as food. Thousands of predators were thus reared and shipped for further releases (Michael, 1995). Despite this success, the method was considered too labour intensive for producing the large numbers of mites required for further introductions. Collecting *A. wallacei* from the field was less expensive, but the laboratory method could be used for mass-rearing other anystids.

The embryos of *A. wallacei* require a minimum of 12 days under moist conditions to reach the prelarval stage, the onset of which is recognisable by a splitting of the egg shell and by the emergence of larval legs (Otto & Halliday, 1991). Eggs kept under conditions of low relative humidity (33% RH) or medium RH (55%) die without hatching. If the prelarva is then maintained under humid conditions (>93% RH), larvae emerge after 4–5 days. Prelarvae kept in dry conditions enter a sort of 'suspended animation', surviving periods of low humidity. Upon being wetted with free water (e.g. rain), they will hatch within a short time. In the field the fully developed prelarva of *A. wallacei* is capable of surviving dry summers, continuing its development in the autumn, after the first rains. A practical application based on this 'suspended animation' is that eggs in the prelarval stage can be stored until required, and exposed to free water to bring about rapid hatching. Development, from larva to adult, required 28 days at 22°C (Otto & Halliday, 1991).

An important, unexplored aspect of the biology of *A. wallacei* is its possible competition with *Bdellodes lapidaria* (Kramer), an effective bdellid predator of the lucerne flea. A significant reciprocal numerical relationship was observed between them following unseasonal early rains (Michael *et al.*, 1991). The two predators may have competed for food before prey emergence, or *B. lapidaria* became the prey. An unexpected result was that reductions in *B. lapidaria* were followed by increases in lucerne flea numbers.

Miscellaneous

Ticks (Ixodoidea) have very few arthropod natural enemies. In small laboratory cages *A. wallacei* attacked tick larvae (Holm & Wallace, 1989). Single *A. wallacei* and *A. baccarum* were placed with several larvae of the paralysis tick, *Ixodes holocyclus* Neumann, or with the cattle tick, *Boophilus microplus* (Canestrini) (both hard ticks, Ixodidae), in small cells. The predators fed on both tick species; the potential of these mites (and other Anystidae) as natural enemies of ticks in the field should thus be further explored.

Very little is known on the impact of other anystids on additional pests. James (1995) reported that *Walzia australica* Womersley was among common predators associated with RLEM in New South Wales. Data on the effect of pesticides on the Anystidae are presented in Chapter 43.

As noted, *Anystis* spp. are usually considered to be generalist predators that will feed on any arthropod that they can catch, as well as on dead prey. Conventional wisdom argues that their non-specificity and cannibalism detract from their value as efficient natural enemies. The observations of Michael (1995), Mostafa *et al.* (1975), Otto & Halliday (1991) and Sorensen *et al.* (1976), however, suggest that *Anystis* spp. have distinct feeding preferences and that such feeding may significantly reduce pest populations.

Chapter 7
Arrenuridae

Diagnosis

Arrenurids are heavily sclerotised water mites, with closely fitting dorsal and ventral plates. The palpal tibia bulges ventrally to form an uncate palp. The coxal and ventral plates are united, and the medial genital aperture is located beyond coxae IV. The genital acetabula are placed on wing-like plates that are incorporated into the ventral plate. Males usually possess a posterior extension that merges into the dorsal plate. The larvae are heteromorphic to the deutonymphs and adults; separate keys are therefore needed for their determination (see Chapter 3). About 20 genera and several hundred species were proposed in the family. The genera of the Arrenuridae were treated by Cook (1974), the North American taxa by Smith & Cook (1991) and the Australian genera by Harvey (1998). Jin (1997) listed the Chinese species.

General biology

The Arrenuridae is a family of freshwater mites assigned to the Hydrachnidia (also called Hydracarina or Hydrachnellae). This is a cohort of aquatic prostigmatid super-families whose members have only three active stages: larvae, deutonymphs and adults; the protonymphs and tritonymphs remain quiescent. Larval arrenurids usual-ly attack insects that develop in freshwater bodies, such as various pestiferous Diptera (including Culicide or mosquitoes; midges or Chironomidae; biting midges or Ceratopogonidae) as well as damselflies (Odonata: Zygoptera) and dragonflies (Odonata: Anisoptera) (Smith & Oliver, 1986). The larvae are attracted to the pupae of dipteran hosts, locating them from short distances by chemical cues (Smith & McIver, 1984a; Baker, 1996), await the emergence of the adults and immediately transfer onto them and begin to feed (Mitchell, 1998). The volume of the mite's body increases many-fold during engorgement, which lasts for about 5–6 days. Detachment of fully fed mites is instigated by the return of the insects to suitable water bodies, and may also be triggered by cues associated with the ovipositional behaviour of the hosts that are perceived by the parasites (Smith & Laughland, 1990; Rolff & Martens, 1997). Mite larvae drop off and resume development in the water. Arrenurid deutonymphs and adults prey on small crustaceans, such as Ostracoda and Cladocera, as well as on small mosquito larvae (Rajendran & Prasad, 1994), and sometimes even on larvae of ceratopogonids and chironomids (H.C. Proctor, personal communication, 2000). Sperm transfer is indirect (e.g. via spermatophores) but requires a meeting between the sexes. *Arrenurus* spp. deposit about 55–130 eggs/female (Cassagne-Méjean, 1966; Stechmann, 1978), their number and size depending on the hosts: arrenurids that attack Odonata produce many, but small eggs, whereas those that parasitise Diptera deposit few, but large eggs (Cook *et al.*, 1989). All eggs are laid in gelatinous masses, attached to aquatic plants or stones or onto the bottom. The sex ratio is usually female biased, although strongly male-biased populations have been noted (Rajendran & Prasad, 1994; Proctor, 1996). A few Arrenuridae develop directly from egg to deutonymph, bypassing the parasitic larval stage (Smith, 1998).

A single gelatinous feeding tube, or stylostome (Fig. 7.1), which begins at the mite's mouth parts, is formed within the body of the attacked insect as a result of reactions between the parasite's saliva and the host's haemolymph. The mite sucks the host's body fluids through the stylostome. Cellular damage caused by the formation of the stylostome as it is being embedded into the insect's thoracic flight muscles injures the host and impedes its flight (Smith, 1988). Stylostomes remain in the insect's body

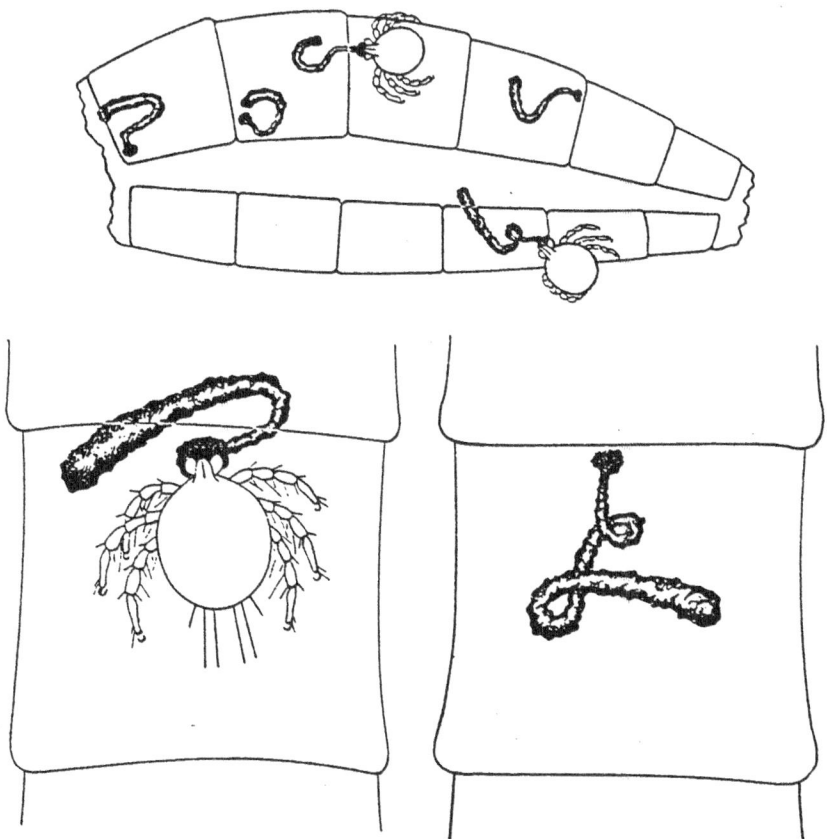

Fig. 7.1 Hydrachnid stylostomes within a mosquito's body. Above: mosquito abdomen with two attached mites and stylostomes; below left: mite with stylostomes; below right: stylostomes remaining in the body of the mosquito after the mite had dropped off. (From Marshall & Staley, 1929.)

after the mite had dropped off, enabling an observer to count them in cleared host specimens, and thus determine the number of mites that had parasitised a given host (see below). As the form of the stylostome seems to be species-specific in *Arrenurus*, it can serve as a taxonomic marker (Lanciani & Smith, 1989). On Odonata the melanised attachment scars left by detached larvae may be counted.

Arrenurus spp. as parasites and predators of mosquitoes

Larvae of *Arrenurus* spp. are the most common acarine parasites of mosquitoes, being associated with several host genera of major medical and veterinary importance (e.g. *Anopheles*, *Coquillettidia*, *Culex* and *Mansonia*). Uchida & Miyazaki (1935) recognised these mites' potential as natural enemies of *Anopheles* spp. (the vectors of malaria), and noted that mosquitoes that bore more than four mites were inactive and did not attack humans. *Anopheles* and *Mansonia* are the most frequent hosts in the tropics and subtropics; at times 80% of their populations may be parasitised (Smith, 1983).

The quantitative effect of water mite larvae (including Arrenuridae) on their insect hosts cannot be directly evaluated in the field. In the case of mosquitoes, the biology of the parasitic mites (which feed on mosquitoes mostly during their first 1–5 days as adults) and of the insects (which begin to search for their own hosts only 3–5 days after emerging) can be utilised to estimate the mites' impact. A common practice is to compare mite abundance and prevalence on freshly emerged insects with those on mosquitoes that had emerged at the same time ('same cohort') but are collected at the later, host-seeking phase. The survival rates of either group, and those of mature (engorged) mosquitoes, are used to provide an indirect estimate of the mites' effect. Rates of past parasitism in engorged mosquitoes are determined by the presence of stylostomes. The data may then be fortified by observations on mite behaviour in the laboratory, including impact on hosts. This topic is discussed in Chapter 44.

Host specificity appears to depend more on the temporal and spatial co-occurrence of mites and hosts than on any specific associations. The common, east Asian *Arrenurus madaraszi* Daday, for instance, attacked several species of *Aedes*, *Anopheles* and *Culex* (Reisen & Mullen, 1978; Rajendran & Prasad, 1994). In contrast, *Arrenurus danbyensis* Mullen seems to parasitise only the North American *Coquillettidia perturbans* (Walker) (Smith & McIver, 1984b).

More unengorged females of nulliparous (without developing eggs) *Anopheles crucians* Wiedermann were parasitised in Florida by *Arrenurus pseudotenuicollis* Wilson than engorged mosquitoes, and the former invariably bore more mites. Lanciani & Boyt (1977) postulated that these 'missing' mosquitoes represented the extent of early, mite-induced, death of heavily infested hosts, which would result in their failing to take their first blood meal. Mosquito females that were parasitised in the laboratory and then either well fed (had imbibed 2.51–4.00 mg blood) or lightly fed (1.01–2.50 mg) deposited significantly fewer eggs with increasing mite loads. The fecundity of the well-fed group was 175 eggs/female with two parasites/host, decreasing to 125 eggs with nine to 16 mites/host. Comparative values for the lightly fed insects were 140 and 55 eggs/female, indicating that the reduction in fecundity in the latter group was significantly greater. The mites also decreased mosquito survival. The longevity of *A. crucians* was reduced after 5 days of parasitism only when attacked by more than ten mites, and more mosquitoes died as parasite load increased. This suggests that the detrimental effect of the mites is due primarily to the extraction of host tissues (Lanciani, 1979a). The reduction in the reproductive potential of young mosquitoes, in their first ovarian cycle, was calculated to be about 35% (Smith & McIver, 1984b).

The average number of stylostomes in field-collected young female and male *A. crucians* was significantly higher than in older mosquitoes from the same cohort, collected 3 days later. This loss of heavily parasitised hosts was attributed to *A. pseudotenuicollis* (Lanciani, 1979b, 1987).

About 42.5% of a group of newly emerged *C. perturbans* were missing from the older, host-seeking populations of that mosquito in Ontario, Canada, a decrease attributed to *A. danbyensis*. In addition, the mosquitoes produced fewer eggs when carrying more than five mites/host: each additional mite reduced fecundity by 4.5 eggs/mosquito. However, even loads of 16–20 mites/host did not decrease the amount of blood ingested by the mosquitoes (Smith & McIver, 1984b).

An *Arrenurus* sp. that parasitised *Mansonia uniformis* (Theobald) in southern India affected neither host fecundity nor the hatching rate of its eggs. However, mite presence significantly reduced the quantity of blood taken and prolonged, by more than 2 days, the period between feeding and oviposition. Rajendran & Prasad (1992) postulated that heavily parasitised mosquitoes have a lower chance of taking a second blood meal, thereby detracting from their potential as disease vectors as well as lowering their fecundity. *Arrenurus* spp. parasitised about half the population of *M. uniformis* in Assam (eastern India), with an average of 7.2 mites/host (Bhattacharyya *et al.*, 1994), values that suggest that the mites could have a substantial effect on that mosquito.

In Pakistan *A. madaraszi* killed preovipositing females of several *Anopheles* and *Culex* spp. and reduced the survival of their males (Reisen & Mullen, 1978). In total, 37 mosquito species (most in *Anopheles* and *Culex*) were hosts for *Arrenurus* spp. in India. Parasitisation in the southern and warmer regions had two annual peaks, but only one in the northern parts (Malhotra & Mahanta, 1994).

Variations in the effect of *Arrenurus* spp. on various mosquitoes, as reported by diverse authors, suggests that each parasite/host association has its own characteristics, dependent, among other factors, on the size of the partners and on intrinsic defence mechanisms (see below). In addition, environmental conditions (e.g. insufficient food) that reduce host fitness may decrease their ability to cope with different mite loads (Smith, 1983).

Aside from reducing host fitness by parasitisation, *A. madaraszi*, may also decrease mosquito numbers by direct predation. In the laboratory its nymphs and adults fed on small mosquito larvae, preferring *Aedes albopictus* (Skuse) over *Aedes vittatus* (Bigot) and *Anopheles* spp., and living longest on the first prey. Rajendran & Prasad (1994) qualified their data by noting that these values were obtained in the confines of a Petri dish and do not reflect field situations. At 29–30°C the mite's life cycle required about 7–8 weeks.

Certain mosquitoes (e.g. *Aedes cinereus* Meigen) may overcome the mite's attack by avoidance responses when contacted by the larvae. The responses consist of vigorous abdominal movements or rapid swimming; both usually result in mite dislodgement (Smith & McIver, 1984a). A unique mechanism of dragonfly resistance to *Arrenurus* is discussed below.

Arrenurus as a parasite of other insects

The pattern of midge (Chironomidae) parasitisation by *Arrenurus crassicaudatus* Kramer was explored in south Wales by Booth & Learner (1978). The mite had one principal host, whose main period of emergence coincided with the parasite's own life cycle, suggesting phenological adaptation. Although both host sexes were attacked, females carried more mites, which were usually attached to their abdomens. When too many mites attacked a given midge, their mutual jostling tore the intersegmental membranes of the insect's abdomen, causing host (along with parasite) death (Mitchell, 1998).

Parasitism of *Arrenurus* on other insects, such as Odonata, can adversely affect host fitness by both lowering the fecundity of young host females and reducing male condition (Forbes & Baker, 1991). Mite-induced decreases in the physical condition of attacked damselflies significantly reduced their flight distance and increased the risk of being caught by predators in flight (Åbro, 1982; Reinhardt, 1996). Heavy loads (>30/host) of *Arrenurus mitoensis* Imamura & Mitchell on the thoraces of the damselfly *Cercion hieroglyphicum* Brauer hindered its movements and may kill this host (Mitchell, 1968).

One species of dragonfly, *Sympetrum internum* Montogomery, has an effective immune response to *Arrenurus planus* (Marshall). The inserted stylostome turns bubble-like and collapses, which results in the parasite's death (Forbes *et al.*, 1999). Although known so far only from a single host species, such a mechanism of resistance could be more common in nature.

In formulating recommendations for future research on the use of water mites to control flies of economic importance, Lanciani (1983) noted the dearth of experiments intended to quantify the damage that is caused by these parasites to their natural host populations. He also noted that although laboratory rearing methods are at hand, information required for biological control purposes, e.g. specificity, power of increase and host searching, are not available. Some progress towards obtaining such data was made by Smith & McIver (1984a, b). They noted that the distribution of an Old World *Coquillettidia* was analogous to that of the New World species (*C. perturbans*), but that it was parasitised by other *Arrenurus* spp. These authors suggested that *A. danbyensis*, parasitic on the New World *Coquillettidia*, may have potential in controlling Old World species of *Coquillettidia*, and that *Arrenurus* spp., attacking the latter hosts, could help to control *C. perturbans* in North America. The Arrenuridae are among the 'neglected predators' in freshwater communities noted by Proctor and Pritchard (1989), but lacking more observations and experimental data, their potential for controlling dipteran pests remains uncertain.

Chapter 8
Ascidae

Diagnosis

The dorsal plate of the Ascidae bears more than 23 pairs of setae, and may be entire or subdivided. The stigmata, placed between legs III and IV, are connected to straight peritremes. Legs I carry ambulacra but there are no leg grooves and the metasternal shields are small. The female genital shield is not flask-shaped and the distal digits of the male spermadactyl are free. There are close to 30 genera and several hundred species in the family, which was defined by Lindquist & Evans (1965); the European species were keyed by Karg (1993) and the Australian fauna was revised by Halliday *et al.* (1998).

General biology

Many Ascidae are free-living predators that inhabit upper soil layers, plants and stored products, where they feed on nematodes and small arthropods. A few species are parasites of cockroaches and moths (Egan & Moss, 1969; Treat, 1975). Ascids are often carried in large numbers by the flies and moths on whose eggs (and young larvae) they feed; others are dispersed by beetles (Kinn, 1983b). A large group of pollen-feeding species (mostly in the genera *Rhinoseius* and *Proctolaelaps*) lives in tropical flowers and is dispersed by hummingbirds (Colwell & Naeem, 1994). Other Ascidae feed on fungi (Ibrahim *et al.*, 1989) or are omnivorous. The females are usually bisexual and mate to produce eggs, but 30% of the species listed by Walter & Lindquist (1995) were thelytokous.

Temperate-zone Ascidae have colonised protected artificial habitats, such as food stores, mushroom-growing facilities and greenhouses, but rarely occur on plants. Under warm and humid conditions, however, they commonly inhabit plants. Ascids were the most numerous predatory mites found on the leaves of rainforest trees and on tree crops in tropical north-eastern Australia (Walter & Lindquist, 1997). One species was the dominant natural enemy of an acarine pest of rice in southern China (Zhang & Lin, 1991).

Blattisocius spp.

Members of the cosmopolitan genus *Blattisocius* occur in food stores and feed on the eggs and young larvae of resident pest moths and beetles, and on various mites. *Blattisocius keegani* (Fox) raised a generation within 1 week at about 27°C and produced about 20 eggs when feeding on beetle eggs (Barker, 1967). *Blattisocius tarsalis* (Berlese) was the second (after the cheyletid *Cheyletus malaccensis* Oudemans) most abundant acarine predator in tropical and subtropical storage facilities (Haines, 1984), and was believed to be the major factor that controlled the almond moth, *Ephestia* (sometimes placed in *Cadra*) *cautella* (Walker), infesting bagged maize in Kenya (Graham, 1970). The predator survived periods of moth absence by feeding on eggs of the red flour beetle, *Tribolium castaneum* (Herbst), but did not control it. Nevertheless, this food source ensured that enough mites would be present in the bags to prevent *E. cautella* from establishing therein. Moth outbreaks, which elevated *E. cautella* to major pest status, were attributed to the effect of the organophosphate OP pesticide malathion (applied against other insects) that killed more mites than moths. Graham (1970) concluded that 'the insecticides ... were helping *E. cautella* to escape control by *B. tarsalis*'.

The mite completed a generation (egg to egg) in about 1 week at 27°C and 73% relative humidity (RH), when offered live eggs of *E. cautella* (Haines, 1981). Eggs killed by chilling provided almost as good a diet. Each paired-for-life female deposited about 32 eggs during 3 weeks, and then lived and fed for another 5 weeks. The sex ratio (SR) was 0.7, and the intrinsic rate of increase (r_m) calculated under these conditions came to 0.2428/day. Juvenile mites consumed four moth eggs during their development, males about 65 eggs throughout their lives, and females 100 (the ovipositing females consumed 2.6 eggs/day). *Blattisocius tarsalis* offered only *T. castaneum* eggs developed more slowly and their mortality (especially in males) was higher, possibly owing to the inability of male nymphs to pierce the eggs' thick shells. In consequence, the SR was 0.9 and the r_m was 0.1760/day. Females offered only wheatgerm, yeasts or no food at all survived for 11 days. They were dispersed by phoresy on moths and beetles (Haines, 1981), preferring female hosts, whose longevity was reduced when nine or ten mites were present on a single insect (White & Huffaker, 1969).

Haines (1981) evaluated the role of *B. tarsalis* as a biocontrol agent. It co-occurs with the moth in most tropical and subtropical regions, is a mobile and voracious predator of pest eggs during all its active stages, including the long postovipositional period, and may reach prey even in the inner layers of the grain bulk. The rapid rate of increase of *B. tarsalis* is in excess of the prey's, and the mite is dispersed while clinging to its food source. As it prefers female moths for phoresy, the predator is more likely to reach the prey sooner. Finally, *B. tarsalis* can survive and develop on another pest of stored products, the beetle. The predator tends to die out within a short time in pure moth cultures, owing to the longer generation time of *E. cautella*, whose populations tend to become synchronised. In consequence, any mite juveniles that emerge within the moth cycle (e.g. while all moths are in their larval stages) cannot not find prey eggs and will starve. The presence of *T. castaneum* eggs, an inferior but acceptable alternate diet, allows the predator to survive and increases its potential to control the pest. The feasibility of augmentative releases of *B. tarsalis* for moth control is dependent on both mass-rearing the predators and knowing when prey eggs would be available in the moths' synchronised life cycle (Haines, 1981).

The use of *B. tarsalis* to control stored product pests has not formerly been a viable option in regions where arthropod contaminants in the grain bulk require additional treatments. However, regulatory pressures in European countries are forcing cutbacks in the use of fumigants within stored product stores, resulting in renewed interest in biocontrol options (see Chapter 38, ABAs as enemies of stored product pests). This is the context in which *B. tarsalis* will be assayed against the Mediterranean flour moth, *Ephestia kuehniella* (Zeller), the major pest in flour mills in Denmark (Hansen, 1998).

Arctoseius spp.

Arctoseius spp. feed on eggs and young larvae of *Lycoriella auripila* (Winnertz) (Diptera: Sciaridae), a pest of cultivated mushrooms. The hatch of pest eggs was reduced by around 85% after placing ten females of *Arctoseius cetratus* (Sellnick) into small pots containing several hundred fly eggs. In the laboratory the mite's life cycle (adult to adult) required about 2 weeks at 22°C. Average daily fecundity was 2.5 eggs/female, produced during 9 days (Binns, 1974). The predation of *A. cetratus* on *L. auripila* was not disrupted by the fungicide (and acaricide) benomyl (Binns, 1973). The predator also fed on cecid (Diptera: Cecidomyidae) larvae, omnivorous mites and nematodes. Dispersal was by mated females that attached to sciarid flies.

Arctoseius semiscissus (Berlese) was dispersed by *L. auripila* within mushroom houses in Poland. Female mites detached when the flies started to lay their eggs, on which they immediately fed. A single mite damaged 20 eggs within 3 days, and young larvae were also taken. The predator was believed to effect the almost total disappearance of sciarids within a few mushroom production cycles (Dmoch, 1995). When supplied with sciarid eggs in the laboratory at 20°C, development from egg to adult required almost 8 days, the SR was 0.67, each female produced about 58 eggs, net reproductive rate (R_0) was 24.5 and r_m came to 0.23 (Rudzińska, 1998).

Lasioseius spp.

Lasioseius parberlesei Tseng is a natural enemy of *Steneotarsonemus spinki* Smiley (Heterostigmata: Tarsonemidae), a major acarine pest of rice (Tseng, 1984). The predator showed a numerical response to changes in pest populations and was believed to hold the tarsonemid's numbers at an endemic, non-injurious level in southern Taiwan. *Lasioseius chaudhrii* (Wu & Wang) is the dominant predator of another tarsonemid pest of rice, *Tarsonemus fuzhouensis* Lin & Zhang, in Fujian Province, southern China (Zhang & Lin, 1991, who placed the predator in the genus *Gnorimus*). The populations of *L. chaudhrii* tracked those of *T. fuzhouensis* at its summer and autumn peaks, and a single predator killed 200–300 pest individuals during its life in the laboratory.

Lasioseius bispinosus Evans was among several predatory mites that were collected in Japan during a survey for natural enemies of the bulb mite, *Rhizoglyphus robini* (Claparède) (Lesna *et al.*, 1995). This small predator fed almost exclusively on the eggs and juveniles of the intended prey, and seldom on its females. Although capable of suppressing prey populations within jars, *L. bispinosus* was not deemed to be very promising. Its small size may facilitate penetrance into bulbs, where it could feed on prey juveniles, thereby having a stronger impact on the pest.

Ascidae as natural enemies of nematodes

Lasioseius dentatus Fox (often identified as *L. scapulatus* Kennett) was reared on the nematode *Aphelenchus avenae* Bastian in Petri dishes (Imbriani & Mankau, 1983). The presence of even a single mite in a dish caused a reduction of around 70% in nematode numbers, and with four mites/dish, prey populations declined by 99% (Fig. 38.1). The mite attacked and consumed several plant- parasitic nematodes and also completed its life cycle on mould fungi (e.g. *Aspergillus* sp.). Prey suitability appeared to depend on nematode size, because larger worms reacted to mite attack by violent movements and subsequent escape. The reproduction of *L. dentatus* was asexual and rapid; a generation was completed in 6–8 days at 24°C. The mite's short life cycle, great appetite and ability to subsist on alternate food (fungi), as well as its density dependence, were considered to contribute to the efficacy of *L. dentatus*. However, the predator probably does not reach all available prey, because it prefers the uppermost soil strata, whereas plant-feeding nematodes inhabit deeper root zones. Imbriani & Mankau (1983) concluded that the use of this predator would be most appropriate for nematode control in special situations, such as greenhouses or organic soils (see Chapter 38, ABAs as enemies of nemotodes). *Proctolaelaps dendroctoni* Lindquist & Hunter, formerly believed to be a predator of pine bark beetles, was reared on an exclusive diet of nematodes. The mite, which is phoretic on bark beetles, starved when confined only with these insects (Kinn, 1983b).

Nematophagous mites could interfere with biological pest control by feeding on purposely released entomogenous nematodes. The risk was assessed by Epsky *et al.* (1988), who exposed the nematodes to several soil mites. Only the ascid *Gamasellodes vermivorax* Walter, a relatively large species, completed its development on these nematodes. However, the mite's large size was expected to restrict its search for entomogenous nematodes to the upper soil layers, thus making it of little danger to entomogenous nematodes.

Miscellaneous

A substantial increase in the number of Ascidae found in the surface litter of an apple orchard was preceded by an application of chicken manure. Childers & Enns (1975b) postulated that these predators could affect pest mites that overwinter in the surface litter. Chicken manure might thus be used to promote pest control.

An important factor that could affect the performance of Ascidae in stored products is the presence of other predators, especially Cheyletidae. A study of the interactions between *Blattisocius dentriticus* (Berlese) and *Cheyletus eruditus* (Schrank) indicated that when both were propagated in the same containers, the former was usually eliminated within 6 weeks (Burnett, 1977) (see Chapter 42). However, observations that *Blattisocius* spp., as well as cheyletids, are abundant in tropical stored products (Haines, 1984) indicate that under natural conditions these predators are able to coexist. The effect of pesticides on Ascidae is discussed in Chapter 43.

In conclusion, although some Ascidae are generalist predators, they could be used for the control of certain pests under specialised conditions.

Chapter 9
Bdellidae

Diagnosis

The Bdellidae have snout-like mouthparts, with long palpi that terminate in strong setae but lack the palpal thumb–claw complex. They are large (up to 4 mm), red–brown or green mites, whose prodorsum bears two prominent sensilli. About 100 species in 15 genera were assigned to the family. The American species were revised by Ateyo (1960), the Australian taxa by Wallace & Mahon (1972, 1976), the Taiwanese fauna by Tseng (1978) and those of Hawaii by Swift & Goff (1987).

General biology

Bdellids are common on plants and in the soil, occurring also in stored products. They are active hunters of small arthropods, locating the prey with their palpi. Once the prey (most often springtails or mites) has been located and sized, the predators tether it with a squirted salivary liquid that hardens upon contact into silken lines. The bdellids then suck out the body contents of the prey. Some species are parthenogenic (Soliman & Mohamed, 1972). Eggs are often deposited in masses within silken nests. Bdellids are efficient natural enemies of the lucerne flea in Australia (Wallace, 1981). Reduction in the populations of this pest by *Bdellodes lapidaria* (Kramer) was probably the first demonstration of pest control by a mite (Womersley, 1933). Studies on this and other bdellids, especially *Neomolgus capillatus* (Kramer), have been conducted in Australia since the early 1930s and are continuing.

Bdellodes lapidaria

The lucerne flea was introduced into Australia in the late nineteenth century and became a major pest of subterranean clover in Western Australia (WA) (Wallace, 1954). In 1931 it was noted that a previously ravaged field, south of Perth, showed little damage. Only few *Sminthurus viridis* were seen during a visit to the site, but large numbers of a 'small red mite', recognised as a predator and identified as *Biscirus lapidaria*, were obtained (Womersley, 1933). (The species was later transferred to the genus *Bdellodes*.) Observations conducted during the following 18 months indicated that the predator was slowly extending the area wherein it was controlling the pest. In one paddock 'there was a very finite line of attack, in which the bdellids were most abundant. In front of this line no mites were to be found, while the *Sminthurus* population was very heavy'. During the ensuing 18-month period it was estimated that an area of about 26 ha was cleared of the pest. The gradual increase in the region colonised by *B. lapidaria* suggested to Womersley (1933) that its spread could be facilitated by transferring it to new areas. Introductions to other plots were then made, consisting at times of no more than a dozen, or up to 50 individuals; all were successful. Currie (1934) revisited some of the same release sites a year later and reported that the predators had spread further. He found a consistent pattern of lucerne flea control, in which the predator's advance along a field resulted in pest population reductions of about 75% (from 20–36 to 5–10/sampling point; Fig. 44.1). Shipments to other states, namely South Australia, Victoria and Tasmania, were then initiated. *Bdellodes lapidaria* hides under any shelter (e.g. branches or logs) available in the field during rains. This suggested placing 'trap' boards on the soil in areas where the predator occurs and harvesting them from the 'traps' after heavy showers. About 1000 mites/h per person were thus collected into glass tubes that had been provisioned with young prey stages. Collected mites were kept in cold storage and sent to the three states, were released and became established. Later reports differed in their appraisal of the effect of *B. lapidaria* on pest populations, as it became known that the predator had reached these states before the releases (Ireson, 1984).

Wallace (1954) demonstrated the predator's beneficial effect in WA by applying the pesticide check method (Chapter 44). The insecticide dichlorodiphenyltrichloroethane (DDT, lethal to the predator but not to the pest) was added to a fertiliser during a top-dressing administered to clover in plots that were infested by the pest, where *B. lapidaria* was also present. In the first experiment (DDT added at concentrations of 0.5%), the predator was totally eliminated from the pesticide plots, wherein the pest became significantly more numerous than in the control plots. The results of the second experiment (with 1.0% DDT) were even more striking. Pest populations declined in treatment and in control plots, but whereas those in the DDT plot declined by about 80%, those in the unsprayed control dropped by more than 98%. Within 10 weeks the lucerne flea population in the unsprayed controls was only around 5% of that in the DDT plots (Fig. 9.1). By the end of the experiment there was about one predator for every three lucerne fleas in the controls, compared with one *B. lapidaria* for every 134 pests in the DDT plots. The perceived need for using selective pesticides (e.g. that affect the pest but not the predator) became clear from this experiment.

A summation of 8 years' research led Wallace (1967) to conclude that whenever an infested field was sprayed with DDT, pest populations would reach levels two to 15 times higher than those found in unsprayed plots. Furthermore, a very significant regression was obtained between the numbers of *S. viridis* and those of *B. lapidaria* that had been recorded 8–9 weeks previously. Wherever predator numbers exceeded $20/m^2$ early in the season, no pest outbreaks would occur later on in the season (attributed in part to the more efficient feeding of juvenile bdellids on young prey stages). In contrast, when *B. lapidaria* density was $<10/m^2$ in the spring, *S. viridis* might increase to pestiferous levels later on. Several other indigenous and introduced bdellids also occurred in pastures damaged by the lucerne flea, but none appeared to affect pest numbers.

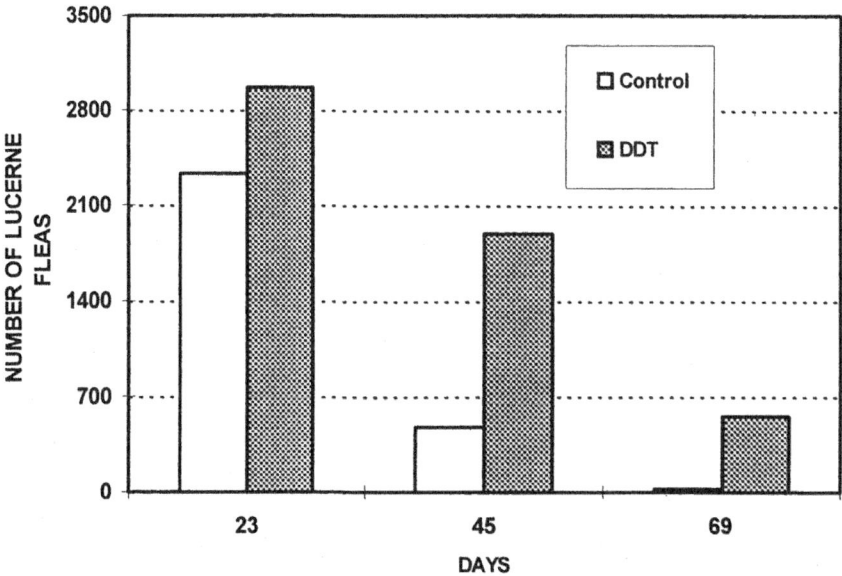

Fig. 9.1   Numbers of lucerne flea (*Sminthurus viridis*) in a clover field treated with DDT and in the control plot, 23, 45 and 69 days post-treatment. (Based on Table 1 in Wallace, 1954.)

Five factors were postulated to affect the pest's year-to-year numerical fluctuations in WA pastures. These included the weather, soil type, vegetation, predation by *B. lapidaria* and consumption of apparently toxic bodies of dead lucerne fleas by their own young. Low temperatures and heavy rainfall during autumn, winter and spring created conditions favourable to juvenile lucerne fleas when suitable plants were available. Wherever *B. lapidaria* was present it fed heavily on the pest. Such control could, however, be indirectly disrupted by year-to-year weather fluctuations. Lack of autumnal rains reduced the number of available pests, thereby engendering irregular oscillations in the predator–prey cycles. The lucerne flea appeared to be the preferred, but not exclusive, diet of *B. lapidaria*, which devoured alternate prey (other collembolans and psocids). Feeding on such prey reduced the predator's fecundity and its numbers declined after more than one season. The predator's populations grew again only in response to increases in lucerne flea numbers; the climatic factors that limited pest numbers thus also affected predator density. The long-term beneficial effect of *B. lapidaria* was evinced by the decline in *S. viridis* from a pest that required annual insecticidal treatments to minor status, with local outbreaks occurring only once in every 3–5 years (Wallace, 1967).

The eggs of *B. lapidaria* are usually deposited close to the soil surface, often stuck to plant debris. Being always in diapause to some extent, their subsequent hatching takes place over a long period. Egg development was most rapid at 30°C, but occurred at a wide range of temperatures (16–38°C) (Wallace, 1971); later temperatures <21°C and wetting then permitted their hatching. These findings suggest that mature eggs hatch only after the arrival of seasonal rains, which also herald the appearance of the prey. Eggs in diapause were successfully used to transport *B. lapidaria*.

Adults of *B. lapidaria* lived for about 9 days at about 18°C in the laboratory (Currie, 1934). A single mite devoured 18 prey nymphs/24 h on 3 successive days; adult spring-tails were attacked but often escaped. Eggs were never taken; Jenkins (1935) noted that the mites would starve but not feed on lucerne flea eggs. In the autumn *B. lapidaria* may be displaced by another predator, the anystid *Anystis wallacei*, owing to competition for food (Michael *et al.*, 1991; Chapter 6).

In north-western Tasmania *S. viridis* was a pest whose control in pastures required chemical treatments, as no significant predator–prey interactions with *B. lapidaria* were observed during several years. This was attributed to unsuitable climatic condi-tions that affected the predator and to the presence of abundant alternate prey species. As no other acarine predators were seen to provide good pest control a search for more climatically preadapted natural enemies was advocated (Ireson *et al.*, 2001, and earlier papers).

The success of *B. lapidaria* in WA prompted its introduction into South Africa to control the lucerne flea. About 80,000 mites were sent during 1963–1966 and released at several sites in the Cape Province (Wallace & Walters, 1974). A total of 30,000 eggs, laid by the mites during the passage and deposited on pieces of cardboard or on tissue-paper strips, was also placed at the release sites. Sampling at two of the sites showed clear interactions between *B. lapidaria* and another pestiferous collembolan, *Bourletiella arvalis* Fitch. Low pest densities followed periods of high bdellid num-bers, and large pest populations (>100/m^2) occurred 4 weeks after low *B. lapidaria* numbers had been recorded. Significant negative interactions between predator and prey populations on the same day were sometimes evident (Wallace & Walters, 1974), attesting to the effect of *B. lapidaria* on *B. arvalis*. The value of the predator as a bio-logical control agent was enhanced by its ability to control more than a single pest springtail.

Neomolgus capillatus

Wallace & Mahon (1971) surveyed the distribution of *S. viridis* and *B. lapidaria* in Australia. Both occurred under almost the same climatic conditions, except for a small but significant difference. The lucerne flea, being slightly more tolerant of dry conditions, had penetrated deeper inland. Consequently, there remained a belt of land, varying in width from a few to over 100 km, where the pest was not controlled by *B. lapidaria*. Another predator, better adapted to dry inland conditions, was thus needed, leading to a survey of bdellids in the eastern Mediterranean, the assumed origin of the lucerne flea. *Neomolgus capillatus*, an abundant species that fed on the pest in pastures and grasslands, from Scandinavia to Morocco, was the most promis-ing candidate (Wallace, 1974). About 70% of the 18,000 mites that were collected in The Netherlands during 1965 arrived safely in WA, but failed to establish. The fail-ure was attributed to the humid provenance of the mites, which were assumed to lack the ability to survive the long and dry summer in WA.

A second effort was made in 1969, using predators collected in the Atlantic coastal districts of Morocco. In that year, the end of the growing season in Morocco happened to overlap with the beginning of the season in WA, enabling the collection and almost immediate release of *N. capillatus*. Mites were collected by a sucking machine, sorted and placed in small cages (see Chapter 39). About 17,500 mites were sent to WA, released at one site and became established there. The mites had laid about 16,000 eggs in transit, which were retained in the laboratory and later placed in the field. Two years passed before a small *N. capillatus* population became established from the egg introduction (Wallace, 1974). Pest populations in fields without predators carried about 3200 lucerne fleas/m^2, whereas in the release plots their numbers were only 730/m^2, a 78% reduction that demonstrated the predator's efficacy (Wallace, 1981). The inability of *N. capillatus* to spread for more than about 70 m/year required their redistribution to additional sites (Michael *et al.*, 1991). About 150,000 *N. capillatus* were collected in 1988 and released in about 100 fields; at one site the predator became established after the release of only 85 mites.

As noted, the situation in Tasmania was ripe for introducing additional predators. The European survey had shown that *N. capillatus* was abundant in diverse climatic zones. The similarities between some areas of predator origin, especially north-west France (Brittany) and north-western Tasmania, the target zone, suggested that populations originating from Brittany might be suitable. A total of 33,500 European mites was released during 1985–1990 (Ireson & Paterson, 1991). Subsequent sampling (Ireson & Webb, 1995) indicated that mite activities had reduced the mean autumnal maxima of *S. viridis* by more than 95%. Predator populations in excess of 25/m^2 early in the season were associated with low, non-damaging pest levels later on. A positive regression was calculated between numbers of *N. capillatus* recorded on certain dates and those of *S. viridis* tallied 2 weeks previously, a relationship that accounted for 80% of the variation in pest numbers. Most control was attributed to the presence of *N. capillatus* in the fields during autumn, when pest eggs were hatching, and the predator's population consisted mainly of young nymphs (Ireson & Webb, 1995).

As noted, *N. capillatus* has a slow rate of spread, being easily hindered by obstacles such as roads or creeks. This made it necessary to redistribute the predator in Tasmania. Vacuum-collected *N. capillatus* were placed into ventilated plastic containers lined with paper tissues, stored overnight at 4°C and transferred to new sites. Mites were released during spring, summer and autumn but not in winter, the period of least predator activity. Over 1.6 million *N. capillatus* were released at 840 sites on all Tasmanian dairy farms where *S. viridis* was considered a pest. Predators were placed into small areas within a relatively high (15–20 cm) growth of pasture plants, and farmers were advised to restrain their cattle from grazing to below 10 cm at these sites, and not to use pesticides there (although *N. capillatus* seems to be tolerant to many chemicals). The overall establishment rate was about 90%, but there is still a need for intrafarm redistribution of the predator (Ireson *et al.*, 2001).

Other Bdellidae

Additional Bdellidae are known to feed on the lucerne flea and/or on another pest, the red-legged earth mite (*Halotydeus destrucor*, RLEM) in Australia (Wallace, 1974; James, 1995). Little is known about their quantitative effect on the pests, but James (1995) believed that these bdellids were part of a complex of insect and acarine predators that provided natural RLEM control.

Other Bdellidae were reported to contribute to the control of miscellaneous pests, mostly mites. *Bdella depressa* Ewing is a voracious feeder of spider mites (especially the clover mite, *Bryobia praetiosa* Koch), and various springtails in Illinois, USA. Development from larva to adult (at 90% relatively humidity) required 21–30 days at 15°C and 14–21 days at 21°C (Snetsinger, 1956). The larvae devoured at least three eggs or two pest larvae, and each nymph required a minimum of two or three prey nymphs or adults to complete its development. The eggs, deposited in dry tree bark crevices, were the overwintering stage. A similar habitat, on grapevines, was the site preferred by *Bdella longicornis* (L.). Sorensen *et al.* (1983) sampled bdellid numbers on vine bark in California, USA, and found 3.4–8.3 mites/25 cm-long bark sections. Most *B. longicornis* occurred in winter and spring. At that critical period, before resident phytoseiid predators (Chapter 26) become sufficiently abundant to control vine spider mites, each bdellid consumed around 1.8–3.3 pests/day. This early season reduction in pest numbers may expedite their subsequent control by phytoseiids (Sorensen *et al.*, 1983).

Bdella distincta (Baker & Balock) fed on eggs and crawlers of armoured scale insects (Hemiptera: Diaspididae) in Florida, USA (Muma, 1975), but no quantitative data are available. A *Spinibdella* sp. was believed by Goh & Lange (1989) to be a key factor (along with and augmenting a phytoseiid) in reducing phytophagous mites on artichokes in California. Unidentified bdellids were considered among the major predators of lucerne aphids in eastern Australia (Milne & Bishop, 1987). Reporting from the island of Reunion, in the Indian Ocean, Vincenot & Quilici (1995) noted that an unnamed bdellid (probably a species of *Bdellodes*) rapidly controlled spider mites on citrus. Members of the family Bdellidae thus join an assemblage of polyphagous predators from various families, which are major biological control agents of pests.

The need to protect bdellids from adverse agricultural practices goes back to the initial discovery of *B. lapidaria* in WA, when Newman & Womersley (1932) wrote 'it is inadvisable to adopt any artificial treatment owing to the risk of destroying the beneficial creature'. Bdellid susceptibility to insecticides is discussed in Chapter 43.

Chapter 10
Camerobiidae

Diagnosis

The Camerobiidae can be recognised by their long, stilt-like legs, with elongated femora and tibiae, ovoid body, fused chelicerae, weak palpi on a ventrally directed gnathosoma, and looped peritremes. It is a small family with approximately 100 species in six genera. The family was revised by Bolland (1986) and the most recent key to its six genera is by du Toit *et al.* (1998). A camerobiid discovered in Baltic amber (dating back to the Upper Eocene) attests to the antiquity of the family (Bolland & Magowski, 1990).

General biology

Camerobiids commonly occur on plants and in plant litter, where they feed on small arthropods. Species of the cosmopolitan *Neophyllobius*, the largest genus in the family, were postulated to contribute to the control of armoured scale insects (Hemiptera: Diaspididae), by preying on their first instar nymphs ('crawlers'). Pence (in McGregor, 1950) observed the mite to attack unidenitified crawlers, injecting them with an 'opiate' and sucking them dry. Meyer (1962) reported that nymphs and adults of *N. ambulans* Meyer fed on crawlers of the California red scale, *Aonidiella aurantii* (Maskell), a major pest of citrus, in South Africa. The predator was scarce, which made it of only minor interest. Richards (1962) raised the possibility that another (apparently unnamed) *Neophyllobius* plays an important role in the natural control of the oyster-shell fruit scale, *Quadraspidiotus ostreaeformis* (Curtis), in New Zealand. Mites were very abundant wherever the scale was common. In the laboratory *Neophyllobius* fed on crawlers, which were initially paralysed and then sucked dry. Other species of *Neophyllobius* feed on eriophyoid or tenuipalpid mites (De Leon, 1967; Bolland, 1983), but their effect on pest populations is unknown.

Chapter 11
Ceratozetidae

Diagnosis

The integument of the Ceratozetidae is generally smooth, lacking rough sculpture; lamellae and cusps are present, with inner margins of the lamellae fastened to each other. Notogaster bears three to seven pairs of porose organs; pteromorphs are directed downwards, hinged or unhinged. The genital plates have six pairs of setae. The family, first defined by Grandjean (1954, 1963), has close to 50 genera and 300 species. The Russian ceratozetids were reviewed by Shaldybina (1975) and Pavlitshenko (1994), those of New Zealand by Luxton (1985), and the western North American arctic and subarctic species by Behan-Pelletier (1985, 1986).

General biology

Ceratozetidae are common soil and litter mites that feed mostly on fungi and lichens (Woodring & Cook, 1962; Mitchell & Parkinson, 1976), and may also occur in suspended soils and on trees (Aoki, 1971; Behan-Pelletier, 2000). Soil-borne species have a stable population structure and a prolonged life cycle, but in the laboratory a generation may be completed within a few weeks. *Ceratozetes cisalpinus* Berlese raised a generation in 32 days at about 25°C and deposited ten to 16 eggs when fed mushrooms and lichens (Woodring & Cook, 1962). Reproduction is usually sexual, insemination being via spermatophores (Alberti *et al.*, 1991). At least 15 species of Ceratozetidae serve as intermediate hosts of tapeworms, whose eggs are ingested by the mites. Denegri (1993) believed that these eggs are too large to be taken by chance, suggesting that the eggs are probably interpreted by the mites as potential or even preferred food.

Diapterobates humeralis (Hermann)

The hemlock woolly adelgid, *Adelges tsugae* Annand (Hemiptera: Adelgidae), a small, sucking insect, occurs on hemlock in eastern Asia and in North America. The insect deposits its eggs within a single woolly ovisac that is placed on young hemlock twigs. In North America *A. tsugae* is a serious pest of eastern hemlock (*Tsuga canadensis*), and unless checked, may kill trees within 4 years. In Japan, in contrast, the insect does not harm its hosts, being regulated by tree resistance and by natural enemies, including *Diapterobates humeralis*. The latter is common in the pests' colonies, and was the main (and often the only) natural enemy collected. The mite did not attack *A. tsugae* but fed on the woolly filaments that make up its ovisacs. Most (>95%) of the pest's eggs were dislodged and lost in consequence, as the emerging adelgid nymphs were unable to infest nearby trees. The mite destroyed 86–100% of the egg masses at all adelgid-infested sites surveyed on Honshu, and high pest populations were seen at sites where the mites were not present (5–9 ovisacs/5000 cm twigs with mites, vs 58–97/5000 cm twigs without mites). As *D. humeralis* is an effective natural enemy, has a broad tolerance for a wide range of environmental conditions and is harmless to other organisms, it was believed by McClure (1995) to be a suitable candidate for controlling *A. tsugae* in North America. However, the mite's long generation time and low rate of reproduction made it unsuitable for mass-rearing. In addition, *D. humeralis*, introduced from Japan, did not show the same affinity for the pest's woolly ovisac filaments on North American hemlock as it had on the Japanese trees (McClure & Cheah, 1999).

Chapter 12
Cheyletidae

Diagnosis

Cheyletids bear a prominent gnathosoma with a strong thumb–claw complex; the palpal tarsus usually carries sickle- and/or comb-like sensilli. The dorsal setae attain many shapes, from simple, hair-like, to clavate, shell-like or staghorn-like. The males of some cheyletids occur in two forms, the homeomorphs, similar to the females, and the dissimilar heteromorphs. The latter possess robust and elongated palpi within their enlarged mouth parts. The heteromorphs may cause systematic difficulties, because some species in the genus *Cheyletus* are known only in this form (Volgin, 1969). The Cheyletidae was reviewed by Volgin (1969), revised by Summers & Price (1970) and summed up by Gerson *et al.* (1999), who listed approximately 400 species in about 80 genera in the family. Regional discussions include Tseng (1977) for Formosa, Gerson (1994c) for Australia, Hoffmann (1998) for Mexico and Corpuz-Raros (2000) for the Philippines.

General biology

Cheyletids may roughly be separated into two groups, parasites and predators. The dividing line between them is tenuous, although the obviously parasitic taxa possess bizarre modifications of their mouthparts and legs, and bear reduced or augmented setation (hypotrichy or hypertrichy). The parasitic taxa occur on mammals, birds or insects; their biology is little known. The Cheyletidae are usually arrhenotokous, but several species are parthenogenic.

The predators are free-living, slow-moving, yellow or orange-coloured mites that usually approach and capture their prey cautiously or ambush it. They occur on plants and in the soil as well as in vertebrate nests and stored products, feeding on mites or small insects. Several species [e.g. *Hemicheyletia bakeri* (Ehara)] immobilise their prey's body, which suggests the injection of a toxin (Laing, 1973). This mode of capture enables even the larvae of *P. bakeri* to feed on prey several times their own size. Hungry cheyletids suck out the bodies of their victims, but otherwise may attack and kill many prey without prolonged feeding (Ewing, 1912), sometimes leaving them only partially consumed (Laing, 1973). *Cheyletus eruditus* (Schrank), a common inhabitant of stored products, house dust and poultry litter, was the first mite used in predator–prey population studies (Gause *et al.*, 1936). To this day it remains the most studied member of the family.

Cheyletus eruditus

The life cycle of *C. eruditus* required 3–4 weeks at 25–28°C, extending to 52 days at 18.5°C. Each female consumed a single prey mite/day when offered the acarid *Lepidoglyphus* (=*Glycyphagus*) *destructor* Schrank, and deposited a daily average of 5.4 eggs at 25°C for 25 days, with a mean total of 133 (maximum 212) eggs (Barker, 1991). Reproduction was mostly parthenogenic, as males were rare. Ovipositing females remained near their eggs and guarded them, but fed on their own juveniles, and even on weaker adults, during periods of hunger. The predator also attacked other mites, small larvae of moths and beetles, as well as psocids (Psocoptera) (Sinha, 1988; Barker, 1991). Pest species, such as *Acarus siro* L. [formerly called *Tyroglyphus* or *Aleurobius farinae* (L.)] and *A. farris* (Oudemans), appear to be the preferred prey in stored products. The predator may devour 0.9–4.5 prey-mites daily at 25°C and 80% relative humidity (RH) (Solomon, 1969). It had a domed functional response to prey density that peaked at three prey/predator, at a density of 12 *A. siro*/cm², and the tritonymphs preferred to feed on the prey's juvenile stages (Berreen, 1984).

The possibility that *C. eruditus* could reduce the numbers of pestiferous acarid mites in stored products was often voiced. Several authors observed their prey–predator interactions (reviewed by Solomon, 1946; Norris, 1958) but came to different conclusions. Some writers discounted the predator's ability to control stored product mites (e.g. Newstead & Duvall, 1918). Others, including Gause *et al.* (1936), indicated that *C. eruditus* can and does reduce the numbers of acarid mites. The factors underlying these inconsistencies were partially resolved by Solomon (1946 and cited papers), who sampled several hundred stored product (mostly wheat grain and flour) sites in various regions in England and tallied the populations of the resident mites qualitatively as well as quantitatively. The extent of the incurred damage was graded and the prevailing physical environmental conditions (mainly temperature and percentage of moisture content) were recorded. With these data Solomon (1946) related several levels of acarid populations to environmental conditions in the stored products. Pests and predators often co-occurred, but the differences in their environmental requirements usually affected the outcome of the interaction, e.g. whether or not pest control took place. The deciding factor was the relative humidity. When it rose in the substrate, in consequence of pest and/or fungal invasions, the numerical ratio between the predator and pests shifted in favour of the latter (Fig. 12.1). This was due to increased pest reproduction, in part compounded by reduced predator fecundity. A seasonal cycle of control would thus take place: the pest (*Acarus siro*) became very numerous under winter conditions, with *C. eruditus* controlling it in summer.

These field data were confirmed and extended by long-term laboratory studies (Solomon, 1969). By separately plotting the increased rates of *A. siro* and *C. eruditus*, along with data showing days to fastest pest control, and the percentage of success that could be expected to occur at different combinations of temperature and RH, and then superimposing them on each other, a general pattern of prey–predator relationships emerged (Fig. 12.2). The predator prefers temperatures of about 20–25°C and 70–80% RH; when these conditions prevail *C. eruditus* rapidly controls *A. siro*. The pest does better under slightly lower temperatures (<20°C) and higher RH values (>80%),

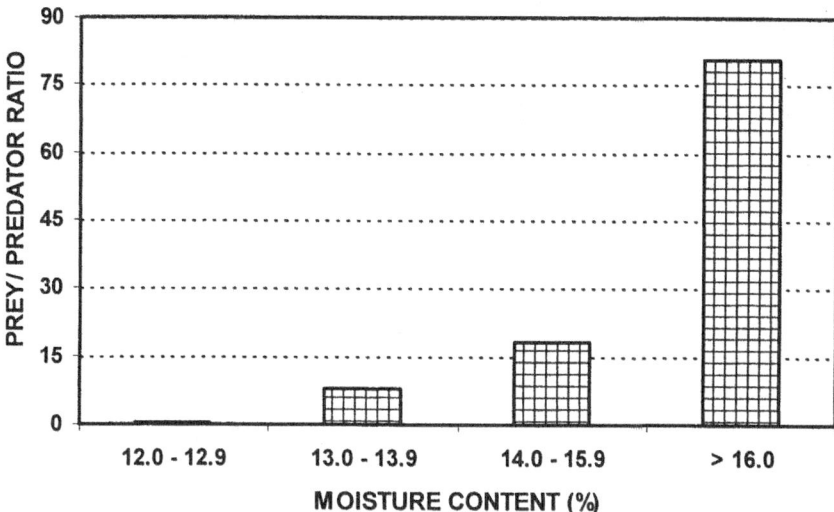

Fig. 12.1 Effect of the moisture content of wheat on the numerical ratios between prey (*Acarus siro*) and predator (*Cheyletus eruditus*), based on a total of 72 samples collected at various sites in England. (Data from Table 2 in Solomon, 1946.)

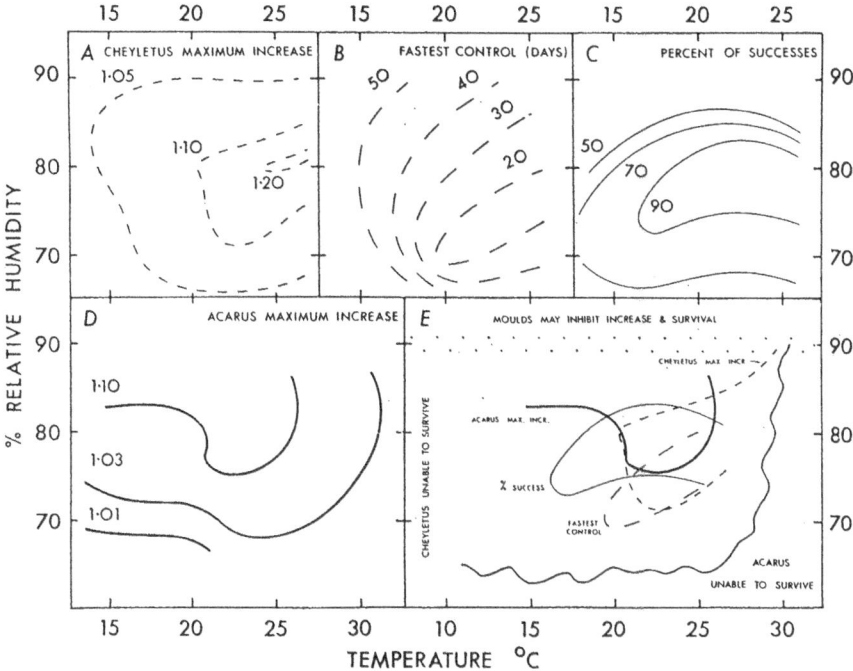

Fig. 12.2 The pattern of prey pest (*Acarus siro*) and predator (*Cheyletus eruditus*) interactions and pest control as affected by temperature and humidity. (A) Rate of increase of *C. eruditus* per day). (B) Days to achieve control. (C) Per cent of pest control. (D) Rate of increase of *A. siro*. (E) Inner curves from A, B, C and D superimposed, and a general indication of conditions limiting the prey and the predator. (From Solomon, 1969, with kind permission from *Acarologia*.)

conditions at which more damage should be expected. This pattern explained the conditions that limit the increase of the two organisms, as well as indicating under what circumstances optimal (or less than optimal) control can be expected.

This seasonal cycle of pest control in the summer and damage in the winter held mostly for bagged wheat, whereas a different pattern emerged from examining wheat in bulk. Continuous sampling in British stores indicated that *C. eruditus* (and possibly *Cheyletus malaccensis* Oudemans, identified as *C. munroi* Hughes by Norris) were numerically dominant in the bulk and controlled the pests during winter. Very few *A. siro* (> 50/100 cm^3 wheat) were seen whenever the cheyletids became dominant, indicating their controlling effect. Predator populations increased in December and January, followed by massive reductions in pest numbers. Norris (1958) noted several cases in which active predators appeared in mid-winter and reproduced in rather humid grains (15–17.5% moisture content). The reasons for the different seasonal activity of *C. eruditus* in bagged and in bulk wheat remain unclear.

Other factors that affected the outcome of the interaction are the nature of the product, its physical state and the means of storage. Gause *et al.* (1936) reported that the substrate on which the pest fed strongly affected its cyclic interaction with the predator. The cycle was curtailed when the pest fed on millet, *C. eruditus* controlling it at the low level of around 20 individuals/0.2 g grain. When feeding on wheat flour, however, acarid numbers were reduced only after they had reached more than 60/0.2 g grain. A mixture of the two substrates resulted in control taking place at around 40 pests/0.2 g. The outcome may be related to lesser prey accessibility to the predator within the finer substrate, compared with the firm surface of millet grains (Gause *et al.*, 1936). This was an early indication that the substrate has a major effect on the ability of *C. eruditus* to control stored product mite pests. Solomon (1946) noted that wheat stored in bags has a bigger exposed surface area than bulk grain. Mite infestations in bagged wheat consequently tended to be more widespread, promoted by higher atmospheric humidities that circulate more freely between the bags. Besides the higher humidity, the activities of *C. eruditus* in such bags is also hindered by its need to enter each bag individually, provided the mites are not excluded by the minute openings in the bagging material. The performance of *C. eruditus* is further influenced by the presence of other predators. During interactions between *C. eruditus* and the ascid predator *Blattisocius dentriticus* (Berlese) (Chapter 8) in the laboratory, the latter was consistently eliminated by the former when they co-occurred in the same container (Burnett, 1977); in a real stored food situation, however, the outcome could differ.

The prolonged survival of *C. eruditus* in laboratory studies was attributed to its delayed numerical response to decreasing prey numbers, and to persisting in searching even at low prey levels (Solomon, 1969). Cannibalism and the ability to endure periods of starvation enable the predator to withstand periods of absolute prey absence. As *C. eruditus* reproduces by parthenogenesis, even very few mite individuals suffice to preserve the population. Armed with these traits, *C. eruditus*, perseveres at very low population densities, surviving until pest numbers begin to increase.

Further indications of the ability of *C. eruditus* to bring about spontaneous control of stored product mites were obtained after such facilities were fumigated with the fumigant methyl bromide. Stores in which predator populations had been dominant were overrun by huge numbers of acarid pests 2–4 months after fumigation, with very few *C. eruditus* being found (Norris, 1958; Pulpán & Verner, 1965). The inference was that before the treatment pest populations were controlled by the resident predators. The fumigation had disrupted this natural control because *C. eruditus* was more susceptible than the pests, as well as being slower in recolonising the substrate. Such observations were the basis of augmentation trials undertaken in Czechoslovakia. Predators were sieved out of infested grain and released into storage bins in which *C. eruditus* was rare. A successful rate of pest (*A. siro* and *L. destructor*) control was obtained in almost 90% of the experiments, sometimes within 2–3 months (Pulpán & Verner, 1965). These authors advocated that *C. eruditus* be collected from stores wherein pest control had already taken place and released (at predator/pest ratios of 1/100 to 1/1000) in spring or autumn, when the prevailing temperatures were around 5–10°C on the grain's surface.

The tedious and labour-intensive task of sieving predators out of grain, and their subsequent collection, was later substituted by mass-rearing methods. Zdárková & Pulpán (1973) found that *C. eruditus* could be kept alive for several months in cold storage (0±1°C and 84–87% RH) without being fed, and then used. This opened the way for the commercialisation of the predator. Mass-rearing consists of placing about 20,000 *A. siro* and 100–200 *C. eruditus* onto 100 g lettuce seeds confined in paper bags. Bags were kept at 25°C and 75% RH for 4–5 weeks, during which the pest first completely consumed the seeds and was then totally devoured by the predator. Within 1 month there were about 2100 ± 600 *C. eruditus* in each bag. The duration of the rearing period is critical because there must always be an abundance of prey- or *C. eruditus* turns on its own. The bags can be used immediately or kept in cold storage for up to 3 months, for later application (Zdárková, 1986). The resulting pure cultures are sold (under the trade name Cheyletin) for release in stored grain bins (Zdárková, 1991). By 1990 two organisations were selling bags of *C. eruditus*, mostly for controlling residual populations of acarid mites in empty stores, in preparation for restocking them with grain (Zdárková & Horák, 1990). Numbers of *Acarus* spp. were reduced almost eight-fold after predator release in empty warehouses and those of *Lepidoglyphus* spp. almost 37-fold. The rate of release was 2000–3000 predators/100 m². Pest acarids were not controlled in damp stores, where the environment favoured the pests over the predators.

At times it is necessary to apply a pesticide to suppress other pests before the cheyletids can be released. The commonly used (in the Czech Republic) organophosphate (OP) chlorpyrifos-methyl was very toxic to *C. eruditus*, leading Zdárková (1994) to recommend that it should be avoided in prerelease treatments. Of seven *C. eruditus* populations that were compared for sensitivity to OPs, the wild strains were the most susceptible and those maintained in the laboratory the least. Zdárková (1997) thus recommended that only mites of the latter populations should be released. More details on the effect of pesticides on this mite are given in Chapter 43.

Various species of psocids (Psocoptera) occur in stored products. Although usually of minor importance, at times they damage commodities and create health hazards by transferring microbiota onto food materials. All stages of psocids were attacked by

C. eruditus, which were the most important factor affecting the populations of one of the pest species in Canadian seed bulks (Sinha, 1988). Cat fleas [*Ctenocephalides felis* (Bouchè)] could be another pest whose population density, especially in nests, might be affected by *C. eruditus*. Williams & Hallas (1988) observed the mite to feed on flea eggs that were deposited in humid litter.

Cheyletus spp.

Cheyletus malaccensis is the most common mite predator in subtropical and tropical stored products (Haines, 1984). Females completed a generation in 20–23 days at 25°C and 75% RH, when offered only adults of the acarid *Aleuroglyphus ovatus* (Troupeau). Each female consumed around 44 prey until it matured (males: 21.4), and then devoured another 238 during its life (males: 38). Mated females produced significantly more eggs than unmated ones (50 vs 25), and also lived slightly longer (Saleh *et al.*, 1986). Prey stage eaten strongly affected the predators' life history: most rapid development occurred when given eggs of *Tyrophagus putrescentiae*, and longevity was prolonged on a diet of larvae. However, fecundity was highest when feeding on that prey's adults (Table 12.1). The predator also affected stored product beetles; Girish *et al.* (1973) obtained reductions in beetle numbers along with increases in predator populations when monitored together in small containers of wheat flour.

Unlike other cheyletids (Chapter 43), *C. malaccensis* is not very susceptible to OPs or prethroids (PYRs), as it had no dose-related mortality to these chemicals when assayed at concentrations of up to 500 mg AI (active ingredient of the chemical) (Takahashi & Wada, 1988).

Cheyletus sp. can limit populations of *Dermatophagoides pteronyssinus* (Trouessart), the house dust mite, which induces human house dust allergy. The predator readily fed on that pest and may eradicate it in mixed cultures (van Bronswijk *et al.*, 1971). Indirect evidence for the predator's effect was obtained by Wassenaar (1988), who vacuumed a domestic carpet six times, every other day, and then left it for 7 weeks. The carpet was cleaned at the end of that period by spray extraction; all mites obtained by either method were determined. The dominant species were the house dust mite and a *Cheyletus* sp. Pest populations were not affected by vacuuming, in

Table 12.1   Effect of prey (*Tyrophagus putrescentiae*) stage eaten on the fecundity of *Cheyletus malaccensis*

Prey stage eaten	Number eaten	Number of eggs produced
Egg	214.9	65.3
Larva	245.5	85.5
Adult	142.1	92.2

Based on data in Yousef *et al.* (1982).

contrast to those of *Cheyletus* sp., which were reduced. House dust mite numbers greatly increased after the 7 week break, whereas those of the predator remained stable. It was postulated that a numerical equilibrium between pest and predator had been in place at the start of the study, an equilibrium disrupted by vacuuming, which removed relatively more *Cheyletus* sp. and thus led to the huge increase in *D. pteronyssinus* (Wassenaar, 1988). The occurrence of high house dust mite populations in better cleaned rooms, from which the predator had been more efficiently removed, was offered as a corroborative observation.

An undetermined *Cheyletus* fed on a false spider mite (Tenuipalpidae) infesting lawn grasses in Egypt. Each female consumed around 223 pests during its life, and each male about 104; Mohamed *et al.* (1982) postulated that the predator had some importance in the biological control of the false spider mite.

Hemicheyletia bakeri

Two separate assessments were made of the ability of this predator to control spider mites (the authors of both studies placed the predator in the genus *Paracheyletia*). Kanavel & Selhime (1967), working in Florida, found the mite on citrus leaves infested with scale insects and spider mites, and reared it on the latter prey. Development required 22–25 days at 24–30°C, seven or eight prey were devoured daily, and average total consumption during a predator's life reached 390–470 spider mites. Each female produced about 65 eggs. According to Laing (1973), reporting from California, development required 47 days at 22°C, fecundity was only 24 eggs/female and spider mite consumption rates were lower. Weekly counts of the predator and of spider mites on strawberry leaves (on which they co-occurred spontaneously) indicated that *H. bakeri* had a rather weak numerical response to pest population fluctuations. Even when *H. bakeri* numbers came to 16 individuals/leaflet, along with only ten pests/leaflet, it could not keep spider mite populations at a low density. The predator was incapable of responding to increasing prey numbers, partially because it foraged only within a limited area (around the leaf's petiole) of the total space occupied by the prey. This, along with restricted mobility, limited reproduction, low rates of prey consumption and generalist feeding habits, relegated *H. bakeri* to a minor role in spider mite control on strawberry. However, as the predator reproduces asexually, separate strains may have been used in these two studies, which were conducted in different regions, using mites obtained from various plants.

Cheletogenes ornatus (Canestrini & Fanzago)

The arboreal *C. ornatus* often occurs within colonies of armoured scale insects (Hemiptera: Diaspididae) infesting fruit trees in many parts of the world. The mite usually hides under the shields of dead scales, ambushing the wandering emerging juveniles (crawlers). The predator caused the death of 37% of the crawlers of the lesser snow scale, *Pinnaspis strachani* (Cooley), that infested Cuban grapefruit trees, being the pest's major mortality factor (Fernández *et al.*, 1996). In Israel *C. ornatus* is a year-round resident of chaff scale (*Parlatoria pergandii* Comstock) colonies on citrus bark, being rare on fruit or leaves (Avidov *et al.*, 1968). Mite populations had an annual cycle; in the spring they recovered from their winter ebb, increased during mid-summer and peaked in late summer, when the population was 13-fold larger than in winter. No reproduction took place during winter. In the laboratory *C. ornatus* fed on one crawler a day; if a second was offered, it was sized, partially consumed and discarded. A third crawler was grasped but released. The predator did not distinguish between the crawlers of different scale insects, as it fed on them indiscriminately, but neither their eggs nor members of other scale insect families were ever taken. Given one crawler a day, *C. ornatus* had a long life cycle during which it consumed few prey (Table 12.2), and survived for long periods without food; almost half of the females were alive after being starved for a fortnight at 28°C and 50% RH. Despite the obvious shortcomings of *C. ornatus* as a natural enemy, Avidov *et al.* (1968) considered it to be a 'second line' predator of the chaff scale, worthy of conserving. This conclusion was based on the mites' year-round residence on citrus bark alongside the pest, on their ability to endure periods of starvation, and especially on the dominance of this predator on citrus bark during late summer, the critical ebb period in the pest's annual population cycle (Gerson, 1967a). In contrast, Zaher *et al.* (1981) reported that the mite fed on the eggs of scale insects and mites, the latter diet promoting highest fecundity. Plant material, such as sap and pollen, was not taken. Rather dissimilar life history data were obtained by Moraes *et al.* (1989) (Table 12.2), who attributed their results to differences in rearing methods, prey suitability and/or predator biotypes. As these authors also noted differences in the chaetotaxy of *C. ornatus*, they may have been working with another species of *Cheletogenes*. The effects of pesticides on this predator are discussed in Chapter 43.

General comments

Notwithstanding their diffident modes of attack, cheyletids feed on and may even eliminate several co-occurring predatory mites. When the phytoseiids *Galendromus occidentalis* (Nesbitt) and *Phytoseiulus persimilis* Athias-Henriot were placed with *H. bakeri* in rearing units, the cheyletid usually killed most other predators.

Table 12.2   Life history parameters of *Cheletogenes ornatus*, reared on crawlers of armoured scale insects in the laboratory, as reported by different authors

Prey scale	Cycle (weeks)	Number of crawlers consumed	Progeny per female	Males present	Source
Parlatoria pergandii	12	90	12	+	Avidov *et al.* (1968)
Parlatoria oleae	8–10	370	48	+	Zaher *et al.* (1981)
Pinnaspis aspidistrae	7	30	8	–	Moraes *et al.* (1989)

Interactions on plants could have a different outcome. Spider mites that infested strawberry plants in a plant growth chamber were rapidly controlled by *P. persimilis*. Low numbers of *H. bakeri*, naturally present on these plants, persisted in the ecosystem for several months, probably owing to their modest food requirements (Laing, 1973). As noted, *C. eruditus* displaced *B. dentriticus* within a few weeks of their concomitant introduction into experimental populations. This is discussed further in Chapter 42.

The meagre data available on the biology and life table parameters of these Cheyletidae suggest that they may comprise two groups with a different pest-controlling potential. Plant-inhabiting species, such as *H. bakeri* and *C. ornatus*, do not appear to have much promise as acarine biocontrol agents (ABAs). They develop slowly, forage only within a limited area, have low rates of increase (Table 12.3) and do not seem to affect the populations of pests (whether spider mites or scale insects). Their value is possibly due to being secondary mortality factors. Members of the other group, represented by *C. eruditus*, develop rapidly, have a relatively high rate of increase and successfully control pests. *Nodele calamondin* Muma, originally described from citrus but later found in stored foods (Barker, 1992), may represent (Table 12-3) an intermediate lifestyle, a plant-inhabiting species becoming adapted to another habitat.

The success of *C. eruditus* suggests that other cheyletids, which also occur in stored products, could be more important in pest control than appreciated. A prime candidate is *C. malaccensis*, which is very prevalent in tropical and subtropical food storage. Additional cheyletids of biocontrol potential in stored products were noted by Kapil & Bhanot (1973) and by Taha *et al.* (1988b). However, *Cheyletus* spp. cause asthma and papular urticaria in humans (Lozano, 1979; Yoshikawa, 1985), and care should be taken in any planned mass-rearing.

Taxonomic investigations of the widespread predators (e.g. *H. bakeri* and *C. ornatus*), supported by studies on the effects of their diets (Table 12-1), could promote the further deployment of these and other cheyletids. The tolerance of the plant-inhabiting Cheyletidae to many common pesticides preadapts them to becoming part of integrated pest management (IPM) programmes.

Table 12.3 Life table parameters of three species of the family Cheyletidae

	Hemicheyletia bakeri	*Nodele calamondin*	*Cheyletus eruditus*
Prey	*Tetranychus urticae*	*Lepidoglyphus destructor*	*Lepidoglyphus destructor*
Temperature (°C)	22	25	25
Net reproductive rate (R_0)	15.63	20.04	74.4
Intrinsic rate of increase (r_m)	0.05	0.048	0.178
Generation time (days)	55	62	24.2
Source	Laing (1973)	Barker (1992)	Barker (1991)

Chapter 13
Cunaxidae

Diagnosis

The Cunaxidae have three to five segmented palpi, which bear strong spines, spurs or apophyses, and usually terminate with a strong claw. The prodorsum carries two pairs of sensilli, and tibiae IV bear one pair. The family includes about 200 species in fewer than 20 genera. It was revised by Smiley (1992), and the Philippine taxa were summarised by Corpuz-Raros (1996 and earlier papers).

General biology

Cunaxids are fast-running, red, brown or yellow generalist predators that attack small arthropods on diverse crops, in stored products and in litter. In the soil they feed on rootknot nematodes (*Meloidogyne* spp.), major pests of many crops. Some cunaxids hunt their prey and fasten it with silken threads that are excreted through their mouthparts (Alberti & Ehrnsberger, 1977), whereas others snare or ambush their prey.

Cunaxids as predators of various pests

Cunaxa capreolus (Berlese) was reared by Zaher *et al.* (1975) on booklice (Psocoptera) and on *Eutetranychus orientalis* (Klein), the oriental spider mite, a pest of citrus. At 30°C a generation was completed in about 4 weeks when feeding on either diet, each mite depositing around 45 eggs. Average prey consumption during development was 50 psocopterans or 110 spider mites, the adults feeding on another 180 of the former quarry or 360 of the latter. The predator did not take prey eggs, nor did it feed on date palm pollen or potato slices. In the laboratory *Cunaxa lukoschusi* Smiley devoured an average of 1.9 false spider mites (Tenuipalpidae) per day (Corpuz-Raros & Garcia, 1995). *Cunaxoides oliveri* (Schruft) fed on the grape rust mite, the eriophyid *Calepitrimerus vitis* (Nalepa), infesting vines in Germany (Schruft, 1971). In India *Cunaxa setirostris* (Hermann) is one of a guild of predators suppressing the white spider mite, *Oligonychus iseilemae* (Hirst), a pest of coconut foliage. At 28°C the predator raised a generation in about a week, its females lived for a fortnight and each laid around six eggs (Sathiamma, 1995).

Cunaxids as predators of nematodes

Coleoscirus simplex (Ewing) was collected in large numbers from various potted plants inoculated with nematodes in Florida (Walter & Kaplan, 1991). The mite raised a generation in about a fortnight at 28°C and each mated female deposited three to six eggs/day. Virgin females did not lay any eggs, but if mated later in life, they produced progeny within a few days. Adult males guarded moulting female tritonymphs, a behavioural pattern indicative of a female sex pheromone. *Coleoscirus simplex* is an active predator that forages at random, catching and paralysing nematodes with an injected toxin and then sucking them dry. In the presence of abundant prey the mite showed a decrease in handling time, as some nematodes were only partially consumed. Galled root segments that contained nematode females or egg masses (in their gelatinous matrices) were not taken. *Neoscirula* sp. had a similar life history (Walter & Kaplan, 1991) with a sex ratio (females/females + males) of 0.7.

Taha *et al.* (1988a) evaluated the effect of acarine versus nematode prey on *Neocunaxoides andrei* (Baker & Hoffmann) in the laboratory at 30°C and 70% relative humidity. The cunaxid was more fecund when feeding on the nematode *Panagrolaimus rigidus* Schneider than on the astigmatid mite *Caloglyphus rhizoglyphoides* (Zachvatkin) (producing 77 vs 68 eggs/female). However, the predator lived for about 10 days longer with mites as the only prey. During its long (50 days) ovipositional period the predator consumed about 645 individuals of *C. rhizoglyphoides*, at an average rate of 13/day.

Walter & Kaplan (1991) concluded that *C. simplex* was unsuitable as a biological control factor of rootknot nematodes, owing to its prolonged colonisation period, cannibalistic habits and preying on other nematophagous mites. However, the qualities demonstrated by *C. simplex* suggest that other cunaxids could be more suitable for nematode control. The strong cannibalism of cunaxids would be a major obstacle to mass-rearing them, but they could be obtained from special trap plants grown in pots (e.g. Walter & Kaplan, 1991).

A *Cunaxoides* sp. is commercially available for the control of scale insects (Anonymous, 2000), but nothing is known about its impact. The effect of pesticides on cunaxids is discussed in Chapter 43.

Chapter 14
Ereynetidae

Diagnosis

Eryenetids bear two pairs of dorsal sensilli, one on the prodorsum, the other on the hysteronotum. They also have two pairs of genital suckers. The anterior tibiae carries the 'ereynetal organ', a structure consisting of a minute internal canal opening near a sensory seta, the tibial famulus (Fain, 1964). The family includes around 30 genera and fewer than 100 species, which are free-living as well as parasitic.

General biology

Fain (1957) divided the Ereynetidae into three subfamilies: Ereynetinae (residents of humid soil habitats, sometimes associated with arthropods), Lawrencarinae (living in the nasal passages of frogs and toads) and Speleognathinae (denizens of the respiratory passages of vertebrates). Species in the latter two groups are wholly parasitic and have an abbreviated life cycle, with fewer discernible instars, whereas the Ereynetinae retain all three nymphal stages.

Riccardoella spp.

Despite being assigned to the subfamily Ereynetinae, *Riccardoella limacum* (Schrank) and *Riccardoella oudemansi* Thor are parasitic on Molluscans. The former attacks snails (Helicidae), whereas *R. oudemansi* is a parasite of slugs (Limacidae). Both species live on the surfaces or within passages in their hosts' bodies, under optimal conditions raising a generation in about 3 weeks (Graham *et al.*, 1996). Dispersal takes place when hosts aggregate during dry periods. The eggs are deposited in the mucus of the host's mantle cavity. Mobile mites may occur away from the slug or snail, relocating hosts by following their slime trials. The status and different hosts of these two species, often confused in the past, were clarified by Fain & van Goethem (1986). It was formerly assumed (Turk & Phillips, 1946) that the mites feed only on the hosts' mucus, thus being harmless, but anatomical and histochemical studies of the food and feeding mechanism of *R. limacum* convinced Baker (1970) that it was not a mucus feeder. The presence of slug amoebocytes in the mite's alimentary canal strongly suggested that it actually fed on host blood. Mites embedded in the slugs' tissues produce a salivary secretion into the host's body, a secretion that promotes further penetration of tissues by forming a feeding tube or stylostome, through which host material can be drawn (Baker, 1970). (A similar structure also occurs in

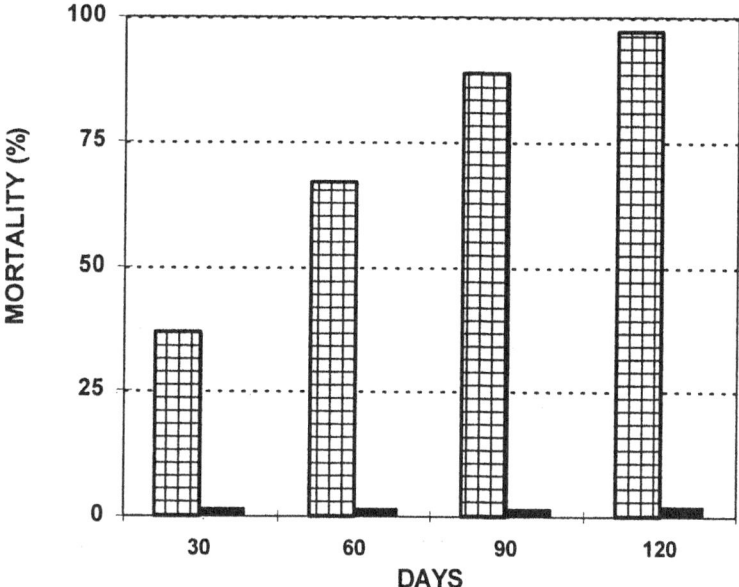

Fig. 14.1   Snail (*Helix aspersa*) mortality rates caused by 150 *Riccardoella limacum* placed with batches of 25 snails, after 30, 60, 90 and 120 days (cross-hatched bars: with mites; dark bars: without mites). (Based on Table 7 in Machado & Thomé, 1994.)

the bodies of vertebrate hosts when parasitised by some other mites; see Chapters 7 and 16.) The presence of many *R. limacum* on the molluscs causes much blood loss and may even kill them (Fontanillas-Perez, 1988; Graham *et al.*, 1996).

In many parts of the world slugs and snails are agricultural pests that damage plant foliage or subterranean parts (Glen *et al.*, 1993). Snails are also reared for gastronomic purposes. *Riccardoella* spp. may thus be natural enemies of pests in some situations, and pests in others.

Machado & Thomé (1994) exposed lots of newly hatched snails (*Helix aspersa* Müller) to various numbers of *R. limacum*. The mites' presence had no influence on snail spawning, average number of eggs/spawn, egg incubation or on hatching rates. However, the presence of mites delayed snail growth and caused high mortality (Fig. 14.1). Heavy infestations of *R. limacum* (20–30 mites/host) reduced the growth of the snails, delayed their reproductive development, and affected the strength and shape of their shells (Graham *et al.*, 1996). Heavy, prolonged mite parasitisation may causes losses of 80–100% in host populations (Fontanillas-Perez, 1988).

Ereynetes sp.

Fungus gnats (Diptera: Sciaridae) of the genus *Bradysia* are common pests in green-houses (Gillespie & Quiring, 1990), requiring control measures. Tritonymphs of an *Ereynetes* sp. were present on *Bradysia impatiens* (Johannsen) in large numbers, almost all examined gnats carrying at least some mites (average load 27/fly) (Zhang & Sanderson, 1993). Heavily infested hosts were irritated and restless, because the mites clustered near their wings, and the gnats' ability to fly decreased with an increase in mite load. Gnats with a load of about 12 mites flew upwards, in a normal pattern. The flight of hosts with about 28 mites/fly was impaired, the gnats being able only to glide, and those that carried as many as 54 *Ereynetes* were disabled and dropped to the ground. This adverse effect on the insect was considered to be beneficial (Zhang & Sanderson, 1993). Tritonymphs of another *Ereynetes* sp. were recovered from adult stable flies, *Stomoxys calcitrans* (L.), a cosmopolitan pest of cattle (McGarry & Baker, 1997). Although the effect of the mites on the flies is unknown, this observation suggests more associations between ereynetids and flies.

Chapter 15
Eriophyidae

Diagnosis

Eriophyids are minute mites (0.15–0.35 mm long) with elongate, annulate, worm-like bodies, two pairs of anteriorly placed legs and, close behind them, a transverse genital aperture. The tarsi lack claws but bear a feather-shaped empodium ('feather-claw'). The chelicerae are stylettiform. About 3400 species in more than 250 genera have so far been named. A major tome on the phytophagous superfamily Eriophyoidea was produced by Lindquist *et al.* (1996). Davis *et al.* (1982) and Amrine & Stasny (1994) presented catalogues to the world fauna. Amrine (1996a) provided a key to all genera, Hong & Zhang (1996) published a catalogue of the Chinese eriophyoids and Baker *et al.* (1996) summarised the species found in the USA.

Life history

The Eriophyidae is the largest family in the Eriophyoidea and some species are serious plant pests; their feeding causes general stunting, brooming, leaf rolling, damage to the reproductive organs and gall formation, as well as fruit and foliage russetting (for their biocontrol, see Chapters 26, 32 and 36). In addition, about a dozen species are vectors of plant viruses. Owing to the damage that eriophyids cause and to their location on plants, they are variously known as blister, rust, gall or bud mites. The indirect injury of the galling species to their host plants may extend beyond the gall itself (as shown in regard to the skeletonweed mite, discussed below). Most eriophyids (and especially those that form galls) are highly specific to their host plants (and at times even to particular tissues or sites), an attribute that makes these mites suitable for weed control. Other traits include dispersal by winds and bringing about slow declines in host vigour, which renders the mites fit for use in weed control along with other agents, without inducing competition (Cromroy, 1979). Eriophyid reproduction is mostly by arrhenotoky. Males deposit spermatophores that the females locate (probably via an unidentified pheromone) and insert into their bodies. Some species, especially those living on evergreens, have a heteromorphic female form, called a deutogyne, which conserves the species during harsh (usually cold) seasons. Deutogynes breed only in the year following their own genesis, when they give rise to the protogynes, the homeomorphic females, which reproduce in the same year.

Eriophyoids occur naturally on many weeds and may seriously damage them, thereby indicating the mites' potential for reducing weed stands. Boczek & Petanović (1996) listed 93 eriophyoids collected on 102 weed species in Poland and Yugoslavia alone, but relatively few have been considered and evaluated for weed control. All are in the Eriophyidae, and only three were successfully introduced across international boundaries (Briese & Cullen, 2001). These are *Aceria chondrillae* (Canestrini), *Aceria malherbae* Nuzzaci and *Aculus hyperici* (Liro). Their history and current status will first be considered, followed by a discussion of other options. Additional eriophyoid candidates for weed control were listed by Boczek & Petanović (1996) and by Rosenthal (1996).

Aceria chondrillae

Rush skeletonweed, *Chondrilla juncea*, is a herbaceous perennial of Eurasian origin that has invaded Argentina, Australia and the USA. Its weed status is due to its being a successful competitor of winter crops and an aggressive coloniser of range lands, and to its clogging up farm machinery. This weed is apomictic, with a number of forms or clones that grow in different parts of its natural range. Different forms of *C. juncea* possess specific chemical and/or mechanical traits that affect their sensitivity to herbivores. A given feeder might thus damage plants of one form, but others could be tolerant or resistant.

Aceria chondrillae occurs in almost all native Eurasian stands of *C. juncea*, being more common in Western than in Eastern Europe, probably owing to the hotter and drier summer in the latter regions. In Western Europe the mite is the most prevalent arthropod found on the weed, as up to 50% of all plants are infested. When feeding on vegetative and flower buds *A. chondrillae* induces the formation of clusters of hyperplastic leafy galls, wherein the mites live and reproduce. Buds are destroyed, seed production is reduced and the plants are much weakened. Long-term mite infestations lead to early plant senescence, stunting and even death (Carèsche & Wapshere, 1974). Buds are reached by mites that 'ride' on new shoots as they grow out of old galls or rosettes [which are the only overwintering sites of *A. chondrillae*, sometimes as deutogynes (Krantz & Ehrensing, 1990)]. The mites raise a generation in about 10 days, and are dispersed by crawling along stems and possibly by winds. *A. chondrillae* fed and reproduced only on *Chondrilla* out of 75 plant species that were offered, and was one of several agents tried for skeletonweed control. The other promising organisms were the gall midge *Cystiphora schmidti* (Rübsamen) (Diptera: Cecidomyiidae) and the rust fungus *Puccinia chondrillina* (Carèsche & Wapshere, 1974).

The mite was introduced into Australia in 1971 (Cullen *et al.*, 1982) and to the USA in 1975–1976 (Sobhian & Andres, 1978). Establishment and success (e.g. damage caused to the weed) seemed to depend on the number of mites present, on the extent of weed stands, on the prevailing climate and on variable interactions between specific weed forms and mite strains. An artificial infestation by large populations of *A. chondrillae* (500–1000/plant) significantly reduced *C. juncea* more than low mite numbers (ten/plant) and 'no mites' controls. Adverse effects included reductions in stem weight, number of growing points, rosette production, number of healthy flowers and dry root weight (Fig. 15.1). The calculated decline in the production of healthy flowers (i.e. reproductive capacity) came to 96% with the high mite treatment and 73% with the low mite treatment. Field observations confirmed that the mite's

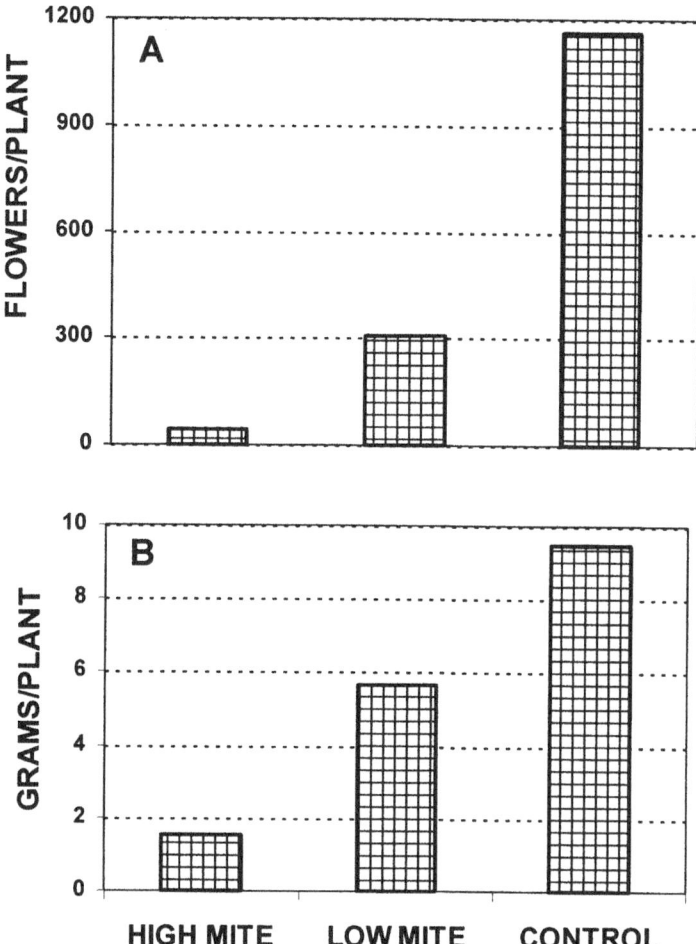

Fig. 15.1 Effects of artificial infestations with *Aceria chondrillae* on (A) the production of healthy flowers by *Chondrilla juncea* and (B) the weight of dry root matter. High mite: 500–1000 mites/plant; low mite: ten mites/plant; control: no mites. (Based on Tables 6 and 7 in Cullen *et al.*, 1982.)

effect is strongly dependent on the size of its populations (Cullen *et al.*, 1982). Mite survival in the field requires the presence of small weed stands wherein it lives during winter. Aggressive weed control that removed most stands of *C. juncea* resulted in mite elimination from such fields. In dense weed stands (100 plants/m^2) *A. chondrillae* increased more quickly, and had a more devastating effect than in sparse stands (Carèsche & Wapshere, 1974).

The failure of the first introduction of *A. chondrillae* into North America was attributed to the harsh (dry) inner Californian climate and to using an unsuitable mite strain (Sobhian & Andres, 1978). Another strain was used in the second introduction and the mite became established (see below). However, no significant correlations between mite attack and weed decline were obtained in California (Supkoff *et al.*, 1988), a result attributed (Rosenthal, 1983) to the area's climate. In Idaho, under more humid conditions, the root and stem weights of the weed were reduced by 56% and 64%, respectively, when it was infested by 890 mites/plant. Lower mite loads (178 mites/plant) brought about smaller reductions, whereas higher numbers (1602 mites/plant) did not increase the damage (Prather *et al.*, 1989). The presence of even a few mites sufficed to reduce seed production; uninfested weeds produced about 250 seeds/plant, whereas mite-carrying plants produced only 23–34.

Recognising the apomictic nature of *C. juncea* and the possible effect of its different forms (or clones) on the efficacy of *A. chondrillae*, Carèsche & Wapshere (1974) assayed mite strains from various Eurasian localities that had presumably become specialised to these forms. This project was undertaken in preparation to introducing the mite into Australia, where the weed occurs in three forms, known as form A (narrowleaf, the most widespread), form B (intermediate leaf) and form C (broadleaf) (Hull & Groves, 1973). Tests with French, Italian ('Vieste') and Greek mite strains on different weed forms showed that only French plants were infested by all *A. chondrillae* strains, and that the Greek strain also established on Australian A and B forms. Form A was the best host for the Greek mites, which were then introduced into Australia. The mites' suitability was demonstrated by reducing flower and seed production of form A by 92.3%, form B by 13%, but form C only by 0.3% (Cullen & Moore, 1983). The Greek strain was also introduced into California; it infested North American skeletonweed but induced no galls on it. In contrast, mites of the Vieste strain, introduced later, formed galls on the North American *C. juncea* and were released (Sobhian & Andres, 1978).

As noted, mites of the Greek strain affected form B plants in Australia, although to a lesser extent. Populations of this strain were maintained on B plants in the laboratory and field for possible adaptation and potential use in field releases. They strongly affected both form A and B plants in intentional infestation trials, and a few of the C form plants also became infested. The great variability obtained led Cullen & Moore (1983) to conclude that there are many environmental factors (e.g. temperature, humidity and especially light intensity) that affect the plant–mite relationship. The occasional severe infestation and strong galling on form B plants should therefore be attributed to serendipitous growing conditions rather than to the selection of mites that are adapting to a new host plant form. This was borne out by failures to establish *A. chondrillae* on form B plants in the field.

The rust fungus (*P. chondrillina*) appears at present to be the most damaging natural enemy that is impacting skeletonweed in Australia and the USA, whereas *A. chondrillae* affects weed populations only in regions where winters are not too cold and there are year-around rains.

Aculus hyperici

St John's wort, *Hypericum perforatum* (called Klamath weed in California), is a European perennial plant that often grows in disturbed areas. Its weed status is due to competition with range plants and to containing the toxin hypericin, which affects the skin of livestock. In Australia St John's wort has been a target for biological control efforts since the 1930s, when several insects were introduced for its control (Briese, 1997). However, only the beetle *Chrysolina quadrigemina* (Suffrian) (Chrysomelidae) has a detrimental, albeit inconsistent, effect on weed populations. Another introduced beetle, *Agrilus hyperici* (Creutzer) (Buprestidae), has not contributed to the weed's control, although it causes significant damage to mature weed plants in Europe (Jupp & Cullen, 1996). The decline of a natural stand of *H. perforatum* in the south of France was associated with a combination of buprestid damage and large populations of *Aculus hyperici*. As the beetle did not attack small plants, *A. hyperici* was considered to be the only natural enemy that consistently caused declines in weed stands. The mite feeds on the rosettes' growing points and flowering stems, causing dwarfing, reductions in plant vigour and consequent mortality. *Aculus hyperici* raises a generation in 3–4 weeks (Mahr *et al.*, 1997).

The mite, first introduced into Australia in 1991, was released at almost 250 sites, and had a preliminary establishment rate of 73%. As its own dispersal (mostly by winds) is limited to about 1–2 km/year, a distribution network was set up to facilitate mite spread in weed-affected areas (Jupp, 1996). St John's wort nurseries were established in south-eastern Australia, at sites where plants artificially colonised by *A. hyperici* were made available to interested persons. A special protocol, which emphasised the optimal conditions for mite colonisation, was formulated to facilitate the release efforts. These conditions included the optimal season (autumn, owing to strong rosette growth), absence of the competing *C. quadrigemina*, plant health (establishment on robust weeds was about twice as successful as on stressed plants), weather conditions (rains washed the mites off) and the density of *A. hyperici* (Mahr *et al.*, 1997). The effect of mites became evident within 2 years; the root systems of heavily infested plants (e.g. 100% of their buds with 1–50 mites/bud) were significantly reduced, stems were lighter and shorter, and they produced fewer seed capsules (Jupp & Cullen, 1996).

Phenotypic variability in weed susceptibility to *A. hyperici* hindered the mite's colonisation in Australia. Jupp *et al.* (1997) infested glasshouse-grown plants of different phenotypes (broad leaves, narrow leaves, tall and narrow leaves, and intermediate) with similar mite loads. The increase in *A. hyperici* populations was monitored by rating numbers of mites in the buds for 14 weeks; the mite effect was determined by weighing the plants. A significant negative relationship was obtained between mite rating and both root weight ($R^2 = 0.8226$) and shoot weight ($R^2 = 0.7697$), demonstrating the effect of *A. hyperici* on *H. perforatum*. The intermediate phenotype was resistant and all plants were alive at the end of the experiment (Fig. 15.2). The other phenotypes had similar mite ratings, but the pattern differed. All tall and narrow, and broad, plants died after 14 weeks, as well as 60% of the narrow phenotype. Mite-resistant and mite-susceptible phenotypes of *H. perforatum* are therefore present in Australia, and the latter were variable in their performance. The presence and potential spread of resistant weed forms could seriously affect its future biological control.

In Australia the mite also survives and reproduces, albeit at low levels, on the native *Hypericum gramineum*, but its effect is milder than on *H. perforatum*.

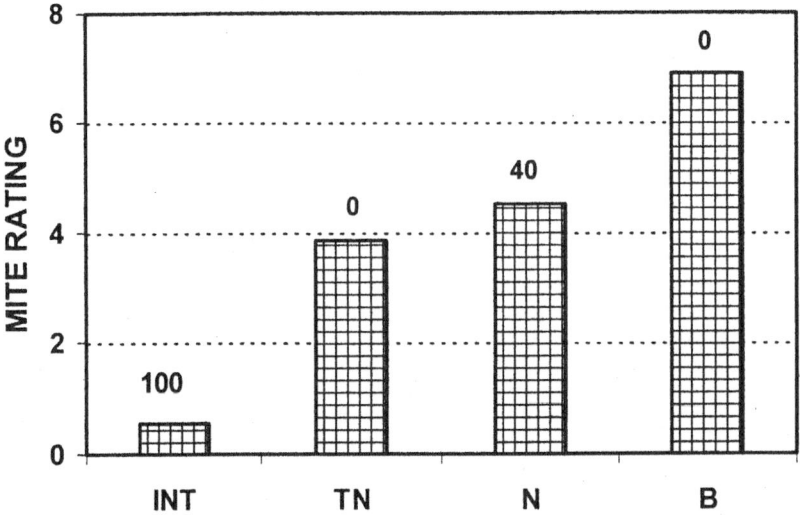

Fig. 15.2 Ratings of *Aculus hyperici* populations in rosettes of four phenotypes of *Hypericum perforatum* (INT: intermediate; TN: tall and narrow; N: narrow; B: broad), and the percentages of weed survival (numbers on top of columns) after 14 weeks in a glasshouse. (Based on Table 2 and text in Jupp *et al.*, 1997.)

Glasshouse infestations of both plants by similar mite numbers resulted in greater damage to *H. perforatum* than to *H. gramineum*; the former grew less and its roots were smaller. *Hypericum gramineum* could thus be affected by *A. hyperici* in the field, but the effect would be much weaker than in the glasshouse. This conclusion was based on low infestation rates on *H. gramineum* in the field, and on its patchy distribution, compared with the large 'monocultures' of *H. perforatum* (Willis *et al.*, 1995).

Although the health of the weed is expected to decline further as a result of mite activity, the overall effect of *A. hyperici* on the distribution and weed status of St John's wort in Australia remains unclear (Briese & Cullen, 2001).

Aceria malherbae

Field bindweed, *Convolvulus arvensis* L., is a perennial growing in most temperate regions. Its weed status is due to competition with cultivated crops, facilitated by its large, deep and persistent root system. Losses due to bindweed in some US states were estimated (Boldt & Sobhian, 1993) to be several dozen million dollars.

In the Mediterranean region bindweed is attacked by many herbivores (Rosenthal & Buckingham, 1982), of which only two, a lepidopterous leaf-feeder and the eriophyid *A. malherbae*, were sufficiently specific and damaging to be considered for introduction into the USA (Boldt & Sobhian, 1993). Feeding by the mite [initially misidentified as *Aceria convolvuli* (Nalepa)] causes leaf furrowing along the veins, which subsequently brings about thickening, wrinkling and twisting; the mites live within these folds. Very small, tubercle-like galls are formed on the leaves, petioles and stems. Galled leaves turn yellow or red. Flowers are not formed on heavily infested stems, and the buds swell to become irregular (Nuzzaci *et al.*, 1985). The annual cycle of *A. malherbae* is well synchronised with that of bindweed, as mites occur on the plants throughout their growing season, from March to November; over-wintering on the rhizomes in the soil (Boldt & Sobhian, 1993). The type specimens of *A. malherbae* were obtained in Italy and Greece, and the material used for the required biological studies originated in the latter country.

Convolvulus belongs to the Convolvulaceae, a family to which sweet potato, *Ipomoea batatas*, is also assigned. Strict host specificity tests, however, consistently showed that the mite could not survive on sweet potato (Rosenthal & Platts, 1990; Craemer, 1995). In contrast, *A. malherbae* reproduced and formed galls on several North American species of another member of the family, *Calystegia*. However, as this plant is not very abundant, its chance of being infested by *A. malherbae* was considered to be low. This assumption, along with the lack of other organisms capable of controlling bindweed, brought about official approval for introducing the mite into the USA. Tests also revealed that various North American geographical bindweed 'ecotypes' differed in their susceptibility to the mite. Bindweed from New Jersey was highly susceptible (96% of plants infested), plants originating in California and Texas were intermediate (87% and 75% levels of infestation) and *C. arvensis* from Nebraska was the least suitable host (53% infestation) (Rosenthal & Platts, 1990).

The mite was introduced into the USA in the early 1980s for quarantine studies, and permission for field release was granted in 1987. In South Africa these events occurred in 1989 and 1994, respectively (Rosenthal & Platts, 1990; Craemer, 1995). The material to be released in the USA was collected in Greece and was immediately air-freighted to a quarantine facility in Texas, where the mites were placed on small potted bindweed plants. Leaves from these plants were used in the field releases (*c.* 15,000 mites). Infested weed crowns were found 4 months later. The mite occurred in over half of the crowns within 3 years and leaves were heavily galled. The infestation spread out from the initial focus to a distance of 9.5 m. These findings constituted proof of the establishment of *A. malherbae* in North America (Boldt & Sobhian, 1993), being the first arthropod that has successfully been established in the USA for weed biocontrol. The mite was secondarily released at other North American sites, but its effect on bindweed populations remains unknown.

During 1992–1995 a separate introduction of *A. malherbae* into Canada was made from Greece. Mites were mass-reared in the greenhouse and infested plants were used in the field. The initial effect was slight, but within 3 years the mite spread from its release focus to a distance of 90 m, covering an area of 3000 m^2. At other sites up to 65% of the weeds were affected. The success of this Mediterranean species in the very cold Canadian continental climate suggests that *A. malherbae* has the potential to establish throughout the entire range of field bindweed in North America (McClay *et al.*, 1999).

At the time of writing *A. malherbae* is the only eriophyid (in fact, the only mite) that is commercially available for the biological control of weeds, being sold by Praxis, located in Allegan, Michigan, USA (Anonymous, 2000).

Aceria acroptiloni Shevchenko and Kovalev

Russian knapweed, *Acroptilon repens* (formerly in the genus *Centaurea*), competes with many crops and may be toxic to grazing livestock. It is attacked by several arthropods, including the gall-forming *A. acroptiloni*. The mite occurs only in the weed's flowering clusters (inflorescences), wherein they raise huge populations, averaging 6000–7000, and up to 13,000/cluster (Kovalev, 1973). The mites spend late summer, autumn and winter as darkly hued deutogynes in the damaged flowers, subsequently raising four or five annual generations. Early-season *A. acroptiloni* initially feed on the leaflets that cover the flowering cluster, cause them to swell and to form bulges, then entering the inflorescence itself, penetrating into and damaging the receptacles. Attack on individual flower clusters is uneven, depending on chance encounters by the mites. Infested clusters become hypertrophied, leaf-like and sterilised; no new shoots are formed and the entire plant becomes stunted. Mites are naturally dispersed within infested plants by the winds and mechanically (during harvest) by farm machinery. Kovalev *et al.* (1974) considered this mite to be one of the more promising agents for the control of Russian knapweed.

Aceria boycei (Keifer)

Several ragweeds (*Ambrosia* spp.) are North American plants that invaded Eurasia and became weeds there. Various arthropods were sent from California to the (then) USSR for ragweed control during 1969–1972. The shipments included ragweed leaves with many galls that contained *A. boycei* (formerly placed in *Eriophyes*). Most of the shipped mites arrived dead and efforts to culture the survivors did not succeed (Goeden *et al.*, 1974).

Aceria centaureae (*Nalepa*) and *Aceria thessalonicae* Castagnoli

Diffuse knapweed, *Centaurea diffusa*, is a Eurasian plant that has spread widely in North America, becoming an important weed of rangelands. *Aceria centaureae* is one of several specific arthropods that attack the rosettes and shoots of *Centaurea* spp., on which it induces blister galls. The mite was introduced in quarantine into the USA in 1987 for host-specificity trials. The finding of a very similar, but vagrant (i.e. not gall-inhabiting and therefore potentially not conspecific) mite on knapweed halted these tests. The unknown eriophyid was shown by Castagnoli & Sobhian (1991) to be a new species, *A. thessalonicae*, differing from *A. centaureae* by an array of morphological characters and by affecting the host differently, causing broom-like deformations. Other injuries attributed to *A. thessalonicae* include reductions in seed formation and killing of leaf rosettes, thus being more harmful to the weed than *A. centaureae*. As these species show strict site preferences, each occupying different plant parts, they might coexist and be concurrently released without risking interspecific competition (Castagnoli & Sobhian, 1991).

Aceria salsolae DeLillo and Sobhian

Salsola kali, Russian thistle or saltwort, of Russian provenance, has become a weed of rangeland and wheat fields in the USA. The plant bears strong and sharp spines that damage farm stock and clutter up machinery. *Aceria salsolae*, of Turkish origin, was assayed in the laboratory by Sobhian *et al*. (1999). Offered several members of the family Chenopodiaceae, it survived and reproduced only on *S. kali*, which it seriously damaged and killed within 60 days. These results, along with its strict host specificity, led to *A. salsolae* being highly recommended for further testing as a candidate for saltwort control.

Phyllocoptes fructiphilus Keifer

As noted, several eriophyids are vectors of plant virus diseases, but none has so far been found to affect weeds. However, an eriophyid-associated disease of unknown aetiology is a natural mortality factor of multiflora rose (*Rosa multiflora*), an Asian thorny shrub that was introduced to the USA as a source for rose root stock and as an ornamental. Its weedy character became apparent by the 1950s, when it spread onto marginal lands and along roads, necessitating control measures (Amrine, 1996b). In the field multiflora rose suffered heavy mortality due to rose rosette disease (RRD) (also called witches' broom of rose), an unidentified agent that can be transmitted by grafting and by the native *Phyllocoptes fructiphilus*. The expression of RRD after grafting took 30–90 days, whereas only 17 days were required for the appearance of the disease after mites, obtained from affected roses, were placed on healthy plants. Both methods provided 100% transmission. In addition, populations of *P. fructiphilus* on diseased plants were many times larger than those on healthy roses. The mortality of infected plants is very high and may reach 98% in a few years.

Grafting RRD-infested buds onto healthy plants is the current method of multiflora rose control, but it is laborious and time consuming. Efforts to manipulate *P. fructiphilus* for weed control are now underway (Epstein & Hall, 1999).

The multiflora rose–RRD–*P. fructiphilus* system represents a case of an endemic herbivore that transmits an apparently native disease to an introduced weed. Under certain circumstances the mite is a pest of ornamental roses, requiring chemical control (Amrine, 1996b).

Eriophyids that hinder pests and/or may serve as alternate diets for predators

Apple rust mite, *Aculus schlechtendali* Nalepa, is a secondary pest of apples in North America, Europe and Japan. Foliage on which the mite had fed and that was subsequently inoculated in the field by the European red mite (ERM), *Panonychus ulmi* (Koch) (Tetranychidae), a major pest of apples, hindered the latter's reproduction

(Croft & Hoying, 1977). Numbers of ERM remained below their damage threshold in an orchard that had carried high rust mite populations. This effect may be due to the minute stylets of the eriophyid, which wound but do not kill the leaf epidermis. The affected epidermis then forms a layer of callus that reduces, or even prevents, the probing, and thus feeding, of the spider mite (Dunley & Croft, 1996). Applications of selective pesticides that only minimally affect apple rust mite populations could reduce ERM numbers through this (postulated) indirect competition.

The conservation of minor-pest eriophyids also promotes the early-season activity of predatory mites, which use such prey as alternate or supplementary prey. The presence of *A. schlechtendali* early in the season, when spider mite numbers are still low, enables the phytoseiid *Galendromus occidentalis* (Nesbitt) (often placed in *Metaseiulus* or in *Neoseiulus*) to build up its numbers. A larger predator population is thus at hand when the pests increase their densities (Hoyt *et al.*, 1979). This mechanism was shown to occur in vineyards as well as in plum and apple orchards (reviewed by Sabelis & van Rijn, 1996).

Eriophyids, serious plant pests in their own right, may be beneficial (1) when they assist in weed control, whether as herbivores or by transmitting diseases to these plants, (2) by inhibiting the development of pest mites, and (3) when they serve as alternate diets for predatory mites early in the season.

Chapter 16
Erythraeidae

Diagnosis

Erythraeids are large, reddish mites with two pairs of prodorsal sensilli, many dorsal setae and a thumb–claw process with long, straight chelae. This family is assigned to the Parasitengona, a cohort of superfamilies whose members have only three active stages. These consist of larvae, deutonymphs and adults; the protonymphs and trironymphs remain quiescent. The larvae are heteromorphic to the deutonymphs and adults, but the postlarval stages of only a few species have been correlated with their larvae. It is therefore necessary to use separate keys for their determination (e.g. Welbourn & Young, 1987). At present the family has about 300 described species in about 20 genera. The family was revised by Southcott (1961), and the interrelationships among the five erythraeid subfamilies were charted by Gabryś (1991), along with a list of all genera. Keys to the North American genera (larvae and post-larval stages) were provided by Welbourn & Young (1987). The larvae of the genera of the Erythraeinae, the largest subfamily, were separated by Southcott (1988).The Middle-Eastern species are being studied by Saboory (2000 and earlier papers).

General biology

The larvae of the Erythraeidae are parasites of various arthropods, including insects and spiders. These larvae may secrete a cementing substance at the attachment site, which later forms a superficial cone. The cone enhances the mites' attachment and surrounds a funnel through which the host's body fluids are sucked (Åbro, 1988). Postlarval erythraeids are predators. In rare cases (e.g. *Balastium*, see below) all stages attack the same animal, which is both host and prey. Fertilisation is by male-deposited spermatophores. Each female usually produces several hundred eggs in batches; most species raise a single annual generation.

Insect parasitism by erythraeid larvae is of considerable geological age. The earliest known cases of animals that parasitise other animals are erythraeids on biting midges (Diptera: Ceratopogonidae) found in Cretaceous Canadian amber (70–80 million years old) (Poinar *et al.*, 1993) and on similarly aged Chironomidae (Poinar *et al.*, 1997). Relicts of a more recent era (found in *c.* 25–40 million-year-old amber) are engorged erythraeids attached to moths (Poinar *et al.*, 1991).

Balaustium spp.

Members of the genus *Balaustium* have diverse feeding habits, often occurring in flowers of annual and perennial plants, browsing on pollen. They may also feed on leaf tissues and even attack humans (Newell, 1963). The larvae are polyphagous predators, like their postlarval stages. Unlike other eryrthraeids, *Balaustium* spp. raise more than one annual generation; two species were evaluated as predators.

The moth *Zeiraphera diniana* Guénée, a pest of pine and larch in Europe, has long-lived eggs whose predators were evaluated in Switzerland by exposing tagged eggs on larch sticks placed on trees in the forest. Eggs were tagged by injecting an innocuous radioactive marker (zinc-65) into moth larvae; the resulting adults deposited eggs that retained their radioactivity even after being ingested by predators. All associated arthropods were collected and assayed for radioactivity; *Balaustium murorum* (Hermann) was the most important predator. Its predation rates remained similar throughout the experimental period, and were higher on the trees' southern aspect. These rates were high on large egg masses and decreased as their density declined. As mite numbers did not change much during the entire period, Delucchi *et al.* (1975) postulated that the functional response of the predators reflected changes in egg population.

Balaustium putmani Smiley was discussed in the Canadian literature (Herne & Putman, 1966) before and after its formal description (Smiley, 1968). Interest in this predator arose from observing its increases in sprayed North American peach and apple orchards (Putman, 1970; Cadogan & Laing, 1977; Childers & Rock, 1981). In southern Ontario the mites overwinter as eggs hidden in bark crevices of fruit trees. The larvae appear in May, adults of the first generation in June, and those of the second generation in late July to August. The predator was often very abundant in apple orchards (ten to 28 nymphs and adults were counted in 1 min on each of six trees), appearing to have a strong impact on the European red mite (ERM), *Panonychus ulmi* (Koch). In the laboratory *B. putmani* consumed about 106 ERM eggs/day; females devoured more than 25 adult mites/day. Some females kept in captivity produced at least 175 eggs, in six batches. These observations suggested that *B. putmani* was an abundant and important member of a guild of predators, including Phytoseiidae and Stigmaeidae, which kept ERM populations from reaching pestiferous levels. Other observations (Cadogan & Laing, 1977) indicated that although the predator would occasionally feed on various other predators, such feeding is minimal except in the absence of adequate alternate diets. In contrast to other members of the genus (see above), pollen was insufficient for survival.

Balaustium putmani, which feeds on many soft-bodied arthropods, also attacked other pests, including all stages of the San José scale, *Quadraspidiotus perniciosus* (Comstock) (Hemiptera: Diaspididae) and apple aphids. Feeding on aphids, however, was dependent on prey size: only insects that were half or less the predator's size were attacked. Larger aphids twitched their bodies vigorously, discouraging mite attack (Childers & Rock, 1981). In addition, *B. putmani* fed heavily on the apple rust mite, *Aculus schlectendali* (Nalepa) (Cadogan & Laing, 1977). It was also reared to maturity on other pests, such as the brown mite, *Bryobia arborea* Morgan &

Anderson, and the two-spotted spider mite (in whose webbing the predator became entangled). When confined without any food, the mite turned cannibalistic (Putman, 1970).

Balaustium putmani was reared in the laboratory within small moistened leaf discs with an abundance of ERM eggs; the provision of sufficient hiding sites minimised their cannibalistic habits. The predator developed from egg to adult in about 50 days at 20°C, or 39 days at 25°C. About 460 eggs/predator were consumed by all mite stages at 20°C and 370 at 25°C (Cadogan & Laing, 1977). *Balaustium putmani* had a functional response to *P. ulmi*, as the number of prey taken increased with the density of eggs offered. The sex ratio of field-collected *B. putmani* was 0.58%.

Miscellaneous

Lasioerythraeus johnstoni Welbourn & Young, a parasite and predator of insects in six families placed in three orders, apparently prefers the tarnished plant bug, *Lygus lineolaris* (Palisot de Beauvois), a pest of field crops and deciduous fruit trees in North America. Bug size usually determined the outcome of the attack: hosts that were only a maximum of four times bigger than the parasite (a ratio common in young bug nymphs) succumbed to *L. johnstoni* (Young & Welbourn, 1987). Most hosts parasitised in the field were first instar nymphs and were killed by the mite. This outcome, along with attack rates that were up to 30–50%, the use of the same host by all active mite stages and its prolonged presence in the field, suggested that *L. johnstoni* may have considerable potential for the control of this pest. The mite occurred in relatively undisturbed, moist, early successional sites in North America (Young & Welbourn, 1988).

Several erythraeid larvae are known to attack the spruce budworm, the moth *Choristoneura fumiferana* (Clemens), a major pest of pine and spruce forests in North America. About 30% of the trapped moths were parasitised, with up to four mites being collected from a single host. Although no data are available on the effects of such parasitism, Houseweart *et al.* (1980) felt that the mites could affect pest flight and fecundity. The larvae of *Callidosoma metzi* Sharma, Drooz & Treat parasitise adults of many moths (Lepidoptera), including several pests. When attacking the apple pest *Platynota idaeusalis* (Walker), parasitisation rates reached 16%. In the laboratory, at 24°C, mite deutonymphs and adults consumed a total of 59 and 230 moth eggs, respectively, but did not eat a variety of other animal foods offered (Sharma *et al.*, 1983). Because of such feeding, and because the body of *P. idaeusalis* is only about ten times larger than that of *C. metzi*, Adler & Browning (1986) postulated that the mite could have an adverse effect on this pest. Erythraeids also attack pestiferous thrips, at times causing up to 20% parasitism (e.g. Goldarazena *et al.*, 1999); their effect on pest populations is not known.

Charletonia taiwanensis Tsai & Chow is a parasite of grasshoppers in Taiwan. Larvae obtained off grasshopper wings completed their development, at 25°C and 80% relative humidity, when feeding on the eggs of various Lepidoptera, Hemiptera and a spider in the laboratory. Eggs and fly maggots were not touched, nor did honey water or pollen grains prolong the mites' lives (Chow, 1996). Egg development required around 68 days, the larva (living on grasshoppers) 12 days and the postlarval stages another 53 days; adults emerged after about 150 days. Young adults tended to aggregate and mate in the dark areas of the container. Many larvae may concurrently parasitise and harm individual grasshoppers, but their impact on the host is not clear.

The available data suggest that Erythraeidae possess a natural tolerance to certain pesticides (see Chapter 43). Welbourn (1983) presented a list of erythraeids and their host, or prey species. Additional unexplored target pests are tsetse flies (*Glossina* spp., vectors of the causal agent of sleeping sickness), parasitised by several species of *Leptus* (Fain & Elsen, 1972). As noted, the juveniles of larger injurious insects (e.g. moths and thrips) are also promising targets, although the univoltinism of the Erythraeidae (except for *Balaustium*) would render them unsuitable for controlling multivoltine pests. However, conservation efforts and mass-rearing for releases on suitable dates against specific pests are possibilities to be tried.

Chapter 17
Eupalopsellidae

Diagnosis

The Eupalopsellidae carry very long palpi and chelicerae. The palpal thumb–claw complex is reduced and the empodia terminate in two capitate raylets that are longer than the claws. They are yellow to orange mites that usually occur on plants. This is a small family with five described genera and about 30 species. The Eupalopsellidae was revised by Meyer & Ueckermann (1984, 1989), the Australian species were listed by Gerson (1994a) and the status of the family was clarified by Gerson & Walter (1998).

General biology

Eupalopsellids are often associated with armoured scale insects (Hemiptera: Diaspididae), serious pests of many perennial crops, on whose eggs or first stage nymphs ('crawlers') they feed. Reproduction is by arrhenotoky and they raise a generation in less than 1 month (Gerson & Blumberg, 1969).

Saniosulus nudus Summers

This mite has been collected from diaspidids in different parts of the world (Gerson, 1994a). It is an agile predator that raises 'families' in the prey's colonies. After catching a crawler, *S. nudus* holds the prey by its anterior legs and sucks out the body fluids. Feeding may last for 30–40 min or more, after which the predator pushes the shrivelled cadaver off its chelicerae (Gerson & Blumberg, 1969). Older scales may be attacked, but do not appear to be harmed.

Male and female *S. nudus* are voracious feeders, devouring an average of 125 and 257 eggs, respectively. Prey consumption depends on which species' eggs or crawlers are being fed upon. About 85 eggs of *Lepidosaphes pallida* (Maskell) were devoured by individual mite females (at 27°C) during 20 days; the comparative value for eggs of the purple scale, *Lepidosaphes beckii* (Newman), was only 40. Feeding on *L. pallida* was also conducive to higher fecundity, as the mite deposited about 34 eggs while consuming its eggs, but only about 19 when feeding on *L. beckii* (Zaher *et al.*, 1984). Oviposition began 1–2 days after emergence. The eggs, usually deposited in groups of eight to ten, were placed underneath the living or dead scales. Each female may lay about 40–50 eggs. At 24°C the mite completed a generation in about 3 weeks, whereas only 2 weeks were required at 28°C.

The population trends of *S. nudus* in Israel were followed by monthly sampling of citrus bark infested with the chaff scale, *Parlatoria pergandii* Comstock. Mite numbers were low during winter and spring, peaked in late summer (as pest populations were also increasing) and then declined (Gerson, 1967a). In India *S. nudus* fed on the sugar cane pest, *Melanaspis glomerata* (Green); the mites were located under the mother scales where they attacked emerging crawlers (Rao & Sankaran, 1969).

In insectary mass-rearings of diaspidids the mite may become a serious nuisance, to the extent that chemical treatments have to be applied for its control (Gerson & Blumberg, 1969).

Eupalopsis jamesi Gerson

This species was associated with another citrus pest, the California red scale, *Aonidiella aurantii* (Maskell), in New South Wales, Australia. It occurred on most fruits throughout the year, with up to 64 individuals/fruit, being usually found around the calyx. The predators aggregated under adult females and attacked their emerging crawlers. The populations of *E. jamesi* followed those of their prey, being higher in the autumn and winter than during spring to early summer. James *et al.* (1997) believed that *E. jamesi* plays an important role in the pest's control, being a member of the guild of natural enemies associated with *A. aurantii*. The mite was absent from citrus groves with a history of insecticide use, suggesting its susceptibility to pesticides.

To conclude, the field observations along with the mites' nuisance role in mass-rearing facilities indicate that eupalopsellids have some potential for reducing diaspidid populations under conditions of abundant prey.

Chapter 18
Galumnidae

Diagnosis

Galumnidae bear large and movable pteromorphs that extend interiorly and posteriorly. The lamellae are weak or absent, the smooth cuticle has four pairs of porose areas and the genital plates bear six pairs of setae. The family consists of about 40 genera and 450 species. Balogh & Balogh (1990) prepared a key to the genera, the New Zealand taxa were reviewed by Luxton (1985) and those of the Iberian Peninsula by Pérez-Iñigo (1993).

General biology

Galumnidae are soil mites with diverse feeding habits. Many are intermediate hosts of tapeworms, as the mites ingest the eggs from which the young worms then emerge. If not eaten by the tapeworms' vertebrate hosts, the mites die from the infestation. Denegri (1993) believed that tapeworm eggs were too large to be eaten by chance; their ingestion could therefore be interpreted by the mites as potential or even preferred food. Reproduction is usually sexual, sperm transfer being via spermatophores (Alberti *et al.*, 1991).

Orthogalumna terebrantis Wallwork

Waterhyacinth, *Eichhornia crassipes* (Pontederiaceae), is a free-floating, perennial plant that forms dense mats on waterways, canals, reservoirs and lakes, and is considered a serious weed in many warm parts of the world (Harley, 1990). The waterhyacinth mite, *Orthogalumna terebrantis*, of South American origin, feeds on some Pontederiaceae, but mainly on *E. crassipes*. The mite deposits its eggs in the weed's false leaves (pseudolaminae) where all immature stages live, excavating numerous longitudinal tunnels (Del Fosse, 1978). Heavily infested leaves may have 25–50 feeding tunnels in a 2 cm^2 area, making up to 250–500 tunnels/leaf; 1 m^2 of the weed would thus bear 20,000 or more mites. Heavily infested leaves show yellowish-brown streaks, which are the excavated galleries wherein the various stages of *O. terebrantis* live. Each of the older, larger leaves may have more than 500 tunnels; these leaves later turn brown and dry out, the mites dying therein (Perkins, 1973). Adult *O. terebrantis* emerge through large holes, mostly on the upper surface of the leaves. At advanced infestation stages neighbouring galleries may overlap, forming large dry patches on the leaf surface. Total leaf collapse may occur within 1 month of infestation (Sumangala & Haq, 1995). However, large mite numbers are required to inflict serious damage. Cordo & DeLoach (1976) estimated that 200,000 galleries/m^2, or as many as 75,000 mites/m^2 or 10,000 galleries/plant, would be needed to kill most waterhyacinth leaves.

The number of tunnels, and of emerging adults, seemed to decrease with rising temperatures: at 10–30°C the mites excavated an average of 1060 tunnels and produced 1400 adults, but at 20–40°C only 580 tunnels were burrowed and 350 adults emerged. A mean of 21–24 eggs/female per week was deposited within the range of 10–35°C, little mite growth taking place at lower temperatures (Del Fosse, 1977a). According to Ganga-Visalakshy & Jayanth (1991), each female deposited about 60 eggs at 26 ± 1°C, a life cycle was completed in around 4 weeks and females lived for up to 11 weeks. The mite raised two or three annual generations in Argentina, where adults overwintered on waterlettuce (*Pistia stratiotes*), possibly subsisting on algae that grow on this plant (Cordo & DeLoach, 1976).

Direct damage, as noted, is caused by mite feeding and tunnelling. Indirect injury to waterhyacinth is due to attracting *Neochetina eichhorniae* Warner, the mottled waterhyacinth weevil, a specific herbivorous beetle (Coleoptera: Curculionidae), to oviposit there. In addition, the mites' emergence holes serve as portals for *Acremonium zonatum*, a fungus that causes zonate leaf spot disease of waterhyacinth and is disseminated by *O. terebrantis*. The weevil produced significantly more eggs in the presence of the mite, a stimulating effect attributed to the release of a kairomone from wounded waterhyacinth tissues (Del Fosse, 1977b). Coexisting mites and weevils lead to higher population levels of both organisms, without harming either. *Orthogalumna terebrantis* starved but did not consume any offered weevil eggs, and there was no evidence of weevil feeding on the mite. When used together, their combined activities reduced waterhyacinth density by about 50% within 1 year, an improvement over the sum of their effects alone (Del Fosse, 1978).

Orthogalumna terebrantis collected in Argentina appeared to differ from mites studied in Florida. The former fed only on waterhyacinth, burrowed diffuse galleries and attacked plants growing in shade as well as in sunlight. The Florida mites fed on another member of the same plant family, *Pontedria cordata*, had more concentrated galleries and restricted themselves to shady areas. These populations, initially considered to be separate strains or even distinct species (Perkins, 1973), were later shown to be only slightly different (Cordo & DeLoach, 1976).

The mite thus affects waterhyacinth by direct feeding, by promoting the activity of another biocontrol agent and by exposing the weed to a phytophagous organism. Despite these attributes, little has been done with *O. terebrantis*, which may be an underestimated biocontrol agent (Charudattan, 1986). However, the mite's association with *A. zonatum*, a pathogen of several *Ficus* spp., has negated its release in Fiji (Kamath, 1979). The mite was introduced into India, Central America, and Africa (Waterhouse, 1994). In Zambia and India *O. terebrantis* became well established (Hill, 1997; Jayanth & Ganga Visalakshy, 1989). At least four other Galumnidae also feed on the leaves of waterhyacinth in India (Sumangala & Haq, 1995).

Waterhyacinth mats serve as breeding sites for mosquitoes, requiring chemical treatments. Five organophosphates assayed had little effect on the mite, but a herbicide was detrimental. This is discussed further in Chapter 43.

Pergalumna spp.

Pergalumna spp. thrive on nematodes and several species devoured a free-living nematode, *Cephalobus*, in 120–180s (Muraoka & Ishibashi, 1976). An undetermined species of *Pergalumna* was cultured on nematodes for over a year (Rockett, 1980). All mite stages (except for the larvae) fed on nematodes, although they showed no special preference for this diet and consumption seemed to depend on chance encounters with the prey. A typical bout consisted of *Pergalumna* grasping a nematode at one end and continuously pulling the prey into its oral cavity. The potential of Galumnidae in nematode control is discussed in Chapter 38 (ABAs as enemies of nemotodes).

Chapter 19
Hemisarcoptidae

Diagnosis

The Hemisarcoptidae are small, whitish mites with a prodorsal sclerite. The genital and anal orifices are located beyond coxae IV, confluent in the females, separated in the males. The latter also have a median ventral sucker anterior to the genital area. All tarsi terminate with an elongate, sucker-like ambulacrum. The hypopodes (facultative deutonymphs), in contrast to the other active stages, are heavily sclerotised and dark brown. They bear a pair of eyes but lack functional mouthparts, and have a caudal, ventral attachment plate. Their tarsal appendages lack the sucker-like ambulacra of the other stages, and the fourth pair of legs terminates with two or three long setae. The family has about a dozen genera, of which several are known only as hypopodes (e.g. OConnor & Houck, 1989). OConnor (1982) defined the family.

General biology

Hemisarcoptidae occur on insects, in nests of vertebrates or in sheltered habitats, e.g. the subcortical regions of trees; the life history of most forms is unknown. *Hemisarcoptes* is the only genus whose biology has been studied. All of its species feed on armoured scale insects (Hemiptera: Diaspididae), which include major agricultural pests, and the hypopodes are phoretic on predatory ladybirds (ladybeetles) of the genus *Chilocorus* (Coleoptera: Coccinellidae), which feed on the same prey.

A brief description of the biology of these pests is necessary in order to discuss the mites' life history. Diaspididae attack perennial crops such as apple, citrus and date palms. They are sessile insects whose body is covered by a hard, detachable shield (the 'scale') and whose bodies can be white, yellow, purple or any combination thereof, and may have a round, pear-like or elongated shape. The shields are formed as the insects increase in size and assume the shapes of their bodies. Diaspidids lay their eggs (or more rarely, give birth to young nymphs, termed 'crawlers') underneath the shield. The crawlers raise a flap in the shield's brim and walk out. At the first moult they lose their legs and the body becomes sac-like. Females undergo only one additional moult to become sexually mature, remaining at their site of feeding for the rest of their lives. Males undergo three more moults, emerging as short-lived, winged adults. In tropical and subtropical regions a cycle may require several months, whereas in colder regions armoured scale insects raise only a single generation (univoltine cycle).

Hemisarcoptes spp. may be parasitic or predatory. As parasites they suck out the contents of the scale's body, with their mouthparts twisted backwards while embedded in the host. Most active instars of *Hemisarcoptes* attack any available host stages

136

and eventually bring about their death. When only a few *Hemisarcoptes* parasitise a female, its fecundity is reduced and may later be totally curtailed; when more than ten mites attack an ovipositing scale, death soon follows (Gerson & Schneider, 1981). The development of immature hosts attacked by the mite ceases and they die. The bodies of the whitish mites acquire the colour of the body fluids of the host scale (Gerson, 1967b; Kaufmann, 1977). When parasitising the chaff scale (*Parlatoria pergandii*), for instance, *Hemisarcoptes coccophagus* Meyer became purplish, but turned yellowish when attacking *Hemiberlesia lataniae* (Signoret), the latania scale. *Hemisarcoptes* behaves like a predator when feeding on diaspidid eggs. Over 90% of the eggs of *Lepidosaphes ulmi* (L.), the oystershell scale, a major pest of apples, were devoured in eastern Canada by *Hemisarcoptes malus* (Shimer) (Tothill, 1918b). As the host dies or the prey is depleted, the mites wander off in search of other scales. Mating usually takes place on the host, prior to dispersal, and both sexes can mate several times (Izraylevich & Gerson, 1995b).

These mites prefer the ovipositing females of their hosts, in the field as well as in the laboratory. Almost twice as many ovipositing oriental scales [*Aonidiella orientalis* (Newstaed)] that infested papaw (*Carica papaya*) in Queensland were attacked in the field by an undescribed Australian *Hemisarcoptes* sp., as were young, non-ovipositing females (46% vs 25%; Gerson, 1994b). Establishment rates of *Hemisarcoptes cooremani* (Thomas) on the ovipositing scales of several hosts in the laboratory were higher than on other stage (Luck *et al.*, 1999a). The mites' preference for such females may be attributed to several factors, the first being easier access to this female phase, compared with younger, pre-ovipositing females. The margins of the shields of young scales are strongly affixed to the plant's surface, at times even embedded within the epidermis. To reach the bodies of such hosts the mites must cut their way in through the protecting shield, using a prolonged slicing process that requires several hours (Houck & OConnor, 1990). This mode of penetration is probably unsuitable for smaller mite stages. As the slicing process would require a substantial input of energy, as well as a cessation of feeding, the fecundity of invading females may be reduced. In contrast, access to ovipositing scales is facilitated by the emergence of the crawlers, which lift the shield's margin to egress, an opening used by the mites to gain access. Another factor affecting the mite's preference for ovipositing females could be the richer, more varied diet that such hosts represent, as they also include developing eggs and crawlers.

The time required by the mite to develop from egg to adult (1–2 weeks at 28°C, on optimal hosts) is usually shorter than its total egg-laying period, which can last for 3–4 weeks. The host, as noted, sometimes survives the attack, in which case progeny of the early mite cohort reach maturity and raise a second generation *in situ*, while their mothers are still ovipositing there. The presence of more than a single mite generation on a given host affects the mite's developing cohort. The older females mate locally, mostly with their male siblings. Migration of younger progeny away from the original, depleted host patch results in shifting the mating structure towards panmixis. The sex ratio (SR) in the field is quite variable (see below).

As noted, in cold climates diaspidids raise only a single annual generation, a situation that reduces the available host stages and phases. In northern China a

Hemisarcoptes sp. attacks *Lepidosaphes salicina* Borchsenius, the willow oyster scale. Second instar scales, which occurred in the late spring, were the stage most heavily attacked, but by summer, when the population consisted mostly of ovipositing females, only a few scales were available to the mite. Under such conditions the mite may not produce its full complement of progeny. This *Hemisarcoptes* does not undergo a winter diapause, as its development continues throughout the year, but only a few mites occurred under the hosts' shields during the long northern winters (Ji *et al.*, 1994).

Hemisarcoptes spp. are very tolerant of temperature extremes, being able to survive and reproduce in very cold or hot climates. Lord & MacPhee (1953) reported that *H. malus* moved about even when exposed to near subzero temperatures in Canada. At the other end of the climatic scale, Kaufmann (1977) observed that a *Hemisarcoptes* sp. (now known to be *H. coccophagus*; Gerson *et al.*, 1990) was the only natural enemy capable of attacking the date palm scale (*Parlatoria blanchardi* Targioni Tozzetti) throughout the year in the dry and hot Sahel region. Mite oviposition decreased, but did not cease, even during April to May, the hottest months of the year. In that area the mite raised at least 15 annual generations. Comparable data are not available from other parts of the world, although fewer annual generations would be expected in cooler areas. The host scale also affects the number of annual generations. The development of *H. coccophagus* on oleander scale (*Aspidiotus nerii* Bouché) required twice as many days as when parasitising the latania scale, at the same temperature (Gerson & Izraylevich, 1997); in the field this would translate to fewer annual generations.

Hemisarcoptes and *Chilocorus*

The association between *Hemisarcoptes* and *Chilocorus*, first recognised in California (Bartlett & DeBach, 1952), engendered studies on factors that induce the hypopodial moult and its subsequent termination, biochemical nature and evolutionary significance. Phoresy enables *Hemisarcoptes* to find new host/prey in unexploited areas, and facilitates its dispersal between patches, over short or long distances. The sclerotised integument of the hypopodes protects them against desiccation and radiation as they become exposed during dispersal.

A major factor that leads to hypopus formation is insufficient food. Some young mites attack juvenile scales, whose small bodies cannot provide enough nutrients for subsequent development. In addition, a second mite generation may be living off the same host, but cannot derive sufficient food. Some mites move away or die, but individuals whose diet deprivation had occurred early (during their larval or early protonymphal stage) are still able to continue their development, owing to a survival mechanism unique to astigmatid mites.

As long as the host furnishes adequate nutrients to the young mites, their development is rapid, moulting directly from first to third nymphal instar (protonymph to tritonymph), skipping the deutonymph. However, if the host becomes moribund during larval or early protonymphal growth, the mites are deprived of enough food, and moult to become hypopodes (Gerson & Schneider, 1982).

Termination of the hypopodial period seems to be triggered by chemical cues. Deutonymphs that had resided on *Chilocorus* could not moult to the next stage unless they had spent at least a few days on the beetles. Hypopodes of *H. cooremani* attached to *Chilocorus cacti* (L.) acquired unidentified chemicals (at least water) from the beetle, probably via the mites' caudal attachment plate. This acquisition, which takes place during the attachment period, appears to be a prerequisite for moulting (Houck & Cohen, 1995). Indirect support for the chemical acquisition hypothesis came from a study on a sex-specific tritrophic effect, which moved up from the food of the beetle larvae through to adult mites. Larvae of *Chilocorus bipustulatus* (L.) were reared on two separate prey, armoured scales and soft scales (Hemiptera: Coccidae), and the emerging beetles were then exposed to an abundance of *H. coccophagus* deutonymphs. Subsequent mite survival rates had no effect on the SR of mites that had sojourned on beetles given only armoured scales. Mite females that had stayed on beetles reared on soft scales, however, suffered higher mortality, resulting in a male-biased SR. In other words, some component of the larval diet was retained in the adult beetle and was passed on to the phoretic hypopodes (Ji *et al.*, 1996).

The end of the deutonymphal period was influenced by the presence of hosts or prey. Hypopodes of *H. coccophagus* left the beetles more quickly, and more moulted, in the presence of scale insects (43 vs 25%; Izraylevich & Gerson, 1995c). The exuviae of the deutonymphs were usually located next to the scales, indicating where the mites had moulted, and that this activity had taken place shortly before attacking these hosts. However, only around 43% of the hypopodes that had spent the required period on beetles subsequently moulted, suggesting that more factors could be involved. Length of the attachment period also affected the ensuing longevity of deutonymphs that had failed to moult for some reason: the longer they remained on *C. bipustulatus*, the longer they survived. When the beetles died, hypopodes left them within 1–2 days.

The number of hypopodes on *C. bipustulatus* that live on citrus trees was highly variable throughout the year, with the largest values being recorded during late summer and the lowest in winter, probably reflecting the beetles' annual cycle (Gerson, 1967b). The distribution of deutonymph populations on *C. bipustulatus* was highly contagious: relatively few beetles carried most of the mites (Izraylevich & Gerson, 1995c). This, however, may have been a special case. Every individual of three beetle species (*Chilocorus australasiae* Kerville, *Chilocorus circumdatus* Gyllenhal and *Chilocorus flavidus* Blackburn) examined by Gerson (1994b) in Australia carried between a few dozen and several hundred hypopodes.

Further evidence for heterogeneity in beetle–mite relationships is the variable number of hypopodes found on the elytra of diverse species of *Chilocorus*. The average number of *H. coccophagus* on *C. bipustulatus* is 30, with the highest being 202 (Gerson & Schneider, 1982). Larger numbers occurred on *C. cacti*, which was often laden with about 300 hypopodes (maximum: 800) of *H. cooremani* (Houck, 1994). An average of 250 (maximum: 615) *Hemisarcoptes* sp. was found on *C. flavidus* and a mean of 217 (maximum: 847) on *C. australasiae* (Gerson, 1994b). The location of the mites on the beetles seems to be determined by the pattern of microsetae, or spines, borne on the elytra's underside. Areas of dense spines (located mostly along the anterior margin of the wings) are unsuitable for the mites, which prefer the smoother surfaces (Houck, 1994; Izraylevich & Gerson, 1995c). The variable sizes of these areas (aside from the abundance of *Hemisarcoptes* in the habitat) thus determines the number of hypopodes found on various *Chilocorus* spp. Houck (1999) documented the densities of these spines on the elytra of 19 species of *Chilocorus* and predicted that taxa with fewer spines would be more suitable as vectors of *Hemisarcoptes* spp. When *C. bipustulatus* carried only a few *H. coccophagus* (up to six), all occurred on the same elytron. However, hypopodes were evenly distributed on both elytra as their numbers increased (Izraylevich & Gerson, 1995c).

As noted, hypopodes of *Hemisarcoptes* are commonly found on various species of *Chilocorus* in different parts of the world (Gerson *et al.*, 1990). The specificity of *Hemisarcoptes* for *Chilocorus* is postulated to be genus rather than species specific. Adults of *Chilocorus nigritus* (Fabricius), imported from India, were exposed to hypopodes of the North American *H. cooremani*, with which they had had no common prior history. The deutonymphs rapidly attached to the beetles and later completed their development on scale insects (Houck & OConnor, 1990). Similar observations in regard to *Chilocorus infernalis* (L.) were reported by Hill *et al.* (1993). Some other genera of the family Coccinellidae also carry these hypopodes, indicating that although *Chilocorus* is the beetle genus preferred by the mites, the association is not exclusive. Records include laboratory cultures of *Zagloba ornata* Casey (Sellers & Robinson, 1950), *Halmus chalybeus* (Boisduval) and *Scymnus fagus* Broun (Hill *et al.*, 1993), all predators of scale insects. Field observations (Charles *et al.*, 1995a, b) suggest that *H. chalybeus* also disseminates *Hemisarcoptes* in the field. *Chilocorus* does not occur naturally in New Zealand and did not become established in New Zealand despite several introductions; the mites' occurrence on non-*Chilocorus* species thus probably signals their efforts to adapt to other coccinellids that feed on the same prey. Another host for *Hemisarcoptes* was *Axion tripustulatum* (DeGeer) (Gerson *et al.*, 1990).

As noted, larvae and adults of *Chilocorus* feed on armoured scale insects, and may at times ingest the mites along with their prey. Nevertheless, there is no evidence of competition between these natural enemies, and the widespread occurrence of various species of *Hemisarcoptes* on diverse beetle species, in different parts of the world, attests to advantages accruing to the mites from the association.

Hemisarcoptes coccophagus

A comparison of *H. coccophagus* parasitisation levels on two diaspidids, latania scale and oleander scale, infesting acacia (*Acacia cyanophylla*) in Israel, indicated that mites mostly attacked ovipositing females of both hosts. About eight eggs/mite were found on ovipositing latania scales, but only 3.5 eggs/mite on young females (Izraylevich & Gerson, 1995b). The fecundity of *H. coccophagus* was also affected by the scale species on which it had fed. When offered ovipositing latania scales, each mite produced 80–90 eggs (maximum: 120), the comparative values for oleander

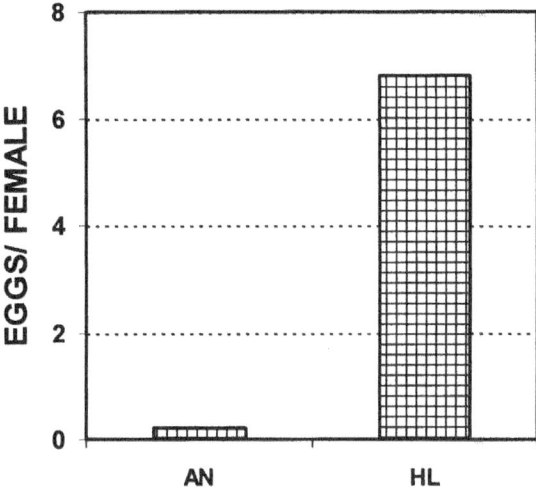

Fig. 19.1 Influence of different host scales (AN: *Aspidiotus nerii*; HL: *Hemiberlesia lataniae*) on the fecundity of *Hemisarcoptes coccophagus*. (Data from Izraylevich & Gerson, 1993a.)

scale being only two eggs/female; low progeny numbers were also found in the field (Fig. 19.1). The higher oviposition of *H. coccophagus* as it fed on *A. nerii* when co-occurring in the field with *H. lataniae* could have been due to the mite's prior feeding on the latter species (Izraylevich & Gerson, 1993a). Host eggs, juveniles and male pupae were rapidly killed by adult *H. coccophagus*.

The plant on which the host scale lived influenced the mite's oviposition indirectly, probably by affecting scale size. Citrus varieties are known to affect the proportions of diaspidids that feed on them (Hare *et al.*, 1990). The dimensions of chaff scales collected off grapefruit bark were 0.55 × 0.67 mm, whereas those from orange bark were 0.37 × 0.44 (Izraylevich & Gerson, 1993a). This difference in host size, and thus in available nutrients, was probably the reason for the significantly different mite fecundity in the field, as well as being the factor that enabled more parasites to develop on the 'grapefruit' scales (Fig. 19.2).

Walking appears to be the mode of dispersal and within-patch host searching of all active mite stages, but the scope of such dispersal is restricted by the mites' limited movement. Nevertheless, in New Zealand *H. coccophagus* spread about 200 m from its point of release within 3 years, whether by self-propulsion, dispersal by *H. chalybeus*, or other means (Charles *et al.*, 1995a, b; see below).

A female-biased SR of approximately 0.65 was initially reported from laboratory studies of these mites (Gerson & Schneider, 1981; Houck & OConnor, 1990). In the field, however, the SR of *H. coccophagus* was variable during the year as well as on its different hosts. It was strongly male biased on latania scale during autumn and winter, becoming female biased in late spring; on oleander scale it was distinctly male biased, and on chaff scale the SR fluctuated throughout the year (see Chapter 40). When the SRs of *H. coccophagus* from all scale species and host plants were plotted

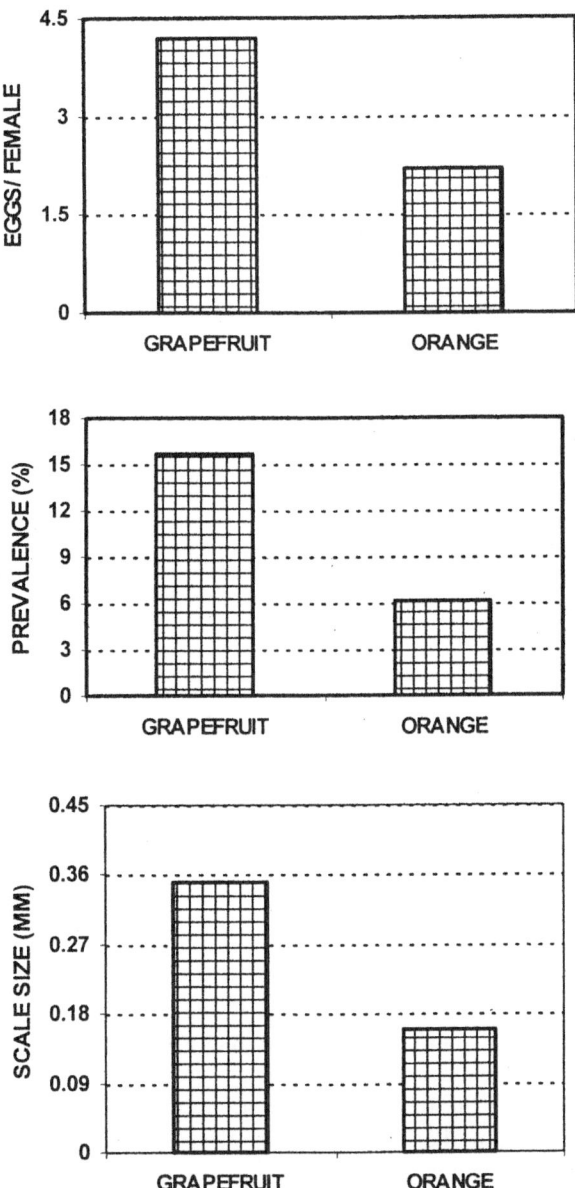

Fig. 19.2 Lower: effect of two citrus species on the size of chaff scale females (*Parlatoria pergandii*). Middle: prevalence of *Hemisarcoptes coccophagus* on these scales. Upper: number of mite eggs. (From Izraylevich & Gerson, 1995b.)

against the mite's mean intensity (mean number of active mites per infested host; Margolies *et al.*, 1982), a significant negative relationship emerged, suggesting that the asymmetric SR ratio seen in the mite's field populations is affected by density-dependent factors (Izraylevich & Gerson, 1995d).

Hemiscarcoptes coccophagus has a diploid karyotype with 2n = 14 in both sexes; the sex-determining mechanism in this species is thus either diplodiploidy or parahaploidy (Izraylevich *et al.*, 1995). The mite's ability to adjust its SR in response to environmental changes, a rare attribute among non-arrhenotokous arthropods (Izraylevich & Gerson, 1996), and the mechanisms of sex allocation, are detailed in Chapter 40.

The parasitisation of *H. coccophagus* on three host scales infesting various host plants was monitored during more than a year's sampling at several sites in Israel. A surge in egg production invariably occurred in the spring (Izraylevich & Gerson, 1993b). In a tally of all mites that had developed on the more suitable latania and chaff scales (comprising 14,315 individuals), eggs made up 46.7% of the population, juveniles 35.6% and adults 17.7%. The latter value (range: 16–23%) was the most consistent feature of all populations (Izraylevich & Gerson, 1993a); deutonymphs were rare on scales in the field. This may then be taken as the overall, annual population structure of feeding *H. coccophagus*. The unsuitability of oleander scale for oviposition resulted in an unstable mite population structure on that host.

The effect of *H. coccophagus* on diaspidids in Israel depended on the latters' stages and, even more, on their specific suitability. The mite's numerical response to the density of the more suitable latania scale was faster than its response to the less suitable oleander scale (Izraylevich *et al.*, 1996), suggesting that latania scale would be more likely to be controlled by *H. coccophagus*. The main impact of *H. coccophagus* on chaff scale infesting citrus appears to be during summer, when pest populations ebb whereas mite numbers surge (Gerson, 1967b). The controlling effect of *H. coccophagus* on the date palm scale in the Sahel was inferred by Kaufmann (1977) from observing that no other active, year-round natural enemies were present and that the mite had a density-dependent response to scale density.

There have been two separate projects to introduce *Hemisarcoptes* spp. into New Zealand. In the first, initiated in 1967 to control San José scale (SJS), *Quadraspidiotus perniciosus* (Comstock), on apples, *H. malus* was imported from Canada to the South Island. The effort was unsuccessful owing to difficulties in rearing host scales in the laboratory, with a consequent dearth of mites for field releases (Hill *et al.*, 1993). The second project, involving *H. coccophagus* and *H. cooremani*, began in 1987 and was aimed at controlling three armoured scale insects infesting kiwifruit (*Actinidia deliciosa*): greedy scale, *Hemiberlesia rapax* Comstock, latania scale and oleander scale. Mite introduction was combined with that of *Chilocorus*, *H. cooremani* being brought in along with *C. cacti* from the USA, and *H. coccophagus* with *C. bipustulatus* from Israel. Beetles and mites were mass-reared in the laboratory on latania scale that was cultured on squash or potato tubers. The single release of *H. cooremani* consisted of around 3000 hypopodes placed in sections of drinking straws that were pinned onto the bark of kiwifruit, but the mite failed to establish. Almost 500,000 *H. coccophagus* were released at 25 sites, mostly in the North Island, either as hypopodes along with the beetles, or as a mixture of feeding stages parasitising scales on laboratory-inoculated squash fruits and potato tubers. These plant parts, placed in plastic bags, were stapled or wedged into forks of branches on the target trees, which were the shelter trees Lombardy poplar, *Populus nigra* var *italica*, and woolly nightshade, *Solanum mauritianum*, all infested by the latania scale (Charles *et al.*, 1995a, b). Releases were made between March and May 1989 at five sites. Regular sampling, which consisted of 40–80 cm^2 of scale-infested bark chips, was initiated 8 months later. Mites were consistently recovered on scales at all sites, serving as evidence for establishment (Charles *et al.*, 1995b). Mite impact on scale populations was assessed as the rate of reproducing host females that were attacked, from all such scales. The mean density (calculated as scales/100 cm^2) of live scales parasitised by the mite rose steadily during the austral fall (February to May) and then declined to a stable 28–34. Differences between mean densities of reproducing scales that infested control trees (without mites), and those on release trees were highly significant. Mite parasitism peaked in May and then steadied at 39–49% (Fig. 19.3). Significantly fewer juvenile scales were found on release than on control trees, suggesting that the mite's impact on pest populations may be permanent.

In analysing the effect of *H. coccophagus* on latania scale, Hill *et al.* (1993) included only hosts that were with mites when sampled. Recently dead scales were therefore excluded from their calculations, possibly causing an underestimation of the mite's impact. Further, all scales on some (3–20%) bark chips taken from release trees were dead, whereas no such mortality was seen in the control tree samples. Such 'local extinctions' of pest populations, along with data presented in Fig. 19.3 and the noted underestimation, indicated the successful biocontrol of the pest by the mite. Hill *et al.* (1993) believed that the most important traits of *H. coccophagus* as a

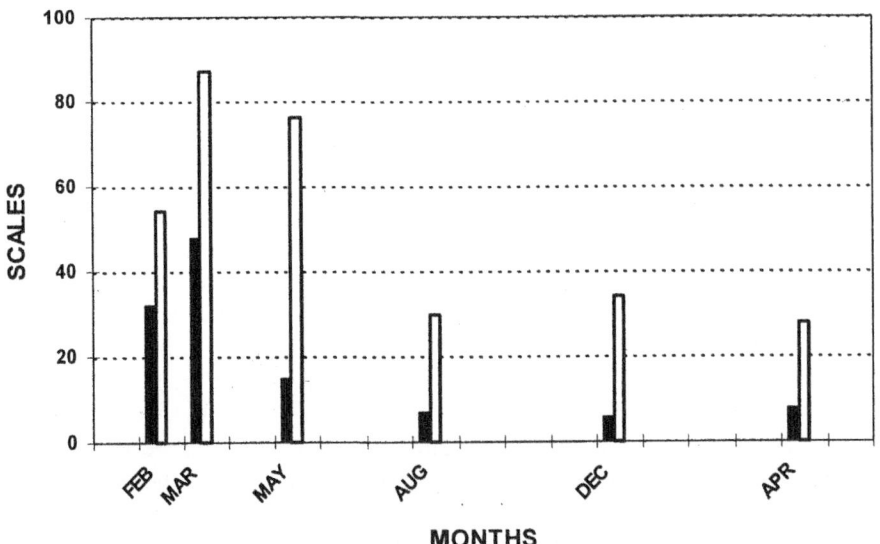

Fig. 19.3. Mean density of latania scale (*Hemiberlesia latania*) on Lombardy poplars with (black bars) and without *Hemisarcoptes coccophagus* (white bars) during a period of 16 months. (Redrawn from data from Table 2, in Hill *et al.*, 1993.)

natural enemy were its host-searching ability and fast rate of increase, which is several times higher than that of the hosts. The percentage of 'local extinctions' increased on the host plants, reaching 53% 4 years later (Charles *et al.*, 1995b).

The discovery of *H. coccophagus* parasitising SJS in one South Island orchard prompted another massive effort to establish the mite on more tree crops there. About 50,000 mites were released on apple, pear and plum trees at various sites, but mites were recovered only from a single site 1 year later, and lost after 6 years. Charles *et al.* (1998) considered this failure to be due to the absence of *Chilocorus* in New Zealand, the mite's inability to survive on univoltine host scales and the presence of competing natural enemies. However, as *H. coccophagus* is well established in the North Island, where it was recovered from at least 11 host scales, the mite could still make its way to the South Island.

Hemisarcoptes malus

As noted in the General biology section, *H. malus* was the first mite ever to be considered capable of reducing populations of a pest, the oystershell scale (Shimer, 1868b). The mite's apparent absence from western Canada suggested that it could profitably be brought there from the eastern states. Introductions to several sites in British Columbia were initiated in 1917 and continued for 23 years, the mites usually being transferred onto apple branches infested by parasitised scales. Although no precise data on its effect were available, Turnbull & Chant (1961) scored this introduction as a successful biocontrol project, an opinion based in part on Tothill (1918a) and Glendenning (1931). Tothill examined close to 18,000 scales on apple trees in western Canada and reported that at some sites 90–95% of their eggs were killed by the mite. This usually occurred on groups of trees, all or most of which were colonised by the mite; when tree colonisation was less than 35%, fewer pest scales were killed by *Hemisarcoptes* (Fig. 19.4). Tothill (1918a) stated that *H. malus* 'effected excellent control' of the pest in western Canada, as its populations were almost completely destroyed on heavily infested trees. At other localities where the mites inhabited fewer trees, fewer scales were attacked.

Clear evidence for the impact of the mite on the oystershell scale was obtained when it rose from being of minor status to becoming a major pest of apples in Nova Scotia, concurrent with changes in fungicide use there. The pest had two major natural enemies, the hymenopterous parasitoid *Aphytis* (then called *Aphelinus*) *mytilaspidis* LeBaron and *H. malus*. Either was deemed capable of controlling the pest. Lord (1947) sprayed infested trees with various fungicides and sampled scales taken from the trees. Numbers of *H. malus* on the sulfur-treated trees were low, in marked contrast to their densities on trees sprayed with fermate, a non-toxic fungicide. Reductions in mite populations coincided with increases in scale numbers. Although this decrease was attributed to eliminating both natural enemies, the role of *H. malus* was clearly seen. Further evidence for mite activity was mortality of juvenile scales, hosts that are not affected by *A. mytilaspidis*. In one case 94.5% of 2800 scale females were attacked by the mite.

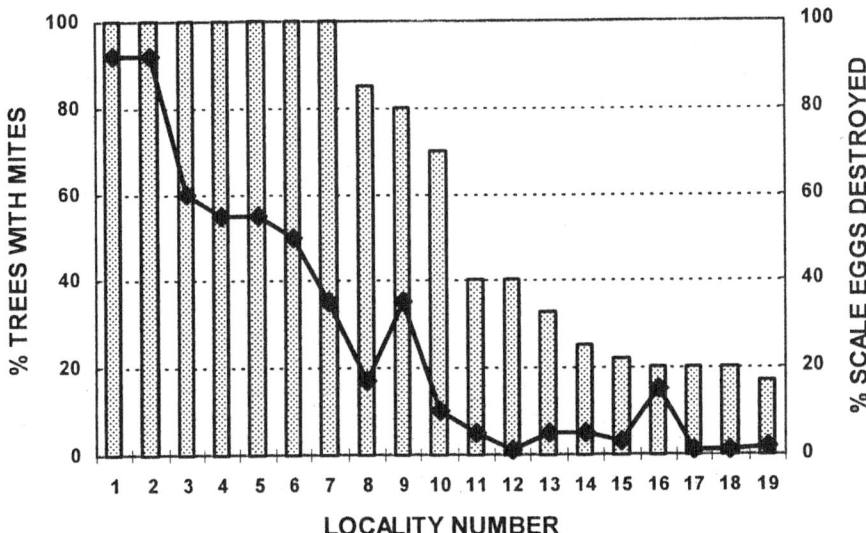

Fig. 19.4 Percentage of apple trees colonised by *Hemisarcoptes malus* (left, columns), and rate of oystershell scale eggs destroyed by the mite (right, line) in 19 groups of trees sampled in various western Canadian localities. (Data from Tothill, 1918b.)

The survival and activity of *H. malus* during the harsh Canadian winter, when no other natural enemies (e.g. *A. mytilaspidis*) were able to survive, further persuaded Lord & MacPhee (1953) that the mite was one of the two major factors controlling the pest. Another approach was adopted by Samarasinghe & LeRoux (1966), who pinpointed the importance of *H. malus* in controlling the pest by applying multifactor analyses of life tables that were constructed for the oystershell scale. Finally, applications of sulfur to apple orchards resulted in large increases in scale numbers. When the treatments were discontinued, the densities of the two natural enemies increased along with a decrease in scale numbers, providing experimental proof that these natural enemies were regulating pest populations (Pickett, 1965).

Miscellaneous

An outbreak of *Lepidosaphes newsteadi* Šulc on cedar trees in Bermuda afforded an opportunity for introducing *Hemisarcoptes* sp. as hypopodes. Mites attached to the bodies of 235 coccinellid beetles, mostly *Chilocorus* spp., were shipped to Bermuda and released, apparently with little success (Bedford, 1949).

Mite tracking of host scale populations onto suboptimal tree aspects was observed in the distribution pattern of *Hemisarcoptes* sp. attacking *L. salicina* on willows in northern China. Mite populations were more abundant on the northern and eastern aspects of the trees than on their southern and western sides, although the latter receive more solar radiation. This distribution was postulated by Ji *et al.* (1994) to be due to the prevailing northern and eastern spring winds that carry the scales' crawlers.

Hemisarcoptes spp. can be mass-reared in the laboratory when cultures of host scales are available. The scales used include the California red scale, *Aonidiella aurantii* reared on green lemons (Gerson & Schneider, 1981), latania scale on squash and potato tubers (Izraylevich & Gerson, 1993a; Hill *et al.*, 1993) and the cactus scale, *Diaspis echinocacti* (Bouché), grown on spineless pods of the cactus *Opuntia* sp. (Izraylevich & Gerson, 1995d). Neither all plant material nor all host scales is uniformly suitable. Only hardy fruit, tubers or pods, which remain turgescent and thus suitable for long periods, are appropriate. As to host scales, the Israeli oleander scale was unsuitable for rearing *H. coccophagus* (Izraylevich & Gerson, 1993a). Houck & OConnor (1996), however, reared *H. malus* and *H. cooremani* for many generations on an American oleander scale, a success that may reflect different host utilisation patterns among the species of *Hemisarcoptes*, or taxonomic uncertainties among the scales (Gerson & Izraylevich, 1997). The different suitability of the various scale stages for parasitisation is another factor that should be considered when mass-rearing the mites (see Chapter 39). The relative ease of rearing *Hemisarcoptes* spp. and *Chilocorus* spp., which feed on the same prey, suggests that collecting, shipping and releasing *Hemisarcoptes* spp. as hypopodes while on the beetles would, as in the past, be a suitable method for further introductions. The effect of pesticides on *Hemisarcoptes* spp. is discussed in Chapter 43.

Prospects

Species of *Hemisarcoptes* kill large numbers of some economically important diaspidids (Table 19.1), but four problems hinder the further use of these mites in biocontrol projects. The first is the lack of taxonomic expertise and keys for the recognition, and thus utilisation, of additional species that are known to exist (Gerson *et al.*, 1990; Gerson, 1994b). The second is the mite's apparent lack of competitive ability. Whenever *Hemisarcoptes* spp. occurred as the only (or dominant) natural enemy attacking armoured scale insects, e.g. in apple orchards in eastern Canada after a cold winter, or on date palms in the Sahel, it provided satisfactory, economic-level pest control. However, in many other cases, when the mites were only one group of several natural enemies attacking a given pest, their effect was seasonal at best. The third constraint is the inability of the mite to track rapid changes in host populations, owing to its limited host-finding abilities whenever and wherever its vector beetles are not present. Lord (1947) noted that at orchards where *H. malus* had brought about 'sensational' scale control during one year, the mite could not be found in the next. These dilemmas could probably be overcome by releasing *Hemisarcoptes* spp. along with *Chilocorus* spp. An understanding of the population structure and suitability of any given target diaspidid for *Hemisarcoptes*, and the mites' specific aggregation patterns, should facilitate future releases (Izraylevich & Gerson, 1995a, b). Finally, *H. cooremani* (and, by projection, other *Hemisarcoptes* spp.) has unique allergens that could affect personnel working with this natural enemy (Arlian *et al.*, 1999).

Table 19.1   Rates of *Hemisarcoptes* spp. attack on various hosts infesting commercial crops in different parts of the world

Host scale	Highest attack rates	Comments	Country	Source
Pseudaulacaspis pentagona (Targioni-Tozzetti)	100	Uneven attack	Bermuda	Simmonds (1958)
Aonidiella orientalis (Newstead)	40		Queensland (Australia)	Gerson (1994b)
Lepidosaphes ulmi (L.)	<90	Uneven attack	Eastern Canada	Tothill (1918b)
Parlatoria blanchardi (Targioni Tozzetti)	75	Uneven, attack mostly on eggs	North Africa	Smirnoff (1957)
Parlatoria pergandii Comstock	60	Uneven	Israel	Izraylevich & Gerson (1993b)
Aonidiella aurantii (Maskell)	75	Uneven	Israel	Gerson (1967b)

The few successes with *Hemisarcoptes* spp., whether as resident agents or when introduced, suggest that in certain cases these mites could be used to great advantage. Three factors should promote their further utilisation: their tolerance to extreme climatic conditions, the possibility of rearing them in large numbers on common hosts grown on inexpensive plant material, and their easy transport to the field with their vectors.

Chapter 20
Hydryphantidae

Diagnosis

Hydryphantidae are reddish freshwater mites usually with a soft, papillate or lined integument, but variously sized dorsal and ventral plates may be present. The palp is chelate and the palpal tibia bears a dorso-apical distal process. The eyes are located in distinct capsules; a median eye occurs when a frontal plate is present. The coxal plates are in four groups and the genital field consists of two movable plates and three or more pairs of genital suckers. Larvae are heteromorphic to the deutonymphs and adults; separate keys are therefore needed for their determination (see Chapter 3). The Hydryphantidae is a large family, with over 50 genera and many hundreds of species. The genera were treated by Cook (1974), the North American taxa by Smith & Cook (1991), the Mediterranean species by Gerecke (1996 and earlier papers), the Chinese taxa by Jin (1997) and the Australian genera by Harvey (1998).

General biology

These freshwater mites are assigned to the Hydrachnidia (also known as Hydracarina or Hydrachnellae), a cohort of aquatic prostigmatid superfamilies whose members have only three active stages. These consist of the parasitic larvae and the predatory deutonymphs and adults; the protonymphs and tritonymphs remain quiescent. The larvae move on the surface film of freshwater bodies, from which they jump onto potential hosts. Dipteran hosts include biting flies, mosquitoes, black flies (Simuliidae) and horse flies (Tabanidae) (Smith, 1983; Smith & Oliver, 1986), families of medical and veterinary importance. The larvae drop off after feeding and develop further in the water. Nymphs and adults may feed on the juveniles of the same species that they had parasitised, or on other insects (Mullen, 1977). Sperm transfer is indirect (e.g. via spermatophores deposited on substrates), and eggs (up to 200/female) are deposited in masses. Many North American hydryphantids develop in temporary pools, springs and seepage areas (Mullen, 1977), habitats that also abound with mosquitoes, whereas the African species live in permanent waters (Smith, 1983).

The pattern of mosquito parasitism by some hydryphantids (see below) differs from that of members of other water mite families, because they attack hosts when returning to water bodies, not as these adults first emerge.

Hydryphantidae as natural enemies of mosquitoes

The life cycle of *Thyas barbigera* Viets required 2 years in the north-eastern USA, each female depositing about 45 eggs. Hatching larvae parasitise several species of the mosquito *Aedes*, most hosts being females that were attacked as they returned to water bodies to oviposite. The overall rate of parasitism was <1%, with usually about 1.7 mites/host. During engorgement the body of the mite increases three- to four-fold. In contrast to other water mites, no stylostome (feeding tube) is formed in the host's body during parasitisation (see Chapter 7).

Submerged nymphs and adults of *T. barbigera* survive for long periods (more than 1 year) without any nourishment. In the laboratory they fed on eggs of *Aedes*, but not on other aquatic organisms, including Ostracoda, worms and juveniles of other water insects. The scant feeding of *T. barbigera* on mosquito eggs probably has no effect the pests' populations (Mullen, 1975, 1977).

Fenley (1966) believed that the flight and host-seeking activities of *Aedes ventrovittis* Dyar were curtailed owing to infestation by *Panisopsis* sp. (an uncertain identification; Smith & Oliver, 1986). Mite-free mosquitoes flew and fed throughout the day, but parasitised hosts were disadvantaged by being restricted to flying only during the evening hours.

Hydryphantidae as natural enemies of water bugs

Hydryphantes tenuabilis Marshall occurred on all five immature stages of the aquatic bug *Hydrometra australis* Say (Hemiptera: Hydrometridae), but large amounts of body fluids were extracted only from the adults. Bug longevity was reduced after 5 days when attacked by two or more mites in the laboratory, the death rate increasing along with parasite load. The rate of increase of *H. australis* was reduced from 0.08 for unparasitised hosts to 0.05 for those with ten *H. tenuabilis*/host. This suggested that the effect of the mite is mostly due to extraction of host tissues, especially during early adult life (Lanciani, 1979a, 1995).

To sum up, mites of the family Hydryphantidae adversely affect pests as well as non-pests, but there is little evidence to show that they seriously reduce populations of mosquitoes or other insects of veterinary or medical importance.

Chapter 21
Laelapidae

Diagnosis

Laelapidae bear a dorsal plate with more than 23 pairs of setae. The female has a flask-shaped genital shield, the metasternal shields are small and the stigmata, with straight peritremes, open between legs III and IV. There are no ventral leg grooves and legs I usually carry ambulacra. The Laelapidae is a large family, with over 50 genera and several hundred species. A recent taxonomic treatment is provided by Casanueva (1993), Radovsky & Gettinger (1999) defined its five subfamilies and Karg (1993) discussed the European species.

General biology

Many laelapids are ectoparasitic on small mammals or associated with various arthropods; pest species include the honey bee parasites *Varroa destructor* Anderson & Trueman (often placed in the family Varroidae) and *Tropilaelaps clareae* Delfinado-Baker & Baker. Other species are free-living soil-borne predators. The provenance of the latter has probably preadapted them for thriving in highly humid artificial habitats, such as greenhouses or mushroom houses; laelapids are also common in stored products (Hughes, 1976). Many species raise a generation in 2–3 weeks. Arrhenotoky appears to be the main sex-determining mechanism (De Jong *et al.*, 1981; Ruf, 1991), but species in some taxa produce only uniparental progeny. The resultant populations may be exclusively female, whereas others consist of males and females (Mihm & Chiang, 1976).

Stratiolaelaps miles (Berlese)

This species (often placed in the genus *Hypoaspis*) is being used in the control of glasshouse pests, including larvae of sciarid flies (Diptera: Sciaridae), which inhabit rotting plants and roots, damaging mushrooms and ornamental crops grown in humid organic substrates. They feed on young roots or fungal mycelia and transmit plant pathogens; adult flies may become nuisance pests of personnel and customers. *Lycoriella solani* (Winnertz) infests mushrooms, whereas *Bradysia paupera* Tuomikoski and *B. tritici* Coquillet are greenhouse pests.

Reared on a diet of *Bradysia* spp. at 25°C, *S. miles* required little more than a fortnight to raise a generation. The mite's average fecundity at that temperature was 57 eggs/female, nearly all eggs were deposited during the first 3 weeks of adult life and the most female-biased sex ratio (SR) (see Fig. 40.3) was obtained (Ydergaard *et al.*, 1997). The mite developed and oviposited within a wide range of temperatures (15–30°C), but the best life table parameters were obtained at 25°C [net reproductive rate (R_0) = 49.7; r_m = 0.133]. Acarid mites, such as *Tyrophagus putrescentiae*, which are common in rotting organic substrates, may serve as food, although they are less nutritious than sciarid larvae. Predators feeding on these acarids had prolonged lives but fewer progeny, of which fewer were females (see Fig. 40.2) (Enkegaard *et al.*, 1997). The mite was mass-reared in a mixture of sterile milled sphagnum and washed vermiculite (ratio 2/1 by volume) in sealed but aerated and wetted plaster-of-Paris containers. The flour mite, *Acarus siro* L., was provided as food and a total of 25 eggs/female was deposited on this prey (Ali & Brennan, 1997). About 50,000–60,000 predators were thus obtained in 3–4 weeks at 28°C, provided cultures were wetted and ventilated (Wright & Chambers, 1994). The highest yield of eggs would be obtained if the SR of *S. miles* were maintained at F2/M1 (Enkegaard *et al.*, 1997). Steiner *et al.* (1999) developed another mass-rearing method, with the ubiquitous *T. putrescentiae* as prey.

Chambers *et al.* (1993) employed *S. miles* to control *Bradysia* spp. infesting potted cyclamen and poinsettia. Habitats of predator and young larvae, the preferred prey, overlap because the pests deposit most of their eggs in the top 1 cm of the substrate, where the mites aggregate (Wright & Chambers, 1994). Optimal release rates were established by placing various numbers of predators in each pot; their effect on the sciarids was determined by tallying flies caught in yellow sticky traps and comparing these values to those from non-release pots. Effective control, indicated by a lack of subsequent pest resurgence, was obtained with at least 55 mites/pot; more predators did not improve the results. Significantly negative relationships between rate of predators released and numbers of flies caught were obtained for both ornamentals. However, the beneficial effect of *S. miles* was restricted to the early sciarid generations, probably because pest populations were limited by insufficient food later in the season.

To assess the efficacy of *S. miles* as a biocontrol agent of *L. solani*, bags of compost used for growing commercial mushroom (*Agaricus bisporus*) were artificially infested by the pest (Ali *et al.*, 1997). Infestations were either with three pairs of *L. solani*, or with three such pairs to which 100 predators were added, or with *S. miles* alone, as well as a control treatment, with neither flies nor mites. Predator effect was evaluated by determining the number of flies that emerged from day 30 to day 70, and by comparing the yields of commercial-rate mushrooms. Mite presence had a significantly adverse effect on the numbers of emerging flies (Fig. 21.1, LS+SM) [the presence of a few flies in the mite-only treatment (SM) was attributed to contamination]. Numbers of *L. solani* obtained from the controls (also ascribed to natural infestations) did not differ from the fly-only (LS) treatment. Of more economic interest, adding *S. miles* significantly improved yields (Fig. 21.2). Several *S. miles* at times fed on a single large *L. solani* larva, which they may have paralysed with an injected toxin (Ali & Brennan, 2000). The predators were always associated with the prey, never occurring on mushroom cups, a factor important for growers and consumers. Placing *S. miles* onto the compost during the early, mycelial colonisation

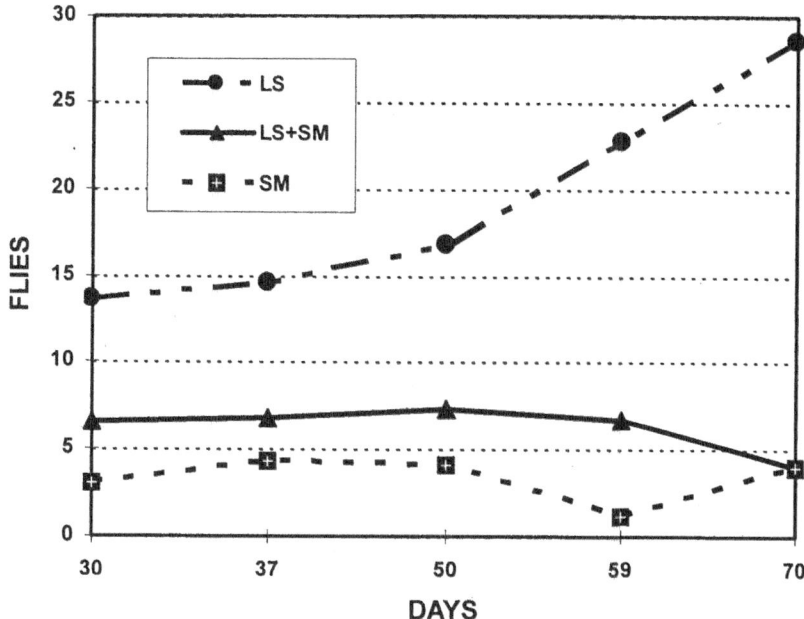

Fig. 21.1 Effect of *Stratiolaelaps miles* on the emergence of adult *Lycoriella solani* from mushroom bags at five sampling dates (LS: infestation by three pairs of *L. solani*; LS + SM: infestation by three pairs of *L. solani* with 100 *S. miles* added; SM: 100 *S. miles* without flies. (From data in Table 2, Ali *et al.*, 1997.)

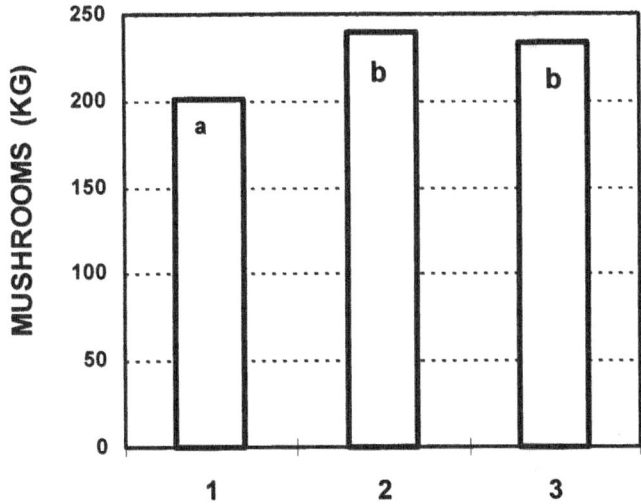

Fig. 21.2 Mushroom yields obtained from 1 t of compost when (1) only *Lycoriella solani* was added to mushroom bags; (2) *L. solani* along with 100 individual *Stratiolaelaps miles* were added; and (3) *S. miles* alone were added. Different letters within columns denote significant differences at the 5% level of probability. (From data in Table 5, Ali *et al.*, 1997.)

phase reduced pest emergence by 87%; later introductions were less effective (Jess & Kilpatrick, 2000).

The predator was assayed against shore flies, *Scatella* spp., which reduce the commercial value of ornamentals by leaving faecal spots on their leaves. The results, although rather ambiguous (Lindquist *et al.*, 1994), suggest another target for the mite. Pupae of the western flower thrips (WFT), *Frankliniella occidentalis* (Pergande), a serious pest of ornamentals, were attacked by *S. miles* in the soil, being assayed in rose pots in Finland. An application of 50 predators/pot reduced damaged leaf area by 41%, and the adult pests by 33%. When another predator, the phytoseiid *Neoseiulus cucumeris* (Oudemans), was concurrently applied to the potted roses, overall control of WFT was much better (damaged leaf area was reduced by 66%), but the populations of *S. miles* were reduced. This was attributed to prey deprivation by the phytoseiid; Linnamäki *et al.* (1998) believed that WFT control would be improved by maintaining more *S. miles* in the soil. However, neither predator succeeded in controlling onion thrips, *Thrips tabaci* Lindeman, on leek (*Allium porrum*) seed crops under protected cultivation, although *S. miles* did quite well in the laboratory (Rat-Morris, 1999).

Symphyla are centipede-like soil myriapods that can raise large numbers in glasshouses, where their feeding causes the death of entire plants. Their control was attained by several releases (in batches of 40–50/m^2) of *S. miles* in a German chrysanthemum glasshouse (Albert, 1999).

Populations of *S. miles* survived without food for several weeks. When subsequently fed, the adults oviposited and lived for over 4 months. This trait, along with the mite's polyphagy, suggests that *S. miles* could be used as a prophylactic, or predator-in-first, biological control agent (Chambers *et al.*, 1993). If introduced early in the season *S. miles* would be on hand before flies or thrips arrived, subsisting on other prey, and could then attack early pest generations.

At present, *S. miles* is offered for sale against flies by one company, to combat sciarid flies by two suppliers, and for thrips control by nine companies (Anonymous, 2000).

Geolaelaps aculeifer (Canestrini)

This species (often placed in *Hypoaspis*) is a soil-borne, polyphagous predator that has been assayed as a natural enemy of nematodes, Collembola, thrips and acarid mites. Sharma (1971) compared the effect of different predatory mites on populations of the plant nematode *Tylenchorhynchus dubius* (Butchli) in pot experiments. Nematode numbers decreased only in the presence of *G. aculeifer*, the predator killing relatively more pest juveniles than the other mites (see Chapter 38, ABAs as enemies of nematodes). The same predator, or a closely related species, reduced WFT populations by around 30% (Gillespie & Quiring, 1990). In addition, releasing 125 mites/plant in selected rows sufficed to control *Bradysia* spp. damaging hydroponically grown greenhouse cucumbers, maintaining their numbers at acceptable levels. Piatkowski (1997) applied three batches of *G. aculeifer* (a total of 65) to each

poinsettia plant in a greenhouse, to control *Bradysia* spp. Pest populations were monitored with sticky traps and compared with numbers obtained from pesticide-treated
plots. *Bradysia* numbers were similar during the first part of the growing season
(April–May), but around 63% higher in the pesticide plots during the second half
(June–July). The efficacy of *G. aculeifer* is probably affected by soil pore size and the
pattern of prey aggregation in the habitat (Axelsen *et al.*, 1997), topics that remain
unexplored in glasshouses.

A search for natural enemies of the bulb mite, *Rhizoglyphus robini*, in various parts
of the world resulted in finding three acarine predators, of which *G. aculeifer* was the
most effective (Lesna *et al.*, 1995). Four populations of this species were collected:
two from sites about 10 km apart in North Holland (herein designated as NH1 and
NH2), one from Bremen in northern Germany and one from Taiwan. Their ability to
reduce numbers of *R. robini* infesting lily scales was demonstrated and compared in
closed-jar experiments, in which the initial predator/prey ratio was 1/6. The predator
reduced bulb mite populations to less than 5% of their original numbers within
4 weeks, whereas in the control jars the pests increased five-fold (Fig. 21.3). A significant interpopulation difference was found only between the NH2 and Bremen
strains. Much greater differences between strains were seen when their rates of
population growth were compared, as mites of the NH2 strain increased their
numbers three times more than those from Bremen (Fig. 21.4).These differences
seem to reflect the intraspecific variability within the acarine entity currently known
as *Geolaelaps aculeifer* (e.g. Ruf, 1991).

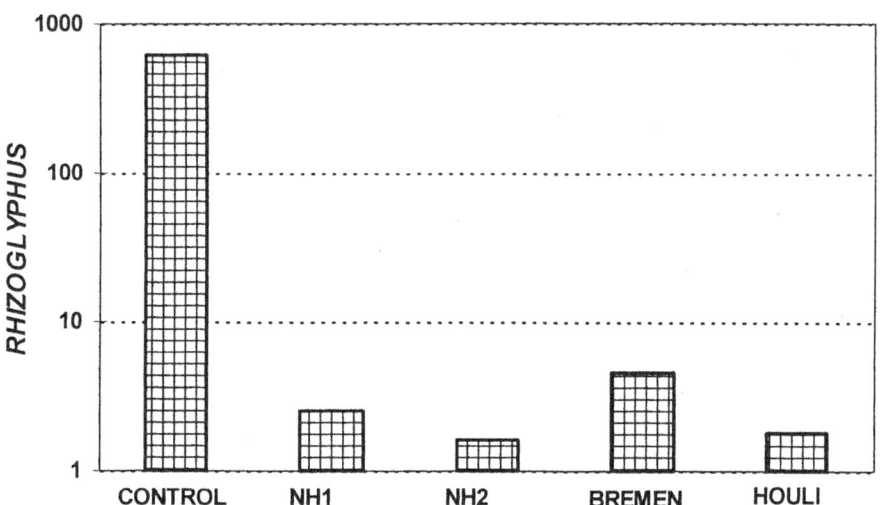

Fig. 21.3 Numbers of the bulb mite, *Rhizoglyphus robini*, found on lily scales in jars 4 weeks
after 120 bulb mites and six *Geolaelaps aculeifer* were placed in each jar. All log values $x + 1$.
Predators originated from four different locations (NH1: 't Zand, north Holland; NH2:
Breezand, north Holland; Bremen: north Germany; Houli, central Taiwan). (Based on data in
Lesna *et al.*, 1995.)

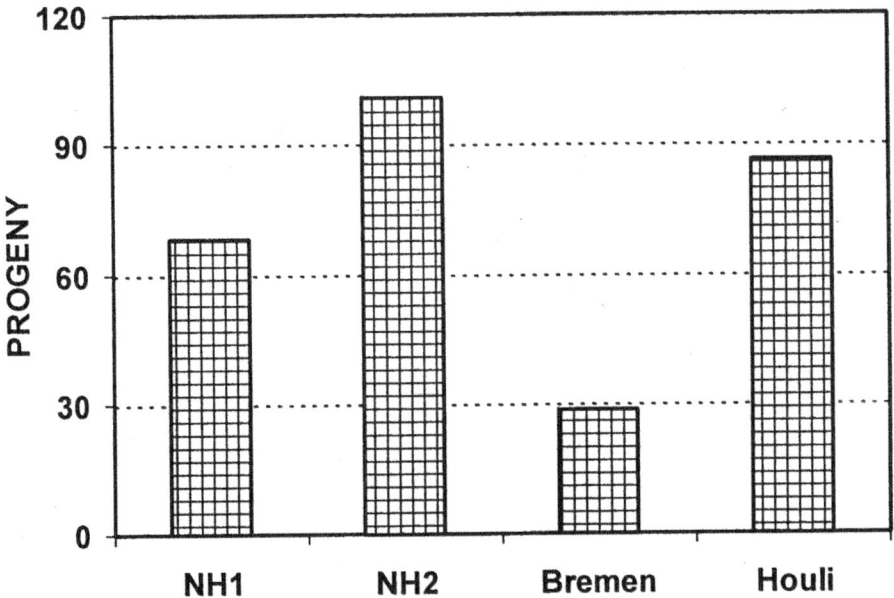

Fig. 21.4 Number of progeny of six females of *Geolaelaps aculeifer*, each representing strains obtained from four different locations (see Fig. 21.3), placed with 120 bulb mites, *Rhizoglyphus robini*, on lily scales within jars for 4 weeks at 25°C. (Based on data in Lesna *et al.*, 1995.)

Lilies are damaged by *R. robini* during their propagation ('scaling'), when bulb scales are separated from the basal plates, inflicting wounds that afford the pest easy access. To control pests and diseases these bulbs are chemically or thermally treated before scaling. Such treatments kill any resident natural enemies (including *G. aculeifer*, often found with *R. robini* in the field), but not all pest mites. The bulbs are then held for 6–8 weeks at 22–23°C in very humid, almost sealed large plastic bags, conditions that promote pest and predator reproduction (Lesna *et al.*, 1996). Placing 15 predators (all stages except for eggs) along with around 300 bulb mites on lily scales within the bags led to a 98% reduction in pest numbers within 4 weeks. Best control (superior to an acaricidal treatment) was obtained when the bulbs were first (prescaling) dipped in hot water, for 2 hours at around 40°C, and predators added during scaling (Conijn *et al.*, 1996).

The predator raised a generation in about 2–3 weeks under optimal conditions. Fecundity depended not only on the particular prey, but also on its specific stage. Offered only eggs or adults of *Rhizoglyphus echinopus* (Fumouze and Robin), each female of *G. aculeifer* produced an average of 71 and 79 eggs, respectively. When given larvae or tritonymphs of the same prey, oviposition came to 114 eggs/female (Ragusa & Zedan, 1988). The effect of other diets, including pollen and bulb mites along with fungi, on the predator's rate of increase is shown in Table 21.1. The females are cannibalistic on their males (Ruf, 1991), but seldom feed on their own eggs or juveniles. The SR is variable within populations, often being at parity, but

Table 21.1   Effect of various diets on the intrinsic rate of increase (r_m) of *Geolaelaps aculeifer* at 24–25°C

Diet	r_m	Source
Tetranychus urticae (live)	0.046	Lobbes & Schotten (1980)
Tetranychus urticae (frozen)	0.071	Lobbes & Schotten (1980)
Pollen of *Vicia faba*	0.062	Lobbes & Schotten (1980)
Drosophila melanogaster (dead)	0.073	Lobbes & Schotten (1980)
Tyrophagus purescentiae	0.121	Lobbes & Schotten (1980)
Rhizoglyphus robini	0.144	Lesna *et al.* (1996)
Rhizoglyphus robini and fungi	0.193	Lesna *et al.* (1996)

some females produce mostly females (Ruf, 1991). The impact of pesticides used in greenhouses, mushroom habitats or product stores on the Laelapidae is discussed in Chapter 43.

Other Laelapidae

Hypoaspis calcuttaensis Bhattacharya significantly reduced numbers of the root-knot nematode, *Meloidogyne javanica* (Treub), attacking okra (*Hibiscus esculentus*) in pot experiments (Walia & Mathur, 1996). Two other laelapids, *Lasioseius athiasae* Nawar & Nasr and *Protogamasellus discorus* Manson, reduced that nematode's density when infesting bean plants in pots by about 75% within 15 days (Amin *et al.*, 1999). The effect on nematode populations in the field has not been determined (see Chapter 38, ABAs as enemies of nematodes).

Hypoaspis spp. are associated with phytophagous scarabaeid beetles (Costa, 1971), including the coconut rhinocerus beetle, *Oryctes rhinoceros* (L.), a pest of coconuts in the Pacific region. In the laboratory the mite reduced the pest's egg hatch by around 70%, and it was released on the Tokelau Islands in the Pacific region as a pest control measure (Swan, 1974).

In mid-western Minnesota, USA, undetermined species of the soil-borne *Androlaelaps* and *Stratiolaelaps* usually reduced populations of the northern and western corn rootworms [*Diabrotica longicornis* (Say) and *D. virgifera* (LeConte), respectively], by about 20% (Chiang, 1970). Predators and prey had overlapping vertical distributions, and the mites fed on rootworm eggs and larvae in the laboratory. In the field mite populations increased three-fold after applications of manure, bringing about around 63% control of the pests. The biologies of both laelapids were studied together (Mihm & Chiang, 1976). A generation (egg to egg) required about 2 months at 20°C and 95% relative humidity, fecundity was 34 eggs/female and some adults lived for 8–9 months. *Androlaelaps casalis* (Berlese), a stored product inhabitant, fed on co-occurring mites and small beetle larvae, producing few (13/female) but large eggs; a generation cycle at 32.8°C was 26 days (Barker, 1968b).

Haemogamasus pontiger Berlese was reared in the laboratory on wounded insect larvae, and produced one or two eggs for 10 consecutive days, completing a life cycle in about 10 days at 27°C (Barker, 1968a). The mite was collected from well-aerated squirrel and bird nests in Scandinavia. It was postulated by Mehl (1977) to be an important predator of flea (Siphonaptera) larvae, to the extent of reducing the numbers of these disease vectors. The possibility that laelapids could reduce flea densities was explored by Ryba *et al.* (1987). Individuals of two flea species, *Neopsylla setosa* (Rotschild) and *Citellophilus simplex* (Wagner), were placed in nests of the European suslik, *Citellus citellus* (L.), a small terrestrial mammal. *Androlaelaps fahrenholzi* (Berlese), *Eulaelaps stabularis* (Koch) and *Haemogamasus nidi* (Michael) were then added to some of the nests. Laelapid-containing nests had significantly fewer fleas, which were more female biased, compared with those devoid of mites. The causes of the decrease have not been determined, but the reduction lasted for several seasons. However, more of the surviving flea females in mite-inhabited nests were engorged and carried mature eggs, a phenomenon attributed by Ryba *et al.* (1987) to mite stimulation of flea feeding and reproduction.

As noted, methods for mass-rearing predatory laelapids are available and their use in glasshouses and mushroom-rearing facilities is of growing interest. An issue of concern in their mass-rearing is the apparently variable sex-determining mechanism in genera such as *Androlaelaps*, *Geolaelaps* and *Stratiolaelaps* (as detailed in Chapter 39).

Chapter 22
Limnesiidae

Diagnosis

Limnesiids are usually covered by a weak integument, although dorsal plates may be present. The palpal telofemur bears a single ventral seta, peg-like or hair-like, often set on a tubercle. The lateral eyes are separated, placed either below the integument or incorporated therein. Coxal plates occur in four groups of two each, or the anterior plates are connected, thus forming three groups. Coxal plates IV are large and triangular, their femoral sockets placed on the posterior lateral margin. Legs IV are without claws, but a few blade-like setae may rise near the tips of the tarsi. The larvae are heteromorphic to the deutonymphs and adults; separate keys are therefore needed for their determination (see Chapter 3). About 30 genera and several hundred species were proposed in the family. The genera of the Limnesiidae were treated by Cook (1974), the North American taxa by Smith & Cook (1991), the Chinese species by Jin (1997) and the Australian fauna by Harvey (1998).

General biology

These freshwater mites are assigned to the Hydrachnidia (also called Hydracarina or Hydrachnellae). This is a cohort of aquatic prostigmatid superfamilies whose members have only three active stages: larvae, deutonymphs and adults; the protonymphs and tritonymphs remain quiescent. Sperm transfer may be indirect (e.g. via spermatophores placed on substrates) or direct (by copulation). Eggs are deposited singly or in batches of up to 50, within a gelatinous mass. The larvae parasitise midges (Chironomidae), biting midges (Ceratopogonidae) or mosquitoes, and the postlarval stages feed on the juveniles (including their eggs) of these and other water insects, and on Cladocera (Smith & Oliver, 1986). Limnesiids are attracted to chemical stimuli that emanate from wounded prey, especially if the victim is already being devoured by conspecifics (Proctor & Pritchard, 1989). A few Limnesiidae bypass the parasitic larval stage in their development, the deutonymph emerging directly from the egg (Smith, 1998).

Limnesiidae as natural enemies of mosquitoes

Adults of *Limnesia jamurensis* Oudemans were found in natural ground ponds located in Papua New Guinea (then called New Britain), along with relatively few larvae of malaria-transmitting mosquitoes. In the laboratory the mite fed on first instar mosquito larvae, daily consuming abou seven small larvae of *Anopheles farauti* Laveran (a vector of malaria) or eight larvae of *Culex pullus* Theobald. When eggs of both mosquitoes were offered together, mites devoured 18 eggs/day. Larvae of *L. jamurensis* also parasitised adult *C. pullus*, but only to a limited extent. Laird (1947) believed that the mite 'plays a useful role' as a predator of the early stages of these mosquitoes. Smith & Oliver (1986) questioned the accuracy of the mite's identification.

Limnesiidae as natural enemies of other water insects

In Florida, USA, *Tyrrellia circularis* Koenike, which attacks biting midges, produced more than 60 eggs/female. They hatched in about 10 days at 23°C and the parasitic larval stage lasted for 4–9 days under the same conditions (Lanciani, 1978). Survival of laboratory-infested hosts depended on mite load. The 50% survival of unparasitised biting midges was about 8 days and those with four to six mites/host slightly more than 2 days (recalculated from Lanciani (1986). The nymphs and adults occurred in moist mats of rotting vegetation along with their prey, the eggs, larvae and pupae of various midges. A single, large mite may consume the entire body contents of smaller prey. Lanciani (1986) believed that evaluations of the impact of *T. circularis* on biting midges populations should include predation as well as parasitisation.

Adults of *Limnesia undulata* (Müller) that attack larval Chironomidae fed at an average rate of 0.5 prey/48 h in the laboratory and were capable of prolonged periods (e.g. several weeks) of starvation (Paterson, 1970). Attack of larval midges (fourth instar) was always initiated by a single predator, which other mites joined when the prey became motionless. *Limnesia maculata* (Müller) consumed a single chironomid third instar larva per day, requiring several hours to suck out the prey's tissues (Timmermans *et al.*, 1992).

The voracity of *Limnesia patagonica* Lundbald was assayed with Cladocera as prey in a temperate lake in the southern Andes. In the laboratory the mite consumed as many as 40 prey/day, a rate at which the predator could suppress the prey's populations (Balseiro, 1992). Although this was not expected to happen at that lake, owing to the coolness of the particular habitat, the data suggest that under suitable conditions *Limnesia* spp. could be important predators. Paterson (1970) believed that water mites (*L. undulata* and a *Piona*, Chapter 27) might be important predators of chironomid larvae in natural southern Canadian environments.

Chapter 23
Macrochelidae

Diagnosis

Macrochelids bear peritremes that are looped around their stigmata, which are located between legs III and IV, and lack ambulacra on the thin legs I as well as leg grooves. Females have small metasternal plates and the spermadactyl digits of the male are distally free. The family consists at present of about 400 species in 15 genera (Krantz, 1998). The genera were reviewed by Krantz (1962), Karg (1993) keyed the European species and the Australian taxa are being revised by Halliday (2000).

General biology

Macrochelidae are fast-moving, free-living cosmopolitan predators found in habitats that are rich in decaying organic material, including manure. Dung-inhabiting macrochelids have evolved two strategies to exploit different types of manure (Krantz, 1983), both dependent on insect phoresy. A few have adapted to large, long-persisting 'domestic' piles (e.g. in poultry houses or cattle sheds) and are dispersed on synanthropic filth flies (Muscidae). Others occur in ephemeral dung pads in the field, being carried between these habitats mainly while clinging to dung beetles (Scarabaeidae). Many of these associations are triggered by specific chemical cues (Krantz, 1998).

Most Macrochelidae are arrhenotokous, unmated females depositing only male eggs. Development is often very rapid, a generation being completed in less than 1 week. Female deutonymphs produce a body-surface, ether-soluble sex pheromone that elicits guarding behaviour in males (Yasui, 1992).

House flies (*Musca domestica* L.), nuisance pests of farm workers that may create public health problems, as well as the face fly, *Musca autumnalis* De Geer, and the stable fly, *Stomoxys calcitrans* (L.), both serious pests of cattle, are considered to be filth flies. All deposit their eggs in and on animal droppings, whether in domestic, pastoral or uncultivated habitats. The eggs hatch in the dung, and the juveniles develop and pupate there. The juveniles of all filth flies are attacked by many predators and parasites, including several species of *Macrocheles*. The mites feed on fly eggs and first stage larvae, and their common association with the pests has suggested that they could be used as biocontrol agents (Pereira & de Castro, 1945).

Macrocheles muscaedomesticae (Scopoli)

This, the best known and most abundant species in cattle and poultry manure, preys on, and is disseminated by, many filth flies. It lives in the outermost layers of manure heaps, the region in which flies usually place their eggs. The predator feeds on fly eggs (which it prefers) and on first instar larvae, as well as on acarid mites and nematodes. Development is rapid, as a generation was raised in 4–5 days at 26.7°C and 55–60% relative humidity (Wade & Rodriguez, 1961). Virgin females produced 50% more eggs than fertilised mites (92 vs 61). Females, which lived for over 3 weeks, produced five to seven progeny/day. The capacity for increase is very high; Krantz (1983, adapting data from Cicolani, 1979) calculated the mite's net reproductive rate as 58.07 (at 30°C), and its intrinsic rate of increase as 0.9057/day. Unfed mites kept on moist paper survived for around 21 days.

Desiccation and feeding by other coprophagous insects (e.g. beetles) quickly reduce the quality of the dung pad. House flies overcome the problem by developing rapidly, leaving the source pad and searching for fresh manure. *Macrocheles muscaedomesticae* reaches that habitat while clinging to the emerging flies.

The sex ratio (SR) of mites developing in poultry manure was 0.7–0.8 (Stafford & Bay, 1994), but habitat and crowding affected the SR (Chapter 40). Data on mite development, population dynamics and feeding rates were incorporated within a simulation model (MACMOD) (Geden *et al.*, 1990). The model included the predator's and prey's non-linear development, longevity, attack rates and *M. muscaedomesticae* fecundity at different temperatures, the effects of crowding and prey type and quantity, but not its spatial distribution. The mite is highly aggregated, because its distribution in manure is affected by prey availability and prevailing humidities (Stafford & Bay, 1994).

The combined effect of two macrochelids [*M. muscaedomesticae* and *Macrocheles merdarius* (Berlese)] and unnamed members of the family Parasitidae (Chapter 25) on house fly juveniles was demonstrated by Axtell (1963). He created uniform arenas with metal barriers that delimited sections of intact calf manure, excluding any incoming mites and flies by applying sticky glue around their outer sides, and placing a screen on top of each cage. Any resident predators were eliminated by an acaricide in half the arenas, which served as mite-free controls (the chemical has no effect on fly juveniles). The other arenas were treated only with water. Batches of 20,000 house fly eggs were then placed into each arena; glue-coated traps within each cage caught the emergent flies. A mean of 83 flies was obtained from the acaricide-treated arenas, compared with 28 in the controls; thus, the mites reduced pest numbers by two-thirds. In an outdoors experiment, with natural manure piles, the decline in fly numbers by resident predators came only to 31–45%. The effect of *M. muscaedomesticae*, along with another macrochelid, *Glyptholaspis confusa* (Foà) (which often co-occurs; see below), was assayed by introducing both mites together (each in batches of 200 females) into fresh manure placed in cages, which were then inoculated with 20,000 fly eggs/cage. Emerging fly numbers were reduced by around 94%, demonstrating the predators' effect (Fig. 23.1). In other experiments fly emergence decreased by 86–94% (Axtell, 1963).

Similar results were obtained by Singh *et al.* (1966), who put mites and eggs of prey [the house fly, the face fly and the lesser house fly, *Fannia canicularis* (L), another nuisance pest] onto fresh manure in boxes. Prey/predator rates were kept at 1/5. The

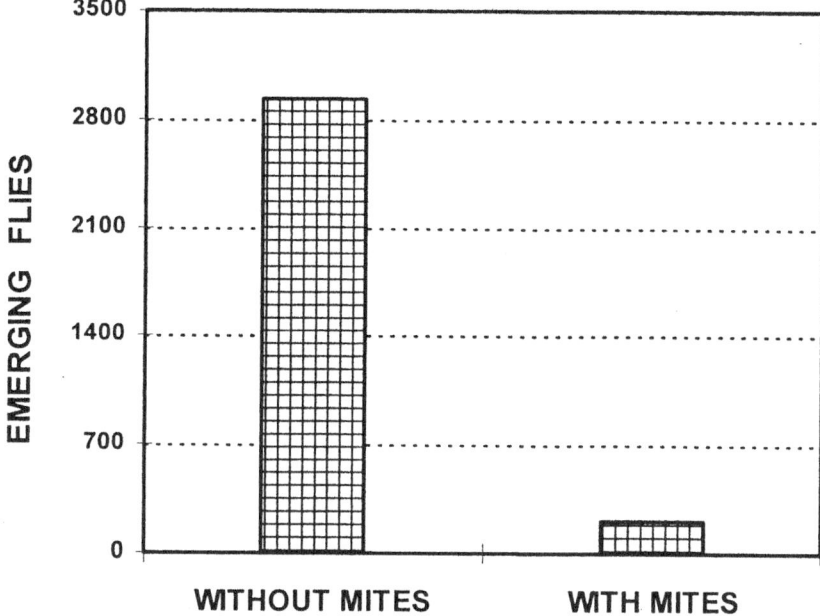

Fig. 23.1   Effect of *Macrocheles muscaedomesticus* and *Glyptholaspis confusa*, each in batches of 200 adult females, on the emergence of house flies developing in fresh cattle manure. (Based on Table 4 in Axtell, 1963.)

mites reduced house and little house fly numbers by 85–92%, but those of the face fly only by 23.4% (Fig. 23.2).

High rates of house fly control (86–99%) under semi-field conditions were reported by Rodriguez *et al.* (1970). However, the ease of pesticide usage, along with the only partial control obtained at times by the acarine predators, indicated the need for an integrated approach to the problem (Axtell, 1986). Cultural control methods noted were keeping dung piles dry and the judicious removal of manure; biological control methods included the use of natural enemies, and acquiring data about their susceptibility to pesticides.

The poultry or cattle manure pile is the universe in which filth flies and their natural enemies interact. Over time this substrate undergoes chemical and physical changes, besides being exposed to various treatments, all of which affect resident predators and parasites. Most arthropod inhabitants of older manure, including predatory mites, occur within a narrow band in or just under the surface, indicating that many parts of the heap are unsuitable. Geden & Stoffolano (1988) speculated that this restricted distribution could be the result of poor gas exchange or prior use by saprophytic organisms, and might also be due to the high degree of compaction in the heap's midst. Whatever the reason(s), optimal predation may thus be seen as a matter of manure management. Manure in the various animal-rearing facilities must occasionally be cleaned out, an activity that is followed by changes in pest and natural enemy numbers. Axtell (1986) advocated the removal of poultry manure during

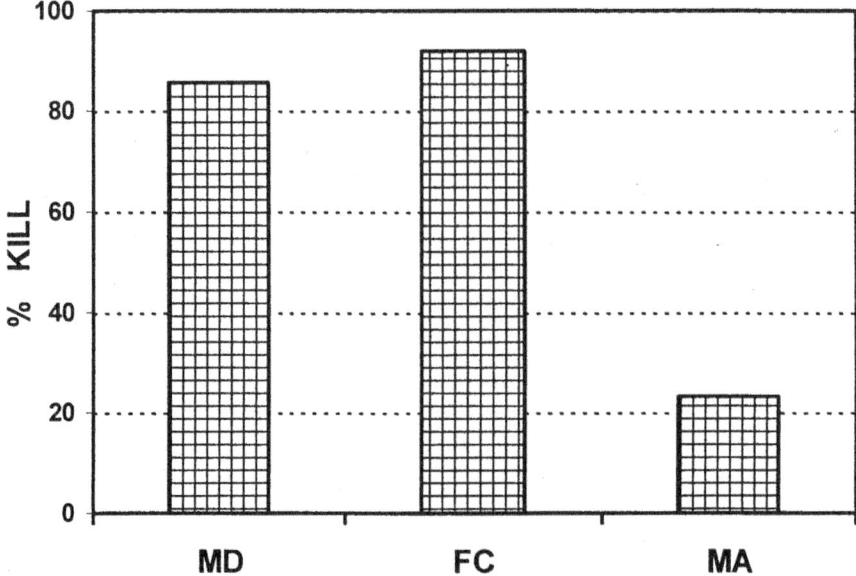

Fig. 23.2 Percentages of the eggs of *Musca domesticae* (MD), *Fannia canicularis* (FC) and *Musca autumnalis* (MA) killed by *Macrocheles muscaedomesticus*, at an approximate mite/egg ratio of 1/5. (Based on Table 1 in Singh *et al.*, 1966.)

the cold season, when fly breeding is minimal. The weekly, or fortnightly, removal of manure in summer resulted in high fly numbers (Peck & Anderson, 1970), whereas relatively few flies emerged from undisturbed manure. Populations of *M. muscae-domesticae* required 3–4 weeks to re-establish themselves in disturbed manure, but when some adjacent undisturbed dung was available their numbers increased soon after heap removal. Reductions in pest control due to cleaning animal stalls could thus be alleviated by leaving refugia of natural enemies within piles, layers or entire rows of droppings (De Jesus & Rueda, 1992). At times there is little benefit from various removal strategies, because some residual manure is often left in place (Mullens *et al.*, 1996).

Manure removal practices differ in different facilities, with various outcomes. In the Philippines cattle, swine and carabao (water buffalo) manure is cleaned out and piled outside the facilities, where flies and mites maintain their populations. In broiler houses manure is almost completely removed after each chicken harvest, greatly reducing the numbers of predatory mites (De Jesus & Rueda, 1992).

Juvenile flies are often controlled with larvicidal pesticides that are sprayed onto the manure. Many compounds kill the predators along with the pests, but as flies recolonise manure heaps more quickly than do the natural enemies (including *M. muscaedomesticae*), the pests resurge rapidly (Axtell, 1968). Damage to natural enemies was reduced by spraying only the upper, inside surfaces of poultry houses and the outside of adjacent buildings, areas where the flies rest. Pest numbers were further reduced by trapping and by manure removal early in the season, before flies became abundant, practices that did not affect resident predatory mites (Axtell, 1970).

The ability of *M. muscaedomesticae* to bring about satisfactory house fly control, even without competitors (see below), also depends on the presence of alternate diets. Such prey (e.g. acarid mites, nematodes and various soft-bodied larvae) help to maintain predator fecundity and longevity when house fly numbers decline. Simulations (Axtell, 1991) showed that ten *M. muscaedomesticae* females could bring about eventual house fly control in the presence of alternate prey, but even 1000 mites would not achieve a similar outcome in its absence. Pesticides that cause 20% mite kill (and eliminate 95% of the flies) would not affect pest control in the presence of alternate prey, but control would not succeed in the absence of other prey. Finally, removal of half of the manure did not affect control in the presence of other prey, but resulted in fly outbreaks when no other animal diets were present. The simulations indicated that pest reductions would be inadequate when 90% of the manure was removed. Further, predation rates were approximately 60% lower when many nematodes occurred in the manure, an outcome attributed to reduced feeding on fly eggs in the presence of many nematodes (Geden & Axtell, 1988).

Consumption of prey depends on the predator's preference for and ability to feed on different fly species, as well as on substrate and diet conditions. Given a choice among three prey species (the horn fly, *Haematobia irritans* L., the house fly and the blow fly, *Chrysomya megachephala* F.) in the laboratory, *M. muscaedomesticae* preferred the eggs and first instar larvae of the house fly, with the horn fly coming second. *Macrocheles limui* Samsinak likewise preferred the house fly, but the blow fly was its next choice, and *M. merdarius* almost exclusively chose house fly juveniles (De Jesus & Rueda, 1992). The latter predator destroyed 3.1 fly eggs/day and was considered to be a good biocontrol agent (Ho & Auemetua, 1990). Feeding on stable fly eggs by *M. muscaedomesticae* was partial at best, a result attributed to the thicker chorion of these eggs (Kinn, 1966). A similar rationale was invoked (Halliday & Holm, 1987) to explain the feeding patterns of nine macrochelids assayed in the laboratory. All devoured eggs of the Australian bush fly, *Musca vetustissima* Walker, but feeding on the eggs of the buffalo fly, *Haematobia irritans exiqua* De Meijere, which have a thicker chorion, was restricted to the larger Macrochelidae. Egg size, and their mode of being placed into the substrate, affected the mites' predation rate. Face fly eggs are hard to access because they are completely inserted into manure during oviposition, which may explain the predator's reduced feeding on them (Singh *et al.*, 1966). *Musca sorbens* Wiedemann, another filth fly, deposits clusters of several hundred eggs into fresh dung and they soon hatch; *M. muscaedomesticae* cannot colonise the dropping rapidly enough to destroy the eggs. In addition, the attacked outer eggs in the cluster collapse, and become hard and dry, forming a barrier on the surface of the cluster that protects the inner eggs from further predation (Toyama & Ikeda, 1976).

Substrate effects on biology are manifested in various ways. When placed in the laboratory on steer manure, *M. muscaedomesticae* killed about 8.5 house fly larvae/ day, on filter paper this value rose to 15, and when put on larval fly media, the mite destroyed about 20 prey/day (Wade & Rodriguez, 1961). The addition of 0.85% ammonium hydroxide increased the destruction of house fly eggs (to 25/day), probably owing to a stimulated biting response (Wallwork & Rodriguez, 1963). The predator's diet during development also affected its rate of predation; mites fed stable fly eggs did not consume as many eggs as did those given house fly eggs (Kinn, 1966). For modelling purposes it was assumed that the predator consumes about 13.9–15.4 house fly juveniles/day (Axtell, 1991).

The outdoor abundance of manure-inhabiting macrochelids changes with the seasons on different substrates. In Philippine poultry houses *M. muscaedomesticae* occurred in largest numbers during May and July (De Jesus & Rueda, 1992), but it was rare in dairy cattle and swine manure (Table 23.1). In North America mites were also abundant in poultry manure during summer and low in winter. In cattle dung, however, their numbers peaked in winter and spring months, ebbing in summer (Rodriguez *et al.*, 1970).

The attachment of *M. muscaedomesticae* to house flies depends on the age of the manure. In drying dung (where few prey are found) the mites are attracted to a water-soluble chemical(s), which is carried by all adult flies. Most mites cling to the flies' weakly sclerotised abdomen and to membranes between head and thorax, and between thorax and abdomen (Jalil & Rodriguez, 1970a). Temperature has a strong effect on mite attachment; significantly more *M. muscaedomesticae* were found clinging to flies at 27°C than at 16°C. When flies reached new animal droppings the mites detached, as they were more drawn to the fresh substrate. The attractiveness of ammonia and skatole (products of protein putrefaction that are common in manure) to the predators probably mediates the detachment of the mites from the flies.

Most studies on Macrochelidae as ABAs have emphasised their effect on prey juveniles, although phoresy on flies could decrease their fitness. Young flies were attacked and killed by clusters of *M. muscaedomesticae* (Wade & Rodriguez, 1961). The mites bit and fed through the intersegmental membranes of the flies, thereby decreasing their weight; mite-free flies lived significantly longer than did those with mites (Jalil & Rodriguez, 1970a). In Germany the lesser house fly was controlled by

Table 23.1 Numbers of the more abundant predatory Mesostigmata per sampling unit, collected from droppings of various farm animals

	Manure			
	Poultry	Carabao	Dairy cattle	Swine
Mite(s)				
Macrocheles muscaedomesticae	1335	6	3	58
Macrocheles merdarius	342	15	53	816
Macrocheles limui	–	260	3	–
Uropodidae	12,422	367	1013	1126
Parasitidae	14	67	54	55

Data recalculated from Tables 1 and 2, de Jesus & Rueda, 1992.

M. muscaedomesticae in an animal stall. The insects, attacked as they emerged, could not flex their wing and many were found with stunted wings. A delay in cleaning these stalls resulted in total fly control (Keilbach, 1978). The presence of even a single *M. muscaedomesticae* on females of *F. canicularis* sufficed to bring about a significant reduction in their oviposition (Borden, 1989). *Macrocheles muscaedomesticae* is one of the few species in the genus that feed on their adult fly vectors (Krantz, 1998).

Macrocheles muscaedomesticae does not occur in carrion, although it fed in the laboratory on eggs of *Phormia regina* (Meigen) (Calliphoridae), a carrion-associated blow fly (Kinn, 1966). The human-bot fly, *Dermatobia hominis* (L.) (Cuterebridae), was attacked by *M. muscaedomesticae* in a laboratory in Brazil. The flies distribute their juveniles by capturing mosquitoes and houseflies onto which they attach their own eggs, thereby reaching suitable hosts where they hatch. Mite-carrying bot flies were unable to copulate and failed to catch carrier flies, thus totally curtailing their reproduction. Moya Borja (1981) advocated studying these mite–fly interactions in the field.

Other macrochelids

A year's census in the Philippines indicated that *M. merdarius* was most abundant in swine droppings, the largest numbers of *M. muscaedomesticae* were obtained from poultry manure, and *M. limui* dominated in carabao barns. The latter species (which is disseminated by dung beetles) was absent from poultry and swine manure (De Jesus & Rueda, 1992) (Table 23.1).

Droppings of cattle and horses, animals introduced into Australia during European colonisation, have accumulated on the soil surface in heaps, fouling vast areas of pasture land and serving as breeding sites for filth flies. The Australian dung beetle project (Bornemissza, 1976) consisted of efforts to introduce dung beetles that specialise in bovine droppings. The beetles burrow through the droppings and comminute them, reducing their suitability for the flies.

The cryptogenic *Macrocheles glaber* (Müller) (later found to consist of the nominal species and of *M. perglaber* Filipponi & Pegazzano) is common in south-eastern Australia on several indigenous dung beetles (Wallace *et al.*, 1979). Fifty *M. glaber* females, added to a 1000 ml dung pad with 300–400 fly eggs in the laboratory, completely eliminated the pest. This result depended on manure quality and on mite arrival time at the pads. Predation rates of *M. glaber* placed on pads within 1 h of bush fly oviposition came to 90%, declining to 60% when mites were added after egg hatch, because the emerging fly larvae managed to escape by burrowing into the dung (Wallace *et al.*, 1979). Early mite arrival depends on rapid vector arrival at the fresh pad, implying that only beetles with a flight pattern that coincides with that of the flies will contribute to mite efficacy. The insufficient fly control by the indigenous mite and by dung beetles led Tyndale-Biscoe *et al.* (1981) to advocate the introduction of additional natural enemies.

Macrocheles peregrinus Krantz was introduced into north-eastern Australia from South Africa (Wallace & Holm, 1983). This predator was chosen from various candidates owing to its wide range of dung beetle vectors and its perceived ability to establish in warm areas, arid or humid. A culture from Pretoria provided the initial 200 mite eggs that served to initiate a large culture in Australia. The introduced species (released in batches of 200/fresh cattle pad) rapidly became established and wide-spread. It colonised at least 180,000 km^2 around the two release sites within 14 months, probably owing to a short life cycle (a generation in 3 days at 27°C) as well as to its ability to use different beetle vectors. However, buffalo fly numbers were reduced only by 33% when *M. peregrinus* acted alone. This was attributed to fly oviposition underneath dung pads, sites not easily accessible to the mite, to the tough chorion of buffalo fly eggs and to the need for a greater predator/prey ratio than that obtained in the field (Roth *et al.*, 1988). Another factor was the preference of *M. peregrinus* for the bush fly. The associations of Australian *Macrocheles* spp. with dung beetles were explored by Wallace (1986) and Halliday & Holm (1987).

Glyptholaspis confusa is a voracious predator that is currently available in commerce (Anonymous, 2000). In the laboratory it consumed four to ten house fly juveniles/day (Halliday & Holm, 1987), but in poultry manure it becomes abundant only in long-standing pads, wherein few fly larvae develop (Geden *et al.*, 1988). In contrast to this discouraging report, Perroti (2001) found that deutonymphs and females of *G. confusa* were ferocious predators of the horn fly in the laboratory. Only around 19% and 28% of fly eggs and larvae, respectively, survived in the presence of the mites.

The association between macrochelids and beetle pests could be used to convey entomopathogenic fungi to these insects. The Pales weevil, *Hylobius pales* (Herbst) (Coleoptera: Curculionidae), a serious forestry pest, carried phoretic mites, including a *Macrocheles* sp. Mites contaminated with the fungus *Metarhizium anisopliae* were collected from healthy hosts and allowed access to the weevils, of which about 80% became infected (compared with 7% in the mite-free weevil control group). All mites died from the disease (Schabel, 1982). The ability of macrochelids to locate their hosts, pests that inhabit hard-to-reach intrabark locations, could make them efficient disease carriers.

Macrochelid efficacy in dung pads may be reduced by competition with other predators. When fly eggs were exposed to the parasitid *Poecilochirus* sp. (Chapter 25) and to *M. muscaedomsticae* (introduced together at a ratio of 4/1), significantly fewer pests were killed than when either predator was assayed alone (Geden *et al.*, 1988) (see Chapter 42). Competition between *Macrocheles robustulus* (Berlese) and *M. glaber* reduced the latter's populations when placed together in cow dung. The smaller, but more aggressive *M. robustulus* attacked and killed *M. glaber* (Wallace & Holm, 1984) (Fig. 23.3).

Macrochelids possess several attributes that could promote their use for pest control. They are easily reared on flies or on nematodes (Royce & Krantz, 1991), raising a generation in less than a week under laboratory temperatures (20–30°C). They produce over 100 progeny/female and live for at least 2 weeks (Cicolani, 1979; Krantz, 1983). Populations of macrochelids could thus be augmented in situations where the local predators had been eliminated. The adverse effect of various macrochelids on their fly carriers (e.g. *Fannia canicularis*) is an aspect that could lead

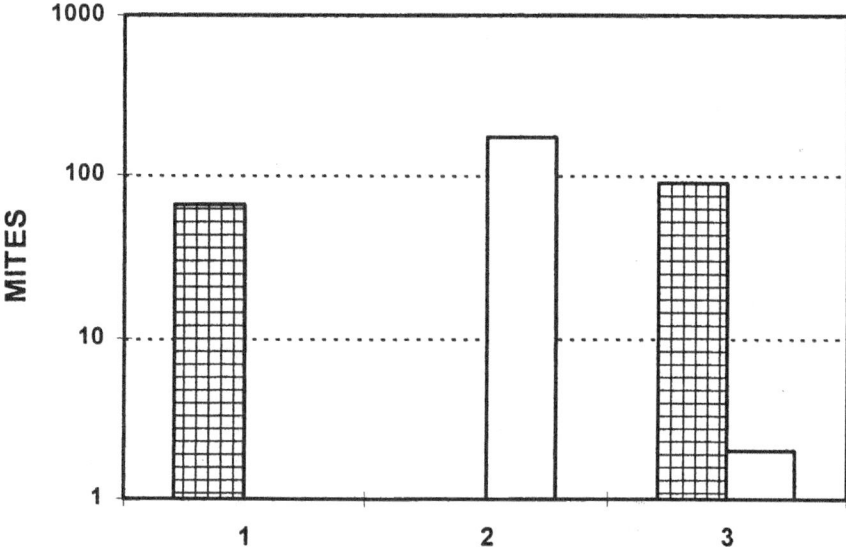

Fig. 23.3   Numbers (log values $x + 1$) of *Macrocheles robustulus* (cross-hatched bars) and of *Macrocheles glaber* (white bars) obtained 2 weeks after 20 females of each species were placed in small boxes; 1 and 2: each species alone; 3: the two species together. (From data in Wallace & Holm, 1984.)

to additional control options. The scope for introductions of exotic macrochelids for synanthropic fly control (in 'domestic' manure piles), however, appears to be limited. Dung pads attract an indigenous fauna that would limit the activities of exotic species. Introductions of such macrochelids could be more promising for the control of 'pastoral' filth flies. That, however, would require more data on species identification, biology (e.g. feeding preferences), ecology (e.g. climatic adaptations, ability to withstand competition) and interactions with potential vectors (Krantz, 1983).

Insecticides are sometimes applied onto manure; their effects on macrochelids are discussed in Chapter 43. Anderson (1983) advocated the selection of pesticide-resistant fly predators that could be incorporated into integrated fly management programmes.

Chapter 24
Otopheidomenidae

Diagnosis

The Otopheidomenidae are weakly sclerotised mites with a reduced number (>21 pairs) of simple dorsal setae; the anus may be terminal. The peritremes are short and the fixed digit of the chelicerae is reduced or missing. It is a small family of about ten genera and 25 species that are external parasites of bugs, grasshoppers, moths and termites, also occurring in the tracheae of katydids (Orthoptera: Tettigoniidae). Zhang (1995a) prepared an overview of the family.

Hemipteroseius indicus (**Krantz & Khot**)

Little is known about the life history of these mites. Most of the known facts pertain to *H. indicus*, which is a parasite of the red cotton bug, *Dysdercus koenigii* F., a pest of cotton in India. Attack begins around the hosts' mouthparts, the mites then settling on the thoracic segments covered by the wings, on the underside of the wings and on the dorsum of the abdomen. The parasites feed mostly on the thorax, preferring to oviposit on the host's first abdominal segment. Each female probably produces fewer than 20 eggs (Prasad, 1975), the emerging juveniles remaining on the host. The sex ratio was 0.83. Large populations may accrue at the site, sometimes numbering over 150–200 mites/host (Krantz & Khot, 1962; Prasad, 1975)

It has not been resolved whether the mite attacks juvenile bugs, and how this would affect the hosts. Banerjee & Datta (1980) reported that wingless bug nymphs were never parasitised, even after the mites had been kept without hosts for 3 days, the mites preferring starvation to attacking nymphs. In contrast, Shahi & Krishna (1981) claimed that all host nymphs, including newborn first instar bugs, were attacked. Moderately infested hosts showed no significant mortality, nor were apparent feeding symptoms seen on examined hosts (Prasad, 1975). According to Banerjee & Datta (1980), however, adult bugs, parasitised in the laboratory for 72 h by ten to 20 adults, stopped feeding, became immobilised and died 1–2 days later. Attacked gravid host females died without ovipositing and nymphs of all instars died without further development (Shahi & Krishna, 1981). Based on these observations, Shahi & Krishna (1981) and Zhang (1995a) advocated further studies on using *H. indicus* to control the red cotton bug.

Chapter 25
Parasitidae

Diagnosis

Parasitids possess straight peritremes that open between legs III and IV, and lack leg grooves. Females bear large metasternal shields that flank the triangular genital plate. The spermadactyls of the male's movable and fixed digits are fused distally. Micherdzinski (1969) revised the family, which has about 400 species assigned to more than 20 genera. The European taxa were discussed by Karg (1993); Tseng (1995) summarised the Taiwanese fauna.

General biology

Parasitids feed on small arthropods and nematodes in organic and forest soils, as well as in animal droppings and seashore wrack, occurring also in the nests of birds and rodents, and even among social insects (e.g. bees). Many species are dispersed during their deutonymphal stage by beetles, bees and flies, whereas others are conveyed as adults. The deutonymphal period can last for several months and its onset, beginning, duration and termination appear to depend on chemical signals from the species' specific environment (Costa, 1969).

Pergamasus spp.

All stages (except for larvae) of *Pergamasus quisquiliarum* Canestrini fed on the garden symphylan, *Scutigerella immaculata* (Newport) (Scutigerellidae), a soil-borne pest of many crops in western Oregon, USA. At 20°C the mite completed a generation in 17 days (whereas the pest required 87 days) and deposited about 33 eggs, usually attached to nearby plant roots. During its life the mite consumed 12–14 symphylans, actively searching for this prey. Although *P. quisquiliarum* appears to be a non-specific predator, its close association with plant roots enhances the probability that it will feed on root pests. This and the predator's rapid rate of increase (five-fold faster than the pest) led Berry (1973) to conclude that *P. quisquiliarum* could be an important factor in regulating the rate of symphylid increases in the field.

Pergamasus spp. feed on stored product pest mites (e.g. *Tyrophagus* spp.) in the soil (Karg, 1961). On the basis of their laboratory results Harris & Usher (1978) postulated that the feeding rate of *Pergamasus longicornis* Berlese on soil Collembola (a few of which are pests) may affect their population dynamics. After catching the prey the mite often completely lifts it off the substrate while sucking it dry (Bowman, 1984).

Parasitus spp.

An exploration for natural enemies of the bulb mite, *Rhizoglyphus robini*, conducted in The Netherlands, Taiwan and Japan, resulted in finding three acarine predators (see Chapter 21), including the parasitid *Parasitus fimetorum* (Berlese) (Lesna *et al.*, 1995). It reduced a bulb mite population from 75 to about seven individuals in 4 weeks at 25°C and 70% relative humidity in the laboratory, whereas parasitid numbers increased from four to 23. This species was also one of several natural enemies of the red-egged earth mite, *Halotydeus destructor*, in Australia (James *et al.*, 1995).

Parasitus bituberosus Karg fed on the juveniles of *Heteropeza pygmaea* Winnertz (Diptera Cecidomyiidae), a pest of cultivated mushrooms. Both preferred the tops of compost bags, a spatial co-occurrence that expedited the mite's predation on the pest in laboratory-scale as well as in commercial-scale mushroom compost units. In the pest's absence the mite persisted within bags by feeding on ever-present nematodes. That trait, along with the predator's polyphagy, suggested to Al-Amidi & Downes (1990) that *P. bituberosus* might also be used in glasshouses. Numbers of another mushroom pest, the sciarid fly *Lycoriella solani* Winnertz, were halved by the predator, yield was increased by 18% and the rate of culled mushrooms was reduced by 50% (Al-Amidi *et al.*, 1991).

Parasitidae in poultry manure

The Parasitidae, along with the Macrochelidae (Chapter 23) and Uropodidae (Chapter 37), have members that abound in poultry manure. In North America Parasitidae (mostly *Poecilochirus* spp.) are the first to appear and the first to decline, being neither as abundant nor as efficient as members of the other families (Axtell, 1970). Despite their early arrival, parasitids never reach large numbers, attaining consistent populations only in undisturbed manure. In the Philippines parasitids were more abundant in carabao (water buffalo), dairy cattle and swine manure than in poultry droppings (De Jesus & Rueda, 1992; see Table 23.1), being most numerous during autumn. *Poecilochirus monospinosus* Wise, Hennessey & Axtell feeds on house fly juveniles in poultry manure. A life cycle required 18 days at 26.6°C, females had an average of 90 progeny and total longevity was 8 days. Deutonymphs killed three to five eggs or four first instar larvae/day; the females nine to 13 eggs or 17–24 larvae/day. The latter prey was clearly preferred over eggs, but larger fly juveniles (second or third-instar larvae) could not be overcome. The predator occurred mostly in late spring and early summer, and was considered to be only a minor, short-term factor in suppressing fly populations (Geden *et al.*, 1988; Wise *et al.*, 1988).

The presence of *P. monospinosus* along with other, more efficient predators in dung caused a net loss of fly control by other predators, whether through preying on their eggs and juveniles, or by disturbing them during prey handling (Ignatowicz, 1979; Geden *et al.*, 1988; and see Chapter 42). Pesticide effects on parasitids are discussed in Chapter 43.

Chapter 26
Phytoseiidae

Diagnosis

Phytoseiids bear an entire dorsal plate with fewer than 23 pairs of setae, their stigmata open between legs III and IV and the metasternal plates are small. They lack leg grooves, legs I carry ambulacra, and the digits of the male spermadactyls are distally free. The systematics of the Phytoseiidae had a convoluted history that has not yet been entirely resolved (e.g. Denmark *et al.*, 1999), but most current students use the system presented by Chant & McMurtry (1994). It divides the family into three subfamilies, namely (in descending number of known species), Amblyseiinae, Typhlodrominae and Phytoseiinae. The subfamilies differ not only in morphology but also in biological attributes, such as patterns of oviposition (Zhang, 1995b). These subfamilies, or groups within them, further differ in their original habitats. Ehara & Amano (1998) noted that the Japanese Phytoseiinae were more closely associated with stable habitats than with any prey. Efficient biocontrol agents occur in all subfamilies.

Close to 1700 phytoseiid species have been described to date, in around 50 genera, but the actual tally may be lower. On the one hand, Chant (1992) estimated that as many as 20–25% of the given names could be synonyms, but on the other hand, some of the widely distributed taxa are now known to consist of two or more sibling species (McMurtry, 1980; Congdon & McMurtry, 1985; Beard, 1999). Many phytoseiids remain to be discovered, especially in the southern hemisphere (e.g. Walter & Beard, 1997; Beard, 2001). Lists of and keys to species were prepared for various regions of the world, including Australia (Schicha, 1987; Beard, 2001), the Philippines (Schicha & Corpuz-Raros, 1992), Korea (Ryu, 1993), Japan (Ehara & Amano, 1998), Central America (Denmark *et al.*, 1999), South Africa (van der Merwe, 1968; Ueckermann & Loots, 1988 and earlier papers), Europe (Karg, 1993), and Israel (Swirski *et al.*, 1998). A catalogue of all then-known species was prepared by Moraes *et al.* (1986), and Kostiainen & Hoy (1996) compiled a comprehensive bibliography (1960–1994) of the family.

General biology

Phytoseiids live on plants and in upper soil layers. They are fast-moving, proactive predators, feeding mostly on mites but also on small insects, nematodes and fungi, and may feed on plants, including pollen and extrafloral exudates. Phytoseiids are the best known and most studied group of predatory mites, owing to their success in controlling spider mites, other mites and thrips (Thysanoptera). Over half of the literature on acarine biocontrol agents (ABAs) pertains to the Phytoseiidae (Chapter 1), and about 20 species are currently being mass-reared (Chapter 39) and sold by about 50 companies world-wide (see Table 46.1).

Supplied with sufficient and suitable food, many phytoseiids develop within 1 week at 27°C and 60–90% relative humidity (RH) (Tanigoshi, 1982). They usually deposit 30–40 eggs/female, although some of the highly fecund species (e.g. *Phytoseiulus* spp.) produce notably more. Most phytoseiids prefer relative humidities between 70 and 90%; drier or more humid atmospheres may reduce the foraging activities (Mori & Chant, 1966), fecundity (Reuveny *et al.*, 1996) and longevity (Boyne & Hain, 1983) of different species. Egg hatch is usually maximal at or very near saturation humidity (Bakker *et al.*, 1993; Croft *et al.*, 1993). *Euseius scutalis* (Athias-Henriot) is an exceptional case, as some of its eggs hatched even at 19% RH (Bounfour & McMurtry, 1987).

Sex determination in most phytoseiids is by parahaploidy or pseudo-arrhenotoky (Schulten, 1985), a situation where mating is required for the production of male and female progeny, but the male embryo becomes haploid because its egg undergoes paternal genome loss during early development. The secondary sex ratio (SR) of Phytoseiidae in the laboratory is variable, from 0.5 to >0.8. Adult (=tertiary) SR in the field is often greater than 0.75 (Sabelis, 1985). In addition, some species are thelytokous (Wysoki & Bolland, 1983). Males locate females (and even their deutonymphs) by a sex pheromone, which invokes a distinctive 'hovering' behaviour in the males (Hoy & Smilanick, 1979).

Phytoseiids survive winters either by going into a reproductive diapause (reviewed by Veerman, 1992) or by being resistant to low temperatures. Diapause is induced by a combination of short daylight and low temperatures (detailed below for some species). Diapause appears to be facultative, and may be modified or eliminated by suitable selection without affecting the predators' performance. Resistance to low temperatures differs even among closely related species. About 50% of a population of *Phytoseiulus persimilis* Athias-Henriot, kept at 4°C and 80% RH, remained alive for more than 60 days; the same percentage of *Phytoseiulus macropilis* (Banks) survived for only a fortnight under similar conditions (Takahashi & Chant, 1994). The ability of phytoseiids to withstand long cold periods is being studied for storage purposes (Chapter 39).

Although the Phytoseiidae, like all Mesostigmata, lack eyes, their behaviour seems to be affected by prevailing light conditions. (Perception of infrared radiation by a parasitic mesostigmatid mite, in another family, was demonstrated by Bruce, 1971.) *Amblyseius largoensis* Muma, for instance, when given an abundance of pollen in the laboratory and placed under regimes of continuous (artificial) light, or continuous darkness, or natural, light/dark L12/D12 h conditions, produced the most eggs under the latter conditions (Kamburov, 1971). Light intensities of 1000 lumens/ft^2 (1000 lumens/30.5 cm^2) in the laboratory hindered the predation rate of *P. persimilis* and reduced its fecundity, both of which were best in total darkness (Babikir, 1978). The movements of *P. persimilis* were affected by light and by ambient humidities, because starved mites, or predators that had been kept under dry conditions, stayed longer in shaded areas, whereas sated females walked towards a light source (Bernstein, 1983). Various species show a distinct pattern of diurnal behaviour, in the laboratory and in the field. The mating frequency of *Amblyseius teke* Pritchard & Baker, kept under constant laboratory conditions of 27°C and 65% RH (light conditions unstated), peaked between 08.00 and 11.00 h, whereas it fed mostly at night; larvae were active only during the morning (Schultz, 1974). *Typhlodromalus peregrinus* (Muma) foraged on the outside of citrus trees in Florida from 1 h after sun set until dawn, the mites on the external leaves usually being much more numerous at night than during the day (Muma, 1969).

Plants damaged by spider mites during the day had a higher production of herbivore-induced volatiles that attract predators (see Chapter 41), compared with the night's output. This enhanced the activities of *Neoseiulus womersleyi* (Schicha) during the day, showing an indirect light effect (Maeda *et al.*, 2000).

Lunar periodicity had a unique effect on the activity of *Typhlodromus pyri* (Schueten). Lowest prey consumption rates were recorded at new moon and full moon periods. As all mites were reared in the laboratory, in the absence of moonlight, their periodicity may have been correlated with the gravitational or geomagnetic influences of the moon, or was endogenous (Mikulecky & Zemek, 1992). Diurnal activity in the field and in the laboratory has implications for sampling and release protocols, but has little been studied. The effect of daylength on diapause was noted above.

Phytoseiids disperse by running on leaves, walking along spider mite webs, crawling on the soil (Takafuji, 1977; Raworth *et al.*, 1994) and while borne on air streams (Johnson & Croft, 1981; Hoy, 1982; Sabelis & Dicke, 1985; Charles & White, 1988). Dispersal by winds is usually associated with prey depletion, but may be delayed for a few days, owing to the presence of prey residues (e.g. web residues; kairomones) on the substrate, and to cannibalism. The initiation of dispersal behaviour depends therefore on a compromise between conflicting stimuli: hunger that drives the mites to forage, and tactiles and/or volatiles that continue to signify prey presence (Charles & White, 1988). Mites that are ready to disperse (mostly young, mated females but also some males and deutonymphs) move onto exposed surfaces and place themselves in a position to be lifted by air currents. Different species probably require different wind velocities to detach from the substrate and 'jump' into air streams (Tixier *et al.*, 1998). After arriving on plants the phytoseiids locate prey by various means. *Galendromus occidentalis* (Nesbitt) (often placed in *Metaseiulus*) reacted strongly to cues (silk and other residues) left on leaves by *Tetranychus* spp. (on which it feeds), but only weakly to other spider mites that are not its preferred diet. The effect of the residues or cues persisted for a few days (Hoy & Smilanick, 1981). Similar results were obtained with *Neoseiulus fallacis* (Garman) (formerly placed in *Amblyseius* or in *Typhlodromus*), where a marking pheromone was postulated to affect the predator's searching activities (Hislop & Prokopy, 1981). In addition, phytoseiids are attracted to volatiles emitted by plants that had been damaged by spider mites. The volatiles differ according to the infesting herbivores and to the damaged plants, and attract different predators (Sabelis & van de Baan, 1983). The attracting chemicals may differ among apple cultivars (cvs) (Takabayashi *et al.*, 1994b) and depend on the age of the injured leaves (Takabayashi *et al.*, 1994a; see Chapter 41).

Dispersal on insects (phoresy) is rare, the only known case is *Kampimodromus aberrans* (Oudemans) being disseminated by the aphid (Hemiptera: Aphidoidea)

Myzocallis coryli (Goeze). The association seemed to be restricted to the aphid's spring generation, when eight to ten mites were sometimes found on each insect caught in sticky traps. No *K. aberrans* occurred on aphids later in the summer, and the insects were unharmed by the mites (Krantz, 1973). Two species of *Asperoseius* were found on nematocerous Diptera (Fain & Krantz, 1990), but the significance of the association remains unknown.

Parrott *et al.* (1906), Ewing (1914) and McGregor & McDonough (1917) provided early reports of the potential of phytoseiids for reducing the numbers of pestiferous rust and spider mites. Smith & Goldsmith (1936) later noted their effect on tarsonemids and MacGill (1939) was probably the first to report phytoseiid predation on thrips. Little attention was paid to these predators until the close of World War II. Spider mite outbreaks that often followed the application of synthetic pesticides on commercial crops (beginning in the mid-1940s) indicated that the effect of a formerly operating inhibiting factor (or factors) had been disrupted. McMurtry *et al.* (1970) concluded that most spider mite outbreaks were caused by a combination of factors, especially the pests' increasing resistance to pesticides and the insecticide-engendered destruction of their enemies, mostly (but not exclusively) resident Phytoseiidae. This inference was supported by the common observation that spider mite populations (especially of the pestiferous *Tetranychus* spp. and *Panonychus* spp.) in unsprayed agricultural systems were usually small. However, after the pests were exposed to certain pesticides, such as organochlorines (OCLs) or pyrethroids (PYRs), they raised very large populations (Huffaker & Flaherty, 1966; Gerson & Cohen, 1989). Other causes of spider mite outbreaks include the use of new, more susceptible plant species and cultivars, fertiliser applications that increase leaf area, and year-round cultivation, which promoted larger, on-site pest populations.

A major attribute that contributed to the use of phytoseiids in biocontrol projects, as well as in integrated pest management (IPM) and integrated mite control (IMC) programmes, was their ability to develop resistance to toxicants (McMurtry, 1982), a topic discussed below and in Chapter 43.

As noted, phytoseiids utilise many types of food, and these specialisations are correlated with major ecological traits such as aggregating ability and prey density dependence (Zhang & Croft, 1994). The diversity of feeding habits, and of life history traits, suggested that they could be placed within four groups or categories, a grouping that is consistent with the lengths of certain dorsal setae (McMurtry & Croft, 1997). This system was later somewhat modified by Croft *et al.* (1998a, b), who proposed experimental methods to assist in the placing of individual species in the various categories, and emphasised the need to study more phytoseiids. Another improvement was a genetic algorithm with an information-based model selection criterion (Luh & Croft, 1999) that was used to identify biological traits in order to classify the lifestyle types of 13 phytoseiid species. Of the 24 traits that were used, seven pertained to developmental rates, six were related to diets (including cannibalism on young stages), and the others bore upon dispersal and, as noted, setal length (see below). Although the categorisation was based on relatively few species, and might not fully accommodate some phytoseiids, it appears to be quite robust. It could be used when selecting phytoseiids for specific biological control purposes, and indicates evolutionary trends in the family.

Type I consists only of *Phytoseiulus* spp., mites that bear long dorsal setae (Fig. 26.1) and are specialised predators of heavily webbing spider mites, mostly *Tetranychus* spp., more rarely *Oligonychus* spp. Type II phytoseiids feed on spider mites, but are not restricted to *Tetranychus* spp. They devour spider mites that produce little webbing [e.g. *Panonychus ulmi* (Koch), the European red mite (ERM)], and feed on other mites as well as on pollen and even on plant exudates; in common with type I species, they carry long dorsal setae. Type III are generalists that often prefer prey other then spider mites (in whose webbing they may become entangled), such as tarsonemid mites and thrips. They bear short mid-dorsal and lateral setae (Fig. 26.2). Type IV species consist only of members of *Euseius*, generalists that appear to develop and reproduce best on pollen; their population trends seem to be more closely associated with the abundance of pollen than with any prey. Type IV phytoseiids likewise carry short mid-dorsal and lateral setae, and do not do well within spider mite webbing.

Sabelis & Janssen (1994) explored the close association of type I and II species with webbing spider mites, noting the importance of being adapted to the transient nature of webbing spider mite colonies.

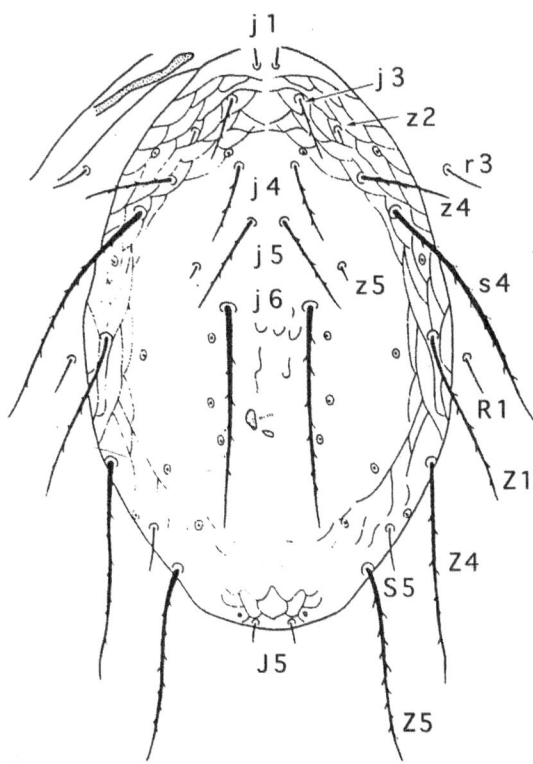

Fig. 26.1 *Phytoseiulus persimilis*. Note long median j4–j6 and lateral s4 setae. (Based on Denmark & Schicha, 1983, with kind permission from Indira Publishing House.)

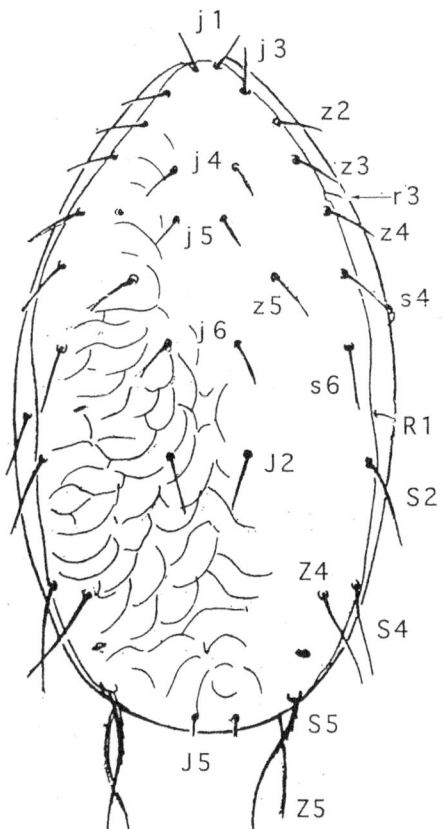

Fig. 26.2 *Typhlodromus pyri*. Note short median j4–j6 and lateral s4 setae.

Type I species

Phytoseiulus persimilis

Members of the genus *Phytoseiulus* live and place their eggs almost exclusively within the webbed colonies of *Tetranychus* spp. The predators have a rapid life cycle (egg to egg in less than 1 week) and are the most fecund Phytoseiidae (see below). The early demonstration that *P. persimilis* controls spider mites on beans (Chant, 1961), peaches (Bravenboer & Dosse, 1962) and roses (Smith *et al.*, 1963), and the realisation that predator numbers will decline along with decreasing prey numbers, suggested the option of augmentation. This, in turn, necessitated the development of simple and inexpensive mass-rearing methods (Chapter 39). Once these were at hand *P. persimilis* was introduced into many parts of the world and became a notable biological control success. It controls spider mite infestations on many crops, in temperate as well as subtropical regions, and despite environmental concerns, did not

Table 26.1 Phytoseiidae that were on sale in the year 2001

Species	Number of companies selling the predator		
	Against spider mites	Against thrips	Against other pests
Amblyseius californicus	23		
Euseius rubini			1 (whiteflies)
Euseius sp.			1 (scale insects)
Galendromus annectens	3		
Galendromus helveolus	5		
Galendromus occidentalis	22		
Iphiseius degenerans	3	10	
Neoseiulus barkeri		6	
Neoseiulus cucumeris	20	29	
Neoseiulus fallacis	6		
Neoseiulus scyphus	1		
Neoseiulus setulus			1 (cyclamen mite)
Phytoseiulus longipes	11		
Phytoseiulus macropilis	3		
Phytoseiulus persimilis	49		
Typhlodromus athiasae	1		
Typhlodromus mcgregori	1		
Typhlodromus pyri	1		
Typhlodromus rickeri	1		

Based on data in Anonymous (2000).

establish on native floras (Walter *et al.*, 1998). Sometimes called the Chilean predatory mite, *P. persimilis* is the best studied ABA as well as being the most advertised for sale (Table 26.1).

Supplied with an abundance of prey, the predator deposited about 78 eggs/female (84% females) in about 15 days at 26°C and 80% RH. Its calculated net reproductive rate (R_0) came to 61.86, its intrinsic rate of increase (r_m) to 0.4282/day and the finite rate of increase (λ) was 1.5345/day (Takahashi & Chant, 1994). These parameters are affected by prey availability and by the host plant. When too few mites were offered, SR was about 0.5 (Toyoshima & Amano, 1998), rising to the usual 0.8 after abundant prey was given. Krips *et al.* (1999) obtained r_m values that ranged from 0.424 to 0.451 when the *Phytoseiulus–Tetranychus* system was maintained on different gerbera cultivars. In addition to its rapid rate of increase, *P. persimilis* is a proactive, voracious forager for spider mites, attributes that enable it to express rapid functional and numerical responses to changes in prey populations. *Phytoseiulus persimilis* swiftly overexploits the quarry, and as it cannot reproduce on other diets, predator numbers then decline.

The usual routine is to release *P. persimilis* to control spider mites on low-growing plants and on shrubs. This practice is widely used in greenhouses on cucumbers, tomatoes and roses (van Lenteren & Woets, 1988), on plastic-covered strawberries

(Cross *et al.*, 1996), and on outdoor crops such as water melons and corn (Pickett *et al.*, 1987). In many cases the predator is used like a pesticide (hence being called a 'living pesticide' or 'biopesticide'). It is meant to provide a rapid solution to the pest problem, but may have only a short-term (or single-season) effect. This often occurs in greenhouses, glasshouses and in annual crop cultivation, where alternate host plants for spider mites (e.g. weeds) have been removed. The scarcity of alternate hosts causes a decline in the spider mites, followed by reductions in predator populations. If spider mite problems then recur, new releases of *P. persimilis* must be made (e.g. Nihoul & Hance, 1993). One approach intended to alleviate this problem was to create asynchronous populations of the two spotted spider mite (TSSM), *Tetranychus urticae* Koch, and of *P. persimilis*. The latter was introduced into one half of a tomato glasshouse, and spider mites in the other part were controlled by spraying with acaricides that did not harm *P. persimilis*. This resulted in prey and predator populations that underwent varying, asynchronous cycles and resulted in satisfactory spider mite control, although relatively few predators (3300/100 m^2) were released (Nihoul, 1993a).

Another approach is the 'pest-in-first' method (Chapter 39, Release techniques), in which spider mites are purposely placed onto the crop (which should, like cucumber or strawberry, be somewhat tolerant to mite damage), and only then are the predators brought in. This method, tried in Finnish cucumber greenhouses (Markkula & Tiittanen, 1976), was considered to be more expensive than predator releases after the pests had been observed. Spider mites that infested field-grown strawberries in south-east Queensland, Australia, were controlled by *P. persimilis* with the 'pest-in-first' method (Waite, 2001).

In the field, where the commercial plant may be surrounded by other spider mite-infested crops or weeds, the predator can persist for longer periods. Infested cucumbers and weeds near a strawberry plot provided off-season prey for *P. persimilis*, enabling it to persist until the following season, when the predator moved onto the crop and controlled the pest (Waite, 1988a). Reductions in spider mite numbers on peaches were obtained through early colonisations of *P. persimilis* that had originated from surrounding plants (Waite, 1988b). Long-term control of *T. urticae* by *P. persimilis* was reported from dense rose hedges in south-east Queensland (Fig. 26.3). Success was attributed to factors that promoted the prolonged prey–predator coexistence: the continuous development of the pest, which under the prevailing subtropical conditions, and with always available fresh leaves, did not undergo a winter diapause; and the dense hedge foliage and its open sides. This density provided refuges to the pests and allowed easy access for the foraging predator along touching leaves. Further, the open sides of the hedges enabled the wind-borne prey and predator to move onto different parts of the thicket, thereby encouraging their population stability and prolonging the long-term interaction. Unsatisfactory spider mite control by *P. persimilis* at another site was attributed to the sparse growth of a different rose cultivar, whose leaves barely touched in the hedges (Gough, 1991).

The predator is seldom used and rarely survives in arboreal situations (although exceptions are known: McMurtry, 1977; Botha & Pringle, 1995). An effort to control spider mites on trees was not economically feasible. Steinberg & Cohen (1992)

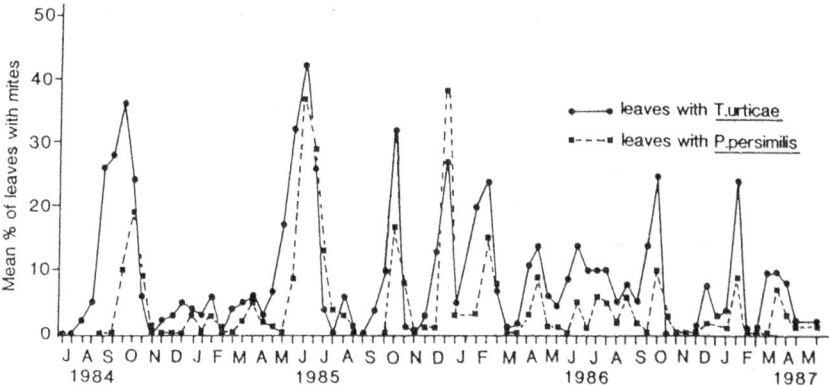

Fig. 26.3   Rate of compound rose leaves with *Tetranychus urticae* and with *Phytoseiulus persimilis* in a greenhouse, demonstrating control of the pest by the predator during 35 months. (From Gough, 1991, with kind permission from Kluwer Academic Publishers.)

released *P. persimilis* in large numbers (1.2 million/ha) on apples; the mites persisted for at least 2 months and controlled *T. urticae*, but the large releases were too expensive. These examples emphasise that each spider mite situation should be considered on its own particular merits.

The specificity of *P. persimilis* for spider mite prey can be a disadvantage if other predators are present on the same plants. When *P. persimilis* and *G. occidentalis* (a type II species) co-occurred on hops that were infested by spider mites, the latter predator provided the better control. This was attributed to the unilateral feeding of *G. occidentalis* on the eggs of *P. persimilis*, which reduced the latter's populations (Pruszynski & Cone, 1972). Placing *P. persimilis* and *Iphiseius degenerans* Berlese (a type III species) together on leaf arenas engendered a similar result for the same reason. In addition, *I. degenerans* laid eggs after feeding on *P. persimilis* juveniles, which the latter could not reciprocate when feeding on the former (Yao & Chant, 1989; see Chapter 42). The oviposition of *P. persimilis* when feeding on the Banks grass mite, *Oligonychus pratensis* (Banks), approximated rates obtained on *Tetranychus* spp. (Pickett & Gilstrap, 1986a). Diets of other spider mites, such as the citrus red mite, *Panonychus citri* (McGregor), and the spruce spider mite, *Oligonychus ununguis* (Jacobi), also allowed a minimal degree of survival and reproduction (Ashihara, 1995).

The existence of 'races' in *P. persimilis* is discussed in Chapter 39. Suffice to note here that since the mite had begun to be widely distributed, 'races' or 'strains' were detected that differ in their life history traits (see Fig. 39.1). Some are more tolerant to heat and to low humidities than others (Perring & Lackey, 1989; Gough, 1991), or show variable resistance to pesticides (Hassan, 1982). The latter trait was observed as soon as the predator began to be widely used (Smith *et al.*, 1963). However, the onset of resistance in *P. persimilis* appears to be slower than in other phytoseiids, and is expressed to lower toxicant levels. Mass-selection efforts to obtain, retain and increase this trait were initiated (e.g. Markwick *et al.*, 1990), and the resulting strains are routinely used in commerce. The introduction of new pesticides, along with the continuing commercial use of the resistant *P. persimilis*, requires ongoing assaying of its reactions to new agrochemicals (Chapter 43). Of 51 insecticides and acaricides assayed by Oomen *et al.* (1991), about half were harmless to *P. persimilis*, as were most of the 33 fungicides, but only half of the 12 herbicides. However, the effects of such chemicals in the field may differ from those obtained in the laboratory. Applications of four pesticides, considered to be harmful to *P. persimilis*, 1 week before or 1 week after its releases, did not prevent the mite from controlling *T. urticae* on strawberry (Cross *et al.*, 1996). Sufficient agricultural chemicals are thus at hand for most IPM programmes that include *P. persimilis*.

The glandular hairs on tomato leaves and stems often entrap predatory mites, detracting from their efficacy. A programme intended to select a strain of *P. persimilis* that will not be hindered by these hairs was undertaken by Drukker *et al.* (1997). A population of *P. persimilis* that had been placed on these plants gradually adapted to its new host and became more proficient at controlling spider mites. This appears to be a unique case of selecting a predator for suitability to a host plant (see Chapter 41).

An emerging problem in the use of *P. persimilis* is susceptibility to a highly contagious disease that reduces its response to spider mite prey (Dicke *et al.*, 2000; see Chapter 39).

Phytoseiulus macropilis *(Banks)*

This is the most widely distributed species in the genus (Takahashi & Chant, 1993) and was the first phytoseiid (when still placed in the genus *Hypoaspis*) to be recognised for its ability to reduce spider mite populations (McGregor & McDonough, 1917). Supplied with abundant prey in the laboratory, *P. macropilis* deposited an average of 56.6 eggs/female (of which 88.4% were females) in 15 days at 26°C and 80% RH. Owing to the comparatively lower fecundity, its life history parameters are slightly lower than those of *P. persimilis*: R_0 came to 50.06, r_m was 0.3862/day and λ was 1.4714/day (Takahashi & Chant, 1994). When supplied with spider mite eggs at 27°C, each female consumed an average of 194 eggs, showing feeding cycles during the third and fourth weeks of life (Shih *et al.*, 1979). The mite is capable of reducing pest numbers on tall perennials as well as on low annuals. An infestation of *Tetranychus tumidus* (Banks) on bananas in Cuba was completely controlled within 3 weeks by *P. macropilis* (released at a prey/predator ratio of 20/1) (Ramos & Rodriguez, 1995). Three female predators/greenhouse bean plant almost eliminated spider mite populations within 5–10 days (Prasad, 1973). In another demonstration of its efficacy, Hamlen & Poole (1980) released *P. macropilis* on the ornamental *Diffenbachia maculata* at prey/predator ratios of 5/1, 10/1 and 20/1. Least damage was seen on plants receiving the 10/1 treatment, on which no spider mites could be found at the end of 12 weeks (Fig. 26.4). However, as with *P. persimilis*, predator declines ensued from prey reductions, followed by pest reappearance which then required a second release of *P. macropilis*. Despite this and other cases of successful spider mite

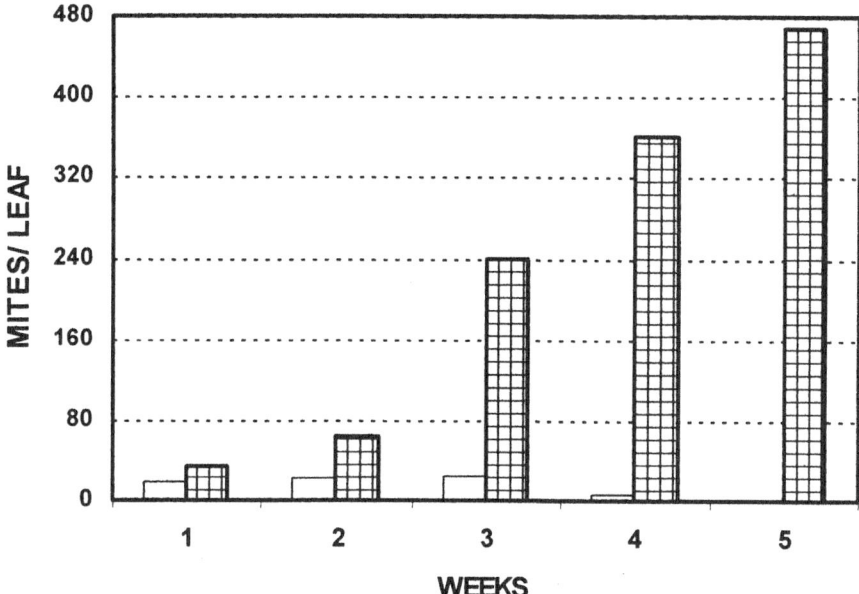

Fig. 26.4 Numbers of spider mites/leaf on *Diffenbachia* plants in the presence of *Phytoseiulus macropilis* (white bars) and in its absence (cross-hatched bars). (Data from Table 1 in Hamlen & Poole, 1980.)

control (e.g. on preserved cut flowers; Price, 1981), *P. macropilis* is commercially available only from three suppliers (Anonymous, 2000).

Phytoseiulus longipes *Evans*

In contrast to the former two species, *P. longipes* (sometimes placed in *Mesoseiulus*) has a limited distribution in the southern Hemisphere (Takahashi & Chant, 1993) and was seldom evaluated as an ABA. Like its conspecifics, it fared best on *Tetranychus* spp., but survived also on *Oligonychus punicae* (Hirst), although producing only half the eggs obtained when feeding on *Tetranychus pacificus* (McGregor) (1.25 vs 2.54 progeny/day) (Badii & McMurtry, 1983). *Phytoseiulus longipes* had the highest r_m (0.549) of all phytoseiids studied to date, as well as the shortest development time (less than 4 days at 25°C), and its other population parameters are comparable only to those of *P. persimilis*. In addition, *P. longipes* has a highly biased SR (0.88), and can withstand long periods of low temperatures (Takahashi & Chant, 1994). These attributes convinced Badii *et al.* (1999) that *P. longipes*, although neglected, is a very promising ABA of spider mites. It is available from about a dozen suppliers.

  The larvae of all *Phytoseiulus* spp. neither feed nor require any food for transition to the protonymph (Takahashi & Chant, 1992).

Type II species

General comments

Type II phytoseiids, like those of type I, live mostly within or very near colonies of their spider mite prey. An adaptation of type I and II species to living and moving in spider mite webs seems to be their long mid-dorsal and some of their lateral setae (respectively, series j4–j6 and J2, and series s-S, Fig. 26.1). These long setae were presumed (Sabelis & Bakker, 1992) to serve as sensory organs when the predators transverse the web mass. Such phytoseiids, which are attracted to their prey by the web, could have evolved in response to spider mite webbing (Gerson, 1985). Only two type II species will be discussed in detail; others are described briefly.

Several common type II species have evolved high levels of resistance to pesticides, an attributed that has made them suitable for augmentation in IPM programmes.

Galendromus occidentalis

Among natural enemies *G. occidentalis* enjoys the distinction of having been 'first' in various acarological, as well as general, areas of biological control. It was the first phytoseiid in which resistance (or tolerance) to pesticides was noted, and the first whose pesticide-resistant strains were introduced into an area where susceptible conspecifics occurred, thereby providing proof of the field efficacy of an artificially improved biological control agent. It was the first pesticide-resistant natural enemy to be transferred between continents, and was the first mite (and probably the first natural enemy) that has been transformed by recombinant DNA methodology. *Galendromus occidentalis* inhabits many wild plants in western North America, which strongly suggests that this is its area of origin. The mite appears to prefer somewhat dry habitats and thrives in young orchards with incomplete canopies.

Galendromus occidentalis consists of several strains that may not be panmictic. Its various geographical populations (or strains) exhibit a diversity of dispersal rates, patterns of pesticide resistance, discrete diapause attributes and different mating habits that, separately or in various combinations, may cause them to be reproductively isolated (Hoy & Cave, 1988). The incompatibility of some strains was explained by the presence of the bacterium *Wolbachia*, known to affect the reproduction of many arthropods (see Chapters 39 and 40), in addition to idiosyncrasies in the mite's reproductive system (Johanowicz & Hoy, 1998). The source of additional interstrain diversity remains unclear. The following summation could therefore pertain to more than a single biological entity.

Provided with an abundance of spider mite prey in the laboratory, *G. occidentalis* attained its optimal development and reproductive rates at 32°C and produced a mean of 33 progeny (60% females) in less than 1 week. The population doubling time was 2.5 days (Tanigoshi *et al.*, 1975). The eggs are usually placed within spider mite patches, because the larva must feed in order to transform to nymphs (Zhang & Croft, 1994). The larvae and subsequent stages spend little time on prey-free leaves and search for food, which include their own eggs and those of other phytoseiids (Croft *et al.*, 1995a); after finding prey the mites tend to stay in the patch. Such behaviour is advantageous for a specialist natural enemy whose strategy consists of quickly finding new prey patches, reproducing there and moving on (or becoming displaced by or fed upon by more generalist predators) (Croft *et al.*, 1995b). However, when other suitable prey is available, *G. occidentalis* remains on crop plants even in the absence of spider mites. In Californian grapevines *Homeopronematus anconai* (Baker) (Tydeidae, Chapter 36) served as a suitable alternate diet; as its populations grew they encouraged the predator's increase (Flaherty & Hoy, 1971).

Galendromus occidentalis is a major ABA of webbing spider mites (especially *Tetranychus* spp.) infesting many deciduous tree fruit crops, hops and roses (McMurtry, 1982). In the USA, its populations followed those of the apple pest *Tetranychus mcdanieli* McGregor in Washington State orchards, had similar distributions and greatly reduced spider mite numbers (Hoyt, 1969a). The predator controlled the pest within a highly successful IPM programme (Hoyt, 1969b); resulting interactions between the prey and predator showed cyclic fluctuations, indicative of pest control. Experimental releases of 32, 64, 128 and 256 predators/apple tree provided satisfactory control of *T. mcdanieli* (whose numbers were reduced from about 120/leaf to fewer than 15/leaf). Optimal results (two to four pests/leaf) were obtained with 128 predators/tree, a compromise between the less effective control with 32–64 *G. occidentalis*/tree and pest overexploitation (with resultant predator–prey imbalances) due to the presence of too many predators on each tree (Croft & McMurtry, 1972). Based on these and other field data, Croft & Nelson (1972) developed an index that was intended to predict the outcome (e.g. chances of successful control) of the interaction between *T. mcdanieli* and *G. occidentalis* on the cultivar Delicious in southern California orchards. The index was based on numbers of prey and predator counted on leaves that were collected weekly. The presence of fewer than 20 *T. mcdanieli*/leaf was considered an effective biocontrol situation, whereas more than 20 pests/leaf represented ineffectual control. However, the 'effective' ratios shifted; at low spider mite levels (nine prey/leaf to 0.05 predators/leaf) the ratio was 179, compared with 20.5 when pest numbers were numerous (18.8 prey/leaf to 0.75 predators/leaf). The prey/predator ratios obtained on any given date would then indicate the situation in the orchard, suggesting whether any interference (chemical or biological) might be necessary. The predictive value of this index requires the continuous monitoring of pest and predator populations, and its application should be re-evaluated in different regions and on other apple cultivars.

Several attributes contribute to the strong functional response that *G. occidentalis* shows to webbing spider mites. Attack rate increases as prey numbers grow because less time is spent in searching for prey. Further, the predator shows wasteful killing even when sated, a trait that causes the functional response to increase even at high prey densities. Two other components of foraging behaviour, e.g. speed of walking and probing activities, are independent of feeding history (Metz *et al.*, 1988). The sum effect of such an aggressive functional response was demonstrated when a single *G. occidentalis* per ten *Tetranychus* spp. (on almonds) sufficed to reduce pest numbers by 23–73 times their former density within 2 weeks (Wilson *et al.*, 1984).

The predator was the most effective of several phytoseiids that were compared for controlling spider mites (the citrus red mite and *T. urticae*) infesting citrus seedlings in the nursery (Grafton-Cardwell *et al.*, 1997). In the presence of several equally suitable spider mites on grape vines, *G. occidentalis* had a greater impact on the more abundant and aggregated prey. When it located a spider mite colony the predator displayed an 'area restricted search' pattern, consisting of slower walking and an increased turning rate. Such searching led to greater kill of the more common and aggregated prey species, and could result in a population shift in favour of less aggregated spider mites among the prey species (Hanna & Wilson, 1991).

When only 37% of a field population of *G. occidentalis* succumbed to the organophosphate (OP) parathion, Huffaker & Kennett (1953) thought that the predator might be tolerant to the pesticide. Resistance to OPs was later ascertained through conventional toxicological tests (Croft & Jeppson, 1970). The susceptibility of other strains of *G. occidentalis* to these pesticides suggested that the OP-resistant (OP-R) mites could play an important role in IPM programmes. They were released in apple orchards where the (susceptible) predators had been eliminated by pesticides, became established and provided satisfactory spider mite control (Croft & Barnes, 1971; Croft & Hoyt, 1983). *Galendromus occidentalis* later became resistant to OP compounds used on many crops in western North America (Croft & Brown, 1975). The resistance ratio (between susceptible and resistant populations) was about 100.

The export of an OP-R strain of *G. occidentalis* from North America to Australia in 1972 constituted the first intercontinental transfer of such natural enemies (Readshaw, 1975). Similar strains were later successfully introduced into other countries. In Australia it was released in apple orchards and, unless disrupted by pesticides, controlled TSSM populations at many sites by 1974. Later (1977) releases were made at 147 New South Wales orchards. A follow-up showed that the predator had become established at most release points and had also dispersed to other orchards. Acaricide use at these sites was reduced by 85%, with substantial gains to growers (Bower & Thwaite, 1982). A still later evaluation (James, 2001) provided a different perspective for this introduction. The predator (along with two other introduced phytoseiids, *T. pyri* and *P. persimilis*) became well established in Australia but did not colonise native habitats and rarely occurred outside crop ecosystems, and required pesticide regimes in order to be maintained, even in horticultural systems. Indigenous phytoseiids, such as *Typhlodromus doreenae* Schicha (see below), later displaced the exotics from orchards kept under low-pesticide regimes. An OP-R strain of *G. occidentalis* was introduced into the former USSR in the early 1980s, in order to control *Eotetranychus pruni* (Oudemans) (sometimes placed in *Schizotetranychus*), a spider mite pest of vineyards. It established and spread widely, reducing pest numbers by about 85–96% (Petrushov, 1987). In New Zealand the predator established but did not provide the expected pest control (Thomas & Walker, 1989), whereas in China *G. occidentalis* survived for a few seasons but did not establish (Z.-Q. Zhang, personal communication, 2001).

The availability of *G. occidentalis* strains that had acquired pesticide resistance in the field served as the basis for a genetic improvement procedure (Hoy, 1985a; see

Chapter 43). After determining that some field populations were resistant to OPs and to sulfur, but susceptible to the PYR permethrin and to the carbamate carbaryl, selection for resistance to these compounds was undertaken. Resistance to carbaryl was obtained by appropriate selections that resulted in a strain (labelled COS-R) resistant to carbaryl, OP and sulfur (Hoy & Standow, 1982). Another strain resistant to permethrin (the P-R strain) was concurrently selected (Hoy & Knop, 1981). COS-R, whose resistance is due to a semi-dominant gene, was released in almond orchards. It established, controlled spider mites, overwintered and survived pesticide applications, providing the first field demonstration of the efficacy of a biological control agent that had been artificially improved in the laboratory (Roush & Hoy, 1981b). The resistance of the P-R strain, however, is polygenic, and might be diluted by large, susceptible resident or migrant populations of *G. occidentalis*. The problem could partly be resolved by the reproductive isolation between the P-R and the resident strains. Another option is to release the improved mites in very large numbers, thus outnumbering and reproductively outcompeting susceptible field populations. In the USA, resistant strains were successfully released in deciduous orchards in Oregon and Washington states (Hoy *et al.*, 1983,). In Russia, Petrushov (1991) developed a PYR-resistant, strain of *G. occidentalis*, which also had multiple resistance to OP compounds.

Galendromus occidentalis was the first mite to be transformed ('engineered') by recombinant DNA techniques (briefly described in Chapter 43). The performance of the wild and transformed populations was similar, indicating that no new biological attributes, adverse or otherwise, were added as a result of this engineering. A transformed strain was released into an experimental site on the campus of the University of Florida in 1996; all transgenes were lost within 7 months and there was no evidence of strain persistence (Hoy, 2000; see Chapter 45). Other strains are ready for release, pending regulatory approval (Presnail *et al.*, 2000).

The predator overwinters on plants. On apples it was found mostly under the bark (Leetham & Jorgensen, 1969) as well as within the fruits' calyx cavities (Gurr *et al.*, 1997). On peaches *G. occidentalis* often occurred in the older (second year and older) fruit stalks (peduncles), which remained after fruit picking (Caltagirone, 1970). In California mite females underwent a facultative (reproductive) winter diapause that was induced by short days and low temperatures. At 19°C and L10/D14 h conditions, 95% of the females were in diapause, a rate that declined when the mites were exposed to longer and warmer days (no diapause at and above 13 daylight hours or 22°C). Diapause was spontaneously terminated (recognised by the resumption of oviposition) in accordance with the natural lengthening of daylight, a process expedited by holding the mites under longer daylight periods (Hoy, 1975). In Australia only 43–59% of the introduced *G. occidentalis* went into diapause under the same conditions, leading James (1988) to postulate that the mite could be adapting to different winter regimes in its new environment. The presence of non-diapausing *G. occidentalis* in orchards would lead to more rapid responses to early-season spider mite infestations, a supposition that led to advocating the protection of these non-diapausing predators.

Being a specialist predator, *G. occidentalis* has some traits that make it a poor competitor with co-occurring phytoseiids (e.g. Downing & Moilliet, 1972). It is not an

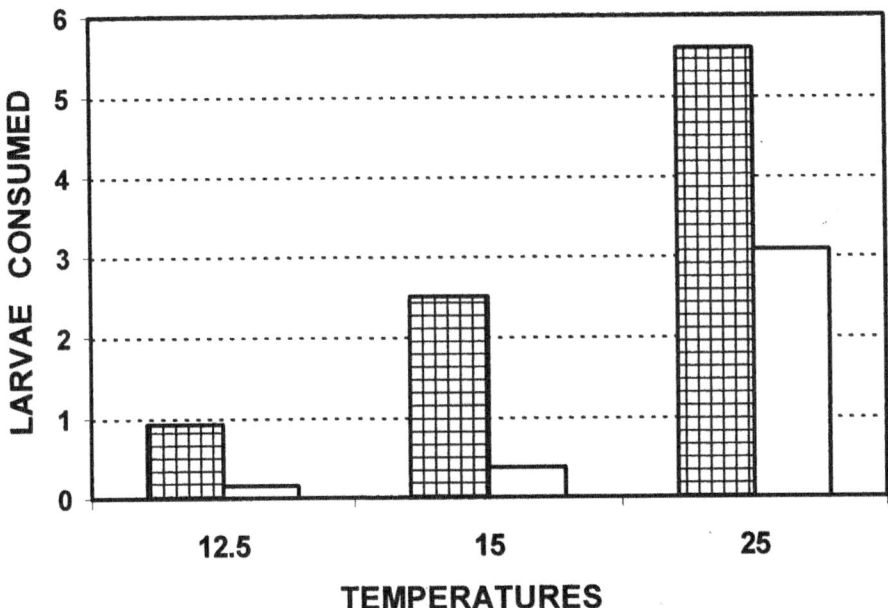

Fig. 26.5 Consumption of *Galendromus occidentalis* larvae by females of *Typhlodromus pyri* (white bars) and of *T. pyri* larvae by females of *G. occidentalis* (cross-hatched bars). (Based on Table 1 in MacRae & Croft, 1993.)

active intraspecific or interspecific predator (except under starvation conditions, as noted), and would thus be more preyed upon than preying on, compared with other ABAs (Croft & Croft, 1996). Unlike *N. fallacis*, a common competitor on apple trees, *G. occidentalis* cannot subsist on pollen (Ahlstrom & Rock, 1973). Further, the adaptation of *G. occidentalis* to warm and dry weather may affect its survival and competitive ability under other climates. The feeding rate of *T. pyri* on larvae of *G. occidentalis* was significantly higher at 12.5 and 15°C than the predation of *G. occidentalis* on *T. pyri* (Fig. 26.5) (MacRae & Croft, 1993). This could explain why *G. occidentalis* is displaced from apple trees in western North America early in the season by the co-occurring *T. pyri*, as the latter is better adapted to cooler and wetter climates. In addition, *G. occidentalis*, most of whose populations had become resistant to pesticides, requires the application of chemicals (that reduce the numbers of other predators) in order to remain dominant (Downing & Moilliet, 1972). In unsprayed or lightly sprayed agroecosystems its status changes as resident competitors or predators [e.g. *T. pyri* and the stigmaeid *Zetzellia mali* (Ewing)] detract from its efficacy (see Chapters 32 and 42).

Methods for mass-rearing *G. occidentalis* in the laboratory and in the field, as well as release tactics, are detailed in Chapter 39. Over 20 companies are currently selling the predator world-wide (Anonymous, 2000).

Neoseiulus fallacis

In the humid central and eastern regions of North America this species occurs on trees and lower vegetation, whereas in the drier western parts *N. fallacis* usually inhabits low and sprawling plants, unless living on irrigated tree crops (Morris *et al.*, 1996). It is the most common phytoseiid on apple trees in many mid-western and eastern commercial apple orchards (Welty, 1995), being an efficient predator of the ERM, TSSM and other phytophagous mites, such as eriophyoids. The availability of the apple rust mite, *Aculus schlechtendali* (Nalepa), as a suitable alternate food is a major factor in maintaining *N. fallacis* populations on apple trees during lack of other prey. It is currently being sold by six companies (Anonymous, 2000).

A generation of *N. fallacis* feeding on TSSM eggs and adults was completed in 4–5 days at 24–27°C, a period that is short among phytoseiids. Each female deposited 30–40 eggs within 2 weeks under conditions of near saturation (Ahlstrom & Rock, 1973; Ball, 1980; Boyne & Hain, 1983). When offered an abundance of spider mites at 26.7°C, individual *N. fallacis* females devoured 10.6 eggs/day or 4.8 females/day. Consumption rose to 12.8 eggs/day as predators were subjected to a diurnal cycle of L10/D14 h, indicating that these were the conditions under which *N. fallacis* would be a more efficient natural enemy. Fewer eggs were eaten by females kept under other photoperiodic regimes (including continuous light or dark) (Smith & Newsom, 1970). Although *N. fallacis* fed and reproduced on other mites, on juveniles of various insects, and even on pollen, survival, activity, fecundity and the development of the second generation (F_1) on these prey were reduced (Ahlstrom & Rock, 1973; Pratt *et al.*, 1999). The disadvantages of *N. fallacis* as a specialist are similar to those of *G. occidentalis*. When inhabiting apple trees with a type III phytoseiid (e.g. *T. pyri*), *N. fallacis* may be displaced because it cannot survive for long without prey, whereas *T. pyri* persists and reproduces on many non-animal diets.

Droplets of an ether extract of *N. fallacis* females attracted its males significantly more, and for longer periods, than male extract or ether alone (Rock *et al.*, 1976). These results, consistent with the production of a sex pheromone, were the first demonstration of such attraction in phytoseiid mites.

The predator overwinters on trees and in the ground cover, but the role of the understorey vegetation in the mite's recolonisation process of apple trees in eastern North America is not clear. Croft & McGroarty (1977) claimed that in Michigan there was a direct relationship between mite numbers in the understorey vegetation (provided it covered at least 20% of the ground) and subsequent biocontrol of ERM. However, no numerical differences were obtained in New York State between predators found on apple trees in orchards whose ground cover had been left intact and those in orchards whose weeds were removed. The ground cover was therefore not considered to play an important role in the dynamics of *N. fallacis* (Nyrop *et al.*, 1994). In peppermint fields in western North America the predator overwintered mostly on dead leaves and the underlying debris, thus being in place to colonise the young foliage and reduce spider mite populations during the following season. It follows that conservation of peppermint debris would enhance pest control (Morris *et al.*, 1996).

After prolonged sampling Croft & McGroarty (1977) devised guidelines for predicting the probability of ERM control by *N. fallacis* on apples (cv. Red Delicious) in Michigan. They were based on an economic threshold of 15 ERM/leaf. Continuous tallying of pests/predator ratios, in 100-leaf samples, provided guidelines for their future relationships. High ERM populations, relative to predator numbers, indicated a low likelihood of control by *N. fallacis*, necessitating acaricide sprays. An intermediate position (with a 50% probability of biocontrol) suggested that the practitioner should wait for 1 week and then, in order to decide whether to spray or to expect the predator to control the pest, sample again. When *N. fallacis* numbers were sufficiently high, relative to the pest, the chances for successful biocontrol were above 90%. Similar decision-making procedures (or indices) for ERM control by other phytoseiids were developed elsewhere (Solomon, 1986; Nyrop, 1988a; Avilla *et al.*, 1993).

As with *G. occidentalis*, populations of *N. fallacis* that survived in heavily sprayed orchards spontaneously developed resistance to many OP compounds, carbaryl and PYRs. By crossing OP-R strains with those that were resistant to carbaryl, Croft & Meyer (1973) obtained a strain that maintained its multiple resistance for 25 generations in the laboratory. The resultant strain (OP × CL) was released (in batches of 100/tree) in apple orchards and became established (Meyer, 1975). Pesticide applications were then needed to reduce the numbers of resident, susceptible *N. fallacis*, which might have diluted the frequency of the Op × CL mites. The availability of resistant *N. fallacis* (as well as *G. occidentalis*) facilitated the implementation of several apple IPM programmes (Croft & Hoyt, 1983). Another IPM programme, based on the OP-R strain, was realised in southern Quebec, Canada, saving commercial apple growers up to 34% of pesticide costs (Bostanian & Coulombe, 1986).

About 300 active stages of the OP-R strain were introduced into New Zealand during 1973–1975, where they served as the nucleus for a mass-reared population (Penman *et al.*, 1979). These predators greatly reduced ERM numbers (from 300 to around 12/leaf) in one season. Notwithstanding this initial success, the long-term efficacy of *N. fallacis* was hindered by winter mortality, displacement by other introduced phytoseiids (*P. persimilis* and *T. pyri*) and the lack of alternate diets. Penman & Chapman (1980) addressed the latter problem by trying a variation on the 'pest-in-first' technique in apple orchards. Understorey white clover was inoculated with TSSM early in the season, providing prey that promoted the increase in *N. fallacis*. This resulted in an earlier (by about 4 weeks) build-up of predator populations on the trees. Nevertheless, owing to its slow rate of population increase in the spring, *N. fallacis* could not prevent pest damage (Thomas & Walker, 1989). The predator was also introduced into Taiwan, where it controlled ERM and TSSM infesting pears in high mountain areas. Releases against Kanzawai spider mite, *Tetranychus kanzawai* (Kishida), infesting mulberry and strawberry, were likewise promising (Lo *et al.*, 1990). In contrast, although *N. fallacis* survived for a few generations in northern China, it did not become established there (Z.-Q. Zhang, personal communication, 2001).

Neoseiulus fallacis controls spider mites on trees as well as on low-growing crops. Spruce (besides apple and pear) may be noted in the former group (Mangini & Hain, 1991), whereas the latter includes lima beans under glass (Hamstead, 1970); peppermint (Morris *et al.*, 1999), corn (Berry *et al.*, 1991), strawberry (Lee & Lo, 1990) and hop (Strong *et al.*, 1997). The effect of hop growth pattern on TSSM control by *N. fallacis*, and the importance of on-plant refugia for this interaction, are discussed in Chapter 39. The sensitivity of *N. fallacis* to low humidity, relative to *G. occidentalis*, suggested to Mangini & Hain (1991) that both might be used to control the spruce spider mite in North Carolina, in tandem. The former would be effective under humid conditions, and *G. occidentalis* could be released during dry and warm spells.

Neither *N. fallacis* nor any other phytoseiids play an important role in controlling TSSM on cotton in the Delta area of Mississippi, USA. The predator could, however, reduce pest populations in another way. Almost 93% of the spider mites were infected and killed by the fungal pathogen *Entomophthora* sp., whose conidia were carried by the predator. As the fungus did not affect *N. fallacis*, it may well have disseminated the pathogen (Smith & Furr, 1975).

Neoseiulus longispinosus *(Evans) and* Neoseiulus womersleyi

The separate status of these closely related, mostly Pacific species was only recently established through biosystematic studies by Ho *et al.* (1995). Both species were kept at the same temperatures and provided with eggs of *T. kanzawai*. The development of *N. womersleyi* females and males was slightly, but significantly, faster than that of *N. longispinosus*, and the former's daily consumption rate was also significantly higher. Reproductive incompatibility between the two species confirmed their separate status, as did their consistent failure to mate when co-occurring on a small island near Taiwan. Although this study indicated that data published earlier about one of these species might be applicable to the other, the similarity in their life histories suggests that most of this information would still be useful. Of special interest could be the correct identification of the '*N. longispinosus*' studied by Hariyappa & Kulkarni (1988), as that predator controlled the broad mite, *Polyphagotarsonemus latus* (Banks) (Tarsonemidae), an unusual prey for a type II phytoseiid.

The distribution of the two species overlaps to a great extent (Ho *et al.*, 1995). The more common *N. womersleyi* occurs in a broad arc from eastern Russia through the Philippines and Australia to New Zealand, whereas *N. longispinosus* is a more southerly species, found in India, eastern China, Taiwan, the Philippines, Malaysia, Indonesia and New Zealand. Both inhabit many wild and cultivated annuals as well as low-growing shrubs, being usually found within colonies of webbing spider mites.

In eastern China *N. longispinosus* is associated with and feeds on mites that infest bamboo, being more efficient at relatively high temperatures (30–35°C) (Y. Zhang *et al.*, 1999). In Taiwan it provided satisfactory control of the Kanzawai spider mite infesting strawberry (Lo *et al.*, 1990). A single *N. womersleyi* (then named *A. longispinosus*), placed with ten TSSMs onto red clover leaves in a northern Japanese greenhouse in the summer, reduced pest numbers and their damage to very low levels within 4 weeks. During autumn, however, the predator failed to control the pest, probably because of winter diapause (Mori & Saito, 1979). TSSM was controlled on strawberry in eastern Australia through a spontaneous colonisation by *N. womersleyi* (or *N. longispinosus*) (Waite, 1988a).

Neoseiulus longispinosus was introduced into Turkey in the early 1990s and assayed against the carmine spider mite, *Tetranychus cinnabarinus* (Boisduval). Best results were obtained when the prey/predator ratio was 5/1, but temperatures around 20°C reduced the predator's efficacy (Colkesen & Sekeroglu, 2000).

Populations of *N. womersleyi* in southern and northern Japan differ in their diapause reactions to combinations of short days and low temperatures. Mites from Okinawa had a very weak diapause response in comparison to the strong reaction of those collected in northern Honshu (Kishimoto & Takafuji, 1997). The former's fecundity was little affected by being kept at different temperatures, indicating that the predator's low diapause rates are associated with an ability to express its full reproductive potential in a wide range of temperatures. These two populations also differed in their response to herbivore-induced plant volatiles. As both were reared for ten generations in the laboratory under similar conditions, Maeda *et al.* (1999) concluded that the differences were hereditary, reflecting the mites' provenances.

Like other widely distributed type II phytoseiids that inhabit many crops, *N. womersleyi* developed field resistance to common pesticides. A multiple-resistant (to OPs, carbamates and PYRs), 'synthetic' strain (called SEL10) was produced in the laboratory by genetic improvement (see Chapter 43) and compared with susceptible strains. Egg hatch and survival rates were similar in all strains, which intermated without reductions in fertility. An unexpected advantage of the SEL10 females was their significantly higher fecundity (Mochizuki, 1996).

Neoseiulus idaeus *Denmark & Muma*

The discussion of this species, one of several neotropical phytoseiids introduced into Africa to control *Mononychellus tanajoa* (Bondar), the cassava green mite (CGM), will be preceded by a few words about the pest. CGM invaded Africa in the early 1970s and soon spread across its 'cassava belt' (encompassing most of tropical Africa), causing widespread damage, mostly in dry regions. Neither chemical control measures nor resistant CVs provided satisfactory control, paving the way for biological control efforts (Yaninek & Herren, 1988). A project, probably the biggest ever undertaken to control a phytophagous mite, was initiated in 1974 by the International Institute of Biological Control. A major component of this project was searching for natural enemies of CGM in the neotropics. More than 50 phytoseiids (many of which were new to science) were collected; ten were chosen after testing (Smith *et al.*, 1996) and sent to Benin to be mass-reared. Meanwhile, Yaninek *et al.* (1989) produced a special training manual, intended to assist personnel engaged in the field research of CGM and its biological control. The manual includes the pest's biological background, describes procedures for fieldwork and notes the natural enemies. In addition, the manual contains illustrated keys to identify common cassava mites, pests and predators. A unique feature of this manual – a procedure never before undertaken in biocontrol projects – was printing it on paper specially treated to be water and tear resistant. This makes the manual a highly suitable tool for perusal and consultation under difficult field conditions.

The imported predators were released in huge numbers (5.5 million individuals at 348 sites in ten African countries), but none became established (Yaninek *et al.*, 1993). Three additional species, namely *N. idaeus*, *Typhlodromalus manihoti* (Moraes) and *Typhlodromalus aripo* De Leon (see below), were later introduced and became established. Details about the mass-rearing, aerial release and establishment of *N. idaeus* in Africa are given in Chapter 39. Like other phytoseiids of wide natural distribution (Brazil and Colombia), *N. idaeus* developed strains that differ in their performance on different prey and in their susceptibility to desiccation. Significantly fewer eggs of a southern Brazilian strain hatched under low humidity conditions. When offered either TSSM or CGM, the predator had a shorter life cycle and produced more progeny on the former prey (Moraes *et al.*, 1994b).

The strain of *N. idaeus* that established in Africa is drought resistant, but has not lived up to expectations in controlling CGM. It spreads slowly, movement into adjoining cassava being measured only in plants/generation and in metres/season (Yaninek & Hanna, 1998). However, it also feeds on another cassava spider mite, *Oligonychus gossypii* Zacher, tracking and subsisting on this prey when it infests weeds during periods of low CGM densities. Weedy cassava fields surrounded by heterogeneous vegetation were therefore believed to be optimal choices as release sites for *N. idaeus* (Yaninek *et al.*, 1991).

Neoseiulus californicus *(McGregor)*

The special rating afforded to *N. californicus* (formerly placed in *Amblyseius*), between types II and III, should be noted before turning to type III phytoseiids. McMurtry & Croft (1997) considered this predator as a type II species, but Croft *et al.* (1998a) changed its rating, because *N. californicus* had more generalist attributes (high fecundity on a thrips diet, less interplant movement, shorter dorsal setae) than a type II specialist such as *N. fallacis*. To this should be added the ability of *N. californicus* to feed on and control a non-spider mite pest, the tarsonemid broad mite (see below). Despite such lesser specialisation, *N. californicus* may control pestiferous webbing spider mites. It tracked TSSM populations on strawberry, albeit with some delay, and its populations were slightly less aggregated, incongruities in the spatial distributions of prey and predator that probably afforded the pests a temporal refuge (Greco *et al.*, 1999). The shelter – as with spider mites on hop, noted above, and on apple, below – reduced prey overexploitation, leading to a persistent prey–predator system; and continued pest control. *Neoseiulus californicus* reaches apple trees by walking on the bare ground (Raworth *et al.*, 1994), although why, and how, the predator is attracted to apples is not clear.

As this case shows, the placement of certain phytoseiids along the type II–type III continuum depends on the availability of sufficient biological data. The choice of a given phytoseiid for a biocontrol project usually depends on its perceived potential to reduce pest populations, which is in turn affected by the particular ABA's lifestyle. Type I and II species, owing to their strong aggregation in prey patches, are best deployed to suppress colonial pests, whereas type III and IV phytoseiids, which hunt for single prey, would usually be more suitable against pests that do not form clusters (exceptions will be noted). This approach could explain some past failures in 'classical' biocontrol projects, adding weight to the argument that life-history studies of exotic natural enemies are a prerequisite for their introduction.

Type III species

General comments

Type III phytoseids, generalists that often prefer to feed on prey other than tetranychids and that readily consume pollen or plant exudates, are still capable of controlling spider mite pests. They subsist on other natural foods, such as hemipteran honeydew, fungi and at times even on plant juices (McMurtry & Rodriguez, 1987), and more often indulge in cannibalism and interspecific predation than do types I and II species. Types III and IV species also produce fewer progeny than the more specialised phytoseiids (Luh & Croft, 1999), and displace the latter when their densities decline owing to a lack of spider mite prey. In addition, type III species do not aggregate within or near prey patches (see below, and Nyrop, 1988b), nor do they oviposit at these sites. Leaf architecture and vestiture (e.g. shelter) affect their behaviour and enhance persistence on certain plants more than the presence of any specific prey (see Chapter 41). McMurtry & Croft (1997) postulated that type III phytoseiids may have evolved in response to the conditions of the plants that they inhabit rather than in consequence of any specific prey. This is supported by many observations, such as the almost exclusive occurrence of *Eharius* spp. on the Labiatae (Moraes *et al.*, 1986), or the restriction of several *Neoseiulus* spp. to certain Australian trees. In the latter case there was no host plant overlap among these mites, although the trees grew in well-mixed stands, a pattern that persisted over several months at several sites in south-east Queensland (Beard & Walter, 2001). The specific preferences of three phytoseiids for grape cultivars affected their persistence on the plants under conditions of prey scarcity, and probably determined their later response to spider mite increases (Duso, 1992).

The more varied diets of type III and IV phytoseiids require a change in format. Only the first three type III species (which are mostly associated with spider mites) will be discussed within their own subdivisions. The other predators, which feed on more than one group of plant pests, will be noted within sections that deal separately with these arthropods [e.g. other mites, thrips and whiteflies (Hemiptera: Aleyrodidae)].

Typhlodromalus manihoti

This neotropical species, a predator of CGM, was initially confused with *Amblyseius limonicus* Garman & McGregor (nowadays studied mostly as a predator of thrips; see below), and some of the early biological data published about *T. manihoti* actually pertain to *A. limonicus* (Moraes *et al.*, 1994a). *Typhlodromalus manihoti* was the most frequently collected phytoseiid on cassava in South America. By using the insecticide check method Braun *et al.* (1989) showed that its removal reduced cassava yields by 30%.

This is another predator introduced into Africa to control CGM (see above). It was first released there in 1988 and has since become established in parts of Benin, Burundi, Ghana and Nigeria, but not throughout the 'cassava belt'. Its quantitative effect on CGM populations seems to be similar to that obtained in Colombia (Toko *et al.*, 1996). The dispersal of *T. manihoti* was measured at several dozen metres/ generation and at 3.3 km/year, and it was five times faster in dry than in wet seasons. By 1998 the predator had become established in about 4300 km², cassava presence being the major factor that explained the predator's rate of dispersal (Yaninek *et al.*, 1998). *Typhloromalus manihoti* occurs mainly on young cassava leaves, its population trends being independent of those of CGM (Bakker & Klein, 1999). When lacking in prey *T. manihoti* subsists on the exudates of young cassava leaves, or even on the mildew *Oidium manihoti* (Bakker & Klein, 1992). However, spider mites (CGM or *O. gossypii*) are needed for reproduction.

The mite required 5.2 days to develop from egg to adult at 25°C and 80% RH in the laboratory. Its fecundity was affected by the cassava cultivar on which it (and the prey) had been reared, as the predator produced most eggs (32/female) on cv Riqueza, the least (20/female) on Olho Roxo. The effect of cassava cultivars on the mite's life table values was very pronounced: R_0 on Riqueza was 15.36 and on Olho Roxo only 9.36, whereas r_m of increase of *T. manihoti* on the Riqueza was 0.256 and on Olho Roxo only 0.119 (Noronha *et al.*, 1995).

Typhlodromalus aripo

This species, the third neotropical phytoseiid established in Africa, became a continent-wide, CGM biocontrol success story. It is now present in 17 west, central and east African countries (International Institute of Tropical Agriculture, 1999), colonising an area of more than 400,000 km². The predator was estimated to have increased yields of fresh cassava root by 30–37%, thereby adding US$70/ha per season to individual farmers (Bellotti *et al.*, 1999). The huge economic success of *T. aripo* engendered the enthusiastic headlines ('Mighty Mite Cuts Cassava Losses' and 'A Mite Strikes Mighty Blow Against World Hunger'), noted in Chapter 1. Other aspects of the CGM project were mentioned above.

The predator occurs mostly on the terminal shoots of cassava, where it lives and shelters during the day. It usually feeds and its juveniles develop on exudates that are secreted at the bases of the youngest leaves. On that diet the mite had 4.5 eggs/female, whereas on GCM its fecundity came to 11.3 eggs/female (Gutierrez *et al.*, 1999b). As the predator develops slowly and utilises an alternate cassava-generated diet, *T. aripo* underexploits its prey, thereby increasing its persistence. Cassava cultivars differ in shoot architecture, which indirectly affects the survival of *T. aripo* and its ability to control the pest.

Other phytoseiids were released in Africa in greater numbers, and in more regions, and yet spread neither as quickly nor as widely as *T. aripo*. During each generation this predator spread by several hundred metres, covering an area of 1500 km² in Benin during a single year (Yaninek & Hanna, 1998). The rapid distribution of the predator is attributed to the custom of African farmers to transfer cassava shoot tips as reproductive material. The predators, living in the tips, are thus transported from field to field and from area to area.

Typhlodromus pyri

Experiments conducted in Europe and North America over more than 20 years, along with anecdotal evidence, suggested that in apple orchards an inverse relationship existed between ERM populations and those of several phytoseiids (at the time identified as *Typhlodromus* spp.). Collyer (1964a), who worked in southern England and conducted some of those experiments, summed them up and concluded that *T. pyri* (called *Typhlodromus tiliae* Oudemans in the earlier literature) (Fig. 26.2) was a major, efficient predator of *P. ulmi*. Her conclusion was based on manipulating prey and predator numbers on seedlings, as well as on field experiments in which the predator was excluded by selective pesticides. Seedlings of *Prunus insititia* were infested with females of *P. ulmi* and used to observe the pest's undisturbed development and the effect of *T. pyri*. Five predator females were placed on each of the latter plants and all seedlings were monitored for 12 weeks. There was a lag of about 1 month until the effect of the predator began to be noted, but by the sixth week the populations of *P. ulmi* were at least 30 times larger in the absence of *T. pyri* than in its presence. This persisted throughout the observational period, regardless of initial ERM infestation rates (Collyer, 1958). Another series of experiments consisted of spraying lime sulfur onto apple trees. The lime reduced populations of both predator and ERM, but in the following 4 years pest densities were much higher, whereas those of *T. pyri* were consistently lower. Two different levels of *T. pyri* on apple trees were also obtained by annually protecting, or by not protecting, the predator's overwintering sites (e.g. bark crevices) from lethal winter sprays. ERM numbers were consistently low on trees whose predators were sheltered and much higher on trees where *T. pyri* had not been protected. The presence of eriophyid mites, while providing an additional diet (for which Collyer coined the term 'alternative food') for the predator, did not detract from its overall efficacy. Collyer (1964a) concluded that *T. pyri*, unless eliminated by pesticides, controlled *P. ulmi* at satisfactorily low levels. Later studies, conducted mainly in apple orchards and grape vineyards in Europe, North America and Australasia, supported this conclusion. In Australia the importance of *T. pyri* as a control agent of apple mites has declined recently, as with *G. occidentalis*, noted above (James, 2001).

Besides requiring a period of some weeks to begin to reduce ERM numbers, *T. pyri* is a persistent, year-round resident of apple trees, and does not disperse much. These features make it a suitable candidate to complement type II phytoseiids (e.g. *G. occidentalis*) in reducing mite populations on apples. Pest outbreaks, especially in warm weather, could be controlled by the latter predator, whereas *T. pyri* is more efficacious in maintaining phytophagous mites at low levels during cool periods (Croft *et al.*, 1995b).

It is a common observation that *T. pyri* occurs mostly on the lower side of apple leaves, where it deposits its eggs, whereas *P. ulmi* inhabits the upper side; e.g. prey and predator populations are randomly or even negatively associated. However, the predator changed its foraging pattern (as evinced by a much more convoluted walking pattern) in response to chemical cues that were either emitted by ERM or associated with that mite. Nyrop (1988b) attributed this searching behaviour to the coevolution of *T. pyri* with prey that is randomly distributed on plants. A predator adapted and attracted to such prey increases the chance of locating it by using a search strategy in which similar amounts of foraging time are allocated to all potential patches. A simulation of the interactions between *T. pyri* and *P. ulmi* indicated that a uniform mixing (e.g. spatial congruity) of their populations would lead to ERM extinction (Walde *et al.*, 1992), whereas the predator could persevere (on alternate diets) at low levels. In other words, the long-term persistence and stability of this predator–prey system depends on their spatial incongruity. The absence of positive spatial density dependence (or lack of prey-induced aggregation) may be common to other type III (and IV) phytoseiids.

While the behaviour that enables *T. pyri* to control ERM on apple in North America appears to be consistent with this premise (and with observations on prey preferences, see below), it is not compatible with reports about the predator's choices when living on grapevines in Europe. The most favoured prey of *T. pyri* on these plants was eriophyids, e.g. the grape blister mite, *Colomerus vitis* (Pagenstecher), and the grape rust mite, *Calepitrimerus vitis* (Nalepa). TSSM (a prey to which the predator did not react; Nyrop, 1988b) was also readily devoured. ERM, however, was of relatively low nutritional value to *T. pyri*, being accepted reluctantly by its juveniles and barely by the adults (Engel & Ohnesorge, 1994). Life-history data (see below) also indicate that grapevine ERM is not an optimal diet for *T. pyri*.

Other (but secondary) apple pests that are held in check by *T. pyri* in North America include the apple rust mite and the brown mite, *Bryobia arborea* Morgan & Anderson. Numbers of the latter, an occasional pest in eastern Canada, were much reduced 4–7 weeks after the predator was added to greenhouse trees, a delay attributed to the restriction of *T. pyri* to leaves, whereas brown mites live on the bark (Herbert, 1962). The rust mite provided for the better reproductive success of *T. pyri*, compared with ERM, although the predator preferred the latter prey. Dicke *et al.* (1990) postulated that this apparent discrepancy might reflect the greater availability of *A. schlechtendali* on apples early in the season. Another explanation could be that ERM contains carotenoids (a dietary component that is rare in apple rust mites) that are required by *T. pyri* for successfully entering winter diapause. This is a facultative (reproductive) diapause induced by short days; Fitzgerald & Solomon (1991) distinguished diapausing from non-diapausing field-collected females by placing them under a daily photoperiodic regime of L8/D16 h. Females that had not oviposited after a fortnight under these conditions were considered to be in diapause.

Compared with many other phytoseiids, *T. pyri* (a northern Hemisphere species) has a low and slow rate of increase. Provided with all stages of ERM, the generation time (T) of the predator was about 15 days at 25°C, R_0 was 11.6 and r_m was 0.075. The latter value came to only 0.036 when the mite was reared on the pollen of *Vicia faba* (Overmeer, 1981). As noted, ERM was not an optimal diet for *T. pyri* obtained from vineyards. The predator developed more rapidly on another grape-inhabiting spider mite, *Eotetranychus carpini* (Oudemans), than on ERM (207 vs 216 h at 26–27°C) and deposited twice as many eggs when offered the grape blister mite than when given enough ERM females (Duso & Camporese, 1991). Zemek (1993) listed various fecundity rates (from 1.5 to 43.4 progeny/female) reported for *T. pyri*, ascribing this variation to different rearing conditions (including diet) and/or to intraspecific variability. Other explanations could include host plant effects and/or the presence of slightly different taxonomic entities. Additional aspects of the feeding of *T. pyri* on various diets are noted in Chapter 41 (and see Porres *et al.*, 1975; Zemek & Prenerová, 1997), as is the effect of pollen on the predator's success in controlling ERM in South Africa (Readshaw, 1998).

OP- resistant *T. pyri* were first noted by Collyer (see Hoyt, 1972) on apple trees in New Zealand (where the mite was inadvertently introduced). Their resistance level was initially low, the median lethal concentration (LC_{50} ratio) between resistant and susceptible mites being only 9.7, compared, at that time, to around 100 in *G. occidentalis* and *N. fallacis*. However, the continuous application of the OP azinphos-methyl in the same apple orchards further selected in favour of the partially resistant predators, which attained higher resistance levels within a few years. By that time an integrated ERM control programme had been initiated in New Zealand (Collyer, 1980). Its core consisted of continuous monitoring of the pest's populations, which allows the practitioner to apply, if needed, one or two annual sprays aimed at the pest's larvae, its most susceptible stage. Milder pesticides greatly reduced ERM numbers while causing only minor damage to the predator. This shifted the previous pest/predator numerical ratio from 2/1 to 1/3–1/10, enabling *T. pyri* to keep the small surviving pest population well below its economic injury level. The need to apply PYRs in orchards then led to a successful 'improvement' programme, in which the OP-R *T. pyri* was selected for resistance to PYRs, and the resistant mites were successfully released and controlled ERM (Suckling *et al.*, 1988).

The New Zealand OP-R strain was introduced into England in 1977 and became established; an indigenous strain, showing stable resistance to OPs and to carbaryl, was later found in nearby apple orchards (Kapetanakis & Cranham, 1983). Other OP-R *T. pyri* strains, collected in southern England, had other patterns of resistance to diverse OP insecticides, differences in total developmental periods, and variations in body and setal size. Each of these strains could therefore respond differently to the applied pesticides, affecting their usefulness as ABAs. Fitzgerald & Solomon (1999) suggested that the term 'OP-R', as used for phytoseiids, might be an oversimplification that should be appraised when using the mites in apple orchards.

OP-R strains of *T. pyri* that also became spontaneously resistant to PYRs (Kreiter *et al.*, 1995) or were purposely selected ('improved') for this trait (Markwick *et al.*, 1990; Solomon & Fitzgerald, 1993) are nowadays widely used. One of these strains, transferred from New Zealand into eastern Canada, survived the prevailing subzero climate and has retained its levels of resistance to PYRs, provided no cross-breeding with PYR-susceptible *T. pyri* took place (Hardman *et al.*, 1997, 2000). High levels of multiple resistances to OPs, carbamates and PYRs were later found in mites living on vineyards in France, as well as resistance to the fungicide dinocap, used against powdery mildew in grapeyards (Kreiter *et al.*, 1995, 1998a). The OP-R strain 'Mikulov', found in the Czech Republic, is tolerant to dry conditions and seems to have developed resistance to other pesticides, such as the acaricide dicofol and the fungicide mancozeb (Pultar *et al.*, 1992). Mancozeb, used to control downy mildew, a fungal disease of grapevines, had probably reduced the numbers of most resident phytoseiids in the early 1980s. 'Mikulov' is now being sold by European companies (Copping, 1998); this and other strains of *T. pyri* with multiple resistance are a major component in apple IPM programmes in many parts of the world (Blommers, 1994). This predator is the only ABA whose susceptibility to new plant protection chemicals is being regularly monitored by the IOBC/WPRS working group 'Pesticides and Beneficial Organisms' in the laboratory and in the field (Sterk *et al.*, 1999). Data on the effects of all 'phytopharmaceutiques', plant protection chemicals used in French vineyards, on *T. pyri* and two other phytoseiids, are annually published by Kreiter *et al.* (1998b and earlier publications).

Typhlodromus pyri is one of several predators that occur naturally on grapevines in continental Europe, but is the only phytoseiid considered to be important for the natural control of ERM and *E. carpini*, their resident pest mites (Schruft, 1985). The elimination of indigenous phytoseiids in northern Italian vineyards by mancozeb was followed by their recolonisation with *T. pyri*. It was released in the early spring by placing around 100 diapausing females (located in crevices and buds on branches taken from unsprayed grapes) onto individual plants. These females remained within their shelter for several weeks, thus surviving extreme climatic changes, and emerged as temperatures rose. They then fed on alternate prey (e.g. tydeids) and on overwintering *E. carpini* females, being on hand to attack larvae of *P. ulmi* as these hatched from their winter eggs. *Typhlodromus pyri* showed a clear numerical response to ERM populations and controlled that pest (Duso, 1989). The successful colonisation (or recolonisation) of vineyards by *T. pyri* was postulated to depend on many factors (Duso & Pasqualetto, 1993; Delbac *et al.*, 1996). These include the prevailing climatic conditions, the specific grape cultivar and strain of *T. pyri,* the availability of alternate diets, the presence or absence of insect enemies of phytoseiids, interspecific competition, reductions in pesticide (including fungicide) use, and the acceptance of higher economic damage thresholds by farmers and consumers. This success highlights the option of augmenting predator populations by diapausing phytoseiids that are released in the orchard before they, and their prey, become active. Control of grape eriophyoid mites by *T. pyri* will be discussed below.

A major factor affecting the natural recolonisation of fruit trees by *T. pyri* (and other phytoseiids) is the surrounding evergreen vegetation. It provides shelter, other prey and pollen, enables the predators to maintain populations away from the crop plants, and provides mites to recolonise crops (Boller *et al.*, 1988). This aspect is discussed in Chapter 41.

The low number of *T. pyri* on pear trees in Switzerland was attributed in part to the scarcity of hairs on the leaves of most cultivars. However, when a benign spray

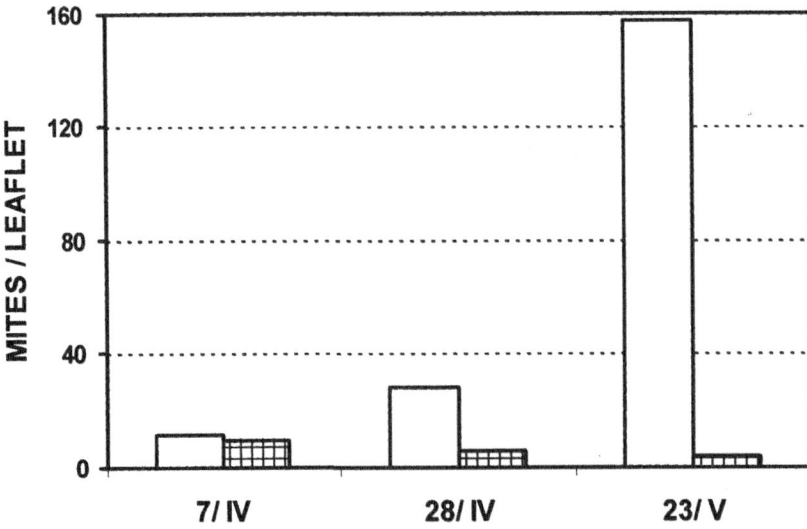

Fig. 26.6   Spider mites (*Tetranychus urticae*) on strawberry leaflets without *Typhlodromus pyri* (white bars) and with *T. pyri* (cross-hatched bars), 3 and 7 weeks after the release of predatory mites on 7 April. (Based on data in Zacharda & Hluchy, 1996.)

schedule was instituted, the predator controlled ERM as well as TSSM. Pest mites were present on less than 5% of pear leaves taken from an orchard that was maintained under an IPM programme, whereas *T. pyri* was found on 30–50% of the foliage and provided good pest control. The presence of *T. pyri* on these leaves was attributed to the small tufts of hairs that grow on their petioles (Baillod *et al.*, 1992). The 'Mikulov' strain was used to control TSSM on strawberry in the Czech Republic. Predators were collected in vineyards, mass-reared on pine pollen and released in numbers that provided a prey/predator ratio of 10/1 on the plants. Pest numbers were greatly reduced within 3 weeks and even more so after 7 weeks (Fig. 26.6). This was a demonstration of the success of *T. pyri* in controlling TSSM on low-growing crops (Zacharda & Hluchy, 1996).

Interactions between *T. pyri*, *N. fallacis* and *G. occidentalis* on sprayed and unsprayed apple trees were discussed above (and see Chapter 42), and interactions of *T. pyri* with the stigmaeid *Zetzellia mali* are noted in Chapter 32.

Phytoseiids that feed on false spider mites

General comments

All members of the family Tenuipalpidae, or false spider mites, are phytophages, but only few are serious plant pests. The tenuipalpids are related to the spider mites by having long, needle-like recurved chelicerae, but are usually smaller, produce fewer progeny and have a longer life cycle (and thus fewer annual generations). Most of pestiferous species are in the very large genus *Brevipalpus*. They damage citrus (directly and by vectoring virus diseases), grapes and numerous ornamentals (Jeppson *et al.*, 1975). Although many phytoseiids associate with tenuipalpids in the field, and feed on them in the laboratory (e.g. Childers, 1994; Gravena *et al.*, 1994), control of pest species by these predators is not common. Available field data indicate that phytoseiid numbers may increase along with decreases in tenuipalpid populations (e.g. Buchanan *et al.*, 1980; El-Halawany *et al.*, 1993), but little is known about their long-range impact. The predators of scarlet mite, *Brevipalpus phoenicis* (Geijskes) (called leprosis mite in South America), which infests tea in Indonesia, include several phytoseiids, but none had a significant effect on the pest under natural conditions. An application of the pesticide dichlorodiphenyltrichloroethane (DDT) onto tea bushes (Oomen, 1982) eliminated competition by other predators, except for one phytoseiid. A subsequent census showed that the latter had reduced pest numbers to only 83% of the predator-free controls, whereas the Stigmaeidae (Chapter 32) reduced pest densities to 27% of the controls.

Typhlodromus doreenae *Schicha*

This species and *Euseius victoriensis* (Womersley) (a type IV species discussed below) often occur together and are capable of controlling various pest mites (James & Whitney, 1991, 1996; Smith & Papacek, 1991). Both are indigenous, generalist east Australian phytoseiids that have successfully colonised exotic crops (e.g. grapevines citrus), controlling some of their acarine pests.

Brevipalpus spp., especially *Brevipalpus lewisi* McGregor (called 'bunch mite' in Australia, because it damages grape bunches), a pest of grapevines, was virtually absent from vineyards in which *T. doreenae* predominated. The predator was found in vineyards that had received inorganic fungicides (e.g. wettable sulfur and copper hydroxide), indicating its tolerance to these pesticides. *Typhlodromus doreenae* also survived applications of certain fungicides, but was very susceptible to insecticides. Bunch mite damage, but no biocontrol activity, occurred in vineyards that were regularly treated with synthetic pesticides, suggesting that *T. doreenae* was the key to bunch mite control (James & Whitney, 1991). The predator underwent a reproductive diapause induced by low temperatures and short days (James, 2000), aggregating on dormant vines. In the spring it fed on the blister mite (as well as on *Brevipalpus* spp.) and later in the season it controlled the grape rust mite (see below).

Typhlodromus doreenae required 8 days at 25°C to develop from egg to adult. Almost all mites that were continuously kept at 30°C completed their development, as did 67% of those held at 35°C. These data, and similar findings about *E. victoriensis*, indicated that both are well adapted to high temperatures, an important factor in their success as ABAs (Taylor & James, 1993). Methods for mass-rearing both species on various diets, including pollen, are in Chapter 39.

Amblyseius channabasavanni *Gupta*

This species, a type III predator, appears to prefer the tenuipalpid *Raoiella indica* Hirst, a pest of areca palm (*Areca catechu*) and date palms in India. The predator's active stages fed voraciously on all pest stages, preferring eggs and quiescent stages. Each female consumed 11–42 eggs, or six to 13 adult *R. indica*/day. Alternate prey, such as spider mites and scale insect crawlers, were also taken. The predator raised a generation in less than 1 week at 27–35°C, producing about 24 eggs/female. Daniel (1981) assumed that the predator's abundance on the trees, concurrent with increases in pest populations, indicated the specificity of *A. channabasavanni* for the tenuipalpid.

Phytoseiids that feed on tarsonemid mites

General comments

Several members of the family Tarsonemidae (Chapter 33) are serious plant pests. They are hard to detect because of their small size and ability to hide in buds, blossoms and narrow plant crevices. Their life cycles are short and they cause much damage before being noted by growers. In addition, pesticides may not reach and control them. Two economically important species, cyclamen mite, *Phytonemus pallidus* (Banks) (formerly in *Steneotarsonemus* or *Tarsonemus*), and broad mite, are the targets of ongoing efforts with ABAs. The former is a major pest of strawberry, whereas the latter attacks diverse crops, including citrus, cotton, jute, papaya, pepper, potato, tea and many ornamentals (Gerson, 1992).

   The body of adult tarsonemids is covered by a thick integument that protects them from many phytoseiids, which in consequence prefer to feed on the juveniles (Croft *et al.*, 1998c). Hariyappa & Kulkarni (1988) noted that *N. longispinosus* never fed on broad mite eggs, probably owing to their cover of studded armature.

Phytoseiids that feed on cyclamen mite

Smith & Goldsmith (1936) were among the first to report that a '*Seiulus*' [later called *Iphidulus*, and probably *Neoseiulus cucumeris* (Oudemans)] fed on cyclamen mite infesting strawberry in California. The predator 'is often responsible for considerable destruction' of the pest and in rare instances brought the cyclamen mite almost to extinction. Owing to its larger size the predator often failed to follow the pest into young, closed strawberry leaflets, which protected *P. pallidus*. Few new leaflets were formed during dry periods, depriving *P. pallidus* of its enemy-free space and exposing it to the predator.

   The dynamics of cyclamen mite control by phytoseiids [*N. cucumeris*, discussed below, and *Neoseiulus aurescens* (Athias-Henriot)] in California were explored by Huffaker & Kennett (1956, and earlier papers). The (then practised) 4 year cycle of strawberry plantings initially affected pest populations, and later those of the predator. Young plants were fully exposed during the first year to prevailing dry and warm climatic conditions, which are unsuitable for the rapid increase of the pest. Predators were still scarce in the second year, whereas *P. pallidus* became numerous within the dense plants, causing most of its damage. In the third and fourth years numerous phytoseiids occurred on the plants and usually reduced pest populations to below injury levels (unless disrupted by pesticides). The predators' ability to feed and reproduce on non-prey diets enhanced their survival during periods of very low (or no) pest presence. Their studies of these pest–predator dynamics prompted Huffaker & Kennett (1956) to augment (called 'stocking' in their assay) phytoseiid populations on strawberry, late in the first or early in the second year. 'Stocking' each plant with two or three phytoseiids late in the first year resulted in a mean of 57 cyclamen mites/sample (e.g. 30 leaflets), whereas pest numbers on predator-free plants came to 1670/sample. Similar reductions in pest densities were obtained as a result of second year stockings, demonstrating that complete cyclamen mite control could be obtained even during the susceptible second year in the strawberry cycle.

In Finland strawberry has a long (4–6 year) growth cycle, and had sustained heavy losses from *P. pallidus* since the cessation of pesticide sprays in 1995. *Neoseiulus cucumeris*, released at the rate of 50 mites/m strawberry row, reduced pest populations within 2 months to less than one-tenth of their numbers in the controls (Tuovinen, 2000).

Moderate TSSM and cyclamen mite populations infesting strawberry were better controlled by *N. fallacis* than by *N. cucumeris*. Fewer individuals of either predator were obtained following releases of both onto plants that were heavily infested by cyclamen mite, but overall pest control was superior and *N. fallacis* performed slightly better. Croft *et al.* (1998c) believed that *N. fallacis* would provide faster control of TSSM and cyclamen mite, whereas in the long run *N. cucumeris* would maintain these mites at lower densities.

Phytoseiids that feed on broad mite

Although no specific acarine predators of broad mite have been identified, several phytoseiids feed on the pest in the laboratory (Badii & McMurtry, 1984; Castagnoli & Falchini, 1993) and contribute to its control in the field (summed up by Gerson, 1992). *Amblyseius ovalis* Evans fed on the eggs and larvae of broad mite and 'checks this mite very effectively' on chilli in Mauritius (Moutia, 1958). Good control of *P. latus* on citrus was achieved with *E. victoriensis* in south-east Queensland (Smith & Papacek, 1985). Wu (1984) reported that *Amblyseius nicholsi* Ehara & Lee consumed 10.5 broad mites/day, providing 'good results' on citrus in Sichuan Province (southern China).

The development of *Neoseiulus californicus* when feeding on *Polyphagotarsonemus latus* was somewhat slower (by about 2 days at 25°C) than on TSSM, but it also increased on the former diet, indicating its potential to control the pest (Castagnoli & Falchini, 1993). *Neoseiulus californicus* was assayed against *P. latus* infesting limes in Florida. Releases in the greenhouse, at the rates of one predator to five to 15 prey, significantly reduced pest densities (from 117–140 to 0.0–2.0/leaf), 14 days post-release, and greatly (10–15-fold) increased leaf area of predator-treated limes. Field releases generally confirmed these results: *N. californicus*, mostly along with *Typhlodromalus peregrinus* (Muma) (an indigenous predator that arrived spontaneously), prevented rapid build-up of broad mite populations on lime fruits. The predators' importance was confirmed with the insecticidal check method. Lime trees were sprayed with an acaricide or left untreated; more fruits with 40–100% of their surfaces damaged by the mite were picked from sprayed trees. Field releases of the predator were therefore considered to be a suitable option for protecting limes in the mild winter–spring season (Peña & Osborn, 1996). During the warm and humid summer–autumn period, however, the predators cannot overcome the pest's very rapid increase.

Satisfactory control of *P. latus* on peppers was obtained by releases of *Neoseiulus barkeri* Hughes in a Florida greenhouse. A single batch of five predators/plant significantly reduced pest numbers, although the peppers suffered some damage. Three weekly releases, however, provided adequate protection until the end of growth. Another strategy, placing ten predators or more on each plant, effectively reduced pest populations from >100/leaf to zero in 1 week, and kept them at that level (Fan & Petitt, 1994). The predator's efficacy against broad mite on other crops in Florida (e.g. beans in a glasshouse and limes in the field), was erratic at best (Peña & Osborne, 1996).

Phytoseiids that feed on eriophyoid mites

General comments

Many Eriophyoidea are pests of agricultural crops, damaging plants by russeting, blistering and galling, as well as by vectoring plant viruses. About 30 of their phytoseiid predators were listed and reviewed by Sabelis (1996). Most are type III and IV species, and eriophyids constitute only part of their diverse diets. Most of the available data on eriophyids as prey for phytoseiids have been obtained under laboratory conditions, often when no other diets were offered.

Sabelis (1996) noted that eriophyoids, which may occur in vast numbers, are very vulnerable to predators and serve as suitable or alternate prey for many phytoseiids. This begged the question of why they often survive in large numbers in the face of strong predation by phytoseiids. Several hypotheses were postulated to explain this apparent paradox. They included the possibility that eriophyoids may be nutritionally deficient or unpalatable; that they are less profitable (for predator reproduction) than other prey, or that eriophyoids defend themselves, escape or hide in refuges. Although these possibilities were mostly refuted, they do explain the survival of galling and blistering eriophyoids, which remain hidden during most of their lives (but could be reached by juvenile phytoseiids). In addition, even if a great many of the gall makers or bud dwellers are destroyed by predators (as might happen when eriophyoids search for new sites), most have a short life cycle and rapidly rebuild their populations. The situation differs in regard to vagrant rust mites, which are always exposed. Owing to predation they probably suffer heavy losses that often lead first to their own local extinction, and then to that of the predators. Mechanisms that bring about the long-term persistence of vagrant eriophyoids may be at the metapopulation level, and explanations should be sought in studying eriophyoid–phytoseiid interactions at different spatial scales (Sabelis, 1996).

The only known prey of *Okiseius* spp. are eriophyoids (Walter, 1999), and *Kampimodromus aberrans* is often found with these mites (Castagnoli & Oldfield, 1996). Several other phytoseiids (e.g. *T. pyri, E. victoriensis*) produce more progeny and develop more quickly when feeding on eriophyoids than on other prey. These mites could therefore be the preferred, or even main, prey of the phytoseiids', and not, as at times inferred, only an alternate diet. Nevertheless, and despite their common co-occurrence and probable long association on plants, phytoseiids do not seem to specialise on eriophyoids, as in the *Phytoseiulus–Tetranychus* association. However, phytoseiids may occur in specific, well-protected sites (e.g. colonies of webbing eriophyoids; Manson & Gerson, 1996), in a manner similar to the occurrence of *P. persimilis* within the colonies of webbing spider mites.

Euseius victoriensis

This type IV phytoseiid (other species in this category will be discussed below) raised a generation in 4–7 days at 22.5–40°C, being still active at the higher temperature (Taylor & James, 1993). The mite reproduced when offered pollen of many plants or eriophyid mites, whereas its oviposition on spider mites was meagre and survival poor (James, 1989). The predator (unless disrupted by pesticides) controls several eriophyid mites that injure crops in Australia. An example is the brown citrus rust mite, *Tegolophus australis* Keifer, a pest in eastern Australia. When predator density exceeded 0.2–0.3/leaf in the Australian spring, hitherto high pest numbers dropped and remained low, with fruit infestation rates of only around 1–2% (Smith & Papacek, 1991). Removal of *E. victoriensis* with an OP was followed by rapid pest increases (from 5/10 cm^2 of leaf surface to 120). Sprayed foliage carried three *E. victoriensis*/100 leaves, whereas unsprayed leaves had about 100. Another co-occurring pest, the cosmopolitan citrus rust mite, *Phyllocoptruta oleivora* (Ashmead), was not satisfactorily controlled by *E. victoriensis*. It overcame small pest infestations only when more than 40 predators occurred/100 leaves. A partial explanation for this could be the predator's tendency to feed only on the eggs and quiescent stages of *P. oleivora*, whereas it devoured all stages of *T. australis*. The association of the latter with *E. victoriensis* is interesting as being a case where an indigenous predator controls an indigenous pest of an exotic crop (Smith & Papacek, 1991).

Euseius victoriensis also controlled grapevine eriophyids. It prefers rust and blister mites over false spider mites and is believed to be the key to the control of several pestiferous eriophyids in south-eastern Australia. Its coexistence with *T. doreenae* on grapevines probably results from the generalist, non-competitive feeding habits of the two phytoseiids (James & Whitney, 1991). *Euseius victoriensis* was considered to be the main natural enemy of TSSM on peaches in the same area (James, 1990). The predator maintained pest numbers at very low levels, probably before they constructed their defensive webbing.

Typhlodromus pyri *and grape eriophyids*

This predator was discussed above as an ABA of various spider mites. Releases of the OP- and sulfur-resistant *T. pyri* (10,000/ha) in Moravia reduced *Calpitrimerus vitis* numbers by about 92% during their August–October peak within 2 years (Hluchy, 1993). In northern Spain Pérez–Moreno & Moraza (1997) followed pest and predator population cycles on grapes for several years. The presence of large *T. pyri* populations at the end of summer resulted in good pest control during the second year, with few rust mites remaining on the vines. In turn, this engendered a reduction in predator numbers, which resulted in poor pest control in the third year. The outcome of this predator–prey association was strongly affected by the weather, because high temperatures and low humidities inhibited predator development while encouraging that of the pest. Other cases of eriophyid control by phytoseiids were noted above.

An interesting new use of phytoseiids in eriophyid control was the application of *N. cucumeris* and *N. barkeri* against *Aceria tulipae* (Keifer), infesting tulip bulbs. Both predators were associated with the pest in the field and reduced its populations in small jar experiments (Conijn *et al.*, 1996).

Phytoseiids that feed on thrips

General comments

Thrips (Thysanoptera) are pests that suck out the contents of plant cells. This feeding results in cell death, extensive silvering, leaf curling and galling. Other, indirect plant injury is incurred by thrips that transmit plant viruses (tospoviruses, the name derived from 'tomato spotted wilt virus', commonly called TSWV). This virus, the causative agent of a serious tomato disease, can only be acquired by the thrips' first instar larvae, but is spread mostly by the winged adults (Bautista *et al.*, 1995). The world-wide increase in glasshouse area has elevated two polyphagous thrips species, *Frankliniella occidentalis*, the western flower thrips (WFT), and *Thrips tabaci* the onion thrips to pestiferous prominence. Both transmit TSWV, and like other thrips (e.g. *Thrips palmi* Karni and *Taeniothrips simplex* Morison) constitute ongoing targets in biocontrol projects using mites (Riudavets, 1995). *Neoseiulus cucumeris*, *N. barkeri*, *I. degenerans* and *A. limonicus*, all type III phytoseiids, are natural enemies of thrips and are being used for their control on many crops (Table 26.2). Several *Euseius* spp., namely *Euseius addoensis* Van de Merwe & Ryke, *Euseius citri* (Van de Merwe & Ryke) and *Euseius tularensis* Congdon, are employed against citrus thrips. However, many pest thrips are not affected by phytoseiids, possibly because of their effective

Table 26.2 Efforts to control thrips pests in greenhouses with phytoseiid predators

Crop	Pest	Predator	Sources
Bedding plants	OT+WTF	NC	Jacobson (1993)
Cabbages	OT	NB+NC	Hoy & Glenister (1991)
Citrus	CT	ID	Grafton-Cardwell *et al.* (1999a)
Cucumbers	OT	NB	Hansen (1988)
	OT	NB + NC	Brødsgaard & Stengaard Hansen (1992)
	WFT	NC	Bennison & Jacobson (1991)
	WFT	AL	van Houten (1996)
	OT+WFT	NC	Gillespie (1989)
Chrysanthemums	WFT	NB+NC	Hessein & Parrella (1990)
Cyclamens	WFT	NC	De Courcy Williams (2001)
Gladioli	TS	NB	Conijn & Groen (1990)
Ornamentals (mixed)	WFT	NC	Sörensson & Nedstam (1993)
Peppers	OT	NB	Ramakers & van Lieburg (1982)
	OT	NB+NC	De Klerk & Ramakers (1986)
	WFT	NC+ID	van Houten & van Stratum (1993)
	WFT	ID	Ramakers & Voet (1996)
Roses (cut)	WFT	NC	Linnamäki *et al.* (1998)

CT: citrus thrips; OT: onion thrips; TS: *Taeniothrips simplex*; WFT: western flower thrips; NB: *Neoseiulus barkeri*; NC: *Neoseiulus cucumeris*; ID: *Iphiseius degenerans*; AL: *Amblyseius limonicus*.

defensive behaviour (below). Sabelis & van Rijn (1997) compiled a list of acarine predators of pest thrips.

Specific problems in WFT control include its feeding within the apical foliage or buds, and the sheltering of pre-pupae and pupae in the soil, often below the plants on which its juveniles had fed. The presence of WFT in hidden plant parts hinders its early recognition and the soil protects the pest from pesticides. The latter may not be an option in floriculture (especially when open flowers are to be treated) owing to phytoxicity. In addition, WFT has developed resistance to several commonly used chemicals. To complicate the issue, many environmental agencies require that some crops (e.g. sweet peppers) be totally devoid of pesticidal residues. Onion thrips is not as fecund as WFT and feeds mostly on the surface of leaves, although it may colonise sheltered sites, e.g. within cabbage heads (Hoy & Glenister, 1991).

Phytoseiids are unable to overcome the large, strong and winged thrips adults, nor can they reach the usually hidden eggs, which leaves them only the first and second stage larvae. The latter try to discourage predation by jerky movements and by ejecting droplets of sticky rectal fluid, activities that are more effective when rendered by the older juveniles. The predators are therefore more successful in attacking the smaller, weaker first stage larvae. In the context of biological thrips control, this suggests that during periods when this stage is scarce, phytoseiids in glasshouses should be supplied with alternate diets (e.g. pollen) to maintain their populations (Bakker & Sabelis, 1989).

Phytoseiids recover more quickly than thrips from low temperatures, and may attack adult prey while these are still immobilised (Grout & Stephen, 1993). Similar preying on second instar or adult thrips could occur during cold periods in the field.

The droplets of rectal fluid that thrips excrete have alarm and defensive functions (Howard *et al.*, 1983; Blum *et al.*, 1992). While droplets deter the phytoseiids that come in actual contact with them, those secreted by WFT are used by its predators as a prey-finding kairomone (Teerling *et al.*, 1993). When offered a choice between leaf discs, either with the kairomone, or only with a solvent-treated wick, *N. cucumeris* spent more time on the former discs. This suggests some specialisation to thrips.

Thrips biocontrol in glasshouses is affected by the crop and by the presence of other natural enemies. For instance, neither *N. cucumeris* nor *I. degenerans* seems to adapt to leek (*Allium porrum*) (Rat-Morris, 1999), although both are common on onion (*Allium cepa*). Other predators, especially anthocorid bugs (e.g. *Orius* spp.), are often concurrently released for thrips control. The co-occurrence of two or more species of predator generates various direct and indirect interspecific interactions that affect the level of biocontrol achieved (Janssen *et al.*, 1998; see Chapter 42). These relationships, as they pertain to individual phytoseiids, will briefly be noted later.

There is considerable commerce in phytoseiids for thrips control. Their mass-rearing and modes of release are based on their reproduction when offered astigmatid mites (e.g. *Acarus*, *Dermatophagoides* and *Tyrophagus*) (see below and Chapter 39). Crops on which phytoseiids are used for thrips control are listed in Table 26.2. In this context *N. cucumeris* is the most 'popular', being available for thrips control by 29 suppliers. It is also offered for spider mite control, but only by 20 companies (Anonymous, 2000).

Neoseiulus cucumeris

MacGill (1939) was apparently the first to note a thrips-killing phytoseiid; the prey was onion thrips and the predator *N. cucumeris* (then named *Typhlodromus thripsi* n.sp.). However, many years were to pass until this mite began to be studied in Dutch glasshouses for thrips control (Ramakers, 1978, 1980). It was mass-produced on flour mite, *Acarus siro* L., in bran (Ramakers & van Lieburg, 1982). Sachets containing bran, flour mite (as sustaining prey) and predator (together comprising a 'slow-release system') are being used to deliver and release *N. cucumeris* in glasshouses (see Chapter 39). This system is currently applied by most UK and Dutch growers of cucumber and pepper (and some flower crops) (Sampson, 1998). The predator's ability to survive prolonged cold storage expedites its use in commerce: about 63% of females kept at 9°C remained alive for 4 months, and oviposited 3 days after being returned to room temperature and offered suitable prey (Gillespie & Ramey, 1988).

The development (egg to adult) of *N. cucumeris* when feeding on thrips larvae took 8–9 days at 25°C (Gillespie & Ramey, 1988; Castagnoli *et al.*, 1990). Adults that were reared on onion thrips consumed more of its larvae (5.4/day) than those of the WFT (3.2/day). In temperate climates *N. cucumeris* undergoes a reproductive diapause

(100% of a population assayed in Europe went into diapause under daily conditions of L10/D14 at 23°C), which greatly detracted from their efficacy in the autumn (van Houten *et al.*, 1995a). However, strains with reduced diapause incidence occur naturally or can be obtained after suitable selection. A New Zealand strain had low diapause rates, which were totally eliminated by selection under long-day conditions. Neither the mites' rates of predation nor their fecundity was affected (van Houten *et al.*, 1995b).

The first instar thrips larvae were too large for young protonymphs of *N. cucumeris*, whose development was delayed in the absence of other mite stages. In the presence of their own adults or of freshly killed WFT, however, the mites' development was quicker than in their absence (3.0 vs 6.3 days at 25°C). The young, 1-day-old nymphs of *N. cucumeris* thus benefit by feeding on prey caught by their own adults (Cloutier & Johnson, 1993). At about one-third of the way through their development the protonymphs become strong enough to kill thrips larvae by themselves. *Neoseiulus cucumeris* usually seems to overcome its prey by force, but may also use another mode. Predators that attacked the large spider mite *Bryobia* sized a leg and hung on until this prey ceased struggling, usually within less than 1 min. *Bryobia* 'gave the impression' that it was paralysed (Burrell & McCormick, 1964); there is no other evidence of toxin use by phytoseiids.

The attack rate of *N. cucumeris* that consumed first instar WFT larvae on cucumber leaf discs was lower than when the interaction took place on sweet pepper leaf discs, and the handling time of each prey was prolonged. Shipp & Whitfield (1991) postulated that owing to the greater pubescence of cucumber leaves (489 trichomes/cucumber leaf disc vs 17/pepper leaf disc) the predator needed more time to locate each prey, leading to a lower predation rate on the former crop. Five to ten times more *N. cucumeris* are required to control WFT on cucumber than on sweet pepper in order to compensate for this plant-generated hindrance of predator efficacy.

The use of *N. cucumeris* to control thrips that infest blossoming ornamentals may be restricted by the refusal of consumers to buy blemished flowers. The predator (released in batches of 100/m² every fortnight) provided good WFT control, and thus satisfactory flower quality, but only during the first two chrysanthemum plantings in a UK greenhouse (Buxton & Finlay, 1993). In contrast, *N. cucumeris* provided satisfactory, season-long control of WFT infesting cyclamen (an ornamental that can tolerate large pest numbers without apparent damage) (De Courcy Williams, 2001).

The predator was assayed against *T. palmi* after this pest invaded Florida. It was released at rates of one *N. cucumeris* for each pest observed on eggplants, totalling 100/plant when thrips numbers were estimated at 99/plant (Castineiras *et al.*, 1997). The predators aggregated where prey densities were high, on fruits as well as on leaves, with numerical increases of *T. palmi* being rapidly followed by those of *N. cucumeris*. Efforts to use the predator in South Africa against the resident citrus thrips (*Scirtothrips aurantii* Faure) failed, because most *N. cucumeris* dropped off the experimental lemon leaves (Grout & Stephen, 1994).

As noted, various natural enemies of thrips, phytoseiids as well as others, may be released on the same plants or else occur there naturally. Interactions of *N. cucumeris* with the often co-occurring *N. barkeri* are discussed below. When *N. cucumeris* was released on potted roses in Finland, along with the soil-dwelling laelapid *Stratiolaelaps miles* (Chapter 21), WFT numbers were significantly reduced only in the presence of both predators. This outcome, which suggests complementary pest control, reflected the different diets of the predators: *N. cucumeris* took young plant-inhabiting WFT, whereas the laelapid fed on thrips pupae in the soil. Indirect competition between the two predators, followed by a decline in *S. miles*, may occur if phytoseiid predation does not leave sufficient prey for the laelapid (Linnamäki *et al.*, 1998). The bug *Orius tristicolor* (White) attacked *N. cucumeris* as well as thrips, but feeding on the latter reduced predation on the mite. The bugs are more mobile and leave the plants after pest density has been reduced, whereas the more sedate *N. cucumeris* remains in place, thus possibly preventing later thrips resurgences. Gillespie & Quiring (1992) concluded that *O. tristicolor* and *N. cucumeris* could simultaneously be used for thrips control in greenhouse crops. Sörensson & Nedstam (1993) reached a similar conclusion about using the mite and another *Orius* on four ornamentals (*Saintpaulia, Impatiens, Gerbera* and *Brachycome*) in Swedish greenhouses. However, Wittmann & Leather (1997) doubted whether thrips control would be improved by having *Orius laevigatus* Fieber along with *N. cucumeris* on the same plants, as the bug fed indiscriminately on predatory mites and pests.

Most biocontrol interest in *N. cucumeris* is due to its ability to affect thrips, although the mite produced significantly more progeny on a TSSM diet than on onion thrips, and was even more fecund when given a mixture of *Quercus* pollen, two mite prey and the thrips (Castagnoli & Simoni, 1991) (Fig. 26.7). Pure diets are rare in nature; the higher fecundity obtained on the mixed diet could thus be a better reflection of the mite's potential fertility (disregarding the fact that under natural conditions the predator will spend more time and energy in foraging). The control of Tarsonemidae by *N. cucumeris* was noted above.

Formerly, there has been some confusion about the correct identity of the entity called *Neoseiulus cucumeris*, which (like other widely distributed phytoseiids) appears to have several strains (van Houten & van Stratum, 1993). The confusion was partially cleared up by Beard (1999, 2001), who provided morphological evidence for separating *N. cucumeris* from the closely related *Neoseiulus bellinus* (Womersley) and other species. However, several taxa may still be masquerading under the name *Neoseiulus cucumeris*, because specimens collected from different parts of the world show some morphological variation, have dissimilar habitats and exhibit different searching patterns.

Neoseiulus barkeri

Formerly known as *Amblyseius mckenziei* Schuster & Pritchard, this species was also noted by Woets (1973) to attack thrips in Dutch glasshouses. Reared on *T. tabaci* at 25°C, it raised a generation in less than 1 week and 63% of the progeny were females. Both genders consumed more than three *T. tabaci*/larvae per day. Other prey included TSSM, broad mites (see above), its own young and the pollen of various

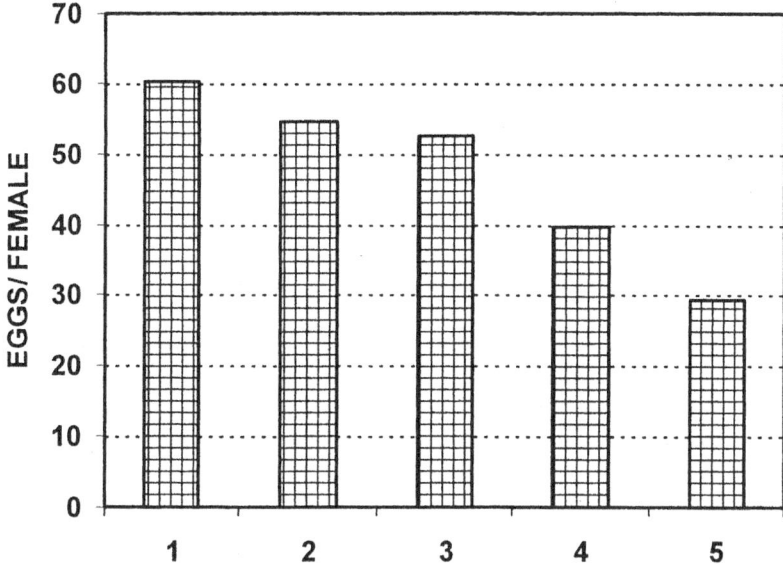

Fig. 26.7 Fecundity of *Neoseiulus cucumeris* reared on five diets (1: a mixture of the mites *Dermatophagoides farinae* and *Tetranychus urticae*, the thrips *Thrips tabaci* and pollen of *Quercus* spp.; 2: *D. farinae* alone; 3: *Quercus* pollen alone; 4: *T. urticae* alone; 5: *Thrips tabaci* alone) at 25°C. (Data from Table 1 in Castagnoli & Simoni, 1991.)

plants (Bonde, 1989). When supplied with TSSM nymphs, *N. barkeri* oviposited about 55 eggs, but only 44 eggs/female when offered the eriophyid *Eriophyes dioscoridis* Soliman & Abou-Awad. The predator could not complete its development on pollen of castor beans (*Ricinus communis*) or date palms (*Phoenix dactylifera*) (Momen, 1995).

Neoseiulus barkeri is not a fully satisfactory control agent of thrips, and is on sale against them only by six companies (Anonymous, 2000). WFT and onion thrips injury to cucumbers in glasshouses was prevented neither by the release of 200 *N. barkeri*/plant early in the season nor by 1000 predators/plant later on. Very large numbers (*c.* 3500 predators/plant) applied early in the season were required to obtain good pest control. Jarosik & Pliva (1995) believed that the weak aggregative response of *N. barkeri* to areas of high prey density, on leaf as well as on plant scale, along with variable developmental rates of the plants and the thrips throughout the cultivation period, decreased the predator's potential. In temperate climates short-day diapause further detracted from this efficacy, although diapause incidence could be reduced by suitable selection procedures without affecting performance (van Houten *et al.*, 1995b).

On the positive side, good onion thrips control with *N. barkeri* was obtained in cucumber glasshouses. Lindqvist & Tiittanen (1989) advocated the weekly release of 15–20 predators/m^2 when 10% of the plants were damaged, with the rate to be increased to 25–30 predators/m^2 when the pest infested 30% of the plants. WFT control was achieved on pelargonium when 15 mites/plant were released for 5 weeks. Glockemann (1992) attributed this success to the co-occurrence of pest and predator on the plant's lower leaves. Turning to another prey, *N. barkeri* was not deterred by the escape activities of *T. palmi*, captured and fed on it, consuming two pest larvae/day (Kajita, 1986).

In northern Europe *N. barkeri* occurs spontaneously with *N. cucumeris* on the same plants. They interact in various ways when feeding on the same prey and the outcome depends on the crop. *Neoseiulus cucumeris* attained higher numbers than *N. barkeri* on sweet pepper, even when the latter was released first or more often. Initially, this outcome was attributed to the reluctance of *N. barkeri* to feed on sweet pepper pollen, a suitable diet for *N. cucumeris* (Ramakers, 1988), but both phytoseiids feed and reproduce on pollen to the same extent. Van Rijn & van Houten (1991) postulated that the greater susceptibility of *N. barkeri* to low humidities and its different behavioural dispersal patterns might affect the outcome. When co-occurring on parthenocarpic cucumbers (which do not produce pollen), *N. barkeri* displaced *N. cucumeris*, either because the latter was an intrinsically inferior competitor on that crop, or owing to the greater mobility of *N. barkeri* (Brødsgaard & Stengaard Hansen, 1992).

Neoseiulus barkeri controlled TSSM on cucumbers for 3 months in Germany with around 100 mites/plant. Pest levels were reduced to about two mites/five leaves, a percentile of the TSSM numbers that were counted on control plants (Karg *et al.*, 1987).

Petrushov (1992) obtained a PYR-resistant strain of *N. barkeri* after exposing it for about 40 generations to pesticides at concentrations that caused 50% mite mortality. The strain was cross-resistant to other PYRs, but no multiple resistance (e.g. to other pesticide groups) was recorded.

Iphiseius degenerans

This species is widely distributed in the southern Palaearctic and in Africa. In comparison to other phytoseiids that feed on thrips, *I. degenerans* showed only intermediate rates of predation and oviposition, but it moves rapidly on and between plants in glasshouses, has no winter diapause and its eggs are relatively tolerant to desiccation. In addition, *I. degenerans* is a frequent visitor to pepper flowers, improving its access to pollen and thrips (van Houten *et al.*, 1995a). One suggested method of introducing *I. degenerans* into Dutch greenhouses was on potted, flowering castor bean bushes, each with around 2000–3000 predators. The pots were placed among sweet peppers and served as 'banker plants', that is, reservoirs for the colonising predators. Some readily colonised the peppers while others remained on the castor beans, continuing their development by feeding on pollen and plant exudates (Ramakers & Voet, 1996).

The predator was the most promising of four thrips-feeding phytoseiids assayed for control of citrus thrips, *Scirtothrips citri*, infesting potted citrus trees (Grafton-Cardwell *et al.*, 1999a). Releases of *I. degenerans* (ten/tree) onto greenhouse trees resulted in pest elimination within 1 week. The mite was also tried on grafted citrus trees in a commercial citrus nursery, where it was released, twice, by placing two predators/five leaves. Thrips numbers on predator-enriched trees were reduced within 5 weeks, engendering better growth and more leaves than control trees, results comparable to those obtained by spraying.

The predator occurs naturally on many plants in association with spider mites. Its pattern of feeding on this prey, however, is different from that of *P. persimilis* and other phytoseiids that specialise on spider mites. *Iphiseius degenerans* obtained less food than *P. persimilis* from each captured prey, but killed more, the two species sometimes sharing meals (Yao & Chant, 1990). A generation required about 1 week at 23°C; each female consumed more than ten prey/day with abundant spider mites and deposited over 60 eggs (Blommers, 1976). The R_0 of *I. degenerans* feeding on *T. pacificus* at 25°C was 50.86 and the r_m/day was 0.248 (Takafuji & Chant, 1976). These values were considerably lower (R_0 22.5–27.9; r_m 0.179–0.208) when the mite was given only pollen (van Rijn & Tanigoshi, 1999b).

Being indigenous to and widely distributed in Africa, *I. degenerans* was an early candidate for CGM control. It did not aggregate in response to high pest densities and had little spatial overlap with CGM; the latter preferred the upper plant stories whereas *I. degenerans* lived in the middle and the lower 'strata'. Skovgård *et al.* (1993) concluded that *I. degenerans* would be able to control CGM on its own provided pest populations were stressed owing to reductions in cassava leaf quality. This uncertainty about the effect of *I. degenerans* on spider mites may be reflected by its being offered for this purpose only by three companies, whereas for thrips control it is on sale by ten suppliers (Anonymous, 2000).

Some interactions between *I. degenerans* and *P. persimilis* were noted above. When placed together on leaf arenas infested by spider mites, the former displaced the latter. Each fed on the other's eggs and juveniles, but the generalist *I. degenerans* devoured three times more *P. persimilis* and reproduced on that diet, a feat not reciprocated by the latter. However, as their diets do not totally overlap, they can coexist in nature (Yao & Chant, 1989). *Iphiseius degenerans* might be preyed upon by *Orius* spp., but to a lesser extent than *N. cucumeris* (Wittmann & Leather, 1997); the predatory bug and *I. degenerans* could thus simultaneously be used to control WFT, with little mutual interference.

Amblyseius limonicus

The systematic status of this New World species, and its separation from *T. manihoti*, with which it had often been confused, were clarified by Moraes *et al.* (1994a). *Amblyseius limonicus* had the highest feeding rate on thrips larvae (seven/day) and fecundity (3.2 eggs/day) of several subtropical phytoseiids compared for thrips control (van Houten *et al.*, 1995a) (Table 26.2). On cucumbers it was more mobile than *N. cucumeris* and did not enter diapause when placed at low temperatures under short daylight conditions. Its main disadvantage is egg sensitivity to low RH conditions (van Houten *et al.*, 1995b). Nevertheless, *A. limonicus* is an efficient predator of thrips. When it and *N. cucumeris* were separately released in a thrips-infested cucumber glasshouse, by the 38th week only five thrips/leaf remained on the *A. limonicus* plants, compared ith 130/leaf on the *N. cucumeris* cucumbers (van Houten, 1996).

Reared on pollen of *Malephora crocea* (then called *Mesembryanthemum* sp.), or on various spider mites, *A. limonicus* completed a generation in 8–9 days at 27.5°C. Citrus nectar or honeydew of mealybugs (Coccoidea: Pseudococcidae), each offered alone, greatly increased the survival of females and enabled them to deposit a few eggs (McMurtry & Scriven, 1965a). Pollen of almond, avocado and corn, moth larvae, juvenile whiteflies and thrips also encouraged the mite's development and fecundity. Rust mites and moth eggs were unsuitable (Swirski & Dorzia, 1968).

McMurtry & Scriven (1971) assayed the performance of *A. limonicus* (a resident of citrus and avocado trees in California) as a predator of the avocado brown mite, *Oligonychus punicae* (Hirst). Pest numbers on avocado seedlings in a greenhouse were reduced by half, the decrease being similar when the phytoseiid was released at predator/prey ratios of 1/7 to 1/23.

Type IV species

General comments

Type IV phytoseiids comprise only members of the genus *Euseius*, arboreal species that live in the warmer parts of the world. They are generalists whose population trends are often associated with the availability of pollen, a diet on which *Euseius* spp. develop and reproduce best, and with plant (e.g. citrus) growth flashes (Grafton-Cardwell & Ouyang, 1996). Recognition of the importance of the pollen diet has greatly facilitated their mass rearing. Yet Congdon & McMurtry (1988), after comparing the attractiveness of several foods for *Euseius tularensis*, commented that although considered 'generalists' that feed on diverse diets, some type IV phytoseiids could prefer specific prey. *Euseius* spp. bear short mid-dorsal and lateral setae, avoid spider mite webs and serve as enemies of other pests, e.g. thrips, eriophyoids and false spider mites. Nevertheless, some species seem to control spider mites.

Euseius addoensis *and* Euseius citri

Injury by *Scirtothrips aurantii*, the South Africa citrus thrips, is due to their scarring the face of young citrus fruits, causing their downgrading, and to the stunting of small trees. The effect of *E. addoensis* on the level of thrips injury was demonstrated by excluding the predator from citrus trees with an acaricidal treatment (Grout & Richards, 1992a). Rates of scarred, culled fruit increased from 15.5 to 25.5% at one grove, and from 19 to 33% at another. Best pest control (<1% cull) was obtained when at least one predator was present on each inner leaf at petal drop in the spring.

Euseius addoensis plays a significant role in the natural control of the citrus red mite, whose scanty webbing does not hinder the predator. When the ratio between active pest stages and predators during the South African autumn (March–April) was 3/1 or less, effective mite control ensued in that season. However, when the ratio was greater than 3/1, large pest populations developed (Keetch, 1972). TSSM outbreaks on table grapes following the elimination of *E. addoensis* by pesticides (Schwartz, 1990) indirectly indicated that this predator could control webbing spider mites, but relevant experimental data are lacking.

The pollen requirements of *E. addoensis*, and the role of suitable windbreak trees (especially *Casuarina* spp.) planted around groves in maintaining predator populations during periods of prey scarcity, are discussed in Chapter 41. McMurtry (1980) established the systematic status of these South African species. The fecundity of *E. addoensis* was stimulated in the laboratory by several systemic OP insecticides (Grout *et al.*, 1997; see Chapter 43).

Euseius citri is widely distributed in Southern Africa and may control the citrus thrips, although to a lesser extent than *E. addoensis*; the former's ability to feed on adult thrips immobilised by low temperatures was noted above.

Euseius tularensis

This West American species had been masquerading, and was extensively studied, under the name *Amblyseius hibisci* Chant (later placed in *Euseius*), until Congdon & McMurtry (1985) ascertained their separate status. *Euseius hibisci* is the rarer species, living mostly on avocado, whereas *E. tularensis* is widely distributed, being the common phytoseiid on Californian citrus. It overwinters on sucker shoots, raises a generation in about 1 week at 26°C and deposits about 25 eggs/female. It is a natural enemy of citrus thrips and of citrus red mite.

Thrips damage is caused only in early summer, as fruits are developing rapidly and any wounds around the calyx could develop to become irreparable scarring. Tanigoshi *et al.* (1985) treated citrus trees in the early spring with the OP malathion, to which *E. tularensis* was very susceptible, and later placed batches of 2000–2500 predators on each tree. The predator controlled the pest throughout the critical damage period, whereas sprayed trees, which had not received predators, suffered serious thrips damage. The presence of at least 0.25–0.5 *E. tularensis*/leaf during that critical period provided effective control. The augmentation of *E. tularensis* (2000 mites/tree in the spring) reduced fruit scarring only by the insufficient 5% (Grafton-Cardwell & Ouyang, 1995). A novel method tried to augment the predators in citrus orchards was intercropping with various legumes and inoculating them with *E. tularensis* (Grafton-Cardwell *et al.*, 1999b). The cover crops were cut and placed within the young trees; the predators moved onto the trees and enriched their resident mite populations.

Jones & Morse (1995) raised doubts about the actual impact of *E. tularensis* on the citrus thrips after studying mite diets by two separate methods, namely isoelectric focusing electrophoresis and staining of a specific enzyme. Positive results (e.g. signs of thrips remnants) were obtained only in 1.3% of adult *E. tularensis* females that were sampled. Adding to these doubts were observations that the predator was inferior to *I. degenerans* in regard to thrips control, when both were release on nursery citrus (Grafton-Cardwell *et al.*, 1999a).

Although *E. tularensis* is common on citrus trees in California, it does not play an important role in citrus red mite control, showing only a limited functional response to the pest (McMurtry, 1969). A non-inclusion virus disease had far more effect on citrus red mite than *E. tularensis* during 4 years of study (McMurtry *et al.*, 1979). The spider mite was the least preferred diet of *E. tularensis*, which chose *M. crocea* pollen over the citrus thrips, and the latter over *P. citri* in the laboratory (Congdon & McMurtry, 1988).

Populations of *E. tularensis* have long been known to be resistant to parathion (Kennett, 1970). The predator has developed resistance to OPs, formetanate and especially to carbaryl, resistance intensity being affected by previous exposure to pesticides (Grafton-Cardwell & Ouyang, 1993).

Phytoseiids that feed on whiteflies

General comments

Whiteflies are small sucking insects whose bodies are covered by (usually white) wax flakes. The nymphs, which cause much of the damage, are sessile and lack the flakes. Whiteflies are serious pests of subtropical fruit trees as well as other crops (e.g. cotton, vegetables and ornamentals), damaging hosts by feeding, by excreting honeydew that is blackened by sooty mould fungi and by transmitting plant viruses (e.g. geminiviruses). One taxon, the polyphagous *Bemisia tabaci* (Gennadius) (believed by some to consist of a complex of biotypes or even species, e.g. *Bemisia argentifolii* Bellows & Perring, the silverleaf whitefly), has become a major pest of diverse crops in many parts of the world. The pest's economic importance is reflected in its numerous vernacular names, such as the tobacco whitefly (TWF), sweetpotato whitefly and cotton whitefly.

Euseius scutalis

This north African and Middle-Eastern type IV species, formerly known as *Amblyseius delhiensis* Narayanan & Kaur, *Amblyseius gossipi* El Badry or *Amblyseius rubini* Swirski & Amitai, develops when feeding on many animal and plant foods, and even on artificial diets (Swirski *et al.*, 1967; Abou-Awad *et al.*, 1992). Its maximal egg complement was produced when offered only *T. pacificus* or pollen of *M. crocea* (Bounfour & McMurtry, 1987). The predator is adapted to high temperatures, as its intrinsic rate of increase and net reproductive rate were highest at 30°C (0.325 and 20.67, respectively). Some of its eggs hatched even under very dry conditions (19% RH).

Teich (1966) was apparently the first to note and quantify the feeding of phytoseiids on whiteflies and to note a TWF field population that was being reduced by these predators. One of the observed species (*Amblyseius swirskii* Athias-Henriot) is a type III phytoseiid that will not be further discussed. The other was *E. scutalis*.

Meyerdirk & Coudriet (1986) evaluated Jordanian and Moroccan biotypes (or strains) of *E. scutalis* as TWF predators. The former strain was somewhat more suitable, owing to higher fecundity when feeding on TWF eggs and juveniles, and because it survived longer on that diet. An Egyptian strain was assayed against the citrus whitefly, *Acaudaleyrodes citri* (Priesner & Hosny) by Donia *et al.* (1995), who left the whitefly species in the genus *Aleurotrachelus*. These authors concluded that the release of 750 predators/citrus branch would bring about satisfactory pest control; the release of 1000 predators/branch did not improve the results. According to Nomikou *et al.* (2001), *E. scutalis*, as well as *N. californicus*, may be suitable for TWF control in European greenhouses. *Parabemisia myricae* Kuwana, the Japanese bayberry whitefly, was also taken, but only its eggs and first instar larvae were apt prey for *E. scutalis*; the second and third instar larvae were unsuitable (Wysoki & Cohen, 1983).

The relatively long s4 dorsal setae borne by *E. scutalis* may facilitate its movement within webs of spider mites; it was the most effective and prevalent phytoseiid predator of these pests on cotton in Egypt. Releases of 15 predators/plant early in the season reduced subsequent pest populations by 50–75% until October (Osman & Zohdi, 1976), but its numbers were much reduced by most pesticides applied on cotton (Elbadry, 1979, and earlier papers). Releases of 40–70 *E. scutalis* adults/citrus tree in eastern Egypt reduced populations of citrus brown mite, *Eutetranychus orientalis* (Klein), by over 80% in 12 months (Abdel-Samad *et al.*, 1996). Fifty predators/tree caused a decline of almost 90% in the tenuipalpid *Brevipalpus californicus* (Banks) (El-Halawany *et al.*, 1993).

Several other phytoseiids attack TWF (and feed mostly on its eggs and first instar larvae) in the laboratory (reviewed by Meyerdirk & Coudriet, 1986), but none seems to be specific to this, or to other whitefly prey. El Badry (1967) described three phytoseiids that fed on *B. tabaci* in Sudan. One of them, aptly named *Amblyseius aleyrodis* El Badry, was reared on several diets (including spider mites and their eggs, WFT eggs, cotton pollen and plant tissue). Although it produced most eggs (1.8/day) when given WFT nymphs, development on this diet was very slow compared with the effect of other foods (Elbadry, 1968).

Concluding remarks

The wealth of information that is available about the Phytoseiidae has placed serious constraints on this presentation, limiting it to the better known or unique members of the family. Many more phytoseiids, probably including ABAs, remain to be discovered, and much is to be learned even about the well-studied taxa, such as whether they constitute clusters of sibling species (e.g. Beard, 1999). In addition, very little (at times only the name and the collection data) is known about obscure forms that could become major predators of certain pests under unforeseen circumstances. Relevant examples are the species that control CGM and vineyard tenuipalpids. These cases indicate that more phytoseiids should be tested for their potential, especially when considering the very small percentage (about 5%) of species that were assayed. No specialised feeders on solitary spider mites (e.g. *Eutetranychus*, *Bryobia*), eriophyoids or other phytophagous mites have yet been found, but their tentative existence should not be ignored (McMurtry & Croft, 1997). Many phytoseiids readily devour these pests and may even control them, but all seem to be generalists or pollen feeders. The litter-inhabiting species (Muma, 1968) are another group about whose habits, diets and possible effects on soil-borne pests practically nothing is known.

Chapter 27
Pionidae

Diagnosis

Pionids have a weak, smooth integument that may be papillate and bear dorsal as well as ventral plates. The capitulum and coxae are separated, with or without anchoring processes. The chelicerae are separated medially. The palpal tibia carries small setae or protuberances on the inner margin and a sclerotised peg on the distal medial end. The coxae are often separated or fused into two or three groups, and the posterior margins of coxae IV usually bear projections or apodemes. The legs bear swimming hairs and tarsal claws, often with clawlets. Pionid deutonymphs and adults are heteromorphic to the larvae; separate keys are needed for their determination (see Chapter 3). The Pionidae, with around 20 named genera and several hundred species, was revised by I.M. Smith (1976), the Chinese taxa were discussed by Jin (1997) and the Australian species by Harvey (1996).

General biology

Pionids are freshwater mites assigned to the Hydrachnidia, a cohort of aquatic superfamilies whose members have only three active stages. These consist of larvae, deutonymphs and adults; the protonymphs and tritonymphs remain quiescent. The larvae parasitise mostly midges (Chironomidae), insects of nuisance value, and mosquitoes. The nymphs and adults crawl or swim, inhabiting surface or interstitial waters, feeding mostly on cladocerans and chironomid larvae, as well as on copepods, ostracods, and mayfly and mosquito larvae (Mullen, 1975; Smith & Oliver, 1986; Proctor & Pritchard, 1989). Pionids predigest their prey by external enzymes that are pumped into a wound in the victim's body (Riessen, 1982). Sperm transfer is direct. Eggs are deposited within a gelatinous mass, placed on various substrates. More than a single annual generation may be raised by some species. A few pionids bypass the larval stage, the deutonymphs emerging from large eggs (B.P. Smith, 1998).

The juveniles of some Pionidae live in temporary water bodies, withstanding long periods of dryness in an inactive phase, returning to full activity when wetted (I.M. Smith, 1976). Species of *Piona* inhabit all available water bodies; as a result they can occur wherever their prey might live.

Pionidae as natural enemies of mosquitoes

Deutonymphs of a *Piona* sp. daily killed about five first instar larvae of an *Aedes* sp. in the laboratory, and its females preyed on late-instar larvae. The prey struggled to dislodge the mite but succumbed within 1–3 min. Mullen (1975) postulated that such cessation of struggle indicated mite-induced paralysis. Male mites rarely attacked mosquito larvae but fed on previously subdued prey (B.P. Smith, 1983).

Parasitism on mosquitoes seems to be rare. The *Hydrochoreutes* reported by Reisen & Mullen (1978) to attack mosquitoes in Pakistan appears to be a member of the family Limnesiidae (Smith & Oliver, 1986).

Pionidae as natural enemies of midges

The larvae of *Piona alpicola* (Neuman) parasitise midges, whereas the deutonymphs and adults feed on midge larvae and on Cladocera. In the laboratory male development from egg to adult required about 5–9 weeks, and the female only 4–7 weeks (Ellis-Adam & Davids, 1970). Predation of adult *Piona carnea* (Koch) on midge larvae within small glass vials depended on the habits of the prey. Larvae of two relatively inactive target species were rapidly killed, but other insects, although finally taken, delayed the process by various means. One was an active swimmer, which was difficult to catch, another formed silken tubes along the bottom of the vials that afforded it partial protection, whereas the third was harder to overcome owing to its large size (Paterson, 1970). During attack the mite used its anterior legs to seize the larva and then attacked its posterior parts. The prey reacted by violent, defensive movements that did not dislodge the mite. As feeding proceeded the larva became immobile and other mites joined to feed on the same prey. A single *P. carnea* killed an average of 0.86 (0.78–0.94) chironomid larvae/day; mites kept without food survived for several weeks (Paterson, 1970). The quantitative effect of pionids on nuisance midges has not been established.

Pionidae as predators of Crustacea

Studies on the population dynamics of zooplankton Cladocera, especially *Daphnia*, indicate the potential predation rates of *Piona* spp. on smaller prey under natural conditions. Nymphs and adults of *Piona constricta* (Wolcott) cruise rapidly in the water, an attribute that increases their rate of prey encounter. As a cladoceran is secured, its exoskeleton is pierced by the mite's chelicerae, which also secrete external digestive enzymes into the prey's body. Feeding time depends on prey size, ranging from 3–5 min for *Bosmina* (mean body size: 0.4 mm) to 35–40 min for adult *Daphnia* (c. 1.5 mm). The mites were voracious predators, eating as many as ten to 15 prey/day. Prey selection depended on predator age and prey size, on the pioniod's feeding experience and on cladoceran ability to flee approaching mites. In addition, predator preference for a particular prey may change with its abundance (Riessen, 1982; Butler & Burns, 1991). The death rate of *Daphnia laevis* Birge, the dominant zooplankton member in a small Argentine lake, was significantly correlated with the density of *Piona* sp., and the latter's density could be predicted by previous variations in prey density. Matveev *et al.* (1989) estimated that the mite contributed 53% to cladoceran death rate, and that in its absence prey populations could increase ten-fold.

The use of pionids (and other water mites) as acarine biocontrol agents (ABAs) is constrained by several factors. First, there are unresolved systematic uncertainties about many genera (B.P. Smith, 1983), which hinder the use of specific species. Second, water mite communities seem to be affected mainly by abiotic factors (Di Sabatino *et al.*, 2000), suggesting that reciprocal interactions with pest or hosts prey are, at best, of minor importance. A third factor is that the difficulty of raising these mites through their complex life cycles in the laboratory makes their application as ABAs rather tenuous (Proctor & Pritchard, 1989). A fourth problem is the strong interference among these predators. The predation rate of *Piona* sp. decreased 20-fold when four mites, instead of one, were placed in a jar with abundant *Daphnia* (Matveev *et al.*, 1989). Finally, water mites can only be used against mosquitoes (and other Diptera of medical or veterinary importance) that breed in habitats where mites abound, such as swamps or marshes. The many pestiferous Diptera (especially mosquitoes) that develop in ephemeral water bodies, such as tree holes or random containers, are often devoid of mites and would not be affected.

Chapter 28
Podapolipidae

Diagnosis

Female podapolipids usually bear two or three pairs of legs (rarely one or four), but are incapable of mobility (except in rare cases); their body is vermiform, pear-shaped or sac-like, with very few dorsal setae. Males generally possess three pairs of legs (seldom four). The chelicerae are needle-like and the palpi rudimentary. The family contains at present about 150 species assigned to around 20 genera. Earlier taxa were listed by Regenfuss (1968) and a key was provided by Husband (1984), who is studying the family (Husband, 2001).

General biology

Most podapolipids are host specific, external or (rarely) internal parasites of insects (especially beetles), the various stages usually occupying specific sites on their hosts. Like other Heterostigmata, they undergo a shortened life cycle, sexually mature males of most genera hatching from eggs and mating with larval females, which moult to adult females. The posterior part of the body of most females becomes distended (physogastric) as a result of containing the developing progeny. Dispersal of some Podapolipidae usually takes place during host mating; such mites could therefore be considered to be venereal diseases (Hurst *et al.*, 1995). Other dispersal occurs when mites wander within a nest (Husband & Husband, 1996). Reproduction is by arrhenotoky, with a very strongly female-biased sex ratio (SR).

Chrysomelobia labidomerae Eickwort

This mite was initially discovered on the milkweed leaf beetle, *Labidomera clivicollis* (Kirby) (Chrysomelidae) in the USA, and did not appear to affect its host (Baker & Eickwort, 1975). It was later found to be wide spread on the pestiferous Colorado potato beetle (CPB), *Leptinotarsa decemlineata* (Say), in Mexico, a host it was not known to attack in the USA. The Mexican mite was considered to be a separate race of *C. labidomerae*, which led to preliminary evaluations before its possible introduction into the USA. In host specificity tests parasitism was limited to hosts of the tribe Zygogrammini in the phytophagous family Chrysomelidae, and none of several beneficial insects was attacked (Drummond *et al.*, 1985). Beetle sex had no effect on numbers of resident mites, which transferred from host to host at mating. Optimal development and survival of *C. labidomerae* was at 30°C, at which a generation was raised in 14.2 days and females lived for about 20 days, each producing around 30 eggs. Small developmental differences were found between cultures of *C. labidomerae* that had different provenances, confirming the existence of various mite races.

In laboratory trials mite load had an adverse effect on beetle longevity (overall 40% reduction) and dispersal. Short-term flight frequencies of CPBs carrying 20–50 female mites/host were only one-third to one-quarter of those without parasites, and long-term flight of parasitised beetles was curtailed even further (Drummond *et al.*, 1989, 1992). However, parasitised CPBs kept within cages during winter in Rhode Island (eastern USA) lived as long as uninfested beetles, and showed no reduction in fecundity, indicating the need for further studies.

Coccipolipus epilachnae Smiley

The Mexican bean beetle, *Epilachna varivestis* Mulsant (Coccinellidae), is a major pest of soya beans in the USA. Other members of this family (known as ladybeetles) are beneficial predators. *Coccipolipus epilachnae*, which is wide spread in Central America, was introduced into the USA to control adult beetles. The risk of affecting beneficial coccinellids was evaluated by host susceptibility tests, which showed that *C. epilachnae* was restricted to the phytophagous subfamily Epilachninae and did not attack other members of the family (Schroder, 1979). *Coccipolipus epilachnae* completed a generation in 16 days at 20°C and 10 days at 25°C, the lower temperature appearing to be more suitable (Schroder, 1981). Fecundity was 36 eggs/female and the SR was almost 0.99. Dispersal, by larval females, took place during beetle mating. The maturing females then settled at the basal part of the hosts' elytra and began feeding, their body size subsequently increasing about ten-fold.

Early observations suggested that the mite could reduce beetle vigour and reproduction (Smiley, 1974). Twenty pairs of mite-infested beetles, kept in the laboratory, deposited a total of 118 egg masses, compared with 267 masses produced by healthy pairs of beetles (a reduction of 56%), and the longevity of the parasitised pests was reduced by 33% (Schroder, 1982). These encouraging results sufficed to try the mite in the field; about 1000 mite-infested beetles were released in 1979 at several sites in the eastern USA (Schroder, 1981).

No similar reductions in pest populations were obtained by Cantwell *et al.* (1985) in laboratory experiments, despite heavy parasite loads (100–400 mites/host), nor did infested *E. varivestis* consume less food than healthy beetles. These negative results were attributed to different handling procedures and to greater crowding in the laboratory. Whatever the reason(s), these data, along with the high mite numbers seen on beetles in the laboratory (which exceeded levels that would establish in the field), argued against using *C. epilachnae* to control the pest. Two possible explanations for these failures were proposed by Hochmuth *et al.* (1987), who also found that mite presence on beetles affected neither their populations nor the leaf area that they consumed. The first possibility was that the parasitic nature of *C. epilachnae* might have decreased during its prolonged (30 generations) rearing in the laboratory. The other was that the beetle, being a reflexive bleeder, could be tolerant to some haemolymph loss when attacked by the mite.

Podapolipoides alatus Husband

This mite (called *P. grassi* Berlese by Gauchat, 1972) is a widespread ectoparasite of the Australian plague locust, *Chortoicetes terminifera* Walker. The mites infest third instar nymphs and subsequent host stages, usually attaching to the thorax or the anterior abdominal segments, as well as to the wing bases and the genitalia (Gauchat, 1972). The parasite spends its entire life on the locust, dispersing during host mating. Wild locust populations were heavily infested (several insects carried more than 100 mites in their genitalia), but did not seem to be affected. In the laboratory mite infestations caused a decrease in locust vigour, manifested as lessened activity.

In conclusion, the specificity of the Podapolipidae could make them suitable acarine biocontrol agents (ABAs) of various pests, but much more information is required about their detrimental effects on their hosts. A separate problem is the potential adverse consequence of their parasitism on beneficial insects, such as predatory coccinellids. Females of *Adalia bipunctata* Mulsant attacked by *Coccipolipus hippodamiae* (McDaniel & Morrill) deposited significantly fewer eggs than uninfected beetles and most failed to hatch (Hurst *et al.*, 1995).

Chapter 29
Pterygosomidae

Diagnosis

These are red, small- to medium-sized mites with external, emergent peritremes, chelate chelicerae and a thumb–claw complex. The bodies of some pterygosomids (e.g. *Pterygosoma*) are twice as wide as they are long. The family Pterygosomidae (at times spelled Pterygosomatidae) consists of over 100 species in around ten genera, to which Davidson (1958) presented an early key.

General biology

Pterygosomids have only three active stages, all parasitic: larva, deutonymph and adult; the protonymph and tritonymph remain quiescent. They attack lizards, scorpions or various insects, such as beetles, cockroaches and blood-sucking bugs. Reproduction is by arrhenotoky.

Pimeliaphilus **spp.**

Several *Pimeliaphilus* spp. attack 'kissing bugs', members of the family Triatomidae (Hemiptera). This family (or subfamily of the Reduviidae) includes vectors of Chagas' disease (which can be fatal to humans), as well as many nuisance insects. Heavy infestations of *Pimeliaphilus plumifer* Newell & Ryckman on various species of the blood-sucking genus *Triatoma* hindered their moulting and killed young bugs in the laboratory (Newell & Ryckman, 1966). Parasitisation rates on bugs collected in the eastern USA reached 30–40% (especially on the fourth and fifth instar insects) in late summer, but this was usually not fatal to the hosts. Nevertheless, as lizard-parasitising pterygosomids transmit diseases between their hosts, these mites could vector pathogens that might affect pest bugs (Anderson, 1968a).

   In the laboratory, at 30°C and 15–20% relative humidity (RH), *P. plumifer* required at least 60 days to complete a generation. The females, which remained on their hosts throughout their lives, deposited an average of 56 eggs during 21 days, and lived for about 54 days (males for 75 days) (Anderson, 1968b). *Pimeliaphilus zeledoni* Newell & Ryckman parasitised 53% of field-collected *Triatoma dimidiata* (Latreille), the main vector of Chagas' disease in Costa Rica. Large (fifth instar) nymphs were the most heavily attacked, mostly by the mites' larvae. The lack of any indication that the insects were harmed led Calderon-Arguedas (1998) to doubt whether *P. zeledoni* would control the bugs.

Other species of *Pimeliaphilus* attack cockroaches in laboratory cultures; they can destroy a colony within 3–4 months (Field *et al.*, 1966). When present in large numbers (25 or more/insect), *Pimeliaphilus cunliffei* Jack kills its host within a few hours (Cunliffe, 1952, who called the parasite *Pimeliaphilus podapolipophagus* Tragardh). The mite deposited its sticky eggs in batches, placing them in various sites; at 32–35°C they hatched within 6–11 days. The emerging larvae, and later the nymphs, sought hosts immediately after moulting and began to feed; the total life cycle took about 4 weeks. The orange–red adults deposited two to three egg batches (each with *c.* 12 eggs) and lived for 2–3 weeks. The cockroaches tried to remove the parasites and often fed on them. The American cockroach [*Periplaneta americana* (L.)], the German cockroach [*Blatella germanica* (L.)] and the oriental cockroach (*Blatta orientalis* L.) were all attacked. Their subsequent death could have been due to a toxin, rather than to loss of body fluids, because dead cockroaches were not exsanguinated (Field *et al.*, 1966). Reduced hatch of eggs that were laid by heavily attacked cockroaches further decreased their numbers.

Pimeliaphilus spp. do not appear to be very specific, an attribute that would allow them to survive on any suitable host within their range (Anderson, 1968a). One species even transferred from a scorpion to a triatomid bug (Berkenkamp & Landers, 1983). This trait could enhance the mites' use for the biological control (directly or by transmitting diseases) of blood-sucking insects.

Newell & Ryckman (1966) cited Undiano & Shictong, who advocated the use of pterygosomids for the biological control of disease-transmitting 'kissing bugs', an option that still remains to be explored.

Chapter 30
Pyemotidae

Diagnosis

The Pyemotidae can be recognised by their milky white, spindle-shaped or rounded, segmented bodies, with stylettiform chelicerae. Legs I end in a claw but usually lack empodia. Only females bear a pair of pseudostigmatic organs (sensilli); legs II–IV are similar in shape, carrying claws and empodia. Trochanter IV is characteristically subtriangular. Legs IV of the much smaller males terminate with a strong claw. The family, with approximately 150 species in 25 genera, was revised by Cross (1965) and further defined by Mahunka (1970).

General biology

Like other Heterostigmata, the pyemotids have a reduced life cycle, with adults being the only free stage. Mated females seek, parasitise and kill insects. During feeding the mite's juveniles develop and mature within the mother's body, whose posterior part becomes greatly swollen (physogastric), attaining a size several times that of a young female. Reproduction is by arrhenotoky, most progeny (>90%) being females. Each may produce 200–300 progeny, but when several mites parasitise the same host, their bodies expand to a lesser degree and they have fewer progeny. Males are born first, remain on the mother's body and mate with several dozen of their female siblings as they emerge. Host seeking is released by copulation, unmated females staying on the mother's body (Moser *et al.*, 1971). Because a single host individual provides enough nutrients for all of a female's progeny, and because only a single mite stage feeds on the host, which then dies, the Pyemotidae are categorised as parasitoids (Lindquist, 1983). A member of the genus *Pyemotes* (then called *Pediculoides*) was the first acarine biocontrol agent (ABA) to be mass-reared in the laboratory.

Species of *Pyemotes* may attack humans (and domestic animals) with which they come in contact. The resultant bites can induce dermatitis (often called 'grain itch' or 'hay itch'), a factor that seriously hinders the use of these mites for biocontrol. However, not all species affect humans in the same way. Moser (1975) monitored the effect of various *Pyemotes* spp. on insects, cross-mated mites of different origins, and used them to conduct dermatological tests on humans. He found that the genus contains two groups of species. Members of the *scolyti* assemblage attack only bark beetles (Scolytidae) and lack venom. As a group they could be among the most successful natural enemies of bark beetles, usually occurring in small numbers, but at times they eradicate entire bark beetle populations (Moser *et al.*, 1978). Their females occur in two morphs, the usual homeomorph (similar to the male) and the heteromorphic phoretomorph. The latter, an adaptation for phoresy, has a more compact body, darker cuticle and stronger forelegs with enlarged claws (Moser & Cross, 1975). Mites in the *ventricosus* group, which do not express phoretomorphy (Moser *et al.*, 1987), are polyphagous and possess a toxin; these are the mites that cause dermatitis. The venom is used to overcome larger hosts (e.g. insect larvae), which are paralysed but not killed, retaining a relatively high metabolic rate (Weiser & Sláma, 1964).

Pyemotes tritici (LaGreze-Fossot & Montané)

This species, the straw itch mite [formerly known as *Pediculoides ventricosus* (Newport)] is the most prominent in the *ventricosus* group. It parasitises about 150 insect species in many orders (Coleoptera; Diptera; Hemiptera; Hymenoptera; Lepidoptera, Neuroptera and Strepsiptera) (Cross *et al.*, 1975; Marei, 1992), is the best studied of several pyemotids considered as ABAs, and was mass-produced on a commercial basis.

In the laboratory the mite was reared on pupae of the cigarette beetle, *Lasioderma serricorne* (Fabricius). A generation required about 4 days and the mite had around 250 female progeny at $26 \pm 1°C$ and 85% relative humidity (RH). The intrinsic rate of increase was 0.63, with a population-doubling time of 1.1 days, thereby exceeding the growth rate of any potential arthropod pest (Wrensch & Bruce, 1991). In addition, the mite had a very female-biased sex ratio (SR): only 2–9% males were normally produced. Gravid females, kept at 15°C for 30 days and then transferred to 22–25°C, gave birth to a full complement of normal offspring, suggesting that during mass-rearing, overflow mites could be stored for 1 month until used.

The occurrence of *P. tritici* on lepidopterous and coleopterous stored product pests suggested that it could be a suitable candidate for controlling these insects. It was assayed against the Mediterranean flour moth, *Ephestia kuehniella* (Zeller), which produces copious amounts of contaminating webbing. About 70–100 mites were added to 400 moth eggs placed on whole wheat flour in plastic jars. The physogastric females, attached to shrivelling moth larvae, were introduced 10 days after the eggs, ensuring that emerging mite females would immediately find their target larvae. Mite impact was assayed 80 days later by tallying the emerging moths, weighing the webbing and determining flour loss. Only a single moth emerged from each mite-containing jar that had very little webbing, compared with a mean of 318 moths from the controls, along with much webbed material. Weight losses were also significantly smaller and gains in undamaged flour larger (Fig. 30.1). Hoschele & Tanigoshi (1993) concluded that about 80 or more mites could completely destroy a pest population that might develop from 400 moth eggs. The introduction of fewer mites provided inconsistent control. Moth diet affected the results. Insects on whole wheat flour developed more rapidly and escaped the mite, leading to pest persistence and postponing the safe handling of the flour, a handicap from the commercial point of view. In contrast, the slow larval development on white flour enabled *P. tritici* to become numerous and kill all *A. kuehniella*. With no moth larvae left, the mites died out.

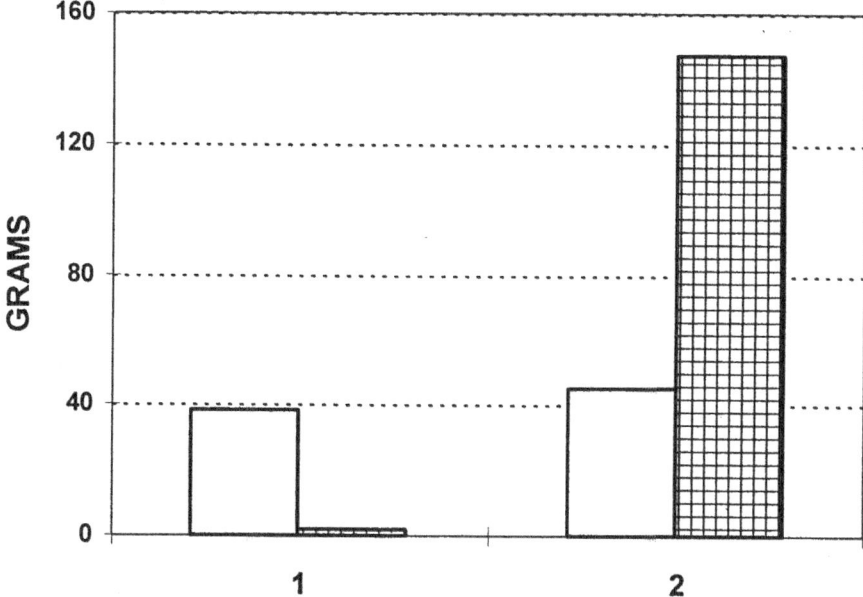

Fig. 30.1 (1) Loss of flour weight (from 150 g) caused by the Mediterranean flour moth, *Anagasta kuehniella* (white bars), and in the presence of *c.* 80 *Pyemotes tritici* (cross-hatched bars). (2) Mean weight of flour that remained undamaged in the presence of *Anagasta kuehniella* (white rectangles) and with mites added (cross-hatched bars). (Data from Table 2 in Hoschele & Tanigoshi, 1993.)

   Nearly half the overwintering larvae of the pink bollworm, *Pectinophora gossypiella* (Saunders), a pest of cotton, were attacked by *P. tritici* in Haryana (India) (Naresh & Balan, 1985, who referred to the mite as *P. ventricosus*). Parasitised hosts died within 4–5 days. Beetle species attacked by *P. tritici* included the groundnut beetle, *Caryedon serratus* (Oliver), a major pest of peanuts in storage. The mite caused 10–47% beetle mortality in the Congo, and was considered (along with larval beetle competition) to regulate the pest's populations (Matokot *et al.*, 1987). The increasing tendency in the UK to replace insecticides that are used to control wood-burrowing beetles with products of lower toxicity is promoting other strategies. Berry (1995) advocated using *P. tritici* as a biological control option. The mite killed larvae of the Colorado potato beetle, *Leptinotarsa decemlineata* (Say), a major pest of vegetables, in the laboratory, but failed to reproduce on that host. *Pyemotes tritici* did not affect pest populations when released (100,000/25 m²) in the field. Nevertheless, Drummond & Casagrande (1989) believed that the mite could be used in inundative releases.
   In Pakistan about 2% of the nests of hornets (*Vespa* spp.), predators of honeybees, were naturally infested by *P. tritici*. Mites were mass-reared on moth larvae and used to infest captured hornets in the laboratory. About 150–180 infested hornets were released at each of three sites at Islamabad, Rawalpindi and Haripur. One month

later 45–83% of the hornet broods were attacked at Islamabad, 81–100% at Rawalpindi 2 months later, and after 3 months all brood was dead at Haripur. No hornets were seen by that time in any of the nests (Ahmad *et al.*, 1985).

Another target for *P. tritici* is ants, especially the imported red fire ant, *Solenopsis invicta* Buren, an agricultural pest in the southern USA. Ant behaviour was used to introduce the mite into the nests. When the pest's tunnels are exposed by scraping the top of the nest, the ants emerge and carry any potential foods into the interior. Parasitised cigarette beetles placed on the nests were immediately taken inside. An important advantage of this mode of release was that it precluded any contact between mites and humans or domestic animals (Bruce & LeCato, 1980). Only about 70% of the nests were subsequently 'inactivated', a failure attributed to the habits of the ants. They forage near the surface on warm spring and autumn days, and thus nearer to the inoculum, but during summer the ants in nests located on exposed ground remain in deeper strata, away from the infested beetles. About 1.4 million mites were introduced into nests in a field trial in central Texas without affecting ant numbers (Thorvilson *et al.*, 1987). Earlier failures in the use of *Pyemotes* to control a cotton pest in Texas during 1902–1903 were documented by Hunter & Hinds (1904).

A *Pyemotes*-containing commercial product was placed on the market by Biogenesis Inc. (Mathis, Texas, USA) (Anonymous, 1985). It consisted of about 700,000 mites, attached to the flour moth *Sitotroga* sp., located in wheat grains in cups. The contents were to be placed onto disturbed nests, where they were presumed to kill all ants within 5 days. The mite is currently available for ant control by three suppliers (Anonymous, 2000).

Pyemotes spp.

Pyemotes barbara Moser, Smiley & Otvos attacked pupae of the Douglas fir cone moth, *Barbara colfaxiana* (Kearfott), a cone and seed pest in British Columbia, Canada. The mite parasitises larvae of several other forest pests in the laboratory and has very potent venom, which paralysed hosts within 15 min. *Pyemotes barbara* was considered a promising candidate for the control of cone and seed pests (Moser *et al.*, 1987). *Pyemotes boylei* Krczal attacks termites, which are household pests (Isoptera) (Vaivanijkul & Haramoto, 1969).

Dendroctonus frontalis Zimmerman, the southern pine beetle (SPB), is a major forestry pest in the southern USA. Its status was postulated to be partially due to immunity from local species of *Pyemotes* (Moser *et al.*, 1978). *Pyemotes dryas* (Vitzthum), a member of the *scolyti* group, was introduced into the USA from Poland and reared in the laboratory. Although it readily attacked SPB brood, *P. dryas* was phoretic neither on this pest nor on six other associated beetles. Lack of phoretic hosts precludes its release in the USA until a suitable vector is found.

In Egypt *Pyemotes herfsi* (Oudemans) killed up to 85% of the pink bollworm's first generation (Tawfik & Awadallah, 1970). In Italy this mite attacked about 30% of a population of *Melanaspis inopinata* (Leonardi) (Hemiptera: Diaspididae) infesting *Cotoneaster* sp. and considered an efficient natural enemy (De Lillo & Porcelli, 1993).

In the laboratory the mite located the eggs of the confused flour beetle, *Tribolium confusum* Jacquelin du Val, at a distance of 5 cm in 2 days, and at a depth of 7 cm within 3 days. Rizk *et al.* (1979) calculated that the progeny of a single mite female could attack all larvae produced by a single beetle within 10 days.

Although they usually parasitise insect larvae or pupae, some *Pyemotes* also attack eggs. An unidentified '*Pediculoides*' fed on eggs of lepidopterous cotton pests in Uganda (Nyiira, 1970), and *P. boylei* parasitised eggs of fruit flies (Tephritidae) in the field (Vaivanijkul & Haramoto, 1969). Feeding on that diet enabled the mites to enlarge their bodies, but not to produce progeny. Such females were too swollen to search for other hosts and died on the eggs that they had parasitised.

Pyemotes spp. attack natural enemies of pests in the field (Traboulsi, 1968; Blahutiak & Alayo Soto, 1982), disturb their activities (Wilson, 1960), and invade pest and natural enemy rearing facilities, causing serious disruptions (Titayavan & Davis, 1988; Marei, 1992; see Chapter 43). Hanks *et al.* (1992) eradicated *P. tritici* with sulfur dustings from logs that contained beetles parasitised by these mites.

The *Pyemotes* toxin and concluding remarks

The gene coding for the paralysing toxin of *P. tritici* (a small protein known as TxP-1) was isolated, cloned and sequenced (Tomalski & Miller, 1991). The gene was inserted into recombinant insect viruses to improve their efficacy as biopesticides, enabling the viruses to paralyse hosts during the infection process (Tomalski & Miller, 1992). Inserted into such a virus and fed to moth larvae within a diet plug, the toxin caused host death within 70–80 h, among the fastest rates recorded. However, the yield of virus progeny was much reduced, suggesting that in the field they would be displaced by wild-type viruses (Burden *et al.*, 2000). The effect of this biopesticide on other natural enemies is not known.

Despite their hazard to humans and domestic animals, *Pyemotes* spp. possess important attributes as ABAs (Bruce, 1983). They are easy to rear and produce many progeny (>100/female) in less than 1 week within a wide range of humidity and temperature conditions (Bruce, 1984). Infesting females, which make up over 90% of the populations, can be stored at low temperatures for later applications. Upon emergence the females immediately seek hosts, which, upon being paralysed, cease to feed. To this should be added the option of rearing *Pyemotes* on synthetic media (Bruce, 1989; see Chapter 39).

Chapter 31
Scheloribatidae

Diagnosis

The integument of the Scheloribatidae is smooth and bears small pteromorphs that may occur as humeral projections. The inner margins of the lamellae are not joined and the cuspis is usually absent. The notogaster has two to four pairs of porose areas and the genital plates carry two to four pairs of setae. The arrangement of cupathia on the palpal tarsus is fan shaped. The tarsi are monodactyl or tridactyl. About 350 known species in about 40 genera are known; the European taxa were reviewed by Weigmann & Miko (1998 and earlier papers) and Pérez-Iñigo (1993). The New Zealand taxa were summarised by Luxton (1985).

General biology

These mites readily ingest various soft plant debris, diverse fungi, algae, lichens and mosses, and even damage live plants (Lan *et al.*, 1986; Stamou & Asikidis, 1992). Their juveniles often burrow in soft tissues, and females deposit their eggs in batches. Reproduction is usually sexual, the females being inseminated via spermatophores (Alberti *et al.*, 1991). The life cycle of scheloribatids may require several months, even in the laboratory (e.g. *Scheloribates laevigatus* Koch raised a generation in 64 days at 25°C; Woodring & Cook, 1962). Many Scheloribatidae are intermediate hosts of tapeworms when they ingest worm eggs. These hatch and develop inside the mites' bodies, the tapeworms arriving in their vertebrate host's body as it swallows the host mite along with other food. As tapeworm eggs are too large to be consumed by chance, Denegri (1993) suggested that the mites could interpret them as potential or even preferred food.

Feeding on pathogenic fungi and gall mites

Scheloribates azumaenis Enami, Nakamura & Katsumata occurs in cultivated soils in northern Japan. In the laboratory it fed on various soil fungi, including *Rhizoctonia solani*, the causal agent of radish root rot. The mite was assayed as a feeder of the fungus. Radish seeds were planted in sterilised soil in jars and then infected by the fungus, with and without added 300 mites/jar, and the germination rate compared with controls. Radish seed germination was reduced in the fungus jars but not when *S. azumaenis* was added (Fig. 31.1). Notwithstanding these promising results, Enami

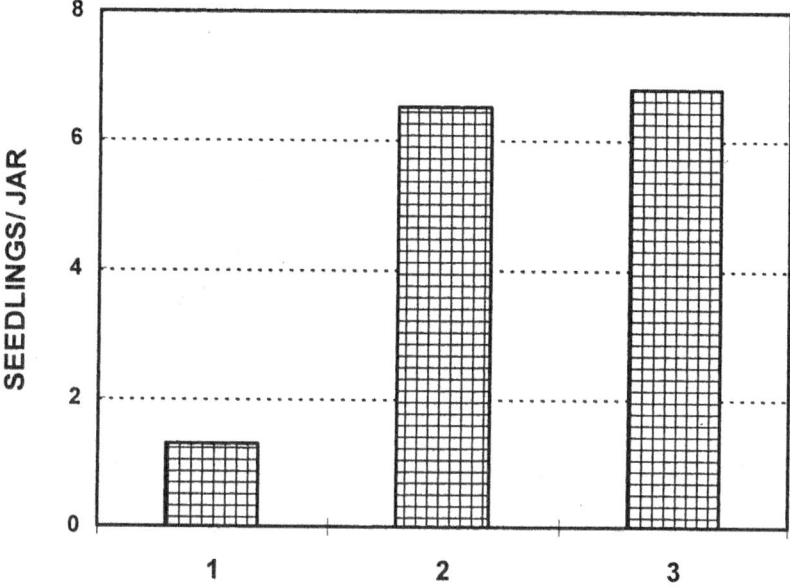

Fig. 31.1 Number of germinated radish seedlings in soil jars: (1) infected with the soil fungus *Rhizoctonia solani*; (2) with the fungus plus *Scheloribates azumaenis*; and (3) control jars. (Based on data in Table 1 in Enami & Nakamura, 1996.)

& Nakamura (1996) noted that in natural soils the mite occurs in smaller numbers, and that its real significance as a biocontrol factor has still to be determined.

Dometorina praedatoria Lan, Xin & Aoki fed on (unnamed) gall mites affecting citrus in Jiangxi province, China (Lan *et al.*, 1986). This observation, along with the apparent preadaptation of Scheloribatidae to animal food (e.g. tapeworm eggs), suggests that additional members of this family could be predators.

Chapter 32
Stigmaeidae

Diagnosis

Stigmaeids are red to yellow, ovoid or elongate mites whose dorsum is often covered by one or more plates. They have no obvious stigmata or peritremes. The chelicerae are short and stylet-like, and a palpal thumb–claw complex is present, its tarsus often bearing an apical three-tined seta. More than 300 species in 25 genera have been named. A key to the genera was presented by Tseng (1982), who also listed the Taiwanese species. Wood (1967) detailed the New Zealand taxa, the Russian species were described by Kuznetzov (1984 and earlier papers), and Ueckermann & Smith Meyer (1987) dealt with the Afrotropical fauna.

General biology

Stigmaeids live on plants and in the soil, often feeding on the eggs and sessile forms of tetranychid, tenuipalpid and other mites that infest commercial crops in many parts of the world. A few species prey on scale insects or parasitise flies (Muma & Selhime, 1971; Swift, 1987; Farag *et al.*, 1990). Species of *Eustigmaeus* (=*Ledermuelleria*) feed on mosses (Gerson, 1972) and the adults of some occur on sand flies (Diptera: Psychodidae), which they wound (Shehata & Baker, 1996). The diets of most species are not known. Many taxa raise a generation in about 1 month and reproduction is arrhenotokous. Stigmaeids (especially in the genera *Agistemus* and *Zetzellia*) are probably the second most important group of plant mite predators (after the Phytoseiidae). [Wood (1967) placed *Agistemus* and *Zetzellia* in synonymy, a suggestion neither universally recognised (Ueckermann & Smith Meyer, 1987) nor accepted here.] Thistlewood *et al.* (1996) believed that stigmaeids (apparently referring to members of those two genera) might be more important in warmer regions, especially as they possess adaptations suitable for preying on rust and gall mites (Eriophyoidea). These include similarities in spatial occurrence, reproductive biologies and population cycles.

Little is known about stigmaeid dispersal (except for *Eustigmaeus*, as noted, and fly-associated species of *Stigmaeus*; Swift, 1987). Delattre (1974) suggested that *Zetzellia mali* (Ewing) might be carried by winds, but the evidence is tenuous (only four out of 12 heavily sprayed apple trees were re-colonised 1 year later by this mite from adjacent unsprayed trees).

The resurgence of the Stigmaeidae

Although often recognised as important predators, the role of stigmaeids in pest control has not always been obvious (e.g. Muma & Selhime, 1971; Amano & Chant, 1990). However, recent reductions in pesticide use in orchards are beginning to reveal their contributions to pest control. Such reductions modify the composition of plant mite faunas, allowing indigenous predators to re-emerge and even to thrive in their former habitats (e.g. Amano & Chant, 1990); the Stigmaeidae are one of those groups (Komlovszky & Jenser, 1992). Comparisons between predatory mites found on sprayed and on unsprayed crops often reveal different faunas. Berkett & Forsythe (1980) sampled mites in sprayed and unsprayed apple orchards in Maine, USA, reporting that five phytoseiids (one of which was dominant) occurred in sprayed orchards, but only very few Stigmaeidae. Twice as many phytoseiids were found on the unsprayed trees, and one (*Agistemus fleschneri* Summers) was the overall dominant predator.

Increases in stigmaeid populations following decreases in pesticide usage are due to several concurrent processes. One is decreased pesticide-caused mortality. The frequency of *A. fleschneri* on sprayed apple trees in Missouri was reduced by far more than that of the phytoseiids (Childers & Enns, 1975b) (Table 32.1). Similarly, phytoseiids in an apple orchard sprayed with a carbamate (CRB) were reduced by 34%, compared with a 93% decrease in stigmaeids (Thistlewood, 1991).

A related process is the decline in the populations of pesticide-resistant phytoseiids, which, favoured by the toxicants, had out-competed the stigmaeids. A third reason is increases in the abundance and composition of the so-called indifferent species (Karg, 1992), which serve as prey when pests are scarce. More Tydeidae and Tarsonemidae, supplementary diets for generalist predators, were collected in abandoned orchards than in commercial plots in Ontario, Canada (Thistlewood, 1991). The greater variety of prey enables more predators to survive in agricultural systems.

A contrary process seems to be taking place in eastern Canada, where *Z. mali* became the dominant predator of apple mite pests. This was probably due to stigmaeid resistance to pyrethroids, which had reduced the Phytoseiidae (Villanueva & Harmsen, 1998).

Table 32.1   Total numbers of Phytoseiidae and of *Agistemus fleschneri* collected in unsprayed and sprayed apple orchards in Missouri, USA

	Unsprayed	Sprayed	Difference (%)
Phytoseiidae	3910	3390	13.3
Agistemus fleschneri	971	439	54.8

Based on Table 4 in Childers & Enns (1975b).

Agistemus exsertus Gonzalez-Rodriguez

Agistemus exsertus is regarded as the year-round, key acarine predator of mite and some scale insect pests on several crops in Egypt (Rasmy *et al.*, 1996). This mite (and other stigmaeids) can be reared on the undersides of excised leaves kept turgid in Petri dishes, where it completed a generation in 3–4 weeks. Eggs are the preferred food, and when offered only this diet (of two prey species), fecundity was twice that when given only spider mite juveniles (80 vs 33–40 eggs/female, 2.9 eggs/day). Longevity was prolonged when fed eggs (Elbadry *et al.*, 1969a, b) (Fig. 32.1).

The specific prey eaten affected the duration of the predator's life cycle as well as its appetite. Development from larva to adult on eggs of the Oriental spider mite, *Eutetranychus orientalis* (Klein), required 4.4 days compared with 6.3 days on eggs of *Tetraychus cinnabarinus*, the carmine spider mite. This could partially explain (aside from specific nutritional values) why *A. exsertus* consumed more of the latter prey than of *E. orientalis* (Table 32.2). Males are often ignored in calculating the potential 'kill' of acarine biocontrol agents (ABAs); those of *A. exsertus*, short-lived and small, killed many prey, their consumption rate being around 30–50% that of the females. They located eggs faster than females and invested a shorter handling time in each individual prey (Yue & Tsai, 1995).

Yue & Childers (1994) and Yue & Tsai (1995) reared *A. exsertus* on eggs of the citrus red mite, *Panonychus citri*, and obtained somewhat different results. Highest fecundity was obtained at 20°C (69 eggs/female) and the best intrinsic rate of increase (r_m) at 25°C (Table 32.3). Higher temperatures reduced predator fecundity (falling to 25.5 eggs/female at 30°C) while increasing that of the prey (Fig. 32.2). *Agistemus exsertus* may thus control this prey only below 30°C.

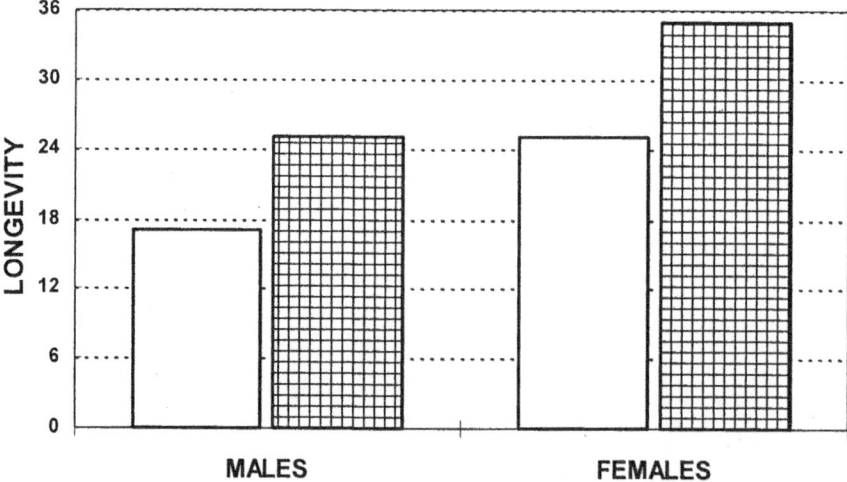

Fig. 32.1 Longevity (in days) of males and females of *Agistemus exsertus* when offered juveniles (white bars) and eggs (cross-hatched bars) of spider mites. (Based on Tables 1 and 2 in Elbadry *et al.*, 1969a.)

Table 32.2   Feeding of *Agistemus exsertus* on eggs and juveniles of two species of spider mites

Prey species	*Tetranychus cinnabarinus*		*Eutetranychus orientalis*	
Prey stage	Eggs	Juveniles	Eggs	Juveniles
Sex of predator				
Female	566.6±27.4	167.3±2.4	400.9±10.1	79.9±2.2
Male	217.8± 8.9	52.8±3.0	132.4± 3.2	34.0±1.0

Based on Tables 1 and 2 in Elbadry *et al.* (1969b).

Table 32.3   Life table parameters of four Stigmaeidae feeding on different prey mites

Predator	*Zetzellia mali*	*Agistemus terminalis*		*Agistemus exsertus*	*Agistemus orbicularis**
Prey	Apple rust mite	Citrus red mite		Citrus red mite	Scarlet mite
Temperature (°C)		19	25	25	17-23
Net reproductive rate (R_0)		10.04	13.5	24.0	21.90
Intrinsic rate of increase (r_m)		0.109	0.149	0.229	0.061
Generation time (days)		21	17.4	14.5	50.6
Source	White & Laing (1977a)	Inoue & Tanaka (1983)		Yue & Childers (1994)	Oomen (1982)

*Oomen (1982) originally identified this species as *Agistemus denotatus* Gonzalez. It is now known as *Agistemus orbicularis* Ehara and Oomen-Kalsbeek (1983).

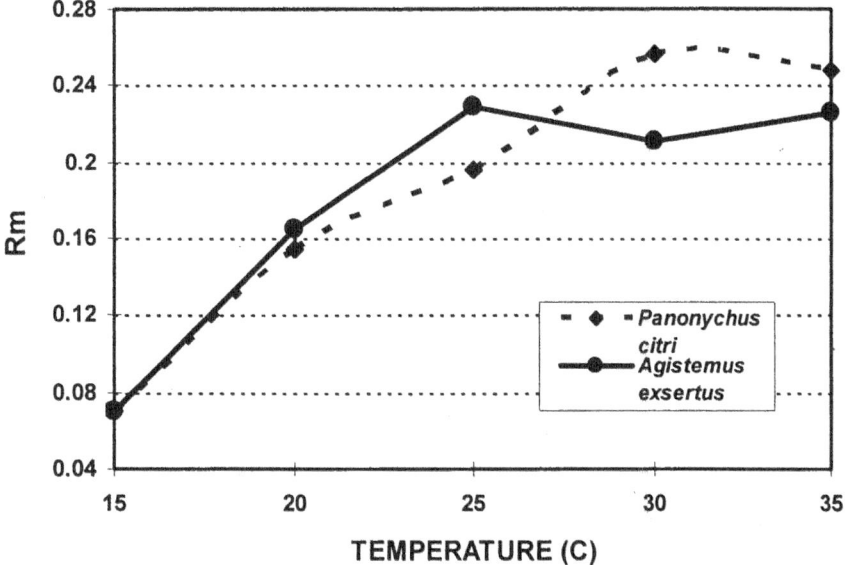

Fig. 32.2   Intrinsic rate of natural increase (r_m) of *Panonychus citri* and of *Agistemus exsertus* at five constant temperatures. (Based on Table 5 in Yue & Tsai, 1995.)

Fecundity was affected by the number of copula as well as by female age at that time. Females that had multiple matings produced more eggs than those that had mated only once (44.3 vs 34.7). A female that had copulated at the age of 5 days deposited 32.4 eggs, compared with 14.6 when first mated at 15 days. The sex ratio (SR) was also affected by female age at first mating. The progeny of early-mated mites had a SR of 0.57, that of late-copulating females being only 0.43 (Abou-Awad & Reda, 1992).

An overabundance of prey induced *A. exsertus* to kill more than it consumed, the predator puncturing most eggs without sucking out their contents (Elbadry *et al.*, 1969b). This depended on prey age, as older eggs were only partially consumed by predators that had totally devoured fresh ones (Salman & Manna, 1991). An average of 60 eggs of the tomato russet mite, *Aculops lycopersici* (Massee), or 45 of its mobile stages, were consumed daily by *A. exsertus* females at 30°C (Osman & Zaki, 1986). The predator also devoured 300 eggs or 280 juveniles and adults of the grape blister mite, *Colomerus vitis*. Eriophyoid eggs were the most efficient prey because they promoted the production of most progeny (Osman *et al.*, 1991). An unexpected factor that enhanced feeding was the presence of insect growth regulators. After such chemicals were applied onto scale insects, an exclusive diet of their eggs increased the predator's consumption rate, from 22.5 to 27–30 eggs/female, and the mite's lifespan was slightly shortened without affecting fecundity (Farag *et al.*, 1990).

Yue & Tsai (1995) infested each of four potted sweet orange trees in a greenhouse with 100 citrus red mites and added ten females of *A. exsertus* to two trees. Pest populations on control trees reached 830 after 4 weeks, in contrast to about 100 on the treatment trees, the predator thus controlling the pest when released at a ratio of 1/10. In southern China the predator was active all year round, and although not being able to subdue *P. citri* in warmer weather (Fig. 32.2), it may help to control the pest along with other predators.

Other species of *Agistemus*

Agistemus terminalis (Quayle) controlled *P. citri* on citrus in southern Japan when pest populations were below 3.4 females/leaf and the prey/predator ratio was 5/1 (Inoue & Tanaka, 1983). The mite was an important natural enemy of scarlet mite, *Brevipalpus phoenicis*, a major pest of tea in Indonesia, feeding on three prey/day and occurring in densities of up to 20/10 cm² of leaf surface. Oomen (1982) considered it one of several stigmaeids capable of keeping the pest's populations at low levels. *Agistemus longisetus* Gonzalez-Rodriguez fed on the eggs and active stages of plant mites infesting apple and other fruit trees in New Zealand. Females, which breed throughout the year, devoured overwintering spider mite eggs in the autumn, remained on the leaves until they dropped, recolonised apples and plums in mid-summer and rapidly raised large populations (100/leaf). Mites kept in the laboratory deposited two to five eggs/day. They reproduced on several prey mites, but feeding on *Tetranychus* spp. and on an eriophyoid reduced oviposition. An advantage held by *A. longisetus* over some phytoseiids was that its populations were not self-limiting at low prey densities. This stigmaeid, along with the introduced phytoseiid *Typhlodromus pyri*, prevented resurgences of harmful mites in New Zealand orchards (Collyer, 1964b).

Agistemus floridanus Gonzalez-Rodriguez was very abundant on citrus leaves in Florida, feeding on several pest mites. It raised a generation within 2 weeks at 25–30°C when offered its optimal diet, the eriophyid citrus rust mite, *Phyllocoptruta oleivora* (Muma & Selhime, 1971). In the orchard, however, the predator seldom co-occurred with the pest on leaves or even in the same groves. This trait, along with susceptibility to pesticides and a low biotic potential, argued against considering it as an ABA on citrus in Florida. *Agistemus fanari* Dosse, another citrus denizen, attacked eggs of spider mites and tenuipalpids in Lebanon; its larvae fed on leaves without causing any damage (Dosse, 1967). Mated or unmated, *A. fanari* produced about 40 eggs/female, lived for around 1 month, and reproduced all year round. *Agistemus hystrix* Gupta is a common predator of mite pests on tea in north-east India, playing an important role in their control (Borthakur *et al.*, 1998).

As noted, *A. fleschneri* was the dominant predator of spider mites in abandoned apple orchards in north-eastern North America. It preferred the mid-rib area on the leaves' underside, feeding until the prey had been depleted, and then foraging on outer leaf margins (Childers & Enns, 1975b). It overwintered in the surface litter, near tree bases. When occurring by itself, the predator was common on main-trunk foliage and rare on leaves of watersprout branches. However, when present with *Z. mali*, the two occupied different tree parts: *A. fleschneri* inhabited inner-tree watersprout foliage, while *Z. mali* lived on fruit-cluster leaves on the trees' outer parts (Holdsworth, 1972).

Zetzellia mali

This mite (formerly known as *Mediolata novae-scotiae* Nesbitt) is considered an important natural enemy of apple mites [especially the European red mite (ERM), *Panonychus ulmi*, and the apple rust mite, *Aculus schlechtendali*] in North America and Europe. It is a sedentary predator that resides in the mid-rib area on the leaves' lower side (Santos, 1976a). The leaves are searched at random, the predator detecting prey by chance encounters, probably mediated by tactile means. After locating the prey *Z. mali* approaches slowly, stabbing and penetrating it in stages, often after several attempts (Clements & Harmsen, 1990). Gravid predator females reacted neither to prey (ERM or apple rust mite), nor towards a leaf without mites, when placed in a Y-shaped olfactometer (Santos, 1991). Similar results were obtained when the mites were offered a choice between infested and uninfested leaf discs. The residency of *Z. mali* on apple leaves infested by various numbers of pest mites changed only in response to mere prey presence or absence (Lawson & Walde, 1993). Numbers of *Z. mali* in the early spring were better correlated with pollen density than with prey abundance (Addison *et al.*, 2000).

Prevailing climates determine the overwintering sites of *Z. mali*. During severe east Canadian winters it hides in the orchard's upper soil layers (Parent, 1967). In the milder western regions the predator remains under loose apple bark or twig scars on the trees, where it aggregates in large clusters, often adjacent to ERM's winter eggs. *Zetzellia mali* is active above 9°C, and is thus capable of feeding on these prey in the autumn, during warm spells in the winter and then in the spring. The predator does not undergo a real diapause (White & Laing, 1977a), which could explain its high winter mortality.

Zetzellia mali also feeds on other prey, such as acarid mites and eggs of various insects. When attacking spider mites, eggs were the preferred prey, followed by quiescent stages and then by active larvae and protonymphs; adults were not taken. The predator devoured its own eggs as well as those of phytoseiids, preferring the latter (Clements & Harmsen, 1993). Feeding on eggs has several advantages for *Z. mali* (and other stigmaeids). Eggs usually make up about 66% of the populations of spider mites (Carey, 1982), providing abundant food, besides being efficient from the reproductive point of view. The fecundity of *Z. mali* was better correlated with number of eggs eaten than with total consumption (Clements & Harmsen, 1990), reflecting the eggs' higher nutritive value as well as the predator's lower energy costs in handling them. Finally, killing eggs decreases rivalry with and attack by other predators.

Feeding on plant substances is required for survival. Larvae of *Z. mali* provided with sufficient eggs but denied access to apple leaves did not moult to their next stage, and the females spent about one-third of their time probing leaf tissues (Santos, 1982). The importance of pollen for the predator's early population build-up, and the feeding of other stigmaeids on plant tissue were noted above.

Host plant effect is further evinced by the pattern of predator and prey preference for various apple culivars (cvs). In eastern Canada *Z. mali* was found (in descending order of abundance) on the cultivars Empire > Red Delicious > McIntosh > Golden Delicious, whereas the prey (mostly *A. schlechtendali*) occurred on Red Delicious > Golden Delicious > McIntosh > Empire (Woolhouse & Harmsen, 1984). Another factor affecting predator–prey interactions is their location on trees. In eastern Canada ERM and rust mites were most numerous on Red Delicious in the lower tree parts (1 m height), whereas *Z. mali* was more abundant at about 3 m (Herbert & Butler, 1973). In France, the predator's distribution was similar at the various apple tree (cv. Golden Delicious) strata throughout the season (Delattre, 1974). On grape vines *Z. mali* was found more often on cobwebbed and woolly leaves then on glabrous ones (see Fig. 41.3), indicating yet another host plant effect. Jenser & Koleva (1996) believed that this explained the preference of *Z. mali* for plants with pubescent leaves. An indirect host plant effect was seen when predation on *A. schlechtendali* was enhanced on trees that had received high levels of a nitrogen fertiliser (Walde *et al.*, 1997).

The optimal reltive humidity (RH) for *Z. mali* was 55–75%, as few eggs hatched and many larvae died at more extreme humidities (33% or 97% RH) (White & Laing, 1977b). The mite deposited an average of 1.7 eggs/female/day; other life-history parameters are listed in Table 32.3. SR in the laboratory was 0.72, consistent for arrhenotokous species. Apple rust mite was the best diet. Predators reared individually devoured around 38 rust mite motiles during development, an adult prey being consumed in 6 min. Given a choice between the two main acarine apple pests, *Z. mali* preferred apple rust mite over ERM, and did not switch from the former to the latter even when reared, from larva through to adult, on *P. ulmi* eggs (Santos, 1976b). However, when the proportions between the two prey organisms were changed, with many more rust mites being offered, a shift took place. At a ratio of 1/4 (ERM/rust mites) *Z. mali* maintained its usual 70% preference for the latter. However, when the offered prey was more than 80% *A. schlechtendali*, *Z. mali* fed almost exclusively on the rust mites (Walde *et al.*, 1995).

Whether or not *Z. mali* controls acarine apple pests is an unresolved question, the answer to which appears to depend on environmental (including pesticide regime) conditions. The predator stayed around the mid-rib on the leaves' lower surfaces, and did not track the apple rust mite onto leaves on which most of its population aggregated (Fig. 32.3). Such lack of spatial congruity, along with its inability to attack adult spider mites, suggested to Santos (1976a) that in the eastern USA *Z. mali* could not maintain apple mite populations below their economic damage levels. In southern Ontario, Canada, *Z. mali* raised about four annual generations and seemed to track the populations of both acarine pests (White & Laing, 1977a). Nevertheless, White & Laing (1977b) doubted whether *Z. mali* could reduce large numbers of pest mites by itself. The recent upsurge in the importance of *Z. mali* in eastern Canada was noted above.

Pesticide-exclusion experiments conducted in the western USA indicated that *Z. mali* could, under certain conditions, reduce *P. ulmi* densities. Parts of an apple orchard were treated with an organophosphate (OP) pesticide that effectively eliminated *Z. mali* (Croft & MacRae, 1993). The resident phytoseiids, *Typhlodromus pyri*

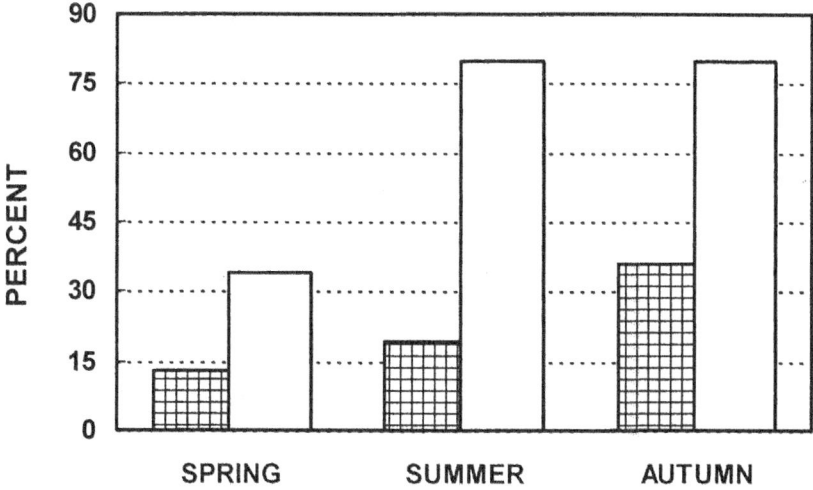

Fig. 32.3 Percentages of apple leaves on which *Zetzellia mali* (cross-hatched bars) and *Aculus schlechtendali* (white bars) occur together during the growing season. (Based on data in Santos, 1976a.)

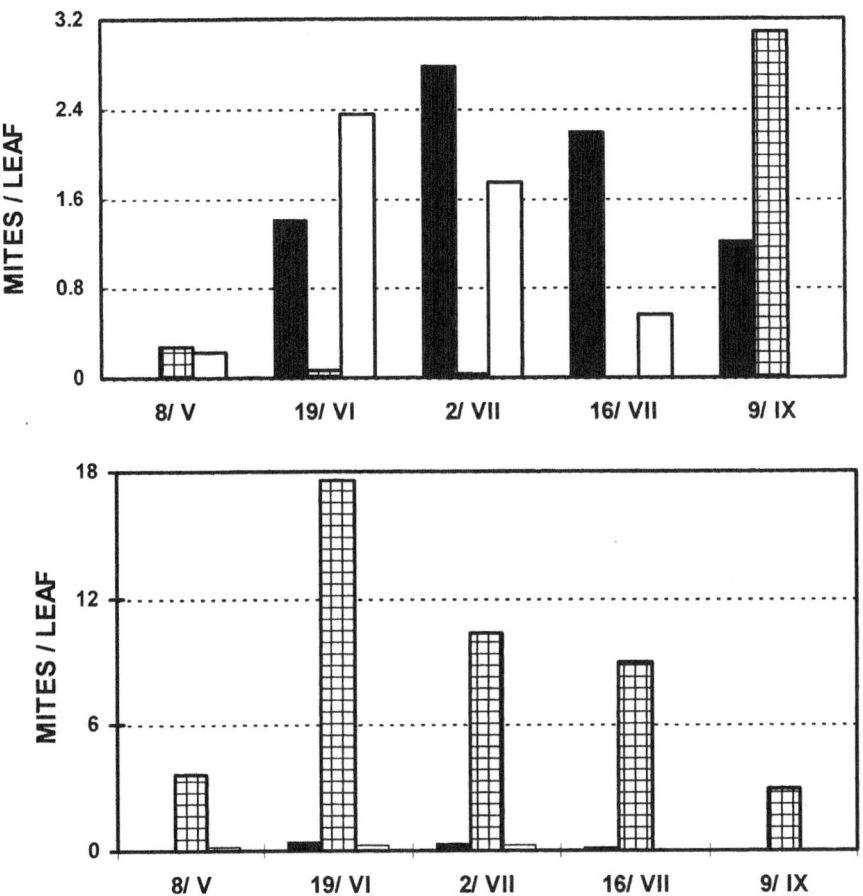

Fig. 32.4 Mites (black bars: *Galendromus occidentalis*; crosshatched bars: *Zetzellia mali*; white bars: *Panonychus ulmi*) obtained from apple plots sprayed (above) and unsprayed (below) with the OP endosulfan. (Based on data in Table 1, Croft & MacRae, 1993.)

and *Galendromus occidentalis*, being somewhat tolerant to the toxicant, were almost unaffected. ERM populations never exceeded 0.3/leaf in plots that contained *Z. mali* during several months post-treatment (Fig. 32.4, above), but the elimination of *Z. mali* allowed *P. ulmi* to reach almost eight-fold numbers (Fig. 32.4, below), and the predator's populations required 5 months to rebound in the sprayed plots.

    These and earlier field experiments indicated to Croft & MacRae (1992) the conditions under which *Z. mali* might control *P. ulmi* by itself. The results also showed what could occur (regarding control) in the apple orchard when the stigmaeid is present along with two competing phytoseiid predators. Where ERM was the target, the company of *T. pyri* enhanced reductions in pest numbers, whereas *G. occidentalis* detracted from overall control. The converse took place in regard to the apple rust mite, a pest better controlled in the presence of *Z. mali* and *G. occidentalis* than of the stigmaeid and *T. pyri*. Whatever the pest, *Z. mali* showed a consistent ability to control either of these mites, provided that its resident spring populations were sufficiently high, and that pest numbers were low. This insight led to advocating that *Z. mali* should be included in guilds of predators introduced for controlling apple mites. Considering the adverse effects of *Z. mali* on both phytoseiids, Croft & MacRae (1993) added the proviso that stigmaeid populations might have to be reduced by a selective toxicant.

The importance of *Z. mali* in apple orchards was demonstrated in France by a pesticide-exclusion experiment; *P. ulmi* increased to over 40/leaf in the predator's absence, but reached only about ten/leaf in its presence (Delattre, 1974). Similar results were obtained in Italy with regard to *A. schlechtendali* (Strapazzon & Dalla Monta, 1988). The adoption of an integrated pest management (IPM) programme in Hungarian apple orchards allowed *Z. mali*, which had formerly been eliminated by insecticides, to re-establish in that habitat. It became the main natural enemy within 1–2 years, regulating spider mite numbers (Komlovszky & Jenser, 1992). In Germany *Z. mali* populations were inversely correlated with those of the apple rust mite. No leaves with at least five stigmaeids carried more than 50 rust mites, whereas large pest populations (>100/leaf) were found on foliage that had only few (two or three) *Z. mali* (Karg, 1992). A website on this mite is accessible at http://www.nysaes.cornell.edu/ent/biocontrol/predators/zetzellia.html

Other species of *Zetzellia*

As noted, scarlet mite is a major pest of tea in Indonesia, where it is attacked by a guild of insect and mite predators. The more important are ten stigmaeids (three *Zetzellia* spp. and seven *Agistemus* spp.), which include *A. terminalis*, *Agistemus orbicularis* Ehara and Oomen-Kalsbeek (initially identified as *Agistemus denotatus* Gonzalez-Rodriguez by Oomen, 1982) and *Zetzellia javanica* Ehara and Oomen-Kalsbeek (initially called *Zetzellia* a). Both latter predators were very common and consumed three to seven scarlet mites/day, *A. orbicularis* being the most abundant stigmaeid on tea leaves. All three were believed to be capable of controlling the pest (Oomen, 1982). The abundance of all stigmaeids was negatively correlated with scarlet mite numbers in seven out of ten censuses taken in the experimental plots. Pest control by the stigmaeids occurred in all plots, albeit to different levels. Competition by other predators (except for one phytoseiid) was eliminated by applying the organochlorine dichlorodiphenyltrichloroethane (DDT) to selected tea bushes. The phytoseiid then reduced pest abundance only to 83% of predator-free controls, the comparative value for the stigmaeids being 27%. Oomen (1982) concluded that stigmaeids are efficient predators of scarlet mite. The predators were common and widespread on tea bushes, accepted the pest as a suitable diet for reproduction for several generations and were usually negatively correlated with scarlet mites.

Zetzellia graeciana Gonzalez-Rodriguez is common in south European apple and grape orchards, where it, along with *Z. mali*, consumed about one-third of the overwintering eggs of ERM (Inserra, 1970).

Interactions with Phytoseiidae

The effects of the interactions between *Z. mali* and phytoseiids on pest control were noted above, and are discussed elsewhere (Chapter 42). Phytoseiids feed on stigmaeids that, in turn, consume the former's eggs. When offered equal numbers of phytoseiid and spider mite eggs, 38% of the diet of *Z. mali* consisted of phytoseiid eggs (Clements & Harmsen, 1990). The greater effect of *Z. mali* on *G. occidentalis* than on *T. pyri* is probably due to the habit of the former phytoseiid to deposit more of its eggs (83%) within the mid-rib area of apple leaves, where the stigmaeid forages. In contrast, *T. pyri* placed only 64% of its eggs there (MacRae & Croft, 1996). Further, Phytoseiidae seldom feed on stigmaeid eggs (Clements & Harmsen, 1990).

The coexistence of the two families on crop plants has raised the question of their comparative ability to control pests. Phytoseiids are usually the better control agents at high prey densities. They kill a larger number of the pests and do so more quickly, are more motile, are usually less affected by spider mite webbing and are more tolerant (or even resistant) to many commonly used pesticides. They are therefore seen as being the 'stronger' predators, with the stigmaeids at times perceived as a hindrance to effective biological control. However, the stigmaeids are more efficient at low prey densities (Clements & Harmsen, 1992). Because they remain on or near trees, they are at hand when pests emerge from overwintering sites or in consequence of new invasions. Stigmaeid preference for eggs enables them to kill pests before these had begun to cause any damage and/or to reproduce. Stigmaeids respond well to prey presence and increase their residence time on infested plants, and persist there longer. During periods of food scarcity they utilise their own eggs. Consequently, a combination of stigmaeids and phytoseiids may reduce pest numbers more than either group alone, over a wide range of prey densities. Stigmaeid–phytoseiid coexistence on trees is facilitated by differences in diet preferences, specific modes of search and attack, and tree residency schedules. Consequently, in temperate climates a combination of stigmaeids and phytoseiids has greater efficacy than either group alone.

This pattern may not hold for the tropics, where stigmaeids play a more important role in plant mite control. Stigmaeids usually have a low rate of increase (Table 32.3), sometimes even lower than that of spider mite prey. The reproductive rate and intrinsic rate of increase of *A. orbicularis*, however, were higher or equal to those of two phytoseiids that co-occurred on tea in Indonesia.

Effect of agricultural chemicals

The pattern of stigmaeid reactions to agricultural chemicals is noted in Chapter 43. Suffice to note here that Stigmaeidae appear to be naturally tolerant to some OPs and to former insecticides (e.g. arsenates, cryolite and DDT). CRBs and pyrethroids are usually very detrimental (White & Laing, 1977a; Wieres & Smith, 1978). As noted, this spectrum of susceptibilities, differing in part from that of the Phytoseiidae, has allowed investigators to pinpoint the potential of either family for pest control.

Taxonomic problems

Apple orchards in North America support more than two dozen phytoseiids (Childers & Enns, 1975b; Amano & Chant, 1990), but only two stigmaeids (*A. fleschneri* and *Z. mali*). The latter is also known from Europe and Iran, which have their own, distinct phytoseiid fauna. These geographical differences in phytoseiid faunas compared with the very wide distribution of *Z. mali* in the Nearctic and Palaearctic raises the question of whether this biological entity is the same species in all regions. It was initially described from apple leaves in Oregon, USA, redescribed under another name in Europe, and then named again (*Mediolata novae-scotia* Nesbitt) from Nova Scotia. Its systematic history and position were clarified by Summers (1960). In North America *Z. mali* is known from wild and cultivated host plants, including pear, apple, box elder tree, linden, plum, oak and willow (Summers, 1960), suggesting a North American origin of *Z. mali*. However, the presence of *Z. mali* on many deciduous and evergreen plants in Hungary (Komlovszky & Jenser, 1992) supports a European provenance for what is regarded as this taxon.

As noted, the evidence that *Z. mali* can be spread by winds is tenuous at best. It is sedentary, remaining on trees or their immediate environment all year round, is not a rapidly colonising species, has no apparent behavioural adaptations for wind dispersal, and is not known to be disseminated by any animals. What then was the mechanism that facilitated the distribution of *Z. mali* throughout northern regions? It could be argued that the mite had become wide spread along with its 'hosts', apple trees. However, the mite was recorded on many other plants in North America and Europe, negating this 'apple' argument. In addition, the faunas of the other apple-associated predators, the Phytoseiidae, differ almost totally between western and eastern North American regions (Amano & Chant, 1990) as well as from Europe (Karg, 1992). Taken together, these morsels of biological evidence suggest that the North American and European populations of what is currently regarded as *Z. mali* may not represent the same taxon.

Agistemus exsertus was initially described from Japan (Gonzalez-Rodriguez, 1965), which lends credence to its presence in China (Yue & Tsai, 1995). The wide geographical separation from the Middle East, whence it has often been recorded, as well as discrepancies in reported biological data, raise doubts in regard to the correct identification of this species.

Future research

Should efforts be made to introduce stigmaeids (especially *Z. mali*, if it turns out to be a species complex) from one part of the world to others? The current policy of preferring specialised natural enemies (to which *Z. mali* does not belong) argues against this option. Conversely, the conservation of stigmaeids would be the best way to maximise their benefits (Thistlewood *et al.*, 1996). Conservation includes selecting the least disruptive spray programmes and understanding the effects of other environmental factors (e.g. available prey, additional diets, other predators, host plant) on the predators. Augmentation (including means of mass-rearing and transportation to release sites) should also be explored.

Stigmaeids can be mass-reared in the laboratory on their prey. Oomen (1982) maintained stigmaeids for several generations on scarlet mites that were kept on tea leaves. Stigmaeids can be reared on alternate foods, such as acarid mites (Rasmy *et al.*, 1987) or pollen (Borthakur *et al.*, 1998). Pollen, however, whether as a short-term or a long-term diet, may have adverse effects. *Agistemus hystrix* completed its life cycle when given only castor bean (*Ricinus communis*) pollen, but lived longer and was more fecund when mite prey were added (Borthakur *et al.*, 1998). The attraction of *A. exsertus* to spider mites was reduced after they were reared for ten generations on the same pollen, and its SR shifted from 0.5 on mite prey to 0.61 on pollen (Rasmy *et al.*, 1996). An artificial diet for *A. exsertus* was developed by Reda (1990). Made up of yeast, milk, amino acids and sugars, it enabled normal mite development, but fecundity was reduced by 30%.

Two kinds of field investigation are needed. The first, which should aim to discover more stigmaeid ABAs, would best be conducted in the tropics. The other required studies should clarify the specific conditions under which stigmaeids could be most efficacious (e.g. Croft & MacRae, 1993).

Interpatch movement of stigmaeids is a poorly understood topic. If some of these mites are wind-borne, is it a passive, random process, or have stigmaeids evolved adaptations for being dispersed by winds, such as moving to outer, higher leaves and positioning themselves against prevailing air currents? If the mites show such behaviour, what are the cues that provoke it? The mites could also be transported by arthropod vectors, but nothing is known about insects that carry species of *Agistemus* or *Zetzellia*.

Chapter 33
Tarsonemidae

Diagnosis

Tarsonemids are small mites (*c.* 0.2–0.3 mm in length) with broad to elongate oval bodies covered by a hard and shiny integument. Females have prodorsal clavate sensilli (pseudostigmatic organs), which are lacking in males. Legs IV differ from other legs: in females they end with apical and subapical whip-like setae, whereas in males they usually terminate with a large claw. The family was revised by Lindquist (1986). It now contains about 700 species in about 40 genera; the Taiwanese tarsonemids were summed up by Tseng & Lo (1980), those of China were catalogued by Lin & Zhang (1999) and those of Costa Rica by Ochoa *et al.* (1991).

General biology

Like other Heterostigmata, tarsonemids undergo a foreshortened life cycle, with only two active stages, larvae and adults. The larvae have a brief feeding period and moult directly to adults, generation time may be 1 week or less, and they reproduce by arrhenotoky. Tarsonemids show the greatest diversity in feeding habits of all acarine families (Lindquist, 1986), their sex ratio (SR) being accordingly quite varied (Kaliszewski & Wrensch, 1993). Some feed on green plants (a few, such as cyclamen mite, *Phytonemus pallidus* or broad mite, *Polyphagotarsonemus latus*, are important pests]; many subsist on fungi, and still others are associated with arthropods, whether as phoretics, predators, parasites [e.g. tracheal mite, *Acarapis woodi* (Rennie)] or parasitoids, or in some yet undetermined relationship.

Iponemus spp.

Iponemus spp. feed on eggs of pestiferous bark beetles (Scolytidae) assigned to the tribe Ipini, mostly of the genus *Ips* DeGeer, on which they also disperse (adhering to the elytral declivity of their vectors). *Iponemus* spp. are parasitoids because a single host individual suffices for the development of all progeny and only a single stage feeds on the host, which is thereby killed (Lindquist, 1983). *Iponemus* females produce 40–80 eggs/each. The larvae stay near the mother's cadaver, developing further without feeding. Males, which neither feed nor disperse, mature first and mate with the emerging females; the SR is very female biased (*c.* 0.95). Under summer conditions development from egg to adult female requires 2 weeks or less

(Lindquist, 1969). The new-generation adults can survive for a long period (from a few weeks to several months) without feeding, awaiting the emergence of new adult hosts. *Iponemus* spp. then board their beetle vectors, whose activities at a new location (e.g. excavation of galleries, copulation, oviposition, moulting to adults) serve as cues for further mite activity. This includes detachment from the beetles, host (egg) questing, feeding, laying eggs and initiating phoretic behaviour (Kaliszewski *et al.*, 1995). As they feed the bodies of the mites become greatly distended (physogastric).

The effect of *Iponemus* spp. on scolytid populations has not been fully evaluated, as authors have estimated rates of egg killing between 1–2% and 10–85% (Lindquist, 1969). Such variation is due to the different species of *Iponemus* and beetles investigated, and to the aims of the investigators. In addition, low (up to 20%) rates of egg predation may reduce intraspecific competition among the beetle's larvae, which would detract from the value of the mites as factors in the pests' control. The role of the mites in the population dynamics of these pests would probably be better understood if continuous life table studies, conducted between and during scolytid outbreaks, could be undertaken (Lindquist, 1969). As noted, *Iponemus* spp. survive for extended periods without feeding, a characteristic that may be of use in manipulating these acarine biocontrol agents (ABAs) for pest control.

Species of *Iponemus* attacking bark beetles were formerly believed to be restricted to the Holarctic region. The finding of a new species on a central African beetle suggested to Lindquist (1996) that the association between *Iponemus* and Scolytidae is probably more widespread and may be 35 million years old.

Miscellaneous

Several tarsonemids are predators of insects and mites. An unidentified tarsonemid (called *Tarsonemus* sp.) fed on the grape phylloxera (*Daktulosphaera vitifoliae*), a world-wide, major pest of grapes. Pest and tarsonemid were reared on grape cuttings in a greenhouse in Germany, and the mite was assumed to cause the death of the former, because no pests survived interactions with the tarsonemid (Forneck *et al.*, 1998). Although mites were seen to follow the prey into their leaf galls, the association could have been accidental, as the tarsonemids also fed on plant leaves and roots. Their eggs were laid there as well as in phylloxera cadavers, hatching after 5–7 days.

Acaronemus destructor (Smiley and Landwehr) fed on eggs of phytophagous mites (Tenuipalpidae and Tetranychidae) infesting pines in California, USA. In the laboratory *A. destructor* consumed one or two eggs/day. In the field mite numbers were highest in September, coinciding with the population peak of its main tenuipalpid diet. However, predator numbers relative to those of its prey were very low (1/65 to 1/100), arguing against important tarsonemid effects on the pests (Smiley & Landwehr, 1976). Two undescribed species of *Acaronemus* were noted from Italy and from Iran (Lindquist & Smiley, 1978). A *Dendroptus* sp. fed on the apple rust mite, *Aculus schlechtendali,* in the laboratory; Villanueva & Harmsen (1996) postulated that the tarsonemid contributed to the mid-summer decline of that apple pest.

Together, these pieces of information suggest that tarsonemids could play a more important role as ABAs. An untested option is to use them to transmit plant diseases to weeds; some species vector plant-pathogenic fungi, for example blue-staining fungus, *Ceratocystis minor,* to pines (Bridges & Moser, 1986).

Chapter 34
Tetranychidae

Diagnosis

The Tetranychidae, commonly known as spider mites, possess needle-like chelicerae, a thumb–claw complex, two pairs of eyes and usually two pairs of duplex setae on tarsus I. At present, the family includes close to 1200 species in about 70 genera. Many species spin webs (hence their vernacular name), which are used for dispersal and protection (Gerson, 1985). Meyer (1987) provided a key to the world genera and a catalogue of the world fauna was published by Bolland *et al.* (1998). Baker & Tuttle (1994) prepared keys to the species found in the USA.

Life history

All spider mites are phytophagous and several are major pests of crops; their biology, natural enemies and control were treated in a multi-author treatise (Helle & Sabelis, 1985). Many species have a high rate of increase, reproduce all year round and can cause severe damage to host plants. These attributes, along with the fact that spider mites infest wild plants, make them suitable candidates for weed control. Although numerous tetranychids are monophagous or have narrow host ranges, the broad polyphagy of the main pest species, their occurrence on some weeds and their uncertain reproductive isolation, diminish their apparent suitability as weed control agents (Hill & Stone, 1985).

Control of prickly pear with spider mites

Tetranychus desertorum Banks (formerly called *Tetranychus opuntiae* Banks) was the first spider mite (and first acarine) used for weed control. A pest of cacti in Texas, *T. desertorum* was accidentally introduced into Australia in the 1920s. It became established in southern Queensland on prickly pear, *Opuntia inermis*, an introduced weed cactus that had seriously depreciated the value of land for grazing. The mite's spread was rapid, its distribution reaching 80 km in all directions from the initial focus within 3–4 years of discovery (Mann, 1970, citing earlier authors). Feeding injured the epidermis of the cactus to the extent that it became corky and whitish-grey, and caused affected segments to dry and drop. Seedlings and young plants were killed, whereas the upper growth of older cacti was destroyed. The mite thinned out 75% of a densely weeded area of about 250 ha in 2 years. Another natural enemy of cacti, the moth *Cactoblastis cactorum* (Bergroth), was later introduced and displaced *T. desertorum* as a controlling agent (Mann, 1970).

Control of gorse with spider mites

Gorse (or furze, *Ulex europaeus*) is a spiny, dense, leguminous, woody shrub of European origin that had been introduced into South and North America, Australia and New Zealand, and became a weed of pastures and disturbed lands. The realisation that chemical and/or cultural means of containing the weed were costly and inefficient led to a long-term gorse biocontrol project in New Zealand. *Tetranychus lintearius* Dufour, the major pest of gorse in Europe, was the most suitable of several arthropods considered and evaluated (Longworth, 1987). The mite lives in large colonies, each containing several thousand individuals, whose members spin profuse amounts of dense webbing that may cover many plants. The colonies slowly move along the gorse shoots, their members feeding, spinning webs and leaving behind eggs and juveniles. *Tetranychus lintearius* raises a generation in 18 days at 25°C, produces about 30–40 eggs/female, lives for about 3 weeks and undergoes no winter diapause (Stone, 1986).

The mite is specific to *Ulex*. This specificity and the large webs were used by van Eyndhoven (1967) to reconfirm the status of *T. lintearius* as a valid species. However, some authors had placed *T. lintearius* in synonymy with the two-spotted spider mite and, as noted, spider mites are often polyphagous. These objections to the introduction of *T. lintearius* into New Zealand required proof of the mite's unique status and of its unqualified host specificity. Further reluctance to approve the introduction was due to concerns raised by beekeepers, farming groups and environmentalists (Longworth, 1987). In consequence, an environmental impact assessment report was drawn up and publicised, asking concerned persons or organisations to comment (Hill *et al.*, 1989). All submissions were examined and objections to the introduction were met by proposing alternative management procedures, and especially by the general recognition that benefits accruing from gorse control (which was not expected to be complete) would outweigh potential losses.

Two issues were of major concern. The first was whether *T. lintearius* is reproductively isolated from two closely related, polyphagous relatives, namely *T. urticae* and *Tetranychus turkestani* Ugarov & Nikolski. This concern was addressed by elaborate crosses and reciprocal crosses among all three species, which clearly established the reproductive isolation of *T. lintearius* (Hill & O'Donnell, 1991a). The other topic was the mite's specificity to *Ulex*. This aspect was explored in England by studying the settlement and reproduction of *T. lintearius* in the field and in the laboratory on many host plants. The mite did not establish on 39 plant species growing outdoors, or on 56 plants in the laboratory. The only exceptions were two other legumes, namely beans and soya beans, on which *T. lintearius* reproduced, but only for one generation. The mite also failed to survive on any of 22 bean cultivars offered. These data, along with no recorded findings of the mite on any plants except for *Ulex* spp., convinced Hill & O'Donnell (1991b) that *T. lintearius* could safely be used as a biological weed control agent.

The mite was released in New Zealand in 1989 on gorse and settled well. Shoots infested with *T. lintearius* were tied onto gorse bushes on their upwind sides at more than 300 selected sites. Although the mite settled throughout New Zealand, establishment was restricted to the cooler and drier regions. Many colonies dispersed from their initial sites by moving along contiguous gorse bushes. Other dispersal was downwind, as mites were found 55 m away from infested sites after 5 months. Some colonies of *T. lintearius* grew rapidly and encased entire gorse plants, causing their shoots to brown and die. Heavy, continuous feeding severely reduced the weed's growth rates, but if the attack ceased for some reason, the plants recovered. Notwithstanding the presence of very large mite populations on gorse, as well as considerable damage, no plant deaths could be attributed to *T. lintearius* (Hill *et al.*, 1991). Further mite populations were imported from north-west Spain and from Portugal and thrived where earlier introductions of *T. lintearius* did not succeed. In a recent evaluation Richardson & Hill (1998) concluded that the mite often fails to provide adequate control of gorse, either because it leaves the plants too soon, or owing to its inability to build up sufficiently large numbers. A major constraint was predation by a coccinellid beetle, which greatly reduced the populations of *T. lintearius*. The mite is now found throughout New Zealand, an establishment that led to its being released in Oregon, USA, where it caused extensive gorse defoliation, and in Hawaii (Markin *et al.*, 1996). A scoring exercise, formulated to predict the suitability of biocontrol agents for weed control, gave the mite relatively high values (Chapter 38, ABAs as enemies of weeds). *Teranychus lintearius* remains to date the only tetranychid ever introduced for weed biocontrol.

Integrated weed control programmes require the continuous evaluation of the effects of one method of control on others. Several herbicides are toxic to *T. lintearius* (Searle *et al.*, 1990), indicating the need for caution.

Spider mites for waterhyacinth control

Waterhyacinth, *Eichhornia crassipes*, is a floating perennial plant that forms dense mats on fresh waterways and lakes and is considered a serious weed in many warm parts of the world (see Chapter 18). Fifteen females of each of 12 tetranychid species were placed on small waterhyacinth plants kept at 25–27°C, and the number of females obtained after 14 days was used as an estimate of plant suitability for the mites. Four species, *Oligonychus silvestris* Gutierrez, *Tetranychus macfarlani* Gutierrez, *T. neocaledonicus* André and *T. tumidus* Banks, produced more than 200 eggs in that period. Their feeding seriously affected the leaves, which turned yellow. Pieterse (1972) concluded that although spider mites would not eradicate dense stands of waterhyacinth, in certain areas they could delay weed growth.

Spider mites as alternate prey and as a 'vaccinating' factor

Two species of spider mites infest grapevines in California: the Willamette mite, *Eotetranychus willamettei* (McGregor), and the Pacific mite, *Tetranychus pacificus* McGregor. The former seldom causes much damage, but Pacific mite is an important pest. Both are attacked by the phytoseiid *Galendromus occidentalis*. Willamette mite, which occurs early in the season and has a scattered distribution, serves as prey that enables the predator to build up and maintain its populations. When Pacific mite then appears, enough predators are present in the grapeyard to effect its season-long pest control (Kinn & Doutt, 1972). Willamette mite thus serves as alternate prey, whose presence is critical for the control of a major pest.

Cotton seedlings, whose cotyledons were damaged by spider mites (or by mechanical abrasion) in the laboratory, or in the field, were less affected by the same pest mite later in the season. This systemic 'induced resistance' (Karban, 1986) was not, however, translated into differences in plant growth or to yields. The potential use of spider mite-induced resistance was later tried against other spider mites affecting grapevines (Karban & English-Loeb, 1990) and against a vascular wilt disease of cotton (Karban *et al.*, 1987).

Independent of predator effect, in Californian vineyards Willamette and Pacific mites seem to be negatively associated (Kinn & Doutt, 1972). Vineyards that undergo early Willamette mite attacks tend to suffer fewer Pacific mite outbreaks later in the season. In the greenhouse Willamette mite had a consistent negative effect on Pacific mites numbers, a result not attributable to natural enemies as these were absent (English-Loeb & Karban, 1988).

Populations of around 1000 *E. willamettei*/plant were placed on grapevines in a commercial vineyard that had a history of Pacific mite outbreaks. By mid-June the populations of *T. pacificus* were about nine times higher on leaves that had not previously been infested by Willamette mite. At a later sampling numbers of both pests were much higher, and the difference, although only three-fold, was still significant (Fig. 34.1). Yields and sugar concentration levels obtained from grapes of both treatments were similar during the first year of the experiment. Comparable differences in mite numbers, and the same similarities in yield, were obtained in the second year, but sugar levels were slightly higher in grapes from 'vaccinated' vines (Karban & English-Loeb, 1990). Commercial growers 'vaccinated' their vines against Pacific mite with large-scale releases of Willamette mite, an effort that has met with encouraging results (Karban & Zalom, 1998).

Hougen-Eitzman & Karban (1995) tested four hypotheses to understand how feeding by one spider mite affects another. The hypotheses included increased predation by a shared natural enemy, direct interaction between the species, local depletion of resources and a systemic reduction in resource quality (or exploitative competition, manifested as induced resistance). Predator absence from 'vaccinated' plants negated the first hypothesis. Direct physical interaction and local resource depletion were discarded because 'vaccination' occurred without any spatial overlap between the mites. The induced resistance hypothesis was accepted because feeding of Willamette mites at the bottom of the plants reduced Pacific mite numbers at the top of the shoots, without their ever meeting. Pacific mite fecundity on 'vaccinated' grape leaves was reduced by 65%, and survival by 56% (Karban & Zalom, 1998).

The effect of 'vaccinating' plants with spider mites against a fungal disease was explored by placing females of *T. turkestani* on cotton cotyledons for a period of 2 weeks. The mites were removed and the plants kept for another fortnight. Spores of *Verticillium dahliae* (strain SS-4), a vascular wilt disease, were then injected into

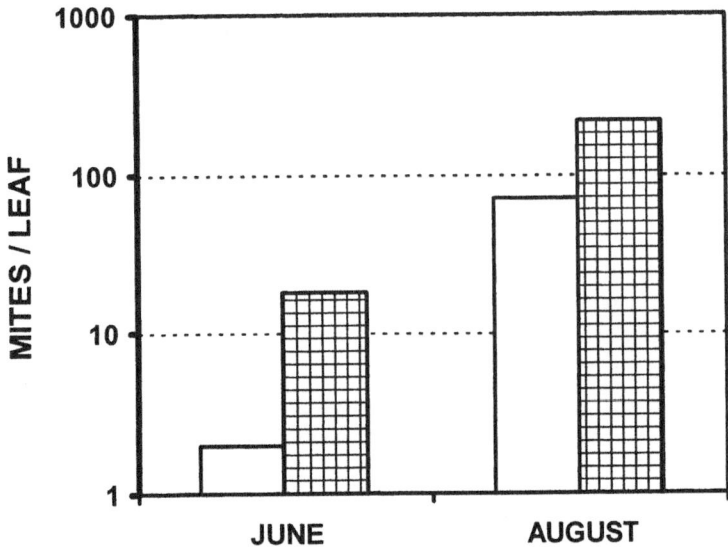

Fig 34.1 Numbers (log values $x+1$) of Pacific mite, *Tetranychus pacificus*, on grapevine leaves formerly infested by the Willamette mite, *Eotetranychus willamettei* (white bars), and on leaves not infested by Willamette mite (cross-hatched bars). Values obtained on the two dates differed significantly from each other. (From data in Karban & English-Loeb, 1990.)

each plant. 'Vaccinated' cotton had significantly less pathogenic foliar yellowing and necrosis. The resistance seemed to be systemic, because plant parts that had not been present during the 'vaccination' were also protected. This defensive result was postulated to be due to the depletion of very rare nutrients that are critical to both mite and fungus, and/or to the activation and/or induction of protective chemicals, such as tannins and isoflavonoids (Karban *et al.*, 1987).

Spider mites thus serve directly as natural enemies of weed pests, as alternate prey for predators, as displacers of other, more injurious mites, and as inhibitors or competitors of plant pathogens.

Chapter 35
Trombidiidae

Diagnosis

Trombidiids are large (up to 2.5 mm in length), usually red mites with a dense coat of setae, short, non-retractable chelicerae and a single set of prodorsal sensilli. This family is assigned to the Parasitengona, a cohort of superfamilies that have only three active stages: larvae, deutonymphs and adults; the protonymphs and trironymphs remain quiescent. The larvae are heteromorphic to the deutonymphs and adults; separate keys are thus needed for their determination. Some genera are known only as larvae whose later stages have not been identified; an effort at correlating larval and adult morphology was made by Zhang (1995c). The Trombidiidae contain about 250 species in over 20 genera, whose larvae were keyed by Zhang & Norbakhsh (1995). The family was revised by Southcott (1987) and catalogued by Makol (2000), but its boundaries are still in flux (e.g. Zhang, 1998).

General biology

Larval trombidiids are parasites of various arthropods (including spiders, insects and other mites), whereas the postlarval stages are predators that forage in the soil or on plants. Trombidiidae are usually univoltine, but a generation may last for more than 1 year. The sex ratio in the field is about 0.5. Adults go through a mating 'dance', the male then depositing spermatophores that are soon taken up by the females (Robaux, 1974), each of which usually lays several hundred to several thousand eggs in the soil. Juveniles need 3–4 months to develop in the laboratory, or about 10 months in the field (Zhang, 1998). The emerging juveniles hunt for suitable hosts; the search mechanism is not known, except that larvae of some species can jump to reach their hosts (Robaux, 1974). During larval parasitisation a feeding tube, or stylostome, is formed within the body of the attacked arthropod, probably as a reaction between the host's haemolymph and the mite's saliva (see Chapter 7). An early appreciation of trombidiids as acarine biocontrol agents (ABAs) was expressed by Howard (1918). Welbourn (1983) prepared a list of Trombidiidae and their host (or prey) species, updated by Zhang (1998).

Allothrombium pulvinum Ewing

The effect of the widespread *A. pulvinum* on aphids and on spider mites has been intensively studied in China. Emerging larvae are positively phototactic and negatively geotactic, traits that guide them from the buried eggs onto the soil surface (Zhang, 1991a). Although host searching appears to be at random, the larvae preferred certain aphid species over others within families, and were more attracted to hosts of some families than to others (Zhang, 1996). In addition, they selected large over small aphids (probably because of less risk of having to change hosts) and parasitised hosts over healthy ones (the former are weaker), and at times preferred winged to wingless aphids. The effect of parasitism depended on aphid size and number of feeding mites. The relatively small bean aphid, *Aphis fabae* Scopoli, succumbed within 3 days when set upon by two or more *A. pulvinum*; development and fecundity were impaired even if attacked by a single mite. The pea aphid, *Acyrthosiphon pisum* (Harris), a larger host, survived the attack of five larvae after 4 days much better (Zhang, 1991b).

Allothrombium pulvinum attacks the cotton aphid, *Aphis gossypii* Glover, an important pest of cotton, early in the season, before other natural enemies arrive. At that time the mite is the most abundant enemy and the pest's major limiting factor (Chen *et al.*, 1994). On continuously grown (monoculture) cotton the predator was twice as abundant as in cotton–wheat intercropping fields, and was far scarcer in formerly flooded fields than under dry conditions, indicating the effect of cultural practices on mite populations.

The deutonymphs and adults of *A. pulvinum* search for prey (spider mites and other small arthropods, including aphids) at random, on the soil surface and on plants. In laboratory studies two adult mites consumed five aphids within 2 h (Zhang & Xin, 1989). Small-scale releases indicated that a mite/aphid ratio of 3/1 could control these pests (Zhang, 1988).

Upon locating a colony of spider mites, the predators easily penetrate the webbing and forage therein. They are very voracious: a single deutonymph killed and consumed around 55 spider mite eggs or up to 36 juveniles/day. Based on his own and others' observations, Zhang (1992) postulated that deutonymphs might be able to control spider mites on cotton at a predator/prey ratio of about 1/50, or if there were about 40 *A. pulvinum* per 100 plants. Indirect evidence for their controlling ability came from observations that cotton plants which carried no predators were significantly more infested by spider mites than leaves with none (Chen & Zhang, 1991). The limited deutonymphal period (June to July), during which they provide effective spider mite control, does not suffice to exclude later pest damage. In addition, the predator is susceptible to many pesticides; fields sprayed with dichlorodiphenyl-trichloroethane (DDT) or an organophosphate compound had few *A. pulvinum* but many spider mites. Optimal predator utilisation within cotton integrated pest management programmes thus depends on applying fewer (or no) pesticides, which should be relatively innocuous, and on spraying only during the predator's periods of inactivity. Another option is to transfer the predator to areas where it does not occur naturally.

Zhou *et al.* (1989, cited by Chen & Zhang, 1991) reported that in western China *A. pulvinum* was an important egg predator of a major lepidopterous pest of peaches.

Allothrombium ovatum Zhang & Xin

When offered a choice among several aphids in the laboratory, the larvae of *A. ovatum* clearly preferred *A. gossypii* over others, a preference reflected in parasitisation rates recorded in the field (H. Zhang *et al.*, 1999). Most *A. ovatum* arrived in cotton fields on winged aphids, the majority (>90%) carrying a single mite. Smaller numbers of *A. ovatum* overwintered in the fields. Their questing larvae walked to and climbed onto plants, parasitising any aphids found there (Zhang & Li, 1996). The mite effect on hosts was similar to that of *A. pulvinum*. Winged aphids died 2 days after being parasitised by five mites, and after 3 days when attacked by three or four mites. In addition, the presence of even a single *A. ovatum* more than halved the aphid's fecundity. Dong *et al.* (1996) and Zhang & Li (1996) concluded that the mite was a very effective early-season natural enemy of *A. gossypii* in China; it attacked most cotton aphids, reducing the fecundity of any hosts that did not die.

Prospects

Several other trombidiids have been shown or inferred to reduce pest populations. For instance, the scarcity of aphids in alfalfa fields in southern France was postulated by Aeschlimann & Vitou (1986) to be due to the presence of *Allothrombium monspessulanum* Robaux & Aeschlimann there. *Allothrombium mitchelli* Davis fed on and seemed to reduce the numbers of the beech scale (*Cryptococcus fagisuga* Lindinger), (Hemiptera: Eriococcidae) in the eastern USA (Wiggins *et al.*, 2001).

Zhang (1998) listed other trombidiids that reduce pest numbers. Before these mites can be used more widely as ABAs, however, more information on various topics should be at hand. This includes a better understanding of their systematics, specific host preferences, the phenologies of the various target pests and the trombidiids' susceptibility to pesticides. Their univoltinism suggests that they would be most appropriate for introductions and in conservation efforts (Zhang, 1991b).

Chapter 36
Tydeidae

Diagnosis

Tydeids are greenish-yellow or pink, small, soft-bodied mites whose idiosoma is striated and sometimes reticulated, with two sensilli and a pair of eyes (and seldom a third median eye). The bases of the needle-like chelicerae are fused or contiguous. The Tydeidae is cosmopolitan, with more than 400 described species, whose taxonomic arrangement is still in flux. The family was revised by Baker (1965), André (1979, 1980), Kaźmierski (1998b and earlier publications) and more recently by André & Fain (2000). However, doubts remain about the definition and boundaries of its major genera (e.g. *Tydeus*, *Lorryia* and *Paralorryia*) and the placement of their species. The Russian tydeids were studied by Kuznetzov (1979 and earlier papers) and those of Antarctica by Usher & Edwards (1986), and those of northern Europe are being revised by Momen & Lundqvist (1996 and earlier papers).

General biology

Tydeids are common, fast-moving inhabitants of plants and soil, where they subsist on many plant and animal foods. They raise a generation in 2–3 weeks and reproduction appears to be arrhenotokous. Sperm transfer is by copulation or via spermatophores (Knop, 1985). Some species (e.g. *Proctotydaeus* spp.) live on insects, but the nature of the relationship is unclear (Kaźmierski, 1998a).

Tydeids as ABAs

Homeopronematus anconai fed on and greatly reduced the numbers of tomato russet mite (TRM), *Aculops lycopersici* (Hessein & Perring, 1986). TRM colonies kept on cut tomato leaflets for 19 days averaged 65 in the presence of *H. anconai* and 427 in its absence. Predator-free seedlings died within 6 weeks, whereas those with *H. anconai* remained healthy. The predator also fed on pollen, fungi and plant tissue, living longer and producing more progeny when cattail (*Typha latifolia*) pollen was added to TRM, and even with pollen as its sole diet (Fig. 36.1).

The mite was reared on various pollens at 40–70% relative humidity (Knop & Hoy, 1983a). The mean generation time was 21 days at 24°C, the lifespan (egg to death) was 40.5 days and around 16 eggs/female were deposited. At 30°C a generation lasted for 12 days, the lifespan was 22 days and around 45 eggs/female were produced. Reproduction was arrhenotokous, with mated and unmated females producing

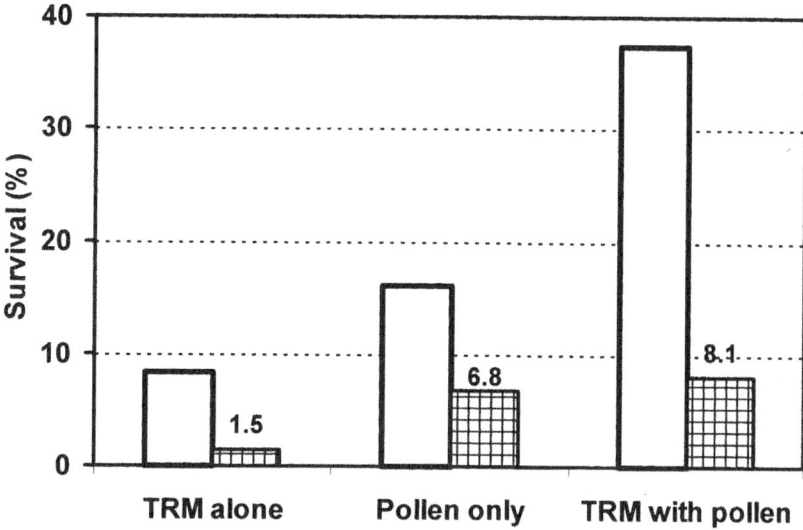

Fig. 36.1 Survival (white bars) and fecundity (cross-hatched bars, with values written on top) of adult *Homeopronematus anconai* when offered tomato russet mite (TRM) (*Aculops lycopersici*) alone, cattail (*Typha latifolia*) pollen alone, or mites plus pollen. (Based on data in Hessein & Perring, 1988b.)

similar numbers of eggs (each of which is deposited on a stalk placed on hairs growing on the leaves' lower side), and the sex ratio (SR) was 0.66. The predator undergoes a reproductive diapause during winter in California, such females forming tight clusters near leaf veins, a form of aggregation also known from other tydeids (e.g. Mendel & Gerson, 1982). Spider mite eggs were preyed upon, such feeding being incidental, subtracting rather than adding to fecundity when pollen was available. The tydeid also fed on eggs of the citrus flat mite, the tenuipalpid *Brevipalpus lewisi* (Hessein & Perring, 1988a), but its quantitative effect is not known.

Pronematus ubiquitus (McGregor) preyed on fig eriophyids in California (Baker, 1968). Feeding on such prey in Egypt, the predator raised a generation in about 30 days at 29°C and produced 25 eggs, their SR being 0.52–060 (Abou-Awad *et al.*, 1999). The quantitative effect of such predation is not known, but reductions in the numbers of another pest, citrus brown mite, *Eutetranychus orientalis*, from 40.8/leaf to four to six/leaf, coincided with increases in *P. ubiquitus* populations on citrus in Egypt (Rasmy, 1970). A *Pronematus* sp. fed on *Brevipalpus phoenicis* infesting tea in India. The predator's larvae consumed two pest eggs/h, five/day, whereas the adults devoured five eggs/h, up to eight/day. It developed in 3–5 weeks, each female laying around 20 eggs. Predator as well as pest preferred young tea leaves, on which the predator fed and oviposited (Borthakur, 1981). Recently, *Pronematus* was transferred to another family (André & Fain, 2000).

In Germany *Tydeus goetzi* Schruft fed on the eriophyids *Colomerus vitis*, the grape erineum mite, and *Calepitrimerus vitis*, the grape rust mite (Schruft, 1972). In south European vineyards *Tydeus caudatus* (Dugés) (sometimes placed in

Orthotydeus) also fed on *C. vitis* (Camporese & Duso, 1995). The citrus rust mite, *Phyllocoptruta oleivora*, a pest of citrus, is controlled in Zhenjiang province (eastern China) by a *Tydeus* sp. Predator populations peaked in early summer and the pest was controlled from May to July. Predator numbers declined later owing to climatic conditions, rising again in the autumn. The development of this *Tydeus* sp. required 20 days at 29°C from egg to adult, each female produced 11 eggs and they lived for 11 days. Each predator consumed more than 50 citrus rust mites (Chen *et al.*, 1988).

Reports on tydeid feeding habits, which range from carnivory through fungivory and omnivory to phytophagy (Hessein & Perring, 1986; Laing & Knop, 1983), add uncertainties about their role in pest control. The doubts were reinforced when *H. anconai* did not develop when given only TRM in the laboratory and did not feed on that pest (Brodeur *et al.*, 1997). Brickhill (1958) provided a possible explanation for observations that linked tydeids with mite or other pest eggs. That author offered eggs of a scale insect (Coccoidea) and of a spider mite (Tetranychidae) to two tydeid species in the laboratory. Reasonable development took place on the latter diet (implying that eggs were a sufficient diet), but most fed-upon eggs subsequently hatched. The effect of tydeids on prey thus remains to be clarified. Further uncertainties are due to using the ambiguous term 'consumption' in regard to the feeding of tydeids. They possess small, delicate chelicerae and may, at best, suck out the body fluids of their prey, but cannot 'consume' an entire egg or a small mite. Thus, it is likely that tydeids were observed to insert their mouthparts into the prey, but without mortality data the results remain to be elucidated.

Tydeids as alternate food for other ABAs

In Californian grapeyards *H. anconai* served as an alternate prey for the phytoseiid *Galendromus occidentalis*, the major predator of pestiferous spider mites (Chapter 26). A dusting of pollen on grape leaves significantly increased tydeid and phytoseiid numbers thereon; Flaherty & Hoy (1971) postulated that the larger populations of the former promoted the numerical increase in the latter. Other phytoseiids, such as *Typhlodromus pyri* and *Amblyseius andersoni* (Chant) [=*A. potentillae* (Garman)], attack the European red mite, *Panonychus ulmi*, in Dutch apple orchards. Although *T. pyri* has a lower rate of increase, it is the more efficient predator, owing to its ability to feed and reproduce on tydeids, prey that encourages its persistence in the orchard (Calis *et al.*, 1988). In contrast, *A. andersoni* neither feeds on tydeids nor survives when spider mites are absent. Tydeids may thus serve as supplemental prey for some predators. Of several phytoseiids inhabiting vineyards in southern Europe, only *Typhlodromus talbii* Athias-Henriot fed on tydeids (Camporese & Duso, 1995).

Tydeids as consumers of fungi

The mycophagous habits of tydeids can indirectly reduce the damage caused by certain pests. Mendel & Gerson (1982) used an acaricide to eliminate *Lorryia formosa* Cooreman from citrus branches that were heavily infested by a pest coccid, the Mediterranean black scale, *Saissetia oleae* (Olivier). The scale's main damage is due to its excreting copious amounts of honeydew, which is colonised by dark fungi (known as sooty mould) that hinder photosynthesis. Sprayed branches were far more blackened by sooty mould than unsprayed parts of the trees, indicating that the mite serves as a sanitising agent in citrus groves.

Grape powdery mildew, *Uncinula necator*, is a major fungal pathogen of grapes in most parts of the world. English-Loeb *et al.* (1999) noted that in the eastern USA *Tydeus lambi* (Baker) (sometimes placed in *Orthotydeus*) commonly inhabited leaves of riverbank grape (*Vitis riparia*) that were rarely infected by the fungus. Batches of *T. lambi* were placed on leaves of the potted grape cultivar 'Cabernet Sauvignon' maintained in a greenhouse. Grape leaves were then infected with the mildew and kept for 3 weeks. Control leaves (infected but without mites) had seven times more mildew colonies than leaves with mites, and cleisthotecia (fruiting bodies that are critical to powdery mildew survival in temperate climates) occurred only on control leaves. Vines on which mites were released had about half the infected leaf area and bore 64% fewer cleistothecia. The mites inhabit small grape leaf cavities (called domatia; see Chapter 41). Leaves with artificially blocked domatia suffered more damage than leaves with intact domatia, and leaves with larger domatia carried more mites and had far less of their leaf area covered by mildew (Norton *et al.*, 2000). These data support the hypothesis that *T. lambi* prevents major mildew outbreaks on riverbank grape, the suppression of cleistothecia being of special significance.

Tydeids as nematophages

The mechanical breakdown and chemical decomposition of plant litter by various organisms releases vital elements for recycling in terrestrial ecosystems. Bacteria, fungi and invertebrates play different roles in this agriculturally important process. Bacteria are very active in the decomposition of creosote (*Larrea tridentata*) litter and are strongly grazed by nematodes, a process harmful to the process. Nematophagous tydeids (unidentified beyond the family designation) were the initial and dominant arthropods that colonised buried litter of creosote in a New Mexico desert. An application of the organochlorine chlordane, which affects neither microorganisms nor nematodes, but kills mites, eliminated the tydeids. The treatment was followed by an increase in the number of grazing nematodes and by a 40% reduction in litter decomposition. Santos *et al.* (1981) postulated that the tydeids reduced nematode densities, which in turn prevented bacterial overgrazing, thereby enhancing litter decomposition.

General comments

The effect of pesticides, to which tydeids are very sensitive, is discussed in Chapter 43. These mites advance pest control via five separate routes. They are predators, serve as alternate food for other predators, 'clean up' after honeydew producers (thus reducing the damage attributable to these pests), reduce infections of plant pathogens and regulate the number of nematodes that are microbial grazers in some soils (hence promoting litter decomposition).

Chapter 37
Uropodidae

Diagnosis

These brown–black, turtle-like mites can be recognised by their leg grooves, by the stigmata that are positioned between legs II and III and by the short, convoluted peritremes. Evans & Till (1979) provided a key to the superfamily Uropodoidea, within which the Uropodidae have not been clearly delineated. Keys to the Middle European and North American taxa were provided by Karg (1989b) and Krantz & Ainscough (1990), respectively. Hirschmann & Wisniewski (1992 and many earlier papers) have been revising the world fauna, which tentatively contains about 100 genera with 1500 species, of which many are known only as deutonymphs.

General biology

The Uropodidae abound in forest soils and other habitats that are humid and rich in organic material, such as insect burrows and animal droppings. A life cycle may require 1–3 months at 20–30°C. They feed on fungi and small animals, including nematodes and juvenile insects; certain species develop more rapidly, are more fecund and survive longer on the latter diets (Nawar *et al.*, 1993). Karg (1986) summarised the feeding habits of the Uropodidae. Reproduction is usually sexual but some species are also parthenogenic. Dispersal is often during the prolonged deutonymphal stage, when the mites attach to various insects (mostly beetles) by an anal pedicel (Faasch, 1967).

Uroobovella marginata (Koch)

This species [also called (*Fuscuropoda vegetans* (De Geer)] is a slow-moving mite that lives in rotting plant material and in poultry and cattle manure. It subsists on many live and dead organic diets, including insect larvae, nematodes, fungi (Faasch, 1967), and even bread and yeast. Reproduction is sexual, eggs being deposited only by fertilised females. The mite completed a generation in 24 days at 27°C and laid seven eggs/day on a diet of nematodes, but only four on house fly juveniles. The females had a long, 7 month postoviposition period (Jalil & Rodriguez, 1970b).

Several acarine predators of filth flies (mostly in the family Muscidae) occur in the manure of animal farms (see Chapters 23 and 25), including *U. marginata*, but the latter plays only a minor role in fly control. Its rate of predation was one-quarter of that of *Macrocheles muscaedomesticae*, the major acarine predator (Axtell, 1991).

Being sensitive to low humidities, *U. marginata* inhabits the below-surface, damper zones of manure heaps, which are the strata preferred by newly hatched fly larvae (maggots). Mite deutonymphs, males and females attack only first instar fly juveniles; *U. marginata* can neither puncture the chorion of the fly egg nor attack its older, larger and more vigorous larvae. The contents of frozen prey eggs, whose chorion had been ruptured, or fresh fly eggs with punctured chorion, were readily consumed (Willis & Axtell, 1968).

The first acarine predator to arrive in accumulating poultry manure heaps is usually *M. muscaedomesticae*, which is disseminated by house flies. Its numbers peak within 2–3 weeks owing to the short life cycle that requires less than 1 week. Populations of *U. marginata*, which arrive later, on dung beetles, undergo a longer cycle, becoming numerically dominant in the heaps only 5–6 weeks after manure accumulation (Willis & Axtell, 1968). In North America their populations peaked in autumn and spring, being scarce during winter (Peck & Anderson, 1969). The effect of *U. marginata* on prey was demonstrated by Rodriguez *et al.* (1970), who obtained 87% mortality of juvenile house flies after adding 200 mites to a container with 1000 fly eggs. In another experiment, 20 predators placed with 250 fly eggs accounted for 27% pest mortality (Willis & Axtell, 1968). Huge numbers of uropodids occur in the manure of poultry, dairy cattle and swine in the Philippines (De Jesus & Rueda, 1992; see Table 23.1); their role was not specified.

Another medical pest affected by *U. marginata* is the lesser house fly, *Fannia canicularis*. Male and female predators consumed an average of 7.0 eggs during 11 days in laboratory experiments conducted at 21°C. This rate of predation was calculated to cause 14–18% mortality in the pest's laboratory populations (O'Donnell & Nelson, 1967). The effect of pesticides applied onto manure heaps is discussed in Chapter 43.

The slug *Laevicaulis alte* (Férussac) is a serious pest of many crops in tropical east Africa, India and Indonesia. It is a nocturnal animal that forages for food in and just after the wet season. During the day slugs remain hidden in the top soil strata. *Uroobovella marginata* feeds on the slugs by night, out in the open, attacking them by day in their hiding places. Humidity is the dominant factor determining mite feeding. Maximal attack (at 25–35°C) occurred at 46–65% soil moisture (percentage of field capacity), but not at 16–25% or 75–85% soil moisture. Raut (1996) postulated that the thick mucus, produced by and covering the slugs, remains impregnable to the mites in dry conditions, whereas in very humid situations the outer surface of *L. alte* may be too wet for *U. marginata*. The mites are gregarious feeders that usually attack the slugs in groups of two to four, or more. They settle on certain areas on the prey's body and feed for hours; as many as 18–22 mites attacked the same slug in the laboratory (Raut, 1996). Distinct wounds, up to 1 mm in depth, appear at the feeding sites, later developing into 'cancerous spots'. By that time the slugs stop feeding and they die within 2–3 days. Wounds that had only been partially damaged healed if no further feeding took place (Raut & Panigrahi, 1991). Not all slug individuals seemed to be attacked and their eggs were not touched.

The effect of *U. marginata* on *L. alte* depended on the injury inflicted by the successive, day-after-day feeding and by the slug/mite ratio. Substantial death of *L. alte* occurred only when this ratio reached 1/5 or more (Fig. 37.1). It follows that a

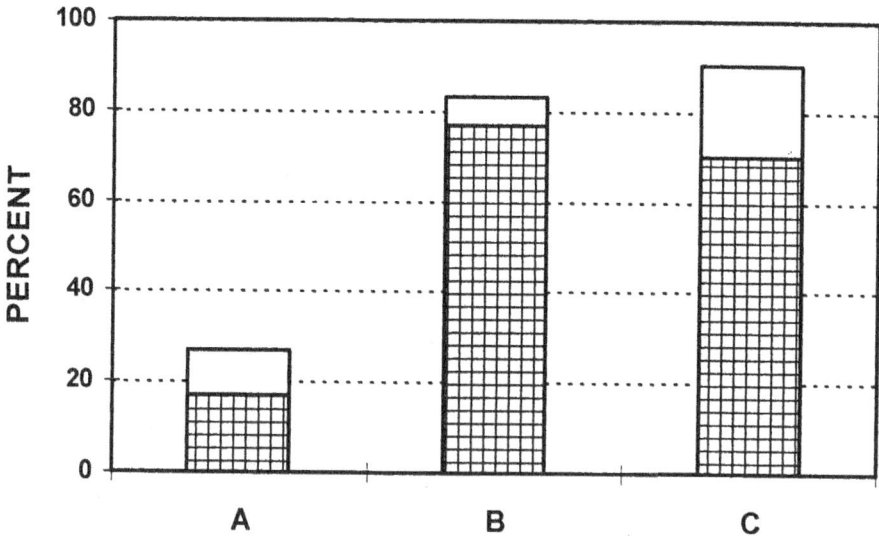

Fig. 37.1 Effect of the ratio between the slug *Laevicaulis alte* and *Uroobovella marginata* on the slug's percentage mortality (cross-hatched bars) and on the mite's percentage attack rate (white bars). A: slug to mite ratio of 1/1; B: ratio of 1/5; C: ratio of 1/10. (Data from Raut, 1996.)

large number of mites would have to be released in the field to achieve the high ratio needed. An untested slug control option is to target their juveniles, which succumb to fewer mites and are more sensitive to desiccation. Both prey and mites can easily be reared in the laboratory, the slug may be reared on potato slices or lettuce (Raut, 1996), whereas the *U. marginata* can feed on the slugs as well as on alternate diets. Royce & Krantz (1991) described a method to mass-rear this species (and other nematophagous mites) on nematodes grown on a fly-rearing medium. The effect of pesticides on uropodids is discussed in Chapter 43.

Chapter 38
Acarine biocontrol agents as enemies of problem-causing organisms

ABAs as enemies of soil-borne pathogenic fungi

Soil-borne pathogenic fungi

Soil-borne pathogenic fungi (e.g. *Fusarium*, *Rhizoctonia* and *Verticillium*) invade susceptible plants via their roots. As they are soil denizens such fungi are difficult to control, and reducing their effect usually requires chemical or physical treatments. Biological control is another quickly expanding option, although very little has been done with mites.

Modes of biological control

Biological control of plant- and soil-borne pathogens has been defined (Cook & Baker, 1983) as reductions in the amount of inoculum (living pathogen units, or propagules, which are available for infection), or in the disease-producing activities of pathogens, that are accomplished by living organisms. The biocontrol of plant-pathogenic fungi might be through the destruction of existing inoculum, by pathogen exclusion from the host or by suppressing it after infection. Destruction of inoculum, through the introduction or encouragement of antagonistic organisms, includes competition, antibiosis and parasitism (or predation). Competition occurs when the controlling agent is more efficient than the pathogen in obtaining nutrients or other essentials in short supply; antibiosis is the secretion of chemicals harmful to the pathogen, whereas predation includes direct exploitation (e.g. grazing) of the pathogen by the antagonist, whether resident or introduced (Cook, 1981). Mite competition with pathogens has not yet been clearly demonstrated, except possibly for the 'vaccination' noted in Chapter 34. Acarine secretions that affect fungi (e.g. antibiosis) have been reported, but predation on disease-producing inoculum by resident fungal feeders seems to be the main effect that mites have on pathogenic fungi. Pathogen exclusion from the host and/or its suppression after infection might also apply in 'vaccination'. An indirect mode of controlling soil-borne plant pathogens is through acarine dispersal of mycoparasites, fungi that parasitise other fungi. Nematodes, bacteria and viruses are also regarded as soil-borne plant pathogens; predators of nematodes are discussed in the next section of this chapter, whereas specific acarine natural enemies of the latter two groups are not known.

Excretion of chemicals: antibiosis

Many mites secrete semiochemicals (externally produced chemicals that affect the same or other organisms in minute quantities; Kuwahara, 1991). Citral, a monoterpene, excreted by the acarid *Carpoglyphus lactis* (L.), suppressed fungal growth in a medium wherein the mite developed. In the absence of mites the medium was overgrown by the mould *Aspergillus* (Okamoto *et al.*, 1978). This is reminiscent of an observation by Oboussier (quoted by Hughes, 1976), that when *C. lactis* fed on a sugary secretion which covered figs, mould growth was prevented by the mite's saliva. Hexyl 2-formyl-3-hydroxybenzoate, isolated from the cuticle of the acarid *Rhizoglyphus robini*, inhibited the growth of *Alternaria alternata*, *Aspergillus niger*, *Fusarium oxysporum* and *Pencillium vermiculatum* (Leal *et al.*, 1990). The extent and effects of such antibiosis in nature are not known (see Chapter 4).

Destruction of inoculum: predation

Much has been written about mycophagous arthropods (e.g. Slansky & Rodriguez, 1987), but insects, and more specifically mites, have seldom been considered as biocontrol agents of pathogenic fungi. The issue is especially intricate in regard to soil fungi, whose abundance is determined by many environmental ingredients. These include abiotic (non-living) factors such as soil chemistry, structure and texture, as well as humidity, temperature and aeration, and diverse biotic components, such as other fungi, additional microorganisms, plant roots and many animals, from protozoa to mammals. Such complex associations constitute almost insurmountable obstacles in the study of the biological control of soil-borne pathogenic fungi by arthropods. A common experimental approach is to extract the animals from infested soils and offer them the relevant plant pathogens in the laboratory. Springtails (Collembola) and mites were often studied in this context (Bollen *et al.*, 1991; McLean *et al.*, 1996).

   The oribatid *Scheloribates azumaenis* reduced the extent of radish root rot, caused by the soil fungus *Rhizoctonia solani*, in jar experiments (Enami & Nakamura, 1996; see Fig. 31.1). In natural soils, however, the mite occurs in smaller numbers and its significance as a biocontrol factor has to be determined. Čatská & Smrž (1989) added *Tyrophagus putrescentiae* and fungal cultures of *Pencillium claviforme* to soil in which apple seedlings were growing. Fungal growth and effect were reduced, but the mechanism was not clear. In the laboratory the mite's feeding on common soil fungi released the sugar trehalose, which served as a nutrient for lytic and antagonistic bacteria. In addition, the mites' excrements increased the pH value of the medium, making it unsuitable for the fungi (Smrž *et al.*, 1987). Certain soil fungi secrete chemicals that attract mycophagous soil mites, a little-studied interaction (e.g. Okabe & Amano, 1990) that may have some significance under natural conditions. Volatiles produced by soil fungi are known to attract collembolan grazers (Bengtsson *et al.*, 1988).

   Other reports provide a less optimistic prognosis. Bollen *et al.* (1991) assayed various soil invertebrates, including the astigmatids *Histiosoma* sp., *Tyrophagus* sp.

and *Rhizoglyphus* sp., against *R. solani* infecting potato seedlings under conditions of high disease pressure. Mites were the least promising of all animals tried. McLean *et al.* (1996) exposed 29 fungal taxa to a collembolan and to the oribatid *Oppiella nova* (Oudemans) in jars to determine whether their grazing could alter the structure of litter fungal communities. No changes in fungal species richness, diversity, dominance or frequency of occurrence were found. This could have been due to several factors, such as only partial consumption of the fungal mass, some fungal parts being inaccessible, fungal growth compensation, or because the animals, which are not specific fungivores, could switch among different fungal species.

Such conjectures suggest that more specific acarine biocontrol agents (ABAs), and/or those that prefer certain fungal structures (e.g. sclerotia), may be more promising. Specificity would also be essential when trying to combine ABAs with other soil-borne antagonistic biocontrol factors, such as fungi and nematodes. A suitable beginning was made by Smrž & Čatská (1987), who found that *Trichoderma*, a mycoparasitic fungus, was barely touched by *T. putrescentiae*, and that no mites survived on this diet. Still, other mycophagous mites could feed on these fungi, negating their beneficial effect. Such generalists would also consume vescicular–arbuscular mycorrhizal (VAM) fungi, a topic about which nothing is known. Given the large number of mites, of most orders, that feed on fungi, it is surprising that no species that strongly prefer, or are restricted to, soil-borne pathogens have so far been found. Finally, a further exploration of the effects of astigmatid secretions on the growth of moulds and other fungi may be rewarding.

Competition or suppression

As noted in Chapter 34, an infection of the soil-borne pathogen *Verticillium dahliae* was suppressed in cotton plants by 'vaccinating' them with an early exposure to spider mites. The mode of action was postulated to be depletion of very rare nutrients that are critical to both mite and fungus, and/or to the activation and/or induction of protective chemicals (Karban *et al.*, 1987).

Transmission of mycoparasites

The mycoparasite *Coniothyrium minitans*, of proven biocontrol activity against the soil-borne plant pathogenic fungus *Sclerotina sclerotiorum* (Budge *et al.*, 1995), was dispersed in sterile and non-sterile soil, to a minimal distance of 55 mm, by the acarid *Acarus siro*. Viable mycoparasite propagules were found on the mite's cuticle (external dispersal) as well as in its faecal pellets (internal transmission). Germination of excreted conidia came only to 11% of the controls, but all faecal pellets contained viable propagules. In addition, except for a slight delay in germination, passage through the mite's gut did not affect subsequent fungal growth. Williams *et al.* (1998) postulated that an application of *C. minitans* with soil animals that transport this mycoparasite could enhance the control of *S. sclerotiorum*.

Acarus siro is not a prevalent soil mite, but similar efforts with common soil Acari, such as *T. putrescentiae* or *R. robini*, may provide better results.

ABAs as enemies of nematodes

Plant-parasitic nematodes

Plant-parasitic nematodes cause widespread damage to many crops by direct feeding, injection of toxins, galling and the transmission of plant viruses. Most plant nematodes live in humid soils; when soils dry up the nematodes can enter an inactive anhydrobiotic state, in which they may be preserved for long periods until reactivated by free water, being it rain or irrigation. Although chemicals are still the common mode of nematode control in some regions, the use of nematicides is being reduced, because of health and environmental considerations. Notwithstanding many early reports on nematode antagonists (especially fungi), it was only when the use of the chemicals was seriously challenged that interest in the natural enemies of nematodes was rekindled.

Biological control of nematodes

The biological control of nematodes has been defined as 'A reduction of nematode populations which is accomplished through the action of living organisms other than nematode-resistant host plants, which occurs naturally or through the manipulation of the environment or the introduction of antagonists' (Stirling, 1991). The realisation that natural nematode control is often associated with certain edaphic conditions has led to the recognition of suppressive soils, such as soils that contain factors that favour natural enemy increases over those of nematodes, leading to improved pest control (Sayre & Walter, 1991).

An early statement about the biocontrol of nematodes was made by Linford & Oliveira (1938). They collected many biotas that were antagonistic to the root-knot nematode *Meloidogyne marioni* (Cornu) in Hawaiian soils. Six (undetermined) mites, all assumed to restrict nematode populations, were among these biota. The many nematophagous mites obtained since are referable to several acarine cohorts. Most of the following records are based on Rockett (1980), Walia & Mathur (1994a) and Walter (1988a).

Mites as nematode antagonists

The mites that feed on nematodes and could reduce their numbers belong to many lineages. In the suborder Astigmata, these are mostly members of the Acaridae, and especially *Tyrophagus* spp. In the Cryptostigmata they are Ceratozetidae (Chapter 11), Galumnidae (especially *Pergalumna*) (Chapter 18) and Haplozetidae. Within the cohort Endeostigmata they include Alicorhagiidae (especially *Alicorhagia*), Alycidae, Bimichaelidae (especially *Alychus*) and Eupodidae, whereas in the Prostigmata they include the Paratydeidae and Tydeidae (Chapter 36). The Mesostigmata contain many families of nematophagous species, such as Ascidae (especially *Gamasellodes* and *Lasioseius*) (Chapter 8), Digamasellidae (especially *Dendrolaelaps*), Eviphididae, Laelapidae (especially *Geolaelaps*) (Chapter 21), Macrochelidae (especially *Macrocheles*) (Chapter 23), Ologamasidae, Parasitidae (especially *Pergamassus*) (Chapter 25), Phytoseiidae (Chapter 26), Rhodacaridae (especially *Rhodacarus*), Veigaiidae and Zerconidae.

Members of the various taxa differ in their mode of feeding. Astigmata, Cryptostigmata and some of the Endeostigmata (e.g. Alicorhagiidae), which possess chelate-dentate chelicerae, ingest the entire nematode, whereas the Mesostigmata puncture and mangle their prey, only ingesting the body fluids. A thick cuticle would thus not be a defence mechanism against the former, but could hinder nematode feeding by some of the latter (Epsky *et al.*, 1988). Predators differ in the stage of prey eaten. *Lasioseius dentatus* (often identified as *L. scapulatus*) fed only on second stage larvae (often termed juveniles) of *Meloidogyne incognita* (Kofoid & White) (Imbriani & Mankau, 1983), whereas the acarid *Sancassania ultima* Samsinak fed on the egg masses, juveniles and females of root-knot nematodes (Sell, 1988, who called the mite *Caloglyphus* sp.). The voracious *Lasioseius* is capable of consuming more than 100 nematode juveniles in a day (Walter *et al.*, 1993). Some nematophagous mites show a functional response to increasing nematode numbers, as the high densities of active prey can stimulate female mites to attack and kill more nematodes than they actually consume (Bilgrami, 1997; Epsky *et al.*, 1988).

Specificity to nematodes

Nematophagous mites may be placed in three functional groups, arranged in increasing order of specificity (Walter *et al.*, 1988). The general feeders do not show preferences and devour any animals that they overcome. They include members of most of mesostigmatid families. The second group are fungivores that also feed on algae and nematodes; they include the Acaridae, the Ceratozetidae and many prostigmatids. The third group are the specialised predators of nematodes, placed in the families Eviphididae, Alicorhagiidae and Bimichaelidae. No member of the latter group is known to be an exclusive feeder of plant-parasitic nematodes, nor were any of these specialists accredited with affecting populations of pest species. The non-specialised predators often (sometimes preferentially) feed on other diets, including fungi, thereby apparently detracting from their pest-consuming potential, while enabling them to survive in the soil for longer periods. Despite their non-specialisation, only members of the generalist and fungivorous groups have been suggested for the control of pest nematodes. These mites are prevalent in the soil and are often very voracious. Some (e.g. *Tyrophagus zachvatkini* Volgin) consume dry, anhydrobiotic nematodes (Walter *et al.*, 1986), a diet that confers the advantage (with regard to biological control) of not being restricted to very humid soil situations. The attraction of *S. ultima* to undamaged roots, where they hunt for gall-making nematodes (Sell, 1988), is an important attribute of that predator.

Mites referable to all above-noted cohorts were present in extractions of arthropods from Colorado grassland soils, with clear dominance by members of the suborders Mesostigmata and Prostigmata (Walter & Ikonen, 1989). Of the 63 species of Mesostigmata obtained, only six did not readily feed on nematodes, indicating that this prey is commonly consumed by mites. Large predators can kill three to five times more prey than small mites (Walter & Ikonen, 1989), but their size does not permit them to hunt within narrow soil channels, thereby restricting their feeding to nematodes that reside in the upper soil layers. Some of the smaller predators (e.g. *Rhodacarus*) have slim, elongated and flexible bodies that facilitate their movement and foraging within the narrow soil pore space wherein many nematodes reside. Smaller mites usually consume relatively few nematodes.

Experimental studies

Most quantitative data about mite feeding on nematodes of economic importance was obtained in laboratory studies, usually conducted in Petri dishes or in pots with soil. Such experiments often show that the mites have a very high rate of predation, seemingly causing close to complete nematode annihilation in the arenas. Imbriani & Mankau (1983) reared *Aphelenchus avenae* Bastian in such dishes and added one, two, four or ten *L. dentatus* for 10 days. The presence of even a single mite in a dish caused a 70% reduction in nematode populations, and with four mites, prey numbers declined by 99% (Fig. 38.1). The very slight increase in survival between the four mites/dish and ten mites/dish could be attributed to mutual interference between predators.

In a similar experiment Bilgrami (1994, 1997) offered an array of detrivorous, plant-parasitic and predatory nematodes to *Tyrophagus putrescentiae* and to *Hypoaspis calcuttaensis* on water agar in the laboratory. Maximal feeding (after 24 h at 28°C) was on the migratory juveniles of sedentary endoparasitic nematodes. The mites killed 82 and 78% of *Anguina tritici* (Steinbuch) and *M. incognita*, respectively, although some nematode species were little eaten. Minimal feeding (but still usually around 40%) occurred on the predatory nematodes, which could have been resistant to predators. Moving to greenhouse experiments, Sharma (1971) examined the ability of three Mesostigmata to reduce nematode numbers in pots planted to grasses and inoculated with 25 adults of *Tylenchorhynchus dubius* (Bütschli). The predators were *Rhodacarus roseus* Oudemans, *Pergamasus truncatellus* (Berlese) and *Geolaelaps aculeifer*. The first two had little effect on final nematode numbers after about 4 weeks (bars 1 and 2 in Fig. 38.2), but the laelapid *G. aculeifer* reduced nematode populations by almost as much as all three predators together (bars 3 and 4 in Fig. 38.2). The predator had another subtle but important influence on nematode populations, as it appeared to prefer pest juveniles.

Fieldwork indicated that nematophagous mites occur in very large numbers in European soils. Arable soils in Germany carried about 37,500 nematophagous

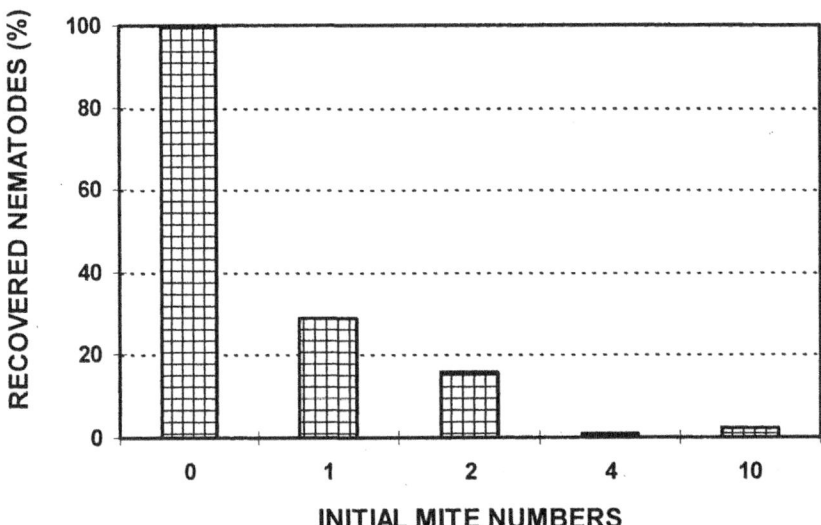

Fig. 38.1 Percentage of nematodes (*Aphelenchus avenae*) recovered from Petri dishes into which one, two, four and ten individuals of *Lasioseius dentatus* were added. (Data from Table 2 in Imbriani & Mankau, 1983.)

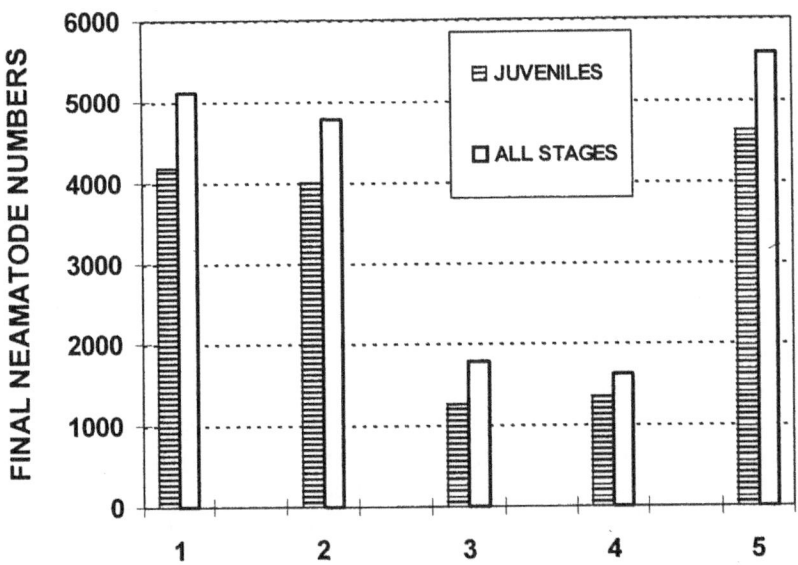

Fig. 38.2 Effect of the predators (1) *Rhodacarus roseus*, (2) *Pergamasus runcatellus*, (3) *Hypoaspis aculeifer*, (4) all mite species together, and (5) no mites, on the nematode *Tylenchorhynchus dubius*, counted 4 weeks after inoculating the pots with the predators. (Based on data in Table 26 in Sharma, 1971.)

Acari/m^2 down to a depth of 15 cm, with the value for compost being ten times higher (Karg, 1983). The presence of such large numbers of predators in the soil suggested that under suitable conditions they might protect field crops from nematode damage. Indirect corroboration for this hypothesis was supplied by Elkins & Whitford (1982), who showed that predatory Acari regulated the numbers of non-pest nematodes in a desert ecosystem. Walter & Ikonen (1989) postulated that in systems where most primary productivity occurs below ground (i.e. grasslands), where nematode biomass is correlated with plant root biomass, and where nematode densities are highest in the root area (rhizosphere), nematophagous arthropods (including mites) could be important predators of plant-feeding nematodes.

Several approaches are available for conducting field experiments intended to evaluate the ability of mites (indigenous and/or introduced) to control pestiferous nematodes. An impressive roster of nematophagous mites, with diverse life history strategies, is already available (many were noted above). The first task could be an assessment of the ability of some predators to affect nematode populations in the field (or greenhouse or mushroom house, Walia & Mathur, 1994b), along with surveying the soil fauna, in which systematics are integrated with biological observations (e.g. Walter, 1987). Information on prey stages taken by the predators should always be included, as some mites do not feed on the adults and eggs of pestiferous root-knot and cyst nematodes (Imbriani & Mankau, 1983). Studies on the soil fauna would probably yield the most data in tropical and subtropical regions, where plant-parasitic nematodes are of more importance and where the acarine fauna is not well known (e.g. Walia & Mathur, 1994a). Rare nematophages that possess high reproductive rates were believed by Walter & Kaplan (1990) to be the most promising ABAs of nematodes. Finally, better water retention and improved soil aeration, by adding organic matter (e.g. manure or compost; van de Bund, 1972), would encourage subterranean natural enemies.

Other agronomic practices that could be used to enhance predatory mite activity include keeping a natural litter layer in place (to reduce soil surface temperatures) and enriching the soil with chemicals that improve porosity and reduce compaction. As noted below, all of these suggestions require vigorous experimental testing. Many plant-protection chemicals (whether applied against weeds, pests or diseases) would be more likely to reduce mite populations than nematode numbers; 'friendlier' treatments could thus affect the balance in favour of the former.

Several major problems have to be addressed before mites can be used for nematode control. The first is the need to find efficient, specific acarine predators that would hunt their plant-parasitic nematode prey within the different soil strata. Part of the latter goal could be attained by bringing the nematodes to the predator instead of the other way around. Some plant-parasitic cyst nematodes produce a sex pheromone that attracts males to their females (Jaffe *et al.*, 1989). Suitable nematode sex pheromones, once isolated, synthesised and formulated on a suitable carrier, could be used to attract nematodes to the proximity of nematophagous mites.

Another issue is the lack of specificity of many predators for their nematode prey, which could have two adverse consequences. The first is that promising predators might be reared and released in the field, only to prefer other prey without affecting pest numbers. The other is the risk that acarine predators would attack beneficial nematodes. The endoparasitic nematode *Controtylenchus brevicomi* (Massey) is known to reduce the fertility of the southern pine beetle, *Dendroctonus frontalis*. Fewer beetles that carried the phoretic nematophagous mite *Dendrolaelaps neodisetus* (Hurlbutt) were also infected by the nematode, implying that the mite had 'cleaned' the beetles of their nematode parasites. This case of mite interference in the natural biological control of a pest was regarded as mutualism between mite and beetle (Kinn, 1980). The contingency that nematophagous mites, released to reduce plant-parasitic nematodes, would attack entomogenous nematodes (used for insect pest control) was explored by Epsky *et al.* (1988). Infective juveniles of two such nematodes were offered to mites referable to three suborders. Although only one (*Gamasellodes vermivorax* Walter) completed its development on a strict nematode diet, its feeding on the juveniles in the laboratory significantly reduced the worms' ability to kill the experimental insect pupae (from about 80 to 25%). Epsky *et al.* (1988) concluded that mites (along with other nematophagous arthropods) could limit the survival of field-applied entomogenous nematodes, even though the predators might devour some of these nematodes to a lesser degrees than others (Bilgrami, 1997).

Also hindering further work on nematophagous mites is the paucity of data about their actual number, distribution and activities in various soils and their layers. Stirling (1991) argued that nematode populations in the soil were actually somewhat small in relation to predator numbers. The more realistic prey/predator ratios that are obtained in the soil are consequently of the same order of magnitude as the daily consumption rates reported for some of these mites in the laboratory. Another consideration relates to the mobility and prey-finding activities of the predators in different soils. Mites are more active in sandy, aerated and loose soils than in heavy, water-logged, compacted earth. Their activities will probably be intermediate in structured soils rich in organic matter. Lack of pore spaces and the presence of free water in heavier soils (especially in deeper strata) would exclude most mites while providing nematodes with refuges. Feeding values (and predictions of control based on them), extrapolated from laboratory studies may thus be misleading.

Little is known about what takes place in the soil when exotic natural enemies are added (or other perturbations occur). The soil ecosystem abounds in predatory and parasitic invertebrates (as well as fungi and bacteria) that could deter any colonisation by newcomers. As noted, studies on the effect of mites on nematodes either were controlled laboratory experiments conducted in pots whose soil had previously been sterilised (therefore containing only a limited flora and fauna), or consisted of field surveys. The latter commonly use ratios between numbers of extracted animal groups to formulate hypotheses about animal activities ('roles') in the soil. As a result, hypotheses on what takes place in the soil when organic matter is added are based on analogies or partial evidence. An example is the assumption that additions of organic matter to the soil cause a reduction in the number of plant-parasitic nematodes by encouraging their natural enemies. Populations of predatory arthropods increased after organic matter (e.g. manure) was added to the soil (Chiang, 1970), but such additions are not necessarily correlated with predation levels on pestiferous nematodes (Stirling, 1991). In fact, increased numbers of non-specific nemato-phagous mites (e.g. generalists) in the soil could result in more encounters among themselves, resulting in intensive intraguild predation. Experimental data (aside from fortuitous observations) are mostly unavailable.

However, mass-rearing nematodes and their acarine predators is simple and cheap (Royce & Krantz, 1991; see Chapter 39), which may encourage and facilitate release experiments. The available evidence suggests that the control of nematodes with mites in greenhouses or in mushroom houses, especially when the plants (or mushrooms) are grown in individual pots or other containers, may be feasible and should be tried.

ABAs as enemies of weeds

Constraints in the application of biological control to weeds

Arthropods, diseases and weeds are the major pests of plants, with weeds being economically the most important. According to a 1978 evaluation (cited in Cromroy, 1983), US$1732.8 million were spent that year on herbicides, compared with US$809.4 million on insecticides. This ratio did not change materially in a later estimate, in which herbicides comprised about 50% of all agrochemicals sold in the world (Woodburn, 1995), whereas insecticides made up only about 30%.

Efforts to use arthropods for biological weed control, which began in the 1830s (Goeden, 1988), pre-date the application of chemicals for that purpose. Despite early and continuing successes, major environmental problems are hindering the wider use of this control option. The problems arise from different perceptions of the pest status of a given weed and from perceived risks due to releasing exotic herbivores into an agricultural and/or native environment. Overlying these dilemmas is the inability to limit or cancel the activities of biocontrol agents after they have been successfully established in a new habitat.

Weeds are indigenous, exotic or cryptogenic plants (Carlton, 1996) that grow where they interfere with management practices, be these horticultural, range maintenance or recreation. However, weeds may be pests in one part of the world and beneficial plants in another. An example is Bermudagrass, *Cynodon dactylon*, which is the preferred grass on many golf courses, but also a major weed in agricultural situations (Cromroy, 1983). Prickly pears (*Opuntia* spp.) provide another example: ornamentals of commercial value in many countries, but weeds of rangelands in Australia.

A given weed may have redeeming features that would argue against limiting its abundance. The introduction of *Tetranychus lintearius* into New Zealand to control gorse was initially contested by beekeepers, farming groups and environmentalists, who contended that the plant could be beneficial. An extended public debate followed, resolved only when all objections were met by proposing alternate management procedures (Chapter 34). A major factor in reaching the consensus was the realisation that benefits accruing from gorse control (which was not expected to be complete) would outweigh potential losses. In contrast, St John's wort (*Hypericum perforatum*), a weed that aggressively competes with range plants and contains the toxin hypericin, was considered suitable for control (Chapter 15).

Except in special cases (as noted), there is usually little opposition to introducing natural enemies to control most exotic weeds, provided suitable precautions (e.g. strict host specificity tests that are conducted abroad or in quarantine) are taken. However, difficulties arise if that weed has indigenous relatives. St John's wort has a native relative in Australia, *H. gramineum*, and the possibility that it might be damaged by the introduced *Aculus hyperici* had to be determined. The mite survived and reproduced on the native plant, albeit at low levels, inflicting only minor damage. Willis *et al.* (1995) concluded that even if *H. gramineum* were to be affected in the field, the effect would be weak (see Chapter 15).

There is strong public objection to introducing herbivores against native weeds (Pemberton, 1985). Some are important components of plant communities and their thinning or removal may have adverse cascading effects on the native wildlife. The ability of biocontrol agents to spread away from the initial introduction area suggests that additional, possibly non-pestiferous stands of the plants may also be affected. Neser & Moran (1985) suggested tactics that could reduce points of contention between conservationists and biocontrol practitioners. These included choosing only plants that were major weeds, selecting natural enemies that would reduce the plants' spread and aggressiveness without affecting their useful properties and, if required, defending the weeds by chemical means. These authors also advocated that consultations be held with environmentalists during the process of deciding on ways and means to control the target weed.

The fear that through 'host shifts' introduced herbivores would infest and damage crop or ornamental plants is another oft-voiced objection. In rebuttal, McFadyen (1998) listed all world-wide known cases (a total of eight) of presumed damage to non-target plants by biocontrol agents. There was only a single case [the moth *Cactoblastis cactorum*, attacking prickly pears in the Caribbean] in which significant injury to a non-target plant was demonstrated. Damage incurred in the seven other cases was minor, insignificant, marginal or unknown. Concerns that introduced herbivores would feed on non-target hosts are routinely alleviated by prolonged laboratory and greenhouse specificity tests, during which relatives of the target weed along with various economic plants are offered as the only food.

Another argument voiced against the introduction of exotic herbivores is that, despite having been strictly tested for their host specificity, they may form totally 'new associations' (*sensu* Hokkanen & Pimental, 1989; see Chapter 39) in the new region and feed on other plants. A well-known case was the shift of a beetle, introduced into the USA to control a weed thistle, onto new hosts, native relatives that were heavily damaged by the insect (Louda *et al.*, 1997). Postintroduction changes in the genetic systems of natural enemies (also called postcolonisation adaptation; see Chapter 39) are on record and their postulated causes were discussed by Murray (1985). Such changes could include 'host shifts', but their actual frequency is not clear. Marohasy (1996) argued that despite over 600 introductions of weed control insects from one geographical region to another, there are very few documented cases of changes in host range. All cases were explained by herbivore behaviour during adaptation, by changes in threshold feeding due to food deprivation and by learning processes, and were not considered to be 'host shifts'. Pemberton (2000) reviewed the 117 cases of biocontrol agents (insects, mites, a nematode and three fungi) established in Hawaii for weed biocontrol and concluded that (except for a single case) there was no evidence for host range changes by introduced species. There is no evidence that biocontrol efforts have ever resulted in the total eradication of a non-target plant species, although (as intended) stands of some weeds have substantially been reduced (Johnson, 1985).

What would happen if the released biocontrol agent was not reproductively isolated from members of a local pest population, and their progeny became more injurious? The application of strict biosystematic tests can go a long way towards dispelling such concerns. One objection to introducing *T. lintearius* into New Zealand was that it may not be reproductively isolated from two polyphagous relatives, namely *Tetranychus urticae* and *Tetranychus turkestani*. The problem was addressed by an elaborate series of crosses and reciprocal crosses among all three species, which firmly established the complete reproductive isolation of *T. lintearius*. This mite's specificity to gorse was assayed in preintroduction tests (conducted in England), which determined that it could not reproduce on diverse hosts either in the laboratory or in the field (Chapter 34).

Other risks associated with biological weed control include the ascendancy of herbivore-resistant biotypes, the danger that other hitherto minor weeds would 'break out', and compensatory growth by the target weed that may affect adjacent plants. The introduction of *A. hyperici* against St John's wort has resulted in reductions in the area occupied by some of the weed's forms in Australia, encouraging others that could be harder to control (see Chapter 15). This risk can seldom be foreseen (e.g. revealed in laboratory studies), and must be dealt with if it develops in the field. Spotted bindweed (*Centaurea maculosa*), a destructive invasive weed in North America, inhibited the growth of a native grass when challenged by a widely-used biocontrol moth. Callaway *et al.* (1999) thought that moth herbivory stimulated the production of defensive chemicals that reduced the growth of the native grass.

A recent evaluation of the risks and benefits of weed biocontrol was presented by McFadyen (1998). Successes in this area have saved many millions of dollars, preserved the lifestyle of entire communities and restored biodiversity to destroyed aquatic ecosystems. If left uncontrolled, weeds cause devastating injury (at times compounded by treating them with chemicals). After weighing the pros and cons, she concluded that classical biological control is the only safe, practical and economically feasible method that is sustainable in the long run.

Methodologies of the biological control of weeds

Biological control of weeds includes several practices. In Chapter 39 the classical biocontrol methodologies are listed, as presented by DeBach (1964) with arthropod pests in mind. Weed biocontrol methods are somewhat different; they are noted here as formulated by McFadyen (1998), with additional notes.

Classical or inoculative method

The introduction of host-specific exotic natural enemies adapted to exotic weeds is the dominant method of weed biocontrol. It seems to be most suitable against perennial weeds growing in dense stands and infesting large land areas. An example (McClay *et al.*, 1999) was noted above.

Inundative or augmentative method

This refers to the mass production and regular releases of native or exotic natural enemies against native or exotic weeds. It requires investments in mass-rearing, and may be most suitable for hard-to-control weeds in annual crops, situations in which rapid and complete control is required (McRae, 1988). The natural dispersal of *A. hyperici* in Australia is limited to 1–2 km/year. A special distribution network was set up to facilitate the mite's spread in weed-affected areas (Jupp, 1996). Weed nurseries were established, artificially colonised by *A. hyperici*, and infested plants were made available to interested persons.

Conservation method

Conservation means protecting the existing populations of natural weed enemies by suitable area management practices, including the use of friendlier chemical treatments, providing overwintering sites and limiting the activities of their own natural enemies. *Aceria chondrillae* was introduced into Australia for the control of skeletonweed, *Chondrilla juncea*. Mite survival in the field requires the presence of at least small weed stands within which it may persist during winter. Farmers were thus advised to avoid too energetic means of weed control, which could result in the elimination of *A. chondrillae* from their plots (Carèsche & Wapshere, 1974). In the past this approach was little emphasised, but Newman *et al.* (1998) advocated that conservation strategies should be integral to any weed biocontrol effort, and especially when using indigenous arthropods against indigenous weeds.

Broad-spectrum method

This approach, added by Wapshere *et al.* (1989), signifies that natural enemy populations are artificially manipulated to restrict their attack on the weed, in order to achieve the desired level of control.

Occurrence of mites on weeds

How common are mites on weeds, how specific is the association and what is the effect of these phytophages? A partial list of generalist and specialist herbivorous mites recorded from major weeds (Cromroy, 1983) suggests that these plants are attacked by many species, some of which are major pests of crops (e.g. *T. urticae*). In addition, it is a common observation that generalist spider mites infest weeds growing in cultivated fields and especially in glasshouses. Weeds are often the first plants to be infested by generalist phytophages in glasshouses and may serve as the mites' 'staging areas' before the commercial crop is infested (Fig. 41.4). There is, however, no clear evidence that these mites seriously reduce the weeds' fitness. Surveys of natural enemies associated with weeds in the Pacific and south Asian regions (Waterhouse & Norris, 1987; Waterhouse, 1994) listed a total of 40 weed species. Only four were infested by generalist mites (e.g. *T. urticae*) and five by specialists. No plant-feeding Acari were associated with the other plants, which comprised around 75% of the weeds investigated. Tseng (1990) sampled 101 weed species growing in paddy and upland rice in Taiwan and obtained only about ten herbivorous species. Thus, despite the occurrence of many phytophagous mites (including diverse eriophyoids; Boczek & Petanović, 1996) on weeds, evidence for their ability to limit the distribution and injury of these plants remains meagre. However, effects could be more subtle, and indigenous plants might become weeds should the limiting effects of diverse indigenous phytophagous mites be removed. Guretzky & Louda (1997) demonstrated that a suite of naturally occurring insects significantly reduced the growth and survival of a native (to North America) potential weed species, and the same could be applicable to mites. In addition, the 'monocultural' stands of many invasive weeds are similar to the fields of cultivated plants, and pests (including mites) probably accumulate there just as they do on crops. For this reason, and owing to the increasing appreciation of the faunal richness of the Eriophyoidea (of which only about 5% have so far been named; Amrine & Stasny, 1994; see Chapter 15), more weed biocontrol agents probably remain to be discovered.

The first mite employed for weed control purpose was the North American *Tetranychus desertorum*, a pest of cacti that was accidentally introduced into Australia in the 1920s. Cacti became major weeds in that continent and the mite was observed to thin out 75% of a densely weeded area of about 250 ha in 2 years. Its importance diminished after the introduction of another, more efficient enemy of cacti that displaced *T. desertorum* as a controlling agent (see Chapter 34).

Only five mite species were deliberately introduced across international borders for weed control (Briese & Cullen, 2001). They include the eriophyids *A. chondrillae* against skeleton weed, *Aceria malherbae* against bindweed (the only acarine natural enemy of weeds currently in commerce) and *Aculus hyperici* against St John's wort. The other two are *Orthogalumna terebrantis* against waterhyacinth (Chapter 18) and *T. linterarius* against gorse (Chapter 34). None has been able to control the target weeds on its own, but they seem to add to the weeds' overall stress and contribute to their decline under specific conditions. The strict specificity of most eriophyids for their host plants and even for certain plant parts is a major advantage when using these mites. The gall formers have the added benefit that when hidden within their galls they are less susceptible to environmental stress, including predation. Studies on *A. chondrillae* also indicate that the damage that these galling mites cause extends beyond the galls themselves.

Predicting the effectiveness of an ABA against weeds

As noted below (see Chapter 39), Harris (1973) proposed a scoring system meant to simplify the process of recognising effective candidates for weed control while excluding unsuitable agents. The system was concurrently modified by Goeden (1983) and Cromroy (1983), the latter using it to predict the potential effectiveness of mites. The Cromroy system has 12 categories, and the candidate ABA is scored by various numerical values. The system (with slight modifications) is used below in order to sum up the scores (marked in **bold**) obtained by the best studied species, *T. linterarius*.

1. Host specificity
   D. Broadly polyphagous 0
   B. Oligophagous (feeding on a limited number of related plants) 1
   C. Restricted monophagous 2
   D. Strictly monophagous **3**
2. Direct damage inflicted
   A. Leaf mining or gall forming 1
   B. Defoliating **2**
3. Indirect damage inflicted
   A. None 0
   B. Reduction of seed production **1**
   C. Disease transmission or increases susceptibility to invasion of other
       adverse organisms 3
4. Phenology of attack
   A. Limited period of attack that does not affect plant's resistance to
       adverse environmental conditions 0
   B. Limited period of attack that, combined with another agent, covers
       the growing season 2
   C. Limited period of attack that increases plant susceptibility to adverse
       environmental conditions 3
   D. Attack during entire growing season **4**
5. Number of generations
   A. Obligate univoltine species 1
   B. Two or three generations/year, according to the climate 2
   C. More than four generations/year, according to the climate **4**
6. Average progeny/female/generation
   A. Under ten 0
   B. Ten to 100 .... **1**
   C. Over 100 2
7. Extrinsic mite mortality factors
   A. Natural mortality mostly due to environmental factors 0
   B. Natural mortality mostly due to competitors for the host 3
   C. Extensive mortality due to specialised enemies and diseases, and
       relative immunity to non-specific enemies **4**

8. Feeding behaviour
   A. Solitary feeders 0
   B. Gregarious or colonial feeders **2**
9. Compatibility with other control agents
   A. Poor, may restrict other agents 0
   B. Good **2**
10. Distribution
    A. Local 0
    B. Covers half the range of the target weed 2
    C. Covers three-quarters of the range of the target weed 4
    D. Covers the full range of the target weed **6**
11. Evidence of effectiveness as a control agent
    A. Failure in previous biocontrol attempt(s) 0
    B. Controls host in native habitat or in one region of introduction **4**
    C. Controls host in two or more regions of the world 6
12. Dissemination
    A. Forms patches or 'pockets' of small populations 1
    B. Forms patches or 'pockets' of large populations 3
    C. Widespread but in small populations 5
    D. Widespread in large populations **7**

According to Cromroy (1983), scores of more than 27 would indicate likely candidates for weed control, and *O. terebrantis* scored 31. When the points for *T. lintearius* were summed up, it scored 40, suggesting that it should be a very effective biocontrol agent. However, as detailed in Chapters 15 and 34, neither species seems to be capable of fully controlling its respective host weeds.

Plant disease transmission by mites

Acari from various lineages are vectors of plant diseases. At least one, the eriophyid *Phyllocoptes fructiphilus*, is used to transmit an apparently viral disease to a weed (multiflora rose) to enhance its control (Chapter 15). Other eriophyids are vectors of plant diseases (Oldfield & Proeseler, 1996) and, owing to their strict specificity, might be considered for weed control. Tarsonemidae (Chapter 33) that are phoretic on pine beetles (Coleoptera: Scolytidae) enhance the transmission of bluestain fungus, *Ceratocystis minor*, between host trees, and infection rates were correlated with mite numbers (Bridges & Moser, 1986). Mites could thus be used to transmit diseases to weeds, promoting their control.

Biological obstacles in the application of biological control to weeds

A problem that could reduce the effectiveness of mites for weed control is pressure from predators. *Arthrocnodax* spp. (Diptera: Cecidomyidae) were considered to 'play the main role' in regulating the abundance of *Aceria acroptiloni*, a potential enemy of Russian knapweed (Chapter 15). Many larvae were seen in mite-infested inflorescences, the predators co-occurring with the mite in most of its natural range (Kovalev *et al.*, 1974). Established colonies of *T. linterarius* in New Zealand rapidly declined as a result of predation by *Stethorus bifidus* Kapur (Richardson & Hill, 1998) (Coleoptera: Coccinellidae). At other times the effect of such predators remains unknown. A phytoseiid lived within *A. chondrillae* galls on skeleton weed in most of the mite's Mediterranean range and even invaded its rearings, but did not seem to have any important effects on its population build-up (Carèsche & Wapshere, 1974).

Cromroy (1983) advocated the preparation of a catalogue of mites found on weeds, which would be readily available to interested scientists. The present authors endorse this suggestion and believe that their natural enemies should also be listed.

ABAs as enemies of scale insects

Scale insects

Scale insects (Hemiptera: Coccoidea) are major plant pests, damaging their hosts by feeding, injecting toxins, transmitting viruses and excreting honeydew on which disfiguring sooty-mould fungi grow. The important families are the Diaspididae (armoured scale insects), Coccidae (soft scales), Pseudococcidae (mealybugs) and Margarodidae. Most scale insects (with the exception of the mealybugs) are sedentary during much of their lives. After eclosion the first instar nymphs ('crawlers') leave their birthplace and, unless blown away by winds, often settle at nearby sites, thus forming colonies. In most cases the settled female spends the rest of her life at that location. The postcrawler instars either lose their legs when moulting or, if these appendages are retained, seldom move. The short-lived, winged males lack mouthparts, serving merely to inseminate the females. Scale insect colonies, while intensifying damage at their sites, also increase the probability of being detected by natural enemies (including ABAs). Another factor that facilitates the biocontrol of individual scale insects is the long life cycle, which prolongs their exposure time. Scale insects usually require several months to raise a generation, a period less than half the time required by most ABAs, which thus produce more than a single generation on a given host scale.

ABAs of armoured scale insects

The body of the Diaspididae is covered by a closely adhering shield that usually protects them from ABAs (thus 'armoured'). However, as the crawlers hatch they raise a posterior flap of the shield in order to emerge. This portal allows ABAs (e.g. *Hemisarcoptes* and *Saniosulus*) to gain access to the hosts. The mites then feed on the body of the female scale, its young and eggs. The Diaspididae is the family that has attracted the most ABAs, being the only coccoid family with a specific genus of associated natural enemies (*Hemisarcoptes*) that can control its populations (Gerson *et al.*, 1990). Most predatory Eupalopsellidae whose prey are known feed on crawlers (although they may be found on the adults), as do some Camerobiidae and Cheyletidae that occur on fruit trees (e.g. Moraes *et al.*, 1989). Pyemotids are generally rare on scale insects, although De Lillo & Porcelli (1993) recorded parasitism rates of about 30% on a *Cotoneaster*-infesting diaspidid in Italy.

Stigmaeidae, especially species of *Agistemus*, may feed on armoured scale crawlers (e.g. Ehara, 1962; Farag *et al.*, 1990), but probably do not affect their populations. Another potential stigmaeid predator of diaspidids is *Eryngiopus*, because three of its species were found under the shields of pestiferous scale insects (Vacante & Gerson, 1989). Several Phytoseiidae feed, reproduce and complete their development when offered only diaspidid eggs and crawlers (McMurtry, 1963; Ragusa & Swirski, 1977). At least one, *Typhlodromus baccettii* Lombardini, seems to have a special relationship with *Carulaspis*, a cypress-inhabiting diaspidid; almost all scales sampled in the spring at Florence, Italy, had the mites under their shields. The predator devoured *Carulaspis* eggs and also fed on available phytophagous mites (Baccetti, 1960). The recent literature on diaspidids as a food source for phytoseiids was reviewed by Schausberger (1998), who demonstrated that three generalists developed and reproduced when given an exclusive diet of crawlers of the San José scale, *Quadraspidiotus perniciosus*. Generalist predators in the families Anystidae, Bdellidae, Cunaxidae and Erythraeidae consume diaspidid crawlers (Gerson *et al.*, 1990), feeding that is probably opportunistic; there is no evidence that it affects the pests' populations.

The citrus red mite, *Panonychus citri*, has a negative, non-destructive association with the California red scale, *Aonidiella aurantii*, a major pest of citrus. Orange leaves were infested with adult mites and about 100 crawlers were added almost daily. Half of the crawlers settled on leaves that had been infested by the mites for 1–4 days, their settlement rate decreasing thereafter. By day 11 no scales were able to settle, suggesting that the mites had rendered these leaves unsuitable for the crawlers (Ebeling, 1948). The mechanism of this inhibition is not known.

ABAs of soft scales, mealybugs and other scale insects

Soft scales do not appear to have any specific acarine predators or parasites, but may be parasitised by opportunistic ABAs, such as *Pyemotes* (Vaivanijkul & Haramoto, 1969). The eggs and crawlers of several soft scale species served as subsistence food for *Amblyseius swirskii* in the laboratory, but only a few mites matured and oviposition was negligible (Ragusa & Swirski, 1977). Mycophagous Tydeidae serve as sanitising agents in citrus groves, because by feeding on the sooty mould that develops on coccid honeydew they reduce the attendant damage (Mendel & Gerson, 1982). Tydeids may serve a similar function in regard to the honeydew of mealybugs, whose eggs and crawlers provided survival food for phytoseiids. Some of the predators fed on this diet and a few eggs were produced, but no progeny developed beyond the protonymphal stage (Ragusa di Chiara & Tsolakis, 1995). Mealybug honeydew provided nutrients for the production of a few eggs by *Amblyseius limonicus* and enhanced female survival (McMurtry & Scriven, 1965a).

Larvae and nymphs of *Leptus* and *Bochartia* sp. (Erythraeidae) attacked the nymphs and adults of the margarodid *Drosicha mangiferae* Green, a pest of mango in India. The mites sucked out the body fluids of the scales, causing them to blacken and shrivel. They killed about 15–20% of the pests, were active from January to May, and later migrated to the soil (Tandon & Lal, 1976). The trombidiid *Allothrombium mitchelli* Davis fed on and seemed to reduce the numbers of the beech scale (*Cryptococcus fagisuga*) in the eastern USA (Wiggins *et al.*, 2001).

This variety of reports suggests that additional natural enemies of scale insects remain to be discovered, especially in the tropics, but available evidence suggests that, with the exception of *Hemisarcoptes*, few ABAs capable of controlling scale insect pests will be found.

ABAs as enemies of stored product pests

Stored product pests

Stored food products provide a benign environment with an abundance of food for many pests, most of which are world-wide in distribution. They consist mainly of the larvae of moths (Lepidoptera) and the larvae and adults of beetles (Coleoptera), as well as a few mites, especially *Acarus siro* in the Acaridae (Chapter 4). Because of health constraints and the emergence of resistance to many pesticides, most stored food pests are difficult to control by chemical means.

Brief history and constraints

The use of natural enemies to control stored product pests (reviewed by Brower *et al.*, 1996; Schöller *et al.*, 1997), advocated since the beginning of the nineteenth century, has largely been 'overlooked and under-exploited' (Haines, 1999). An early effort was an in-depth study on stored grain and flour mites in England during World War I (Newstead & Duvall, 1918), whereas a later enterprise was the taxonomic treatment of these Acari (Hughes, 1976). The best known ABA of stored product mites is *Cheyletus eruditus*. Studies on this species include an investigation of its quantitative relationships with stored product mites (Gause *et al.*, 1936) and commercial efforts to control pest mites by augmenting its populations (Ždárková & Horak, 1990). Bruce (1983) reviewed other efforts in which members of the families Ascidae and Pyemotidae were assayed against various stored product insects; data on these groups and on the Acarophenacidae are given in the relevant family chapters.

Application of biocontrol methods in food stores has often met the objection that even if pests were to be totally controlled by natural enemies, the bodies of the latter would still contaminate the product. The main problem would be in foods where consumers demand absolute purity, whether for health, aesthetic or religious reasons.

Effect of pesticides

The role of predatory mites in spontaneously controlling stored product mites began to be realised in Europe when acarine numbers were tallied after fumigations. Numbers of *A. siro* greatly increased, whereas *C. eruditus* almost disappeared. Stores in which the predator had been dominant were overrun by huge numbers of the pest 2–4 months post-treatment, with very few *C. eruditus* being found (Norris, 1958; Pulpán & Verner, 1965). This suggested that before the fumigation *A. siro* may have been controlled by the predator. However, as *C. eruditus* was more susceptible to the fumigant (methyl bromide) than *A. siro*, as well as being slower to recolonise the substrate, the pest had escaped the regulating effect of the predator. Such observations were the basis of efforts by Pulpán & Verner (1965) to augment *C. eruditus* in stored grain in Czechoslovakia. White & Sinha (1990) believed that the high toxicity of the organophosphate chlorpyrifos-methyl to predatory mites could result in increased pest populations in granaries.

Recent developments

Many of the objections to using natural enemies to control stored product pests have weakened in recent times. First and foremost, the ongoing, urgent need for basic foods in many regions makes for less fastidiousness. Any remaining low levels of arthropod contaminants left by successful natural enemies would be a vast improvement over the serious losses (10–30%; Schöller *et al.*, 1997) incurred by the pests. The presence of low levels of arthropods (or their remains) could and would be tolerated in such regions (Rees, 1994).

In other parts of the world ongoing pressure to reduce pesticide usage in stored foods is also putting new life into using biocontrol (Haines, 1999). As noted by Cox & Wilkin (1998), representatives of the UK grain industry would be interested in using biocontrol agents if other control measures became unviable and if the agents would be equally effective and less expensive. In addition, beneficial insects and mites have the advantage over microbial products (such as the insect pathogen *Bacillus thuringiensis*) of not requiring registration.

In Denmark, reductions in the use of the fumigant methyl bromide are necessitating the implementation of other control strategies. Biological control is prominent among the available options, and the ascid *Blattisocius tarsalis* is one of the two biocontrol agents to be assayed against a major moth pest in flour mills (Hansen, 1998).

Increasing demands for health foods are also promoting biological control in stored food. The claim by producers that their products are totally 'natural', grown and preserved without any chemicals (including pesticides) is integral to their industry. A balance could therefore be found between a few arthropod remnants in the product and the assurance that it is totally pesticide free. The success of *C. eruditus* in eliminating pestiferous mites in stores, before the placement of grain there (Ždárková & Horák, 1990), is an option that should be further explored. Finally, stored foods intended as animal feed are appropriate for biological control.

A promising new development is the addition of biocontrol to integrated pest management programmes within the context of stored product protection (Schöller *et al.*, 1997). Indigenous ABAs, as noted, already play a part in these endeavours. A further step could be the introduction of ABAs from one part of the world to another. No such efforts have hitherto been made, although numerous suitable candidates are available. Haines (1999) and Tseng (1979) provided lists of predatory mites found in many parts of the world, and they could be used (following suitable screening) if and when interest in such projects matures. In the cooler parts of the world, where acarine pests such as *A. siro*, *Glycyphagus destructor* (Schrank) and *Tyrophagus* spp. predominate, cheyletids would probably be of major importance (but see Hansen, 1998). In tropical countries, where mites are only minor pests, owing to the great damage caused by beetles and moths (Gahukar, 1994) the Acarophenacidae, Ascidae and Pyemotidae could be of greater use.

Chapter 39
Introducing, culturing and establishing acarine biocontrol agents

Introduction

The finding, culturing, releasing and evaluation of introduced natural enemies (also termed 'classical importations' and 'classical biological control') comprise the mainstay of biological control; a large section of DeBach (1964) was devoted to this topic. Introductions were of major importance in the past, but the growing awareness of environmental hazards owing to the transfer of living biocontrol agents between regions has reduced this facet of biocontrol (Chapter 45). With the application of all necessary precautions (e.g. Knutson & Coulson, 1997), the use of exotic acarine biocontrol agents (ABAs) should remain a major component of biological control.

The search for ABAs

General comments

Searching for the natural enemies of a given pest usually begins when growers and pest control practitioners are unable, or unwilling, to use other methods of control. Decisions on the feasibility of using biocontrol are based on the pest's own characteristics and on available knowledge about suitable agents and their attributes. Introduced pests of perennial crops that feed in exposed positions, whose natural enemies are known, proven and available, would be the most suitable targets for biocontrol efforts (Barbosa & Segarra-Carmona, 1993). In addition, defensive characters of pests might be good predictors of successful biocontrol (Dyer & Gentry, 1999). However, when the pest is not well studied and/or efficient enemies are not at hand, the practitioner has to decide on the most suitable natural enemy (or enemies). Several guidelines, theoretical as well as practical, have been proposed for this purpose.

Rosen & Huffaker (1983) and Hokkanen (1989) listed the attributes of effective natural enemies in general. These included searching capacity (considered to be the most important trait), density dependence, high specificity, the ability to increase at least as rapidly as the prey, and adaptability to the pest's environment. Luck (1990) suggested that an effective natural enemy might either show a strong attack rate on the pest (e.g. act as a 'biotic acaricide') or bring about a stable interaction with its prey, but not achieve both. Murdoch *et al.* (1985) formulated two predation strategies that could lead to pest control, concurrent with the persistence of the natural enemy.

'Lying in wait' was the term given to polyphagous predators that could prevent pest outbreaks. Specialist natural enemies, considered to be capable of 'curing' prey eruptions, were called 'search and destroy' species. Bakker (1993) preferred these categories to 'older' terms such as 'generalist' and 'specialist' natural enemies. However, as is implicit in this text, it is impractical to draw a clear line between these categories in regard to ABAs, which exhibit much variation in their modes of pest control. Many efficient ABAs (e.g. Anystidae, Ascidae, Laelapidae and some Phytoseiidae) are polyphagous and yet 'search and destroy' their prey, whereas some monophages (e.g. Acarophenacidae, Hemisarcoptidae) are slow-acting parasitoids that do not 'cure' host eruptions. Another way to categorise ABAs could be as reactive natural enemies, those that wait for their prey to arrive (some Cheyletidae), and as proactive ABAs, those that actively hunt for the prey (e.g. specialists such as *Phytoseiulus persimilis*). However, even these categories (at least in regard to ABAs) are not rigid, and each case has to be evaluated on its own merits.

In regard to phytoseiids, it is important to establish their feeding category, as well as dispersal abilities, voracity and survival during prey rarity (McMurtry, 1982). Other major attributes include climatic requirements (including day length), environmental constraints on survival in new habitats, including in the presence of other natural enemies, and tolerance or resistance to insecticides. Another criterion for selecting a suitable natural enemy is its ability to move within the prey's habitat. Gerson (1985) suggested that the ability of predators of unknown feeding habits to survive within spider mite webs could be an indication of their specialisation for such prey (see also Sabelis & Bakker, 1992). Host (or prey) selection by natural enemies is usually determined by the suitability of the host for the reproductive success of the natural enemy. Luck (1990) argued that behavioural observations on host finding by the enemies, along with data on the survival of their progeny and fecundity, could be used to rank potential host resources. The availability of such resources in the field might therefore help in predicting the success of a natural enemy. The attributes of efficient ABAs are discussed in Chapter 44.

As noted, specific conditions in the pest's environment often decree the attributes required from the ABA. *Neoseiulus cucumeris*, used for thrips control in Dutch greenhouses, undergoes a winter diapause. This led to a search for predators that could be used in winter when peppers are grown. A strain of *N. cucumeris* with low diapause incidence was assayed first, followed by a comparison between five subtropical phytoseiids under laboratory conditions (van Houten & van Stratum, 1993; van Houten *et al.*, 1995b). The candidates, besides feeding and reproducing on thrips and pepper pollen, were also expected to lack a short-day diapause and to tolerate relative aridity. *Euseius scutalis* and *Euseius tularensis* were eliminated owing to low predation and oviposition rates when feeding on thrips (Table 39.1). *Amblyseius limonicus* had the highest predation and oviposition rates, and did not enter diapause, but its eggs were sensitive to low relative humidity (RH). This environmentally driven process of elimination left *Iphiseius degenerans* and *Euseius hibisci*, species that showed median rates of predation and oviposition, had no diapause, and whose eggs were insensitive to desiccation. The final choice of the most suitable natural enemy was thus a matter of compromise, taking several attributes into account.

Table 39.1 Predation (on thrips larvae), oviposition (on thrips larvae and on pepper pollen), incidence of diapause and drought tolerance (expressed as critical humidity) of five phytoseiids tested for their suitability for western flower thrips control in Dutch glasshouses during winter

Species	Predation /day	Eggs/day on thrips prey	Eggs/day on pollen	Dispause incidence	Critical humidity
Amblyseius limonicus	6.9	3.2	1.5	0	0.88
Iphiseius degenerans	4.4	1.4	1.4	0	1.48
Euseius hibisci	3.5	1.7	2.8	0	1.66
Euseius scutalis	1.3	0.3	nt	nt	nt
Euseius tularensis	0.5	0.2	nt	nt	nt

Data from van Houten *et al.* (1995b); SE values omitted. nt: = not tested.

Another example was the search for suitable predators of cassava green mite (CGM), during which a major consideration was that repeated releases were not feasible over the large African areas that were to be treated. The candidate enemies were therefore expected to be capable of permanent postrelease establishment, to be able to survive in humid and dry environments, as well as to have almost mutually exclusive attributes: good searching ability along with subsisting on alternate food sources (Janssen *et al.*, 1990). As cassava in South America and in Africa is grown in humid as well as dry areas, the candidate phytoseiids were assayed for their sensitivity to various RHs. A study of about a dozen species and strains showed that much of the interspecific, and even interstrain variation in tolerance to desiccation could be explained by differences in the average RH prevailing at their collection sites (Bakker *et al.*, 1993). Suitable predators were also preselected by testing their responses to plant-pest volatiles (Janssen *et al.*, 1990). The attraction of 11 candidate phytoseiids to odours of cassava leaves infested by CGM was compared; only four showed a significant reaction, but their prey preferences were not in concordance with olfactometry results. Smith *et al.* (1996) concluded that more than a single assay method should be used to evaluate prey preferences fully.

Moving to biological weed control, Harris (1973) proposed a scoring system meant to recognise suitable candidates, while excluding unsuitable ones. He included many of the traits noted above, adding continuous enemy attack during the growing and/or reproductive season, having many (at least four) annual generations and many progeny/generation, a large size, gregarious feeding, being compatible with other agents and effectiveness in two or more regions. Goeden (1983) proposed a revised three-part system that was to be used in sequence. In the first, a candidate was rated for its effect on the weed in its native range. During the second phase it was evaluated for ease of culture and host plant specificity, and in the third its colonisation history, ecoclimatic similarity to the target region and evidence of field efficacy were assessed. A variation of the Harris system for evaluating ABAs against weeds was presented in Chapter 38 (Predicting the effectiveness of an ABA against weeds). Much of the needed information becomes available only after field and laboratory studies, of which host range testing, discussed below, is a central part.

To reiterate, the decision as to which ABA to select for further study depends on the pest problem at hand. Considering the provenance of the pest and of its natural enemies could thus be a starting point. Four situations are possible, calling for different strategies.

1. Endemic or cryptogenic (species of unknown provenance, Carlton, 1996) pests that have efficient indigenous natural enemies whose efficacy had diminished. Such pests are sometimes believed to have been in 'natural balance' with their natural enemies, a 'balance' that had often been disrupted by pesticides. The solution in such cases could be modifying the disruptive practices by using friendlier spray programmes and/or inundative mass-releases of pesticide-resistant natural enemies. An acarine example is using the native, pesticide-resistant *Galendromus occidentalis* in Californian almond groves for spider mite control (Chapter 26).

2. Endemic or cryptogenic pests without effective natural enemies. The release of exotic ABAs is indicated in these cases. Lockwood (1993), who advocated the total restriction of natural enemy introductions, termed this mode of biocontrol 'neoclassical biological control'. An acarine example is the introduction of *P. persimilis*, of Mediterranean origin, into Australia to control the two-spotted spider mite (TSSM) (Chapter 26).

3. Exotic pests that had arrived without their natural enemies; released from their regulating enemies, they might raise large populations and become pestiferous. If local natural enemies cannot control them, then exotic ABAs from the pests' region of origin should be introduced. This is 'classical biological control'. Acarine examples are the introduction of neotropical predators of CGM, also of neotropical origin, into Africa (Chapter 26).

4. Exotic pests that had acquired indigenous ABAs, whose effectiveness, as in the first case, had been reduced. The decision whether to augment local ABAs, import new ones or employ both options depends on the case at hand. An acarine example is the augmentative release of the indigenous *Typhlodromus athiasae* Porath & Swirski to control the exotic European red mite (ERM), on apple trees in Israel (Mansour *et al.*, 1993), along with the introduction of additional predators against this pest (E. Palevsky, unpublished).

Sources of ABAs

Whatever the pests' provenance, there are three main sources for their natural enemies. The first is the pests' area of origin. 'A basic tenet in biological control is that natural enemies capable of attacking and destroying the pest species in its new home are best sought in the land of pest origin' (Bartlett & van den Bosch, 1964). In that region the enemy may be assumed to be fully adapted to the pest, reflected by holding it at non-pestiferous levels. As noted, this is usually understood to be 'classical biological control'.

If suitable ABAs are not available in the pest's area of origin, exploration may be conducted in areas where the pest (or closely related species) occurs and/or in climatically similar regions. When searching for natural enemies of the lucerne flea, Wallace (1973) surveyed the pest's distribution in western Europe and north Africa. The obtained data were then used, along with suitable climatic maps, to locate Mediterranean zones in which to look for suitable predators (Chapter 9). *Bdellodes lapidaria*, intended for introduction to South Africa in order to control the same pest, was collected in Western Australia in areas of comparable Mediterranean climates (Wallace & Walters, 1974). A formalisation of such efforts was the computer-based system CLIMEX (Sutherst & Maywald, 1985, 1991), intended to provide rapid predictions of an animal's distribution and abundance around the world. It furnishes a single value, the ecoclimatic index (EI), which describes the suitability of various locations for specific cold-blooded animals. It is calculated on the basis of weekly humidity, temperature and daylength readings, taking into account also various limiting factors, such as cold, warm, dry and humid stresses. Although unsuitable where animal occurrence and numbers are not determined by climatic factors, CLIMEX provides the main features of the target's distribution and abundance. It can also be used to explore an animal's responses to various environments, including the potential of its population to increase or decline.

The use of CLIMEX to predict the most likely sites for the establishment of *Anystis wallacei*, introduced into Australia to control the red-legged earth mite, was described in Chapter 6. McDermott & Hoy (1997) used CLIMEX to assess the potential of the transgenic *G. occidentalis* to survive in the field under Florida conditions (Chapter 45).

The second source comprises the usually endemic ABAs observed to feed on the pest (whatever its origin) in its current habitat. Suitable ABAs were often discovered because they occurred with the investigated pests [e.g. armoured scale insects (Gerson, 1967b), grapevine mites (James & Whitney, 1993a), the hemlock woolly adelgid (McClure, 1995) and citrus thrips (Tanigoshi & Nishio-Wong, 1982)]. In other cases they were found during special searches. Lesna *et al.* (1995) explored outbreak sites of the bulb mite in The Netherlands, Taiwan and Japan, and found efficient predators of this pest (Chapter 21). Goodwin & Steiner (1996) looked for natural enemies of the exotic *Frankliniella occidentalis* in Australia and collected around 40 indigenous ABAs. A variation on this method is exposing pest cultures ('trap plants') in the field in order to attract natural enemies. Several phytoseiids were thus obtained in British Columbia from bean plants heavily infested by TSSM (Gillespie *et al.*, 1997).

Native ABAs can also be collected from specialised habitats. Looking for nematophagous mites, Walia & Mathur (1994a) screened soils beneath annual field and ornamental crops, perennial fruit orchards and forests. Dung pads, the breeding sites of filth fly juveniles, provided mesostigmatid ABAs that feed on fly maggots (Chapters 23, 25 and 37). The use of indigenous natural enemies, much undervalued in the past (Way, 1986; Waage & Greathead, 1988), is currently receiving greater interest (Chapters 45 and 47). As Huffaker *et al.* (1970) stated: 'Destruction of predators has been shown to cause increase of mites to high densities ... Thus endemic predators certainly hold much promise. We need, therefore, to investigate their roles in spider mite suppression ... and to test their abilities to keep high densities from developing ...'.

The third source is the natural enemies that had only recently come in contact with the target pests, an approach known as the 'new associations' hypothesis (Hokkanen & Pimental, 1989). Underlying this theory is the belief that long-persisting ('old') exploiter–victim associations might have reached some level of balance (or evolved commensalism) that causes the natural enemies to reduce their virulence. This balance enables pest and natural enemy to coexist, thereby reducing the enemy's efficacy. The corollary is that other natural enemies, which had not evolved with prey or host, would be more virulent towards the latter, and be better biocontrol agents. As stated by Hokkanen & Pimental (1989) and backed up by their statistical analyses, such new associations have brought about the majority (c. 75%) of biocontrol successes, and seem to offer a large untapped pool of biocontrol agents that are especially suitable to control native pests. Support for the 'new associations' theory was presented by Dennill & Moran (1989), who advocated its use in weed biocontrol, on top of the 'classical' method.

The 'new associations' approach has been criticised on statistical and conceptual grounds. Recalculations of the success rate of 'old' versus 'new' natural enemies furnished different results from those provided by Hokkanen & Pimental (1989), 'old' associations offering a significantly higher probability of establishment and success than 'new' associations (Waage & Greathead, 1988; Waage, 1990). In addition, it is unlikely that complex coevolutionary processes, leading to the reduced virulence of natural enemies towards pests, would have taken place on commercial plants within the historical period of crop domestication. In fact, a major advantage of the 'classical' approach to the control of weeds (and possibly other target pests) is that the natural enemy can coevolve with its host, thus alleviating the development of resistance to the agent (McRae, 1988). Long-term delays in applying 'classical' biological control were attributed to mistakes in the identifications of pest and/or natural enemy, leading to costly, misdirected explorations (Rosen, 1986). Had the 'new associations' theory been fully applicable, such mistakes would have increased, rather than decreased, the success rate of these efforts. Furthermore, a major aim of classical biocontrol is the reduction of pest populations to low, subeconomic, self-maintaining levels, with enough pests remaining in the ecosystem to provide hosts (or prey) for their enemies. A stable association would thus emerge as the predator keeps its prey at a low level. In other words, it is in the interest of long-term biocontrol that pest and predator should coexist. 'New association' natural enemies, because of their greater virulence, would wipe out pest populations along with their natural enemies, thus acting like pesticides and leading to new problems during the next population cycle.

The phytoseiids that controlled CGM in Africa were found on cassava plants; that is, they were 'old' associates. Lesna *et al.* (1995) found the more suitable predators of bulb mite at sites where the pest occurred, and many other examples were described in the family chapters. Thus, in agreement with DeBach & Rosen (1991), natural enemies which had evolved with the target pest in its native region should be the first choice for importation, and predators associated with the pest there should be tested before undertaking costly explorations. This does not imply that natural enemies of related species, the 'new associates', are to be ignored. Finally, the universal adoption of this new concept, which holds that natural enemies can 'switch' to other prey or hosts, would sound the death-knell for any importations of exotic natural enemies in the present eco-political climate.

Implicit in explorations for and collections of natural enemies, from whatever source, is that minimal data on the biology and ecology of the pest, the natural enemy, their host plants and their taxonomic placement should be available.

Number of sites to sample and collecting ABAs

Having settled on exploration strategy and search location, the practitioner has to decide how many sites should be inspected and the number of individuals to be collected. As noted below ('Founder effect' and genetic drift), it is best to collect from as many sites as possible and theoretical considerations suggest that samples of ABAs in the range of 20–100 individuals/site would suffice. In addition, sufficient numbers of both genders should be included in the samples (see Chapter 40).

Müller-Schorer *et al.* (1991) proposed a method for determining the minimum sampling effort required for discovering most of the potential weed control agents occurring in the native range of a given plant. Rarefaction (dilution) curves were used to estimate the number of herbivore species expected to occur on a given weed, using presence/absence data and species frequencies. The shape of the resultant curves could indicate sites that still contain undiscovered herbivore species and also whether sampling at additional sites would reveal more species. As ABAs of pests are often associated with specific plants, this method could be useful in the present context.

There are two main methods of obtaining ABAs in the field: taking those that feed on or are immediately associated with the pest, or collecting at random, not clearly associated with any prey. The collected natural enemy is placed in small vials or other containers (which should be suitably cooled, humidified, aerated and labelled), and brought to the laboratory for determination and subsequent rearing.

Currie (1934) provided an early description of collecting ABAs in the field. *Bdellodes lapidaria* hides under bark or fallen branches; placing boards in the field as mite 'traps' enables large-scale predator collections (1000 mites/person per hour) (Chapter 9). Working in another era, Wallace (1972) used a portable, motor-operated sucking machine to obtain large quantities of that mite in the field. Lange *et al.* (1974b) employed three different methods to obtain *Anystis baccarum* for TSSM control. They sieved out egg clusters from the litter in which they had been deposited, collected moving mites with a sweepnet, and trapped adults that ascended or descended from trees. Lesna *et al.* (1995) obtained bulb mite predators that were associated with the pest within infested bulbs, Ramakers (1978) observed phytoseiids attacking thrips in the greenhouse, and Gerson (1967b) collected *Hemisarcoptes* off parasitised scale insects. The advantage of collecting pest-associated ABAs is that the parameters of the collection clearly establish the ABA–prey (or host) affinity. In other cases (especially with phytoseiids), predators may be obtained by beating on (or otherwise processing) plant material and collecting the falling mites.

Precise labelling of all collected mites is an absolute prerequisite for their future use. Information on the specific habitat, time of day and other relevant details should also be noted. It is essential to identify all collected animals and to keep voucher specimens; the same should be done with all reared ABAs. The routine practice of comparing mass-reared mites with those collected in the field helps to preserve the integrity of the rearing.

Choosing the ABA: empirical or predictive approaches

Having collected several natural enemies, should one concurrently introduce all of them, without prior knowledge of their abilities, or first study each and choose only those that appear to be the most promising? DeBach & Rosen (1991) stated that 'Multiple importation, either simultaneously or sequentially, of diverse natural enemy species is the only practical manner of obtaining the best natural enemy for a given habitat, or the best combination of natural enemies for this habitat, or the best combination for the entire host habitat'. They thought that importing similar natural enemies, which compete with each other, would cause the most effective species to win out, thus producing better pest control. If competition persists, the combined effect of two or more natural enemies would lead to even better control. Mori *et al.* (1990) used the indigenous *Neoseiulus longispinosus* along with the introduced *P. persimilis* to obtain better spider mite control on cucumbers than when either predator was assayed alone.

The other view (the 'predictive approach') holds that each (or the more promising) species should first be studied and its interactions with other natural enemies (whether additive, neutral or adverse) determined. What approach should the practitioner chose? Environmental opinion currently leans towards the second option. In fact, during the recent CGM project (Chapter 26) many predators were collected in Brazil, and the more promising were studied in the laboratory before release (Bakker *et al.*, 1993; Smith *et al.*, 1996). An entire guild of natural enemies might be associated with the pest, and their interspecific relationships could be disruptive to overall control (Rosenheim *et al.*, 1995). Such disruption might be due to a very assertive species, or to one with a wider range of diets, as well as to certain unpredictable host plant attributes. Brødsgaard & Stengaard Hansen (1992) compared the performance of two predators of thrips in glasshouses; the better ABA, *N. cucumeris*, was displaced by the less efficient *Neoseiulus barkeri*. Other cases of intraguild predation are noted in Chapters 26 and 42.

Could there be an intermediate option (Ehler, 1990)? Serious invading and/or outbreak pests might best be dealt with by 'empirical' importations of diverse enemies, although some prior knowledge must be at hand in order to avoid ecological catastrophes (see later, Use of various strains of ABA). The 'predictive' approach, in contrast, would be suitable for importing enemies against pests of less immediate concern, or to control serious pests which, owing to specific ecological considerations, require more study. At this time it seems that fewer and fewer biocontrol practition-ers subscribe to the 'empirical' approach.

Host range testing

Having collected promising ABAs, the host range should be tested. In the practice of weed biocontrol strict tests for specificity have always been the norm (Chapter 38, ABAs as enemies of weeds), for instance the screening undertaken by Hill & O'Donnell (1991b) before introducing *Tetranychus lintearius* into New Zealand (Chapter 34). Such testing has, however, not been the norm with arthropods, as prac-titioners seldom worry about any non-target (or even beneficial) organisms that the exotic enemy may attack or displace. In fact, a wider diet spectrum was (and is) often taken as an important trait that enables the enemy to survive periods of pest scarcity. Van Driesche & Hoddle (1997) advocated estimating the host ranges of biocontrol agents by looking at records of hosts and closely related non-target species that coex-ist with the candidate enemy, conducting physiological and ecological host range test-ing in the laboratory and then postrelease assessments. The last step is critical because host range estimations in the laboratory can, at best, suggest what would happen outside; only field tests will confirm or reject former predictions. Modest expansion of host range testing for projects on continents was recommended, with more extensive testing being proposed for enemies that were intended for islands.

Use of various strains of ABA

Biocontrol practitioners may collect and release various strains of the same agent, obtained from different areas. These strains (also termed 'races' or 'ecotypes') are intended to increase the genetic diversity of the species in the release area, thereby enhancing the odds of success. In their review of the pertinent literature Clarke & Walter (1995) found little evidence that introductions of more than one 'strain' led to improved success rates in the classical biological control of pests (they did not refer to mites).

The data available on ABAs do not support this hypothesis. Strains from different areas differ in biological attributes that affect their success rate. An Israeli strain of *P. persimilis* was more tolerant to low humidity than a Californian strain (Perring & Lackey, 1989), a trait that affects field survival. Galazzi & Nicoli (1996a) compared the life histories of populations of *P. persimilis* that originated from different parts of Italy; some lived significantly longer (Fig. 39.1) and were more fecund. Pesticide-resistant strains of phytoseiids have usually led to improved pest control (Chapter 26).

Once the various strains commingle in the field they would probably lose their unique features (unless these confer selective advantages). However, use of the unique attributes of different strains to promote pest control in specific situations remains a valid option.

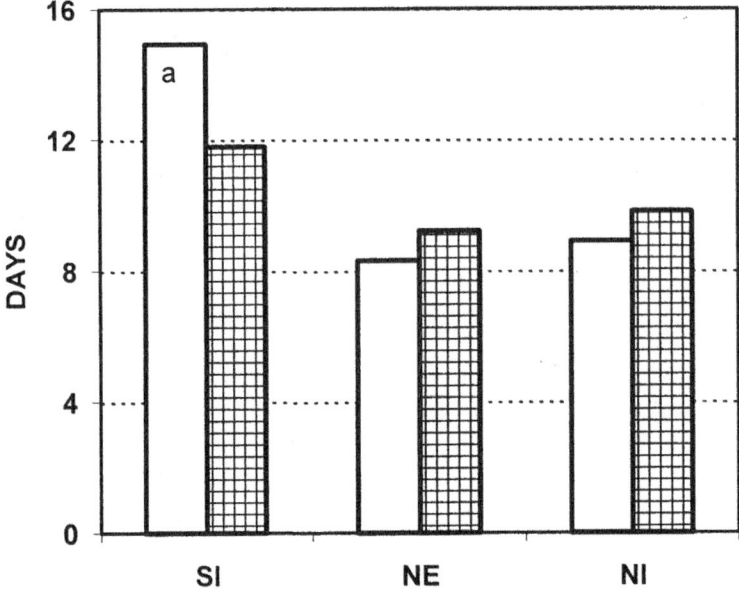

Fig. 39.1 Longevity of *Phytoseiulus persimilis* females originating from three populations (SI: field-collected in Sicily in 1992; NE: a laboratory strain from a British commercial producer; NI: collected in northern Italy in 1991), all kept at 25°C. White bars: longevity at start of experiment; cross-hatched bars: longevity 1 year later; letter denotes a significant difference. (Gleaned from Tables 1 and 2 in Galazzi & Nicoli, 1996a.)

Rearing ABAs

Introduction

ABAs are reared in the laboratory, or at the mass-rearing facility (MRF), for three different purposes: initially, during the postcollection period, to ensure that no contaminants or disease organisms have arrived as stowaways, in order to meet quarantine requirements; secondly, to study the optimal conditions for their mass-culture; and finally, they are cultured in bulk for release (including commerce) purposes.

Rearing the host plants

The mass-rearing of natural enemies of plant pests usually requires the mass-rearing of the prey pests and their host plants; growing healthy plants is thus an essential first step. Basic knowledge about the plants' soil, water, temperature and light requirements is necessary for this purpose.

Legumes (beans, groundnuts or soya beans) are very suitable as hosts for spider mites. They are usually grown at 20–30°C, 50–80% RH and 12–16 h light (>1000 lux) and with the appropriate fertilisers. Cultivars (cvs) of the same host plant may differ

Table 39.2   Effect of four different soil-less substrates on cucumber height, interplant temperatures and numbers of spider mites developing on them

Substrate	Height of plants (cm)	Average temperature (°C)	Average number of mites (seven counts)
Rockwool ('Grodan')	290.5	24.3	173.4
Rockwool ('Flormin')	248.8	24.2	18.6
Polyurethane foam	262.6	23.8	49.0
Peat	267.7	23.0	32.3

Data from Szwejda & Nawrocka (1996).

in their suitability for the prey. Ashihara (1995) tested 13 soya bean and 11 bean cultivars to obtain the most suitable host for three species of *Tetranychus*. Plants are mostly grown in soil, but other substrates can be used. Szwejda & Nawrocka (1996) compared the development of spider mites on cucumbers that were grown on three soil-less substrates ('Grodan' and 'Flormin', two types of rockwool with polyurethane foam, and peat). Plants were tallest, ambient temperatures highest and mite numbers largest on 'Grodan' (Table 39.2).

Unless soils are fumigated, practitioners should avoid their repeated use, which could lead to suboptimal yields and increases the risk of diseases and contaminants. Regular colony turnover is recommended to deal with such dangers. Special care must be taken with plants intended for mass-rearing ABAs for weed control, as these herbivores are intended to damage the plants. They must be in prime condition to support their own pests, and competitors as well as natural enemies should be removed. During the introduction of *T. lintearius* into New Zealand gorse was grown in pots in a shaded house and periodically sprayed with a pyrethroid (PYR, which does not harm this mite), in order to eliminate any encroaching predators (Hill *et al.*, 1991).

Rearing the pests

Spider mites are usually reared under the same conditions as the plants. Other pests, including mites, insects and nematodes, can be obtained in large numbers directly from the field.

Scarlet mite, *Brevipalpus phoenicis*, a pest of tea in Indonesia, was reared on tea leaves whose petioles were kept in plastic tubes with their bases placed in water-filled trays (Oomen, 1982). The predators were reared on the same leaves (Chapter 32). Bulb mite as prey was collected in large numbers from lily fields (Lesna *et al.*, 1995). *Tyrophagus* as prey for various ABAs is cultured on bran, often enriched by wheat germ and yeast (Ramakers & van Lieburg, 1982). Flour mites (*Acarus siro*) were grown on lettuce seeds in small flour bags to serve as prey for *Cheyletus eruditus*. About 20,000 *A. siro* and 100–200 cheyletids were placed in each bag; after 4–5 weeks most prey were consumed and each bag provided about 2100 predators (Chapter 12).

Collembolan prey for lucerne flea predators were mass-reared in small containers padded by moistened plaster-of-Paris and supplied with unspecified organic material (Michael, 1995). Armoured scale insect prey for *Saniosulus nudus* were reared on washed green lemons that had their stems dipped in hot paraffin (Gerson & Blumberg, 1969). When such insects were used as hosts for *Hemisarcoptes* spp., they were reared on potato tubers, squash fruit or pods of smooth cacti (Izraylevich & Gerson, 1995d). In contrast to other plant substrates used, cactus pods have a long shelf-life when infested by scales, are easily grown and maintained, and are seldom attacked by other pests or by moulds. Weiser (1963) and Bruce (1983) used moth, beetle or ant larvae and pupae as hosts for *Pyemotes*, for stored product pest control.

Slug pests (e.g. *Laevicaulis alte*) as food for *Uroobovella marginata* were reared on potato slices or lettuce (Raut, 1996). Nematodes, intended as an alternate diet for ABAs that control nuisance flies, are reared on an organic substrate containing dung and/or on cereals fortified by yeast, on a fly-rearing medium (Royce & Krantz, 1991) or on nutrient agar plates. The latter are then seeded with fungal or bacterial cultures (Imbriani & Mankau, 1983; Bilgrami, 1994) on which the nematodes feed.

Rearing the ABAs

Rearing in arenas

ABAs must usually be reared on an enclosed substrate, to hinder their escape and facilitate their harvest. McMurtry & Scriven (1965b, 1975) described arenas that consisted of dark-painted construction paper overlaying water-saturated foam plastic, itself placed in a 20×20 cm pan filled with water. Strips of wet tissue, placed around the construction paper, served as barriers as well as a water source. This basic unit was later modified, the substrate being changed to black filter paper, metal or plastic tiles (for ease in watching predators). A variation of the barrier involves cutting a ring-like gutter around the arena and filling it with a repellent. Moisture is provided by drilling a hole through the middle of the arena and passing through a wick, whose base rests in water. Food, be it spider mites, other small arthropods or pollen, is added as needed. Friese *et al.* (1987) harvested several thousand phytoseiids during a 4 week cycle with this method. As the mites rest in wedges and avoid direct light, they should be provided with shelter, such as strands or tufts of cotton wool or plastic overlaid with glass cover slips, in which eggs are often placed. Removal of egg-filled cotton tufts from phytoseiid cultures increased their oviposition (Kostiainen & Hoy, 1994). Enlarged units based on the arena model can be used for obtaining more predators with about the same amount of labour (Friese *et al.*, 1987).

Rearing phytoseiids in arenas is especially suitable for predators that fully reproduce on pollen, because the amount of pollen can be controlled and the pollen from various sources may be compared (Ouyang *et al.*, 1992). Rearing on this diet is convenient for routine checks of mite health and sanitation, and for research, because the mites, and their activities, can be observed. An added advantage is that any diet prey does not have to be continuously cultured. Pollen can be collected when available and stored for longer period in a freezer, usually without losing its suitability. An option is to grow suitable plants (e.g. broad bean) for their pollen. It should be noted that different pollens affect phytoseiids in various ways. Pollen of *Typha orientalis* was very suitable for *Typhlodromus doreenae* and *Euseius victoriensis* (James & Whitney, 1993b), but no eggs were produced by *T. athiasae* when they were offered only pollen of *Typha domingensis* (Reuveny *et al.*, 1996). These authors noted that apple pollen, a diet very suitable for *Euseius tularensis*, was unsuitable for *T. athiasae*, (Ouyang *et al.*, 1992). *Agistemus exsertus* produced more eggs on pollen of date palms than on corn or castor bean pollen, whereas pollen of cotton was totally unsuitable, possibly owing to its size and protective spines (Abo Elghar *et al.*, 1969). Pollen-feeding phytoseiids must be provided with urestricted drinking water.

Prolonged or exclusive feeding on pollen could lead to unexpected and sometimes adverse responses from ABAs. The attraction of *A. exsertus* to spider mite prey decreased after being reared for ten generations on castor bean pollen, and its sex ratio (SR) shifted from F1/M1 on mite prey to F1/M1.6 on pollen (Rasmy *et al.*, 1996). Different results were obtained by Dicke *et al.* (1989), who reported that *Amblyseius andersoni* and *Typhlodromus pyri* responded to kairomones of a larger array of prey species when reared on broad bean pollen than when fed only spider mites. This diet-dependent difference in response was attributed to carotenoid deficiency in broad bean pollen, because after adding β-carotene to the pollen the predators responded to prey as did conspecifics reared on spider mites. ABAs must individually be tested for their optimal pollen diets; decreases in predator quality due to mass-rearing in the laboratory are discussed below.

During mass-rearings care must be taken to distinguish between the characteristic growth responses of various ABAs (and even of conspecific populations of different provenance). Phytoseiids assayed for CGM control showed three different responses after 3–4 weeks of rearing (Mégevand *et al.*, 1993). Females showing the first type made up about 25% of the population, other active stages 50% and the eggs 25%. In the second, the ratios were 38, 54 and 8%, respectively, and in the third type the population consisted of 48% females, 36% immatures and 16% eggs. Populations showing the two latter response types had a lower rate of increase compared with those with the first. This could have been caused by suboptimal food and/or to more mutual interference between the larger number of females; whatever the cause (or causes), it should be considered when planning mass rearings. Bearing in mind such growth responses of a species facilitates the rearing of even-aged predators, required for research as well as for mass releases (Drukker *et al.*, 1993).

In order to culture predatory ABAs that must feed on plants, arenas with holes can be used. Plant roots are pressed through the holes to reach an underlying nutrient solution; the leaves remain on the arenas and are thus accessible to the mites. Tedious and labour-intensive variations are to transport the ABAs onto fresh leaves at fixed intervals, or to brush the prey onto the arenas. Alternately, spider mites can be reared in stacks of leaves. The upper leaves, after becoming heavily infested, are transferred to the predator-holding arenas. However, it is difficult to harvest the predators when arenas become heavily stacked with leaves, a problem resolved by using cages (see below).

Rearing on detached leaves or on leaf discs

At times it is difficult to determine prey and predator numbers in arenas, as well as to maintain reasonable levels of sanitation; both problems can be resolved by rearing in detached leaf cultures (Overmeer, 1985). These are leaves that are pressed into wads of cotton wool or filter paper and constantly wetted, usually in an underlying Petri dish. Many spider mite species and their predators were kept for days on such leaves, which can easily be inspected and changed, as needed. The relatively long-lasting leaves of bean and of castor bean are often used, although citrus, avocado and other leaves have also been employed. Nakao *et al.* (1990) used kidney bean petioles, each carrying several leaves, for at least 14 days. Leaves were initially infested by spider mites and *P. persimilis* was added 10 days later; approximately 1000 predators in all stages were collected after another 2–3 weeks.

Leaf discs do not survive for long, and thus are usually unsuitable for mass rearing. In order to rear thrips-feeding phytoseiids Brodeur & Cloutier (1992) placed cucumber leaf discs (lower side up) on an agar-solidified nutrient solution fortified by a plant growth fertiliser and a fungicidal solution. The discs supported the thrips larvae and their predators for more than 15 days; mite escape was minimal and various foods could be added as needed.

Rearing on alternate prey and in cages

As it became feasible to use phytoseiids to control greenhouse thrips (Ramakers *et al.*, 1989), and as these insects are difficult to mass-produce, the predators were reared on other prey, the storage acarids *Acarus* spp. and *Tyrophagus putrescentiae*. The acarids were mass-reared on many substrates, wheat bran being often preferred, as it is cheap and accessible to prey and predators (Ramakers & van Lieburg, 1982). Bran aeration is required for removing excessive humidity and for reducing levels of the carbon dioxide produced by the Acari and the microflora. The mite-containing bran may be applied in the field, or the predators may be extracted by drying the substrate, which causes them to escape, or with water and subsequent sieving.

Some phytoseiids cannot be cultured in relatively open systems, such as arenas or detached leaf cultures, because they tend to escape, becoming entangled in the surrounding barriers, or drown in the surrounding water (Overmeer, 1985), and must be reared in cages. Tanigoshi *et al.* (1975) constructed invertible phytoseiid-rearing units from two cardboard cartons with their bases removed and a window screen fitted between them. The two parts were then taped together, end to end, and a hole (covered only with cloth to permit predator passage) was cut in the screen between them. Leaves infested with spider mites were placed in the upper chamber, which was then inoculated by the predator. After 3 days the unit was inverted and fresh spider mite-infested leaves were placed in the upper chamber. Owing to the phytoseiids' negative geotaxy, they moved upwards, aggregated in the lids and then proceeded through the cloth into the upper chamber. They were then obtained from the lids; each unit yielded 50–100 predators in 3 days. A similar unit, composed of a series of plastic cylinders, was developed by Fournier *et al.* (1985), who obtained 500–2000 *P. persimilis*/day when culturing the predator in a single cylinder. Production schedules can be modified by altering conditions (*c.* 25°C and 80–90% RH) in the rearing units.

Mégevand *et al.* (1993) developed a unit for growing cassava plants within a small space, in order to mass-rear phytoseiids specific to CGM. The unit (a 'cassava tree') consisted of a plastic sleeve filled with rockwool, placed on a metal frame. Sixty-eight cassava cuttings were planted in the rockwool through holes in the sleeve, irrigated and fertilised by a central system. Two weeks later, when they carried five leaves, the plants were infested by CGM and after another 2 weeks several hundred predators were introduced. They were later harvested, each 'tree' providing around 10,000 predators/week. The space required was 56 m^2 in a greenhouse, plus 2 m in an insectary.

Rearing on artificial diets

Rearing generalist phytoseiids on pollen and other non-prey foods suggests that artificial diets could be developed for their mass-culture. McMurtry & Scriven (1965b) offered small droplets of various food solutions (sucrose; molasses; yeast + sucrose; yeast + molasses), placed on excised, washed avocado leaves, to *E. hibisci, A. limonicus, G. occidentalis* and *Typhlodromus rickeri* Chant. The first two, being generalists, developed to maturity and laid a few eggs, but suffered heavy mortality. The last two, which require mite prey in order to reproduce, did not develop beyond the protonymphal stage. Shehata & Weismann (1972) reared *P. persimilis* on a diet that included honey, vitamins, ascorbic acid, choline chloride and distilled water, presented in a membrane. Mite juveniles developed, but the females were smaller and shorter lived, and produced only a few viable eggs. Kennett & Hamai (1980) concocted a diet from honey, sugar, yeasts, casein, fresh egg yolk and water, offering it as droplets dripped onto and coated by parafilm (the 'paraffin-diet droplet'). Given to nine phytoseiids, the diet sufficed for all but two (*G. occidentalis* and *P. persimilis*), but the fecundity of the surviving species was reduced, and viability declined within 2–3 months. A similiar diet, offered via parafilm, was devised by Ochieng *et al.* (1987) for *Amblyseius teke*, and sufficed for several generations. Itagaki & Koyama (1986) reared *Amblyseius eharai* Swirski & Amitai on a chemically defined diet developed for planthoppers (Hemiptera: Delphacidae). It included all essential amino acids as well as some vitamins. Although the rate of adult emergence was high, development was slow.

The overall conclusion is that, except for the diet proposed by Ochieng *et al.* (1987), no artificial diet is yet available for mass-rearing phytoseiids. However, the preparation of such diets, which could be suitable for generalists, would (along with automated rearing technology) reduce the production costs of natural enemies. Such diets might be useful during shipment and could eliminate quarantine problems that arise from the inclusion of alternate prey in the shipments. Artificial diets would be best suited for augmentation purposes (Cohen *et al.*, 1999).

Pyemotes was probably the first ABA to be mass-reared in the laboratory, when assayed in Mexico against the cotton boll weevil at the beginning of the twentieth century (Chapter 1). Weevil-infested bolls were brought into the laboratory, air-dried (to protect against moulds) and exposed to the mites. Rangel (1901) was the first to suggest the use of alternate hosts for rearing an ABA. Weiser (1963) and Bruce (1983) used washed insect larvae and pupae to mass-rear *Pyemotes*; hosts were exposed to mite females, which rapidly colonised them and raised a generation at 25°C in 1 week. Bruce (1989) later developed an artificial diet for *Pyemotes tritici*, consisting of yeast and casein hydrolysates, salt and vitamin mixtures, cholesterol, streptomycin sulfate, an RNA solution, sucrose, gelatin, Tween and distilled water. This stock was kept refrigerated until needed, when egg yolk was added. Diet bags were prepared from folded sheets of a plastic wrap film, the edges of which were heat-sealed. Bags were put in a Petri dish and ringed with a repellent; female *P. tritici* placed on the bags produced around 140 progeny each.

Macrocheles muscaedomesticus and *Uroobovella marginata*, intended for filth-fly control, were routinely reared on a substrate consisting of a 3/1 mixture of a commercial medium used for house flies (CSMA) and fresh cow manure (Willis & Axtell, 1968). Reda (1990) developed an artificial diet for *Agistemus exsertus*, which was similar to that used for phytoseiids. It supported normal development, survival and reproduction, but fecundity was reduced.

Rearing in greenhouses, orchards or the field

Large numbers of predators can be obtained from rearings in greenhouses. Hoy *et al.* (1982) used trays placed on benches in a greenhouse (*c.* 24°C, 35–99% RH, natural daylight) to mass-rear *G. occidentalis*. Beans were grown in sterilised soil in the trays, then infested by spider mites, and predators were added a few days later. A tight schedule of planting, infestation and harvesting was maintained in order to keep the prey/predator ratio between 20/1 and 40/1. If the predators increased too much (which reduced their fecundity), spider mites were imported on infested leaves. When spider mite numbers endangered the plants, an acaricide was applied at rates non-toxic to the predators. More than 1.5 million pesticide-resistant phytoseiids were produced with this schedule during 4 months on approximately 45.5 m² of bench space. The system allowed continuous April-to-October production, preparing mites for field releases. However, the rearing became contaminated from other phytoseiid strains kept there, and the system was less efficient than the field plot method (below). James (1993) increased the output of *E. victoriensis* by dusting pollen of *T. orientalis* over beans grown in a greenhouse.

Field *et al.* (1979) used apple trees that were attacked by spider mites, as well as specially grown and infested soya bean plants, to mass-rear *G. occidentalis*. The latter were released on TSSM-infested trees and, as predator numbers reached 1.2–3.0/leaf, shoots were removed to new orchards and placed on individual trees. An outdoor variation was to 'seed' soya beans with spider mite-infested apple shoots, adding predators later. As the predator populations increased the plants were distributed to orchards. This low-input method sufficed to seed approximately 3000 ha of apple and peach with predators. Hoy *et al.* (1982) reared *G. occidentalis* on soya beans in a special plot of 15 rows. Plants were inoculated by spider mites and then by about 180,000 predators. An average of 8.3 females/leaflet was obtained after about 6 weeks, when there were about 68 leaflets/plant and 30 plants/m per row. It was estimated that at that time about 32 million females plus 30 million juveniles and 38 million eggs of *G. occidentalis* were in the plot, representing a 178-fold increase in predator numbers. Disadvantages of this method are that it does not allow for flexibility in supplying predators whenever they would be needed and that the plot may become contaminated by other species.

To reiterate, a major requisite for maintaining mass-rearings of ABAs is the setting up of tight work schedules. They should include protocols and timetables for preparing substrates, growing host plants, culturing the prey, inoculating and harvesting the predators, and either releasing or storing the predators. Once the operation is running it should be monitored for times actually spent on the various tasks, data that could be used for comparing various rearing methods. Hoy *et al.* (1982) tallied all periods required to mass-rear *G. occidentalis* feeding on spider mites reared on bean plants in a greenhouse. They concluded that the total labour time needed to care for 24 trays was 11.8 h over a period of 42 days; optional tasks such as spraying and introducing more prey required another 0.8–3.8 h.

Large numbers of ABAs may be directly obtained from field populations. During a redistribution project of *Neomolgus capillatus*, intended for lucerne flea control in Tasmania, Ireson & Webb (1996) vacuum-collected large numbers (about one million) of mites from pastures where the predator had already become established.

Harvesting and storage of ABAs

Aspirators can be used to collect phytoseiids; one person aspirated around 1000 mites/h from a good culture (Theaker & Tonks, 1977). Leaves with spider mites and predators can be kept in plastic bags until the plants wither; the predators then move to the top of the bags, from where they may be aspirated or collected through low-vacuum suction pumps. Individuals of fast-running species can be picked up after cooling at 4°C (McMurtry & Scriven, 1975). *Phytoseiulus persimilis* aggregates around prey; leaves with spider mites are placed in predator cultures. As the predators gather on the leaves they are collected (Nakao *et al.*, 1990). This mite has an affinity for threads; Jedlickova (1992) placed dry cotton wads in arenas with starved *P. persimilis*. They rapidly moved into the wads, which were then transferred into vials, sealed and shipped.

Humid granular media are used to collect larger numbers of mites, usually with bran, vermiculite or corncob grits (Pickett *et al.*, 1990). The international shipment of bran is nowadays restricted, and it tends to become mouldy during shipment and storage. However, it is more suitable for predators such as *N. cucumeris*, because their acarid prey is sent along too. The medium is poured and stirred; predators trapped therein are then collected, the remaining mites being used for further rearings. The predator-containing medium is stored at about 10°C and high humidity (95% RH), to prevent drying, and is later used for releases (see below). Estimates of predator numbers can be obtained by taking aliquots of the medium and counting the mites therein. Harvesting methods have a direct bearing on storage, as the same media may be used. Several other commercial methods for harvesting ABAs are in the realm of 'trade secrets'.

Storage of predators has two separate (or combined) purposes: to maintain a stockpile until needed (e.g. by growers), and to maintain them in good condition during shipment and release (below). In commercial parlance, finding the best conditions for storage means ensuring good shelf-life of ABAs.

Cheyletids can be used immediately after being collected or kept for 3 months at $0 \pm 1°C$ and 80–90% RH; mites intended for storage should be cooled gradually, over a period of 3–5 days. The RH must be kept within these limits: higher humidities cause moulding, whereas at 64% RH there was much mortality after 30 days (Zdárková & Pulpán, 1973). Bdellids were kept in a refrigerator (at 5–10°C) for 2 weeks without any damage (Currie, 1934). Gravid females of *P. tritici*, kept at 15°C for about 30 days and then placed at 22–25°C, gave birth to their full complement of offspring (Bruce, 1984). Phytoseiids can seldom be kept for such long periods. Lo *et al.* (1984) stored *N. longispinosus* at 5°C and 85–95% RH for at least 11 days without any adverse effects, but one-third of the mites died after 15 days. An exposure to 9°C caused 37% mortality of *N. cucumeris* after 10 weeks, whereas most mites died when kept at 2°C during the same period (Gillespie & Ramey, 1988). Nakao *et al.* (1990) stored *P. persimilis* at 8–10°C for at least 10 days without loss of vigour. Morewood (1992) found that *P. persimilis* can be maintained at 7.5°C for 4–6 weeks when given only food and moisture. Granular media employed for field distribution were to be added only prior to shipping.

The survival of adults and deutonymphs of *Neoseiulus idaeus* during storage (at *c.* 26°C) was compared by Drukker *et al.* (1993). Adults fared better than deutonymphs, although it was not clear whether this was due to differential susceptibility to dehydration, cannibalism or both. It was concluded that these predators must be stored under cooler conditions, that some nutrients should be offered to them, and that mostly adults were to be stored.

Personnel aspects

Questions of personnel are of major importance in the maintenance of MRFs, but only two aspects will be noted: employee selection and health hazards.

The selection and training of suitable employees, and raising their motivation (as well as safety considerations; see below) were believed by Leppla & Fisher (1989) to be more important than all other aspects of insect mass production. Qualified and dedicated personnel can turn a marginal rearing operation into a success, commercial or otherwise. Running an MRF requires employees who combine a feeling for and an understanding of biological phenomena with technical expertise; this favours personnel willing to take risks, try new methods, and accept responsibility for their role in the operation. It follows that management, at all levels, must exercise great care in chosing employees, providing them with a fitting and pleasant workplace environment and suitably compensating them for their input.

The care of employee health must be a major concern, not only for the usual safety reasons (e.g. hazardous chemicals) but also because some ABAs and/or their prey are allergenic or cause dermatitis. Among ABAs, *N. cucumeris*, *P. persimilis* and *Tyrophagus* are allergenic (Dyne *et al.*, 1996), as is *Hemisarcoptes* (Arlian *et al.*, 1999). Some Cheyletidae induce skin lesions (Yoshikawa, 1985), and *Pyemotes* causes dermatitis (Hughes, 1976). In addition, mites found in hay (including *C. eruditus*) were recently suspected as being vectors of scrapie (Wisniewski *et al.*, 1996). Although this is a disease of sheep and goats, a certain amount of risk may be involved. Among prey, spider mites (including TSSM) are allergenic (Astarita *et al.*, 1994).

Quality control of mass-reared ABAs

Introduction

A major consideration at the MRF is the standardisation of production in order to obtain as many functioning ABAs as possible at lowest cost, within the most suitable time frame. The obtained enemies should consistently be able to control the pest to the desired levels at acceptable prices. An important concern of MRF personnel is to ensure that the ABAs perform as claimed; monitoring their characteristics, correcting any deviations as soon as possible and maintaining the required standards constitute quality control. Ensuring that the mass-produced ABAs are the species that they are supposed to be – a simple but sometimes overlooked precaution – requires the routine collection and examination of voucher specimens, as noted above.

During the routine operation of an MRF, where continuous artificial conditions prevail, the erosion of desired traits may go unnoticed. Erosion could take place at two different phases: the first (discussed below as genetic drift) shortly after an ABA is brought in and is still adapting to MRF conditions, and the second (directional selection) some time later, in response to such conditions (see below). Unless close attention is paid to preserving the desired traits of the natural enemy, it may not perform as expected in the field. Quality control assessments thus have several strategic aims. The first is identifying problems in the production process; early recognition would lead to early rectification. The second is a means of fortifying the commercial repute ('label') of the product in the marketplace. The third is to reduce any negative reactions that customers may have to faulty biocontrol products (Steiner, 1993). Growers buy ABAs to control pests; if performance does not meet the claims the mites will be seen to be low-quality, disappointing products, discouraging consumers from buying them again. This undermining of consumer confidence, rarely acknowledged by commercial producers, could be a serious, long-term hindrance to the increased use of ABAs. Quality control should therefore be considered a major component in the promotion and maintenance of biological and integrated pest control.

Quality control in insect mass-rearing has been addressed in managerial terms (Leppla & Fisher, 1989) which, despite the usually smaller MRFs of predatory mites, are also appropriate for ABAs. The process was envisaged as consisting of three components, or levels, of control: production, process and product. Production control includes running the rearing operation at the required level, by continuous monitoring of procedures according to preset schedules, as well as by routinely checking performance and providing necessary maintenance. Process control is the regulation of performance by comparing the obtained product (e.g. ABAs) with established standards (e.g. egg hatch, longevity). In product control the ABA to be released is checked for its suitability for use (SR, dispersal, etc.).

The International Organisation for Biological Control (IOBC), recognising the importance of this topic, has set up a working group ('Quality Control of Mass Reared Arthropods'), which publishes its proceedings. Recommendations for testing the quality of various beneficial organisms (including *P. persimilis* and *N. cucumeris*) by the producers were made by van Lenteren & Steinberg (1991) and van Lenteren (1993). These tests (part of 'Product control') were to be conducted after completion of all handling procedures, just before shipment. The number of *P. persimilis*/package had to be checked weekly, and was to conform to that specified on the container. The SR had to be over 45% in a weekly sample of 500 mites. Laboratory tests were to be conducted under conditions of 22–23°C, 70% RH and a photoperiodic regime of L16/D8, on bean plants infested with TSSM. At least 80% of the predatory females had to live for more than 5 days. Fecundity was to be more than two eggs/female per day for at least 5 days, and the test was to be repeated over the season. As to *N. cucumeris*, numbers of mites had to conform to the value specified. Three samples from different packages were to be tested weekly per batch. These procedures, meant to provide growers with an ongoing representation of the situation in the MRF, dealt neither with problems that arise in the MRF nor with those that would be seen upon arrival in the field. Dicke *et al.* (1991) suggested using phytoseiid responses to volatiles emanating from spider mite-infested leaves (synomones) as another quality-control parameter. However, phytoseiid reactions to the volatiles could be affected by disease (Dicke *et al.*, 2000), detracting from the usefulness of this character.

Most MRFs (especially larger ones) routinely test the natural enemies that they sell. However, such tests may be insufficient. Steiner (1993) assayed the quality of *P. persimilis* and *N. cucumeris* obtained from three suppliers. Mite numbers in the shipped containers drastically diverged from the stated amount (326–8461 instead of 1000), few females were gravid on arrival, and both longevity and fecundity were below standard. Gravid *P. persimilis* were starved upon arrival and appeared emaciated. The situation in regard to *N. cucumeris*, sent in bran along with prey, was better. However, amounts of mites/container were inconsistent with stated values and fecundity was low. The overall quality was disappointing. Price & Nguyen (1997) tested the dispersal potential of transported *P. persimilis* by placing individuals at the bottom of an 80-mm-high container at 27°C and in daylight conditions. Mites that had not reached the top within 2 h were assumed to be unfit for dispersal. Such fitness was usually reduced after 5 days, indicating that releases should be made as soon as possible.

Steinberg *et al.* (1999) described product control of an ABA at an MRF by a commercial producer. The numbers of *P. persimilis* in shipment containers ('bottles') decreased (by up to 35%) during the packaging process owing to mite escape from the carrier. The situation was rectified by adding extra mites (20% of the required amount) to each bottle.

Lack of adequate nutrition

ABAs, whether predatory or parasitic, are adversely affected by inadequate food, with a resultant decline in colony vigour and field performance. Various acceptable diets, even if offered in abundance, may result in reduced progeny production or in a biased SR (e.g. see Fig. 40.2). Many phytoseiids and some stigmaeids can be reared on various pollens, which differ in their suitability (see Table 41.1) and may affect diapause induction in some species as well as their response to prey kairomones (Dicke *et al.*, 1989). The effect of pollens on *A. exsertus* was noted above. The lack of dietary supplements, such as live prey or moulds (James, 1993) can inhibit the development of certain phytoseiids; an adequate water supply is essential. *Hemisarcoptes coccophagus*, reared on oleander scale, produced only a fraction of the progeny it had on latania scale (see Fig. 19.1), indicating the former's unsuitability for mass-rearing the mite. Culturing *H. coccophagus* on moribund host scales caused the immatures to moult to non-feeding deutonymphs (hypopodes) (Gerson & Schneider, 1982), which do not immediately attack the hosts. Upon providing ABAs with their optimal diets, most of these aberrations can be corrected within one or two generations.

Reduced plant health and phytophagous organisms

It is essential to maintain healthy plants in order to sustain prey, and thus ABA quality. Various herbivores (such as other mites, aphids, whiteflies or thrips) and pathogens may invade the MRF and affect the health of plants. Well-sealed containers, separation of rearing activities in time and place, vigorous isolation and cleaning procedures, and restricting the movement of personnel between rearing rooms (Nakao *et al.*, 1990, Mégevand *et al.*, 1993), as well as continuous inspections, will reduce such invasions. If the invading phytophages are amenable to biological control, their natural enemies could be introduced (van Lenteren & Woets, 1988). Pesticides (including fungicides) that are harmless to the reared ABA may sometimes be required.

Competing predators and contaminants

Unwanted phytoseiids that invade spider mite cultures can be eliminated with the carbamate (CRB) carbaryl or by PYRs, which do not affect this prey (Hill *et al.*, 1991). Ho & Chen (1992) removed invading phytoseiids by dipping infested soya bean plants in hot (50°C) water for 60 s. The invaders were wiped out, whereas the spider mites recovered after a brief hiatus, as did the soya bean seedlings. Hoy *et al.*

(1982) eliminated unwanted phytoseiids in a culture of the pesticide-resistant *G. occidentalis* by spraying with pesticides that affected only the susceptible contaminators. Contamination by other biotypes (or strains, see above) of the cultured predator is a predicament in MRFs when several populations of the same species are being reared; the problem may be alleviated by implementing intrafacility quarantine measures. Unwanted strains can be recognised by using specific markers to check the genetic integrity of the organism in question (e.g. Navajas *et al.*, 2001).

'Founder effect' and genetic drift

The founder effect is caused by very small initial introductions, which may have missed alleles for essential traits (e.g. host finding and specificity) from the feral gene pool, thus reducing from the efficacy of the mass-reared ABAs. The very low numbers that were used to initiate the vast rearings of *P. persimilis* and of *N. cucumeris* (<10 and <50, respectively, van Lenteren & Woets, 1988) suggest that this effect might not be as important as once thought, at least in regard to phytoseiids. Guidelines on numbers to collect were offered above.

Random changes in allele frequencies, known as genetic drift, matter most when the mating populations are small during their first few generations, and in species with biased SRs. The danger declines as the population increases. In the laboratory genetic drift can be monitored (Unruh *et al.*, 1983) by measuring the frequencies of neutral allozymes (slightly different enzymes that fulfil the same function). Roush (1990) suggested maximising the mean number of progeny/parent in order to reduce the impact of genetic drift. Another option is to maintain several subdivided cultures of the founders, so that important alleles would be preserved in at least one. Mégevand *et al.* (1993) proposed maintaining sufficient ABA numbers in the cultures in order to avoid genetic drift.

Inbreeding depression

This refers to the expression of harmful, recessive alleles that become homozygous owing to the increased sib-mating that occurs within MRFs. Such depression brings about declines in vigour, often translated to failures in field performance. Like the former hazards, it affects mostly small populations.

Poe & Enns (1970) explored the extent of this danger by rearing the progeny of single pairs of *Neoseiulus fallacis* and of *P. persimilis* for 1 year. Fecundity of both species decreased by around 50% in each generation, fertility declined by two-thirds within three generations and many of the deposited eggs collapsed. The SR of *P. persimilis* became very female biased (F9/M1), most females being sterile. Hoy (1977), in contrast, reported that sib-mating of 19 lines of *G. occidentalis* for nine generations caused only minimal colony depression. Longevity, fecundity and SR were not affected by the prolonged inbreeding. The lack of deleterious effects could have been due to the natural occurrence of sib-mating in this species as well as to subsequent colony selection for inbreeding. Roush (1990) argued that natural enemies might be less affected by inbreeding depression than other animals. Their numbers in the field fluctuate strongly in response to host/prey abundance, causing the populations to undergo periodic bottlenecks, promoting occasional inbreeding and thereby affecting the selective elimination of deleterious recessive alleles. Any inbreeding effect would be even smaller in haplo-diploid mites, because the heterozygosity of the (haploid) males cannot decline.

Inbreeding depression can be reduced by frequent replenishing of the gene pool from diverse populations and by maintaining variable conditions in the MRF. Other alternatives (Roush, 1990) include maximising the number of progeny/parent and keeping several subdivided, isolated cultures of the founder population, followed by matings between distant individuals. Another possibility is to maintain the initial ABA populations at a constant size, which allows selection to remove harmful alleles as they become homozygous. Incipient depression could then be remedied by hybridising cultures before field release.

Directional selection

The benign conditions prevailing in the MRF (e.g. constant temperatures and humidities, regular light and dark regimes, abundant food, suitable host plants and presence or absence of other organisms) impose strong directional selection on the ABAs. Such conditions may select in favour of natural enemies more adapted to the MRF than to the field, resulting in 'spoilt' mites, unable to survive and perform in the field. If reared with abundant prey, such ABAs could become more density dependent and/or lose their array of reactions to cues that promote prey finding and dispersal in the field. The laboratory population would become more uniform, losing some of its initial variability, whereas the frequency of the hardier genotypes, selected to overcome random stress in the field and with superior prey-finding abilities, may be reduced. After a few hundred generations in the laboratory the population may become 'domesticated', with detrimental behavioural patterns.

Roush (1990) proposed keeping MRF conditions as in nature and releasing natural enemies after as few generations as possible. To preserve important common alleles in the population, Roush & Hopper (1995) suggested fixing these alleles (making them homozygous) in several independent individual family (isofemale) lines (25 for arrhenotokous species, 50 lines for diploids). Mégevand *et al.* (1993) added the frequent renewal of laboratory stock with field-collected mites and ascertaining that MRF conditions did not act as selecting agents. Galazzi & Nicoli (1996b) came to a different conclusion after comparing the quality of *P. persimilis* populations obtained from several sources. The various MRF conditions had no effect on the predator's potential for increase after 1 year (*c.* 36 generations). However, a precondition was that large mite stocks were to be maintained, since in small rearings (500–1000 individuals) predator quality declined rapidly.

Diseases

Causal disease agents, whether viruses, bacteria, fungi or protozoa, reach the MRF undetected, within (or on) the bodies of introduced ABAs or their prey. MRF conditions encourage disease outbreaks by two different modes. First, the stress caused by crowding and inbreeding preconditions susceptible individuals to infectious diseases (Steinhaus, 1958), which then cause epizootics. Second, crowding and unsanitary conditions enhance the spread the disease agents within the population. Little is known about this problem, which is seldom addressed by MRF personnel (an exception is Kluge & Caldwell, 1992) and is usually detected only after declines in natural enemy quality are noted. Chronic, undetected diseases that cause slight but accumulative detrimental effects in ABAs could be a major factor in colony deterioration. The diseases of mites were reviewed by van der Geest *et al.* (2000).

Stress increased the susceptibility of *G. occidentalis* to a weak pathogen, the bacterium *Serratia marcescens* (Lighthart *et al.*, 1988). Juvenile and adult mites were exposed to several stressors (high temperatures, high humidities and crowding), applied singly or in combination. Crowded adults became more susceptible after the high-temperature treatment, which when combined with high RH, significantly increased mortality after 6 days. Females were more affected than juveniles, but fecundity remained unaffected even after inoculation with high (10^8) bacterial doses.

The productivity of phytoseiids mass-reared for thrips control declined as a consequence of a microsporidian infection (Beerling & van der Geest, 1991). *Phytoseiulus persimilis*, obtained from a European MRF, was infected by another microsporidian, which reduced mite appetite, fecundity and longevity (Bjørnson *et al.*, 1996). Efforts to keep disease-free colonies failed and all predators became infected (Bjørnson & Keddie, 1997). An apparently harmless rickettsia infected *P. persimilis* (Šutakova & Rüttgen, 1978). Acaropathogenic fungi (e.g. species of *Neozygetes*) reduce spider mite populations in the field; phytoseiids can transmit the disease into the MRF (Smith & Furr, 1975).

Two distinct forms of microorganisms were isolated from *G. occidentalis* (Hess & Hoy, 1982). One (type A) was benign and seen in all individuals, whereas the other (type B) was found in two-thirds of the mites. It occurred in many individuals within their malphigian tubule lumens and in rectal plugs, and was associated with pale mites and with the death of older females. The prey of these predators, *T. putrescentiae* and *A. siro*, were also diseased (Beerling *et al.*, 1993).

The uncertain as well as often intracellular nature of most disease agents leaves the MRF operator few options except for continuous monitoring and trying to keep colonies healthy and reasonably free of undue stress through frequent subculturing and appropriate diets (Hess & Hoy, 1982). Lighthart *et al.* (1988) recommended that assays of bacterial infection (at least for *G. occidentalis*) should routinely be conducted. Beerling *et al.* (1993) developed monoclonal antibodies to screen for the presence of microsporidia in cultures of *N. cucumeris* and *N. barkeri*.

Abdominal discoloration, sometimes associated with poor field performance, was reported in *P. persimilis* obtained from 14 MRFs, located in North America,

Australia, New Zealand, Europe and Israel (Bjørnson *et al.*, 1997). All samples included individuals with abdominal crystals located within the malphigian tubules; at times the whitish discoloration (healthy *P. persimilis* are orange in colour) extended to the anterior legs. In advanced cases there was a centrally located white spot or U-shaped discoloration in the posterior part of the body. White material often hardened within the anus, forming a plug that inhibited excretion. Affected mites contained numerous birefringent, dumbbell-shaped entities of unknown origin. Some individuals excreted the material and reverted to their original, normal colour; suggesting that affected mites were those that had not been able to excrete the plug. The symptoms did not consistently correlate with any pathogens, and their occurrence in adults and juveniles indicated that they did not signify senescence. Other individuals of *P. persimilis* lost their ability to discriminate between spider mite-infested and uninfested plants. These predators had low fecundity and high mortality, and carried birefringent crystals in their legs (Dicke *et al.*, 2000). When placed (squashed or live) together with healthy mites for 2 days, the fecundity of the latter was reduced and their mortality increased, leading to the assumption that a contaminating microorganism was involved. These findings suggest that some failures to establish ABAs at new release sites and/or control pests with predatory mites of proven ability (not only *P. persimilis*) should be attributed to disease factors.

Wolbachia, a group of intracellular bacteria that may alter the SR of their arthropod hosts, pose a potential problem. The bacteria are widespread in phytoseiids as well as in spider mites (Breeuwer & Jacobs, 1996) and could engender reproductive incompatibility. An infection of *G. occidentalis* was correlated with a decrease in fecundity and a lower SR, although the latter was rectified by applying a heat treatment to infected mites in the MRF (Johanowicz & Hoy, 1998). In addition, *Wolbachia* does not seem to spread rapidly through populations (Johanowicz & Hoy, 1999). Regarding spider mites, *Wolbachia*-carrying females of *Tetranychus kanzawai*, which showed intraspecific incompatibility, were treated with antibiotics. The bacteria were eliminated but the incomplete compatibility was not corrected. Based on these and on crossing experiments, Gomi *et al.* (1997) and Gotoh *et al.* (1999a) inferred that the bacterium did not cause the incompatibility. A similar conclusion was reached by Gotoh *et al.* (1999b) in regard to TSSM. To sum up, *Wolbachia* does not at present appear to constitute a grave danger to ABAs in MRFs.

A method for general quality control

A perusal of the possible reasons for quality decline in mass-reared ABAs suggests that operators of MRFs will seldom be able to monitor all causes and correct them. Methods used for overall quality control of mass-reared insects (Clarke & McKenzie, 1992) consist mostly of process control and product control parameters, such as egg viability and immature weight, adult emergence, fecundity, SR, competitiveness and longevity, as well as more sophisticated techniques, such as electrophoresis. The calibration and routine application of most of these methods are expensive and labour intensive, and could be highly specific while not sufficiently sensitive.

Mégevand *et al.* (1993) summed up efforts to resolve quality control problems within the CGM project. All phytoseiids were reared on their original prey and host plants or, if on an alternate diet, often 'reconditioned' with original prey. 'Mother cultures' were maintained as sources of pure inoculum, using their individuals to start new rearings. Sufficiently large predator numbers were kept to avoid genetic drift, stocks were frequently renewed with field-collected (and verified) mites, it was ascertained that the rearing methods did not become a selective force and new relevant findings were incorporated into the procedures. It is evident that much expertise has to be available to maintain the quality of such predators, an investment that could be beyond the means of small MRF operators.

What is thus required is a set of tests or measurements that can routinely be undertaken by the operator. They should be limited in number and simple, and have a bearing on field performance. Such tests for quality would have to be conducted at three different stages, namely in the MRF, after delivery and in the field. As noted, the IOBC has begun to formulate tests for the first two parts, and, as more ABAs come on the market, others will be drafted. Reductions in pest numbers and/or damage will always remain the best (and economically the most meaningful) measure of the field performance of ABAs, and will be discussed below.

A method that could be applied for evaluating the quality of mass-reared ABAs is the use of fluctuating asymmetry (FA). This is a measure of the differences between the two halves of bilaterally symmetrical organisms, and is the expression of hereditary and environmental stress during development (Parsons, 1990). The completely symmetrical form of a given organism represents the successful stabilisation of all structures during ontogeny by the genome, whereas loss of genetic variation (with attendant loss in fitness) results in increasing FA values (Leary & Allendorf, 1989). Any stresses incurred by a developing organism would increase its FA, a value that has the advantage of being recognisable in the parents before any deleterious changes are noticed in their progeny. A clear relationship between FA and phenotypic quality was demonstrated for many organisms (Polak & Trivers, 1994). It was more consistent than other quality-control parameters assessed when studying the effect of stress conditions on blowfly [*Lucilia cuprina* (Wiedemann); Diptera: Calliphoridae] development. Clarke & McKenzie (1992) advocated FA as a sensitive tool for appraising the fitness of various fly strains.

A plot of the random departures from the expected bilateral symmetry in body characters, which are expressed by a numerical value (e.g. number of setae), obtained from around 50–100 individuals, generates a normal distribution. Mean asymmetry values can then be calculated as the absolute differences between the left and right sides of an individual, summed across characters, and divided by the number of individuals.

Uses that were envisaged for FA in regard to mass-rearing blowflies (Clarke & McKenzie, 1992) could be applied to ABAs. Besides serving as a general tool for assessing quality control during rearing, FA could be used for monitoring colony variance. Species that are newly being reared in the MRF would have high FA, which declines as the ABA becomes adapted to prevailing laboratory conditions. Such adaptation, as noted, may reflect deviation from desired attributes. Charting changes in FA over time would regularly assess population quality and provide criteria for deciding on colony improvement or replacement. FA could also be used as an early-warning system for recognising incipient problems as they arise. More asymmetrical ABAs are probably weaker and thus more likely to become diseased. Another option is to use FA as a means for determining the optimum conditions under which to rear ABAs, as well as for comparing different strains.

An advantage of FA is that its use requires neither costly equipment nor highly trained personnel, except for a dissecting microscope and some experience, making FA more suitable and cost-effective for MRF operators. Practitioners should how-ever be aware of statistical pitfalls in calculating FA, especially in measurement errors. Palmer (1994) and Swaddle *et al.* (1994) addressed such problems and sug-gested protocols to avoid pitfalls.

No single method will always provide the MRF operator with all the information needed to maintain the desired quality of reared and shipped natural enemies (Clarke & McKenzie, 1992). Undue fluctuations in FA cannot reveal to the operator the causes of the adverse change, nor would it be clear whether FA readings relate to field performance. Nevertheless, FA could be a simple, sensitive and relatively inexpensive early-warning method that could continuously be applied to assess the quality of mass-reared ABAs. No data about its applicability to mites are available.

Releasing ABAs and determining their establishment

Strategies, timing and amounts to be released

There are two basic release strategies of natural enemies: inoculation (or seeding) and augmentation. Inoculation refers to releases of exotic agents that are not yet part of their new habitat. They are expected to become established in the new environ-ment (thus 'seeding'), breed there and control the pest. Augmentation alludes to sin-gle or repeated releases of efficient natural enemies, whose (often) short-term effect resembles that of pesticides (see Chapter 46). Parrella *et al.* (1992) noted that the term 'augmentation' has several different meanings. 'Inundative releases' take place when the released enemies themselves, not their progeny, control the pest. They are released in large numbers to effect rapid pest control, acting as 'biological acaricides', and are not expected to survive after controlling the pest. Their ability to breed in the new environment is of minor importance. In 'inundative biological control' natural enemies are periodically released in large numbers, and 'seasonal inoculative biolog-ical control' involves the periodical release in small numbers of enemies to control multivoltine pests in perennial crops (van Lenteren, 1993). In practice these modes tend to intergrade; Parrella *et al.* (1992) used the term 'augmentation' for all, and this is followed here. Inoculation and augmentation may intergrade, as the inoculated enemy may require augmentation, and the augmented enemy may be seeded at new sites.

Timing of releases (whether inoculative or inundative) depends on the related fac-tors of pest (or damage) presence, natural enemy specificity, season and even time of day. As damage may be more readily seen than the pests that cause it (e.g. spider mites), Nihoul *et al.* (1991) defined a tomato leaf damage index meant to warn practitioners of the extent of TSSM injury. Using the index led to releasing only one or two *P. persimilis* on tomato leaves with little damage; two or three predators released onto foliage showed more injury. The index was, however, appropriate only for young plants, which bore small pest populations. By inducing asynchrony in TSSM populations (Chapter 26) and along with the damage index, Nihoul (1993a) obtained good control with relatively few (6570/100 m² in three batches) predators in a tomato greenhouse.

Mass-releases of specialists such as *P. persimilis* are not warranted during periods of prey scarcity, as the predators cannot survive in the field. Pickett *et al.* (1987) released around 15,000 *P. persimilis*/ha to control corn spider mites but obtained poor colonisation, probably because of low prey density at that time. In contrast, generalists (e.g. type III and IV phytoseiids) could be released even before the pests appear, as they would subsist on alternate foods (or, as with thrips predators, on flour mites that are released simultaneously). Further, alternate diets enable them to survive during periods of pest scarcity. Strong & Croft (1996) argued that numbers (and sites) of phytoseiids to be released against TSSM that infested hops in Oregon would depend on the prevailing climate. Hops grow slowly in cool years, slowing pest dispersal through the plants. Predator introductions can thus be gradual, in several doses, at breast height. During warm years the pest rapidly colonises all plant levels and it would then be better to release all predators early in the season, high up on the plants.

Experience with many phytoseiids in Africa led Yaninek *et al.* (1993) to release phytoseiids for CGM control only on cassava plants that were 3–6 months old. They were suffciently grown to support increasing CGM numbers during the dry season, as well as young enough not to be harvested prior to the following wet season. When during the day should ABAs be released? Pickett & Gilstrap (1986b), working in Texas, preferred the morning, in order to minimise mortality from abrupt exposure to high field temperatures, enabling the predators to find shelter (and perhaps begin feeding) before the warmer and drier hours. Others conduct releases in the afternoon, for approximately the same reasons.

How many ABAs should be released and would larger numbers promote better establishment? Practitioners usually strive for larger numbers, a practice supported by observations that the establishment rate of natural enemies usually rises with numbers released (Williamson & Fitter, 1996), and that small numbers/release are correlated with failures. Other failures are due to mistakes in identification (of pest and ABA), as well as lack of adaptations to and adverse conditions at the release site. Hopper & Roush (1993) added the Allee effect (reduction in the fitness of a species due to undercrowding). Reviewing the available data (albeit on hymenopterous parasitoids) led these authors to conclude that a minimal threshold of about 1000/release increases the chances of an exotic natural enemy becoming established.

However, as stated earlier, the vast commercial cultures of *P. persimilis* and *N. cucumeris* were initiated with fewer than 50 individuals and that one of the phytoseiids that afforded CGM control in Africa was released in relatively small numbers. Womersley (1933) established colonies of *B. lapidaria* in western Australia by transferring only about 12–50 mites to new fields. Further, the release of too many predators may be counterproductive. As noted in Chapter 6, 100 *Anystis wallacei*/m² killed about 4.4 pests/day, a value decreasing to 2.2/day with 400 predators, probably owing to intraspecific inhibition (and/or cannibalism). Some biocontrol practitioners release ABAs in vast numbers, believing that more is better and also costing the grower more without suitable returns. TSSM on strawberry was controlled to similar levels after either 125,000 or 250,000 *P. persimilis*/0.4 ha were released, but neither number of predators improved yield or fruit size (Chapter 46, Economics and pricing).

Host plant (Chapter 41) and crop attributes should also be considered when deciding on numbers to be released. Most predators can be expected to find their prey more quickly on plants of a simple architectural structure than those on complex trees, and crops with continuous foliage allow free predator movement between pest patches. In either case fewer ABAs would be needed to control the pests. Small plants, or shrubs and trees planted apart, would need more.

Would the establishment chances of a natural enemy increase with a few (or a single) releases, each consisting of many individuals, or with many releases, each with far fewer enemies? Modelling (Grevstad, 1999) suggests that a single large population is to be preferred in a constant environment, whereas small releases have a better chance in highly variable environments. When introductions of exotic enemies are seen as planned invasions (Chapter 45), it is of interest that the probability of an invader becoming established in a new region increases with the number of introduction events (Kolar & Lodge, 2001).

Transport to the field: long distance

A prerequisite for all modes of transporting ABAs to their release point, whether between continents or from an MRF to nearby fields, is that the mites be as fresh as possible. They are usually stored at low temperatures (to slow down metabolism and reduce intracontainer movement; see below), often within a humid carrier (e.g. bran or tissue paper, or on a host), and conveyed to their destination by the fastest means obtainable.

No data are available about the method used by Riley to transport '*Tyroglyphus phylloxerae*' Riley to Europe in 1873, in the first ever intercontinental shipment of an ABA (Howard, 1930). When *Pyemotes* was introduced into the USA from Mexico, the introducer 'brought with him a supply of the parasites' (Hunter & Hinds, 1904), probably with an insect host, because *Pyemotes* is an obligate parasite. Reports on other early efforts were also vague. Tothill (1918b) wrote that 'As hundreds of the mites [*Hemisarcoptes malus*] can be sent through the mail on an apple twig it should be possible to colonize it in scale-infested places and countries where it may prove to be absent from the local fauna'.

Moser *et al.* (1978) sent *Pyemotes dryas* (to control the southern pine beetle, *Dendroctonus frontalis*) from Poland to the USA in bolts of Norway spruce infested by bark beetles carrying the mite. *Hemisarcoptes malus* was shipped to Bermuda in the hypopodial stage, under the elytra of coccinellid beetles, in an effort to control a pest of cedars there (Bedford, 1949).The same method was used by Hill *et al.* (1993) for introducing *H. coccophagus* into New Zealand. *Macrocheles peregrinus* was shipped from South Africa to Australia as surface-sterilised eggs (Wallace & Holm, 1983). Phytophagous mites intended for weed control must be introduced along with their host plants. Rosenthal (1983) used a special cage for introducing *Aceria malherbae* into the USA from Europe, in order to control field bindweed. The cage, enclosing mite-infested plants, prevented their escape by means of a continuous air-flow, away from its door and through a filter that trapped the eriophyids at the rear of the cage.

Predators are usually shipped alone. Consignments of *B. lapidaria* were made in open glass tubes plugged at both ends with sphagnum moss wrapped in cellophane; the predators, bits of dry bark and young collembolan prey were then added. The tubes were sent in parcels by railway or air and usually arrived in good condition at their destinations (*c.* 90% of the mites sent to South Australia and to Victoria survived). Wallace & Walters (1974) modified the shipping method when transporting bdellids from Australia to South Africa. About 100–120 mites were placed in small plastic tubes, with a holed cap over a plug of saturated tissue at one end and a dry plug of tissue covered by cellophane at the other. Cardboard was placed in the tubes to provide extra surface area and oviposition sites. Moist as well as dried and well-aerated habitats were thus made vailable to *B. lapidaria*. The tubes were packed into cardboard boxes surrounded by loosely packed newspaper and sent by aircraft, attaining a survival rate of 95%.

For overseas transport Mégevand *et al.* (1993) aspirated phytoseiids into disposable pipette tips. The tips were sealed with paraffin at one end, with mite-proof gauze at the other. Upon arrival at African release sites the tips were tied onto the cassava plants and the gauze and paraffin removed.

Transport to the field: short distance

A commonly used method of transporting phytoseiids to the field is to collect predator-rich leaves, place them in paper bags (or other aerated containers, such as cardboard boxes) and transport them in a cool receptacle. McMurtry & Scriven (1962) used an agar-based method for maintaining phytoseiids for 2–3 weeks during transport. Mites were kept on water agar (2%) poured into small vials or other containers and closed at one end with fine mesh nylon cloth.

Mass-reared phytoseiids are collected with wet (or subsequently humidified) bran or vermiculite packed into small containers, each usually containing approximately 2000 mites (in commerce often noted as 1800–2500). Owing to their negative geotaxis, the predators move up and aggregate in the upper parts of and around the containers' lids. This may be corrected by refrigeration *en route* and by rotating the containers around their axes (Fournier *et al.*, 1985). *Hemisarcoptes coccophagus* was released in New Zealand as hypopodes on their coccinellid vector (Chapter 19).

Plastic drinking straws were later used for transporting *Hemisarcoptes* spp. and phyto-seiids to the field (Thomas & Chapman, 1978; Charles *et al.*, 1995b, respectively). Predators were vacuum-sucked into straws, sealed at one end by appropriate gauze and by a stopper at the other, and opened at the release site. Light and permeable paper bags containing *C. eruditus* were used for keeping and transporting this predator to its release sites in food stores (Zdárková, 1986). Wallace (1974) placed diapausing eggs of *N. capillatus* in pastures where their lucerne flea prey was abundant. Potted, flowering castor bean bushes were used to mass-produce and intro-duce *I. degenerans* into sweet pepper holdings (Chapter 26). Argov *et al.* (2002) used citrus seedlings infested by the citrus rust mite to mass-rear phytoseiids introduced into Israel. The seedlings were placed on target trees in the field after the predators had raised large populations. As noted above, Field *et al.* (1979) transported *G. occidentalis* on their apple shoot habitats into new orchards, and Solomon (1986) transferred mass-reared pesticide-resistant phytoseiids to orchards on apple rootstocks. Such plant material, being the predators' natural host, is conductive to better establishment.

Mites for weed control are transported from the greenhouse on parts of infested hosts. Shoots of laboratory-grown gorse infested by *T. lintearius* were placed onto target weed bushes in the field (Hill *et al.*, 1991). *Aculus hyperici*, introduced into Australia to control St John's wort (Chapter 15), has a slow rate of spread. To facili-tate its further dispersal, nurseries of the mite-infested weed were established and the plants made available to interested persons.

Release techniques

Modes of releasing ABAs while on plants or hosts were noted above and most of the following techniques refer to proven predators obtained in commerce (thus being augmentations). If the ABA is released by hand (hand-strewing), the usual practice is to remove the containers from the cooler, roll or agitate them (to homogenise predator distribution) and shake out discrete amounts ('drops') of the carrier on plants or along crop rows. Quantities to be dropped onto each plant depend on the crop, its mode of cultivation (e.g. protected or in the open) and its level of infesta-tion. Although commonly used, this method is imprecise and inefficient. Giles *et al.* (1995) compared the efficacy of a mechanical release system for *P. persimilis* (see below) with hand releases. The former was twice as efficient as hand releases, prob-ably as a result of worker factors (individual habits and fatigue) and inadequate mix-ing within containers, which caused too many mites to be strewn in the first discharges. Bennison & Jacobson (1991) developed a 'controlled release system' of *N. cucumeris*. It consisted of a bran-based predator culture within waxed paper packs of special porosity, each containing around 300 mites in 300 ml culture material (which also contained prey mites). Packs were singly hung on the plants for about 10 days and provided better distribution of *N. cucumeris* within the crop. The predators bred and emerged for at least 6 weeks, the pack also affording them protection against nicotine smokes used in glasshouses for aphid control. Chambers *et al.* (1993), who reared *Stratiolaelaps miles* on flour mites in a medium of sieved peat and vermiculite, released the predator (against sciarid glasshouse flies) on potted cyclamens by sprinkling a little of the medium on the compost surface of each pot.

Giles *et al.* (1995) devised and applied a mechanical system for releasing phytoseiids, based on the concept that the release of biological materials (e.g. ABAs) is similar to the broadcasting of agricultural chemicals, as both demand dependable, controllable and repeatable means. Predators (*P. persimilis*), mixed with vermiculite, were poured into a reservoir and their movements therein curtailed by chilling. The mites dropped from the reservoir into cells within a rotating metering plate which, upon moving, arrived at an exit port, allowing the mites to fall down. An important advantage was that the dropped mites seemed to be undamaged. Field releases were within 5–12% of the intended mite amounts, even in large-scale experiments. The device thus met the design goals of providing viable, mobile predators accurately and uniformly. Takano-Lee & Hoddle (2001) used a mistblower to blow *Neoseiulus californicus* (carried on corn grits) onto avocado trees.

Pickett *et al.* (1987) pioneered the aerial release of phytoseiids (*P. persimilis*) when trying to control spider mites infesting corn in Texas. The intention was to release the predators rapidly at a controlled rate, at low cost, over large areas, using light aircraft. Corn grits were the most suitable carrier, being sufficently heavy to drop in a predictable pattern from an aircraft, large enough to provide the mites with a surface to hold onto, yet light enough to prevent crushing them, and they were available at low cost. Predators were mixed with the carrier and placed in a mechanised, refrigerated cone-shaped delivery system, from which they were metered out onto a rotor. They were then conveyed pneumatically to the outside of the aircraft. Notwithstanding uneven control results, this delivery system provided a nearly uniform distribution of predators on the treated plant canopy. Prevailing winds were a major concern and their effect is to be considered during the planning stage. The smaller cassava fields in Africa necessitated another approach, developed within the CGM control project. It encompassed a unique Airborne Insect Release System (AIRS) and a special packaging and delivery system (Drukker *et al.*, 1993). Mites to be released (*N. idaeus*) were placed within plastic pipette tips (see above), placed in a container. A stopper was placed on top of each container, connected by a cotton string, wound around the plug. The containers were then inserted in place within a release cassette, which held about 360 chambers. During flight the pilot pushed a button that blew each single container out of the cassette. As a result of deceleration brought about by the design of the AIRS, the container exited the aircraft at a velocity of only one-third of the aircraft's speed. At ejection the stopper left the release chamber first, dragging out the parafilm plug and pulling the container out of the cassette chamber. This ensured that predator egress from the containers would only take place in the field (although some might be lost during the fall). The string served the further purpose of increasing the chances that the container would be snared by a plant instead of falling to the ground. In addition, it was devised to orientate the container upwards, offering the mites direct access to the plant via the string. The release system functioned satisfactorily: during ejections most containers opened as expected, were usually snared by the cassava canopy and became orientated with the tip pointing upwards. Drukker *et al.* (1993) advocated that flights be made at the lowest altitudes possible, because this increased the precision of the releases.

Should predators be released by hand or by mechanical means? The former requires more time (and thus longer storage periods that could incur mite mortality) and more labour (limiting the area treated), and the efficacy of release varies between workers. However, owing to greater precision in application, for example pinpointing the predators at infestation sites, hand releases use fewer ABAs/area. Thus, they may be suitable when fewer predators are available and/or labour is affordable. Mechanical release requires less labour, enabling larger areas to be treated more quickly and requiring less storage, but it needs special field data, equipment and personnel, and requires more ABAs/area. For these reasons it is mostly applicable when the added expense is not a limiting factor.

Another method, the 'pest-in-first' approach, may be used to ensure the availability of abundant prey before introducing the predator, especially in inoculation releases or early in the season (see Chapter 26, *Phytoseiulus persimilis*). With this method (first suggested in regard to mites by Huffaker & Kennett, 1956) the pest is placed onto the crop and the predators are added only after the pest has built up small, non-damaging populations. This ensures that immediate, easy-to-find prey will be available to the predator, encourages the establishment of a breeding population, and ensures more even control. Wallace & Walters (1974) liberated a large number of lucerne fleas at the site where *B. lapidaria* was to be released in South Africa, just before releasing the predators. Jedlickova (1992) tried yet another approach, namely continuous introduction. As the plants were never free from spider mites, *P. persimilis* was released in batches of about 500/400 m^2 at weekly intervals, and usually provided satisfactory control throughout the season.

Whatever the method, the release site or sites should be clearly marked in order to facilitate subsequent evaluations. Mahr *et al.* (1997) placed a concrete pad with an arrow at each release site of *A. hyperici*, to mark the spot where a photograph of the weed-infested field was taken initially, and then every year when evaluation was conducted. The photographs provided a visual record of the mite's impact over time.

Establishment in the field

To become established ABAs must first arrive at the release site in good condition and then survive in the target area. The viability of *B. lapidaria* shipped from Australia to South Africa was evaluated by supplying arriving mites with an abundance of prey. Immediate, voracious feeding denoted high viability; restricted feeding indicated moderate survival, whereas no feeding meant low endurance (Wallace & Walters, 1974). Other tests to evaluate the quality of shipped ABAs were discussed above.

If a predator is easily recognised in the field, its establishment can be observed on the plants or in the pest's habitat. Field *et al.* (1979) monitored the establishment rate of *G. occidentalis* on Australian apple trees by placing black cloth bands (into which predators move to overwinter) around their trunks. Preferred spider mite hosts (e.g. beans) can be employed as 'trap plants' to monitor the presence and spread of colonising ABAs; Raworth *et al.* (1994) used this method to monitor the activity of *N. californicus* in apple orchards during the dormant season. Predators labelled with fluorescent powder before release are easier to recognise at the target field (Herren *et al.*, 1987). Charles *et al.* (1995b) examined the bark of kiwifruit and shelter trees infested by scale insects for presence of the introduced *H. coccophagus* in New Zealand. Dung beetles disperse *Macrocheles* spp.; Wallace & Holm (1983) confirmed the presence of the introduced *M. peregrinus* in Australia by trapping beetles in dung-baited pitfall traps a few weeks after release. The exotic *A. malherbae* was considered to be established in Texas, USA, after it reproduced on the host plant and survived in the field for more than 3 years (Boldt & Sobhian, 1993).

Samples of all introduced ABAs (especially when more than a single species is released) should be collected from commercial plants as well as from neighbouring weeds (Yaninek *et al.*, 1993), especially after aerial releases. They should be identified and voucher specimens must be kept (Huber, 1998). If appropriate morphological markers are unavailable, molecular means could be employed to verify what species (or strain) had been recovered (Navajas *et al.*, 2001).

Establishment rates of ABAs at their release sites can also be improved by modified agrotechnical routines. These include crop-specific practices, for example advising farmers not to graze pastures too low where *N. capillatus* had been released (Ireson & Webb, 1996), and hops management. This includes avoiding leaf removal and cutting new growth early in the season, practices that hinder the establishment of predators because they and their prey abound on the cut foliage. Placing the cut leaves back onto the plants may alleviate these adverse effects (Strong & Croft, 1996). Establishment rates of ABAs for weed control sometimes depend on target plant health. Jupp & Cullen (1996) monitored release sites of *A. hyperici* on St John's wort in Australia, and reported high establishment rates (78%) on healthy plants, compared with 40–42% on plants in poor to fair health. The commonsense avoidance of pesticides during releases increases the chances of establishment.

After ascertaining the safe arrival of ABAs at the release site, their efficacy should be appraised. Efficacy can be separated into establishment and pest control. Roush (1990) argued that assessing the effect of releases on target pests over a wide geographical area might require years, and that a more objective standard, such as rate of establishment, should be used. Williamson & Fitter (1996) calculated that approximately one in three of all introduced biocontrol agents managed to establish. The rates were greatly affected by the habitat of the target pest. These rates came to around 20% for enemies of root feeders, compared with 61% for those released against weeds, the overall average being 35.5%. Hall & Ehler (1979) arrived at a similar conclusion, namely that the world-wide rate of successful establishments of all introductions of natural enemies was around 34%. Regarding the ability to control, Williamson & Fitter (1996) estimated that about one-third of established species effectuated pest control. Determining the impact of released ABAs on pest populations is discussed in Chapter 44. Suffice to say that at times reports are only of a qualitative nature; for instance, Glendenning (1931) claimed that *H. malus* (introduced into British Columbia to control an apple pest; see Chapter 19) 'has effected excellent control'. Nowadays quantitative proof is required of the 'before' and 'after' changes in pest numbers, weed stands and yields.

Beirne (1975) concluded that establishment rates for natural enemies (mostly hymenopterous parasitoids) introduced into Canada were 43% in orchards and orna- mentals, 23% in forests and only 16% on field crops. A perusal of the 22 exotic ABAs that became established outside their area of origin (Chapters 6, 9, 11, 15, 18, 19, 21, 23, 26 and 34) indicates that best establishment was in pastures (50%), followed by orchards (25%). Although this total (22) is only about 10% of the Canadian roster, and does not include ABAs established in greenhouses (which were not included by Beirne), it seems that the environment may also affect establishment rates for mites.

It is a regrettable, albeit common, practice not to dwell on biocontrol failures, and efforts with mites are no exception. However, only a critical analysis of past failures can lead to future successes. Yaninek *et al.* (1993) failed to control CGM after introducing phytoseiids from dry neotropical regions into Africa. This brought about a change in policy and, by searching in more humid areas, similar to those where severe CGM damage occurs in Africa, highly successful predators were located. Fauvel & Gendrier (1992) failed to introduce phytoseiids to control ERM in south-eastern France. This was attributed to pesticide susceptibility, inappropriate timing, competition with resident species, lack of alternate diets and low summer humidities. Although covering many factors, the need to assess them was believed to be an important further step. Many ABAs, although introduced and released in vast numbers at many sites, never become established. Such failures occur even when all needed life history data are at hand, but more often when not enough biological information concerning the candidate ABAs is available.

Strict bookkeeping and standardised reporting of ABA amounts released and established are basic requirements in such projects. The impact of the introduction may then be estimated by effects on pest numbers and damage, and on yield. Comparisons with similar results and costs from chemical control are also warranted (as noted in Chapters 26 and 46).

Postcolonisation evolution

The effect of exotic ABAs could be affected by postcolonisation evolution (PE) (also called postcolonisation adaptation); the postulated causes of the few known cases were discussed by Murray (1985) and Secord & Kareiva (1996). The conditions into which natural enemies are thrust, for example initial low numbers with reduced genetic variability impacted into a new environment with abundant food and diverse competitors, may promote evolutionary change. The rise of inherited pesticide resist- ance (discussed in Chapter 43) was taken as evidence of PE (Roderick, 1992). Postcolonisation adaptation of herbivores intended for weed control has always been of concern, but its actual frequency appears to be low (Chapter 38, ABAs as enemies of weeds). Rapid evolution was noted in many interspecific interactions, but Thompson (1998) concluded that changes in the specialisation of natural enemies are usually limited to a small subset of their prey or hosts. Nevertheless, it is necessary to monitor PE by studying genetic variation in small and large introduced populations, along with their patterns of dispersal and mate finding (Hopper & Roush, 1993). PE may also occur in genetically modified ABAs (Chapter 45).

Chapter 40
The sex ratio of acarine biocontrol agents

Introduction

The sex ratio (SR, females/females+males) of arthropod natural enemies is a major issue of continuing interest to practitioners of biological control (Luck *et al.*, 1999b; Heimpel & Lundgren, 2000). Numbers of natural enemies will increase at a rate that is proportionate to the frequency of females in the population (Waage, 1992), because they are the sex that has the greater regulating effect on the pests. The SR determines whether sufficient numbers of both sexes are produced in the field and in the mass-rearing facility (MRF; see below) and affects genetic variability, impacting on the chances of introduced acarine biocontrol agents (ABAs) to adapt to new environments. The diversity of SRs among the Acari was surveyed by Wrensch & Ebbert (1993). The SRs of the Eriophyidae and Tetranychidae were reviewed by Sabelis & Bruin (1996) and by Sabelis (1991), respectively, and will not be further discussed here.

The SR of a given species is primarily affected by its genetic system of sex determination. Three systems of sex determination are recognised in sexually reproducing mites.

- *Diplodiploidy*. Males as well as females develop from fertilised eggs, formed as a result of parental mating. All individuals are biparental and the SR is usually around 0.5.
- *Arrhenotoky*. Males are haploid, being produced from unfertilised eggs, whereas females develop from fertilised eggs; the SR is variable, often around 0.7.
- *Parahaploidy* (also called pseudo-arrhenotoky). Females and males develop from fertilised eggs, but males, which lose their paternal genome, become haploid and transmit only the maternal genome (Norton *et al.*, 1993). The SR is also around 0.7.

All three systems occur in ABAs (Table 40.1). Diplodiploidy occurs in Astigmata, in some Cryptostigmata and some Prostigmata, whereas arrhenotoky is common in the Prostigmata, and may occur in the Cryptostigmata and Mesostigmata. Parahaploidy has so far been shown only in the mesostigmatid superfamily Ascoidea (e.g. Ascidae, Laelapidae, Phytoseiidae), where it is expected to be the rule (Norton *et al.*, 1993). Mated arrhenotokous and pseudo-arrhenotokous females have an advantage over diploid females because they can determine the sex of their progeny by influencing the fertilisation of each egg (Nagelkerke & Sabelis, 1991). In other words, they can adjust the SR of their offspring to an optimal ratio in reaction to environmental factors.

Variations in the relative frequency of sexes may occur at conception (then termed primary sex ratio) at adult emergence (secondary sex ratio) and after adult emergence

Table 40.1   Genetic systems found in the ABA families

Suborder	Family	Genetic system
Mesostigmata	Ascidae	AR; HD
	Laelapidae	AR; HD
	Macrochelidae	AR; HD
	Otopheidomenidae	HD
	Parasitidae	DD
	Phytoseiidae	HD
Prostigmata	Arrenuridae	DD
	Camerobiidae	HD
	Cheyletidae	HD
	Cunaxidae	HD
	Eriophyidae	AR; HD
	Eryenetidae	HD
	Erythraeidae	DD
	Eupalopsellidae	HD
	Hydryphantidae	HD
	Limnesidae	DD
	Pionidae	DD
	Podapolipidae	AR
	Pterygosomatidae	AR
	Pyemotidae	AR; HD
	Stigmaeidae	HD
	Tarsonemidae	AR; HD
	Tetranychidae	AR; HD
	Trombidiidae	DD
	Tydeidae	AR; HD
Astigmata	Acaridae	DD
	Hemisarcoptidae	DD
Cryptostigmata	Ceratozetidae	HD
	Galumnidae	DD; HD

Based on Tables 1.1, 1.2 and 1.3 in Norton *et al.*, (1993).
AR: proven arrhenotokous; DD: proven diplodiploid; HD: haplodiploid karyotype (male/female).

(tertiary sex ratio). Primary SR depends on population structure (see below) and other environmental factors, and may be altered in response to certain life history situations. Secondary SR is affected by sex-specific survivorship during development to adulthood, as well as by the primary SR. Biased tertiary SRs result from differential longevity and/or dispersal of the sexes and from bias accumulated through the skewed primary and secondary SRs.

Although the SR of the progeny is usually fixed in diploid species, its expression may be affected by various factors (see below). As to arrhenotokous and parahaploid species, whose SR is usually around 0.7, diverse increases above this value are influenced by their parents' and their own environment. This chapter looks first at field data, then addresses individual factors, and finally discusses the effect of the SR on the mass-rearing of ABAs.

Factors affecting the sex ratio

SR values of ABAs in the field often fluctuate during and between seasons. Irregular sampling may confuse the picture, because censuses taken only once or twice a season would not detect shorter living or earlier emerging males, or members of either sex that move to other habitats (e.g. Bader, 1980; Proctor, 1992). The SR of 15 species of Phytoseiidae collected from many plants in the eastern USA was from 0.52 to 0.97, each having its own specific value (Dyer & Swift, 1979). The SR of these species (the tertiary SR) was affected by density-independent factors such as temperature, humidity and wind velocity. It was not changed by mite numbers, a density-dependent factor, or by the vegetation, a density-independent cause, nor were there any clear seasonal trends in the SRs of the various populations. Explanations for such changes are sought in laboratory studies, in which the benign prevailing conditions and the care taken with individual rearings tend to lessen developmental mortality, reducing the expression of secondary and tertiary SRs.

Lifestyle

ABAs may control pests by parasitism (including parasitoidism) or by predation. Most parasitic ABAs live and mate only on their hosts or nearby. Males emerge first and inseminate the females as they appear. In these families (Acarophenacidae, Otopheidomenidae, Podapolipidae, Pyemotidae and some Tarsonemidae) males are very potent and the SR is very strongly female biased (Acarophenacidae: 0.99). It is likely that other parasitic ABAs (e.g. Eryenetidae) have a similar SR. Mites in most of these families are dispersed by their hosts, only the mated females usually being phoretic. An exception is the Hemisarcoptidae, whose members can be parasitic as well as predatory, and both sexes are phoretic.

Age, age at mating and number of copulations

Takahashi & Chant (1994) tallied the SR of progeny that were daily produced by four species of *Phytoseiulus*. SR was initially below 0.5, then rose and remained steady, above 0.75, from about day 8 until day 20, when this value declined in two species and remained steady in two others. Abou-Awad & Reda (1992) kept virgin females of *Agistemus exsertus* for 5, 10 and 15 days and then exposed them to males. Early mating resulted in a significantly higher SR, as well as in the production of more progeny (Fig. 40.1). Momen (1994) reported similar results for *Neoseiulus barkeri*, whose SR shifted from 0.56 for females mated at days 1–5 to 0.4 for females placed with males only on day 30. The SR of the progeny of *N. barkeri* females that had

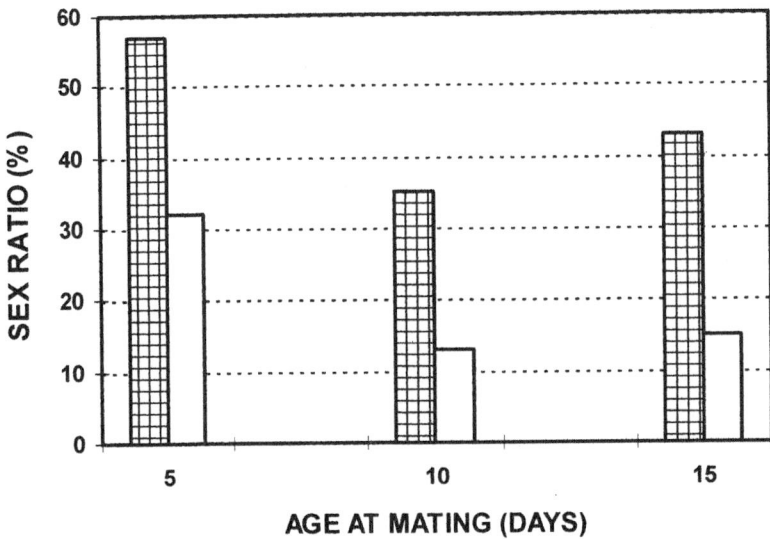

Fig. 40.1 Sex ratio (percentage of females, cross-hatched bars) of the progeny of *Agistemus exsertus* when mated at three ages, and the number of subsequent progeny (white bars). (Based on data in Table 3, Abou-Awad & Reda, 1992.)

several matings increased from 0.64 to 0.68 (Momen, 1993), possibly owing to the presence of more sperm. The reduced SR of the offspring of three other phytoseiids, whose matings were interrupted (Castagnoli & Liguori, 1991), supports this explanation.

Diet and tritrophic effect

The amount, quality and suitability of diets affect investments that females make in the progeny, including their sex. When too many parasitic ABAs develop on the same host, none receives enough food and their female offspring tend to be smaller. Such females produce more sons, a topic discussed in the section on density dependence.

The SR of phytoseiids increased when they were given more of the same prey in the laboratory (Friese & Gilstrap, 1982; Momen, 1994; Toyoshima & Amano, 1998). The SR of *Phytoseiulus persimilis* was affected by the plants on which its prey had fed, being 0.65 on soya bean-reared spider mites and 0.54 on chrysanthemum prey (Popov & Khudyakova, 1989). If based only on field samples, the SR of specialists such as *P. persimilis* may be biased because females tend to leave plants as their prey is depleted. Male phytoseiid juveniles whose mothers did not receive enough food died earlier than their sisters (Toyoshima & Amano, 1998), further skewing the SR.

Different prey (e.g. quality apart from quantity) also affect the primary SR. *Stratiolaelaps miles* produced significantly more females when reared on sciarid flies then when offered astigmatid mites (Fig. 40.2). The offspring of *Blattisocius tarsalis*,

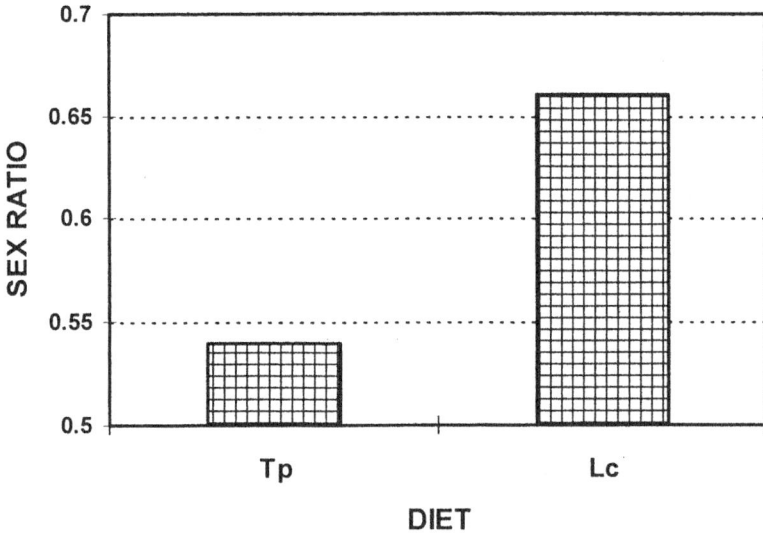

Fig. 40.2 Effect of two prey (Tp: the mite *Tyrophagus putrescentiae*; Lc: the sciarid fly *Lycoriella solani*) offered to *Stratiolaelaps miles* on the sex ratio of its progeny. (Data from Enkegaard *et al.*, 1997.)

reared on moth eggs, had an SR of around 0.7, which rose to 0.99 when the predator was fed beetle eggs. This higher SR was secondary, possibly owing to the inability of male nymphs to pierce the eggs' thick shells (Haines, 1981).

The progeny of *Hemisarcoptes coccophagus*, when parasitising *Hemiberlesia lataniae*, were strongly male biased during autumn and winter, and mostly female biased in late spring. On another host, *Aspidious nerii*, the SR was male biased, whereas on a third scale, *Parlatoria pergandii*, this value fluctuated throughout the year (Izraylevich & Gerson, 1995d). Other aspects of the diet effect (host quality) are discussed below. (section 2.5).

The hypopus of *H. coccophagus* must remain for at least 5 days on *Chilocorus* beetles in order to develop further (Chapter 19). The beetles feed on various scale insects. Larvae of *Chilocorus bipustulatus* were reared on soft scales (Coccidae) and on armoured scales (Diaspididae), and the emerging beetles from either group were exposed to hypopodes from the mite's mass culture. Survival of hypopodes that had stayed on either group of beetles was similar, but the SR of mites that had stayed on 'soft scale beetles' was male biased. Although the precise mechanism of this effect remains unclear, it is a tritrophic effect that influenced the mite's SR (Ji *et al.*, 1996).

Temperature and humidity

Low or high temperatures and humidities influence the mate finding, copulation and sperm vitality of various ABAs, according to their provenances and affecting the SR. The influence of such climatic factors on the SR of phytoseiids in the field was

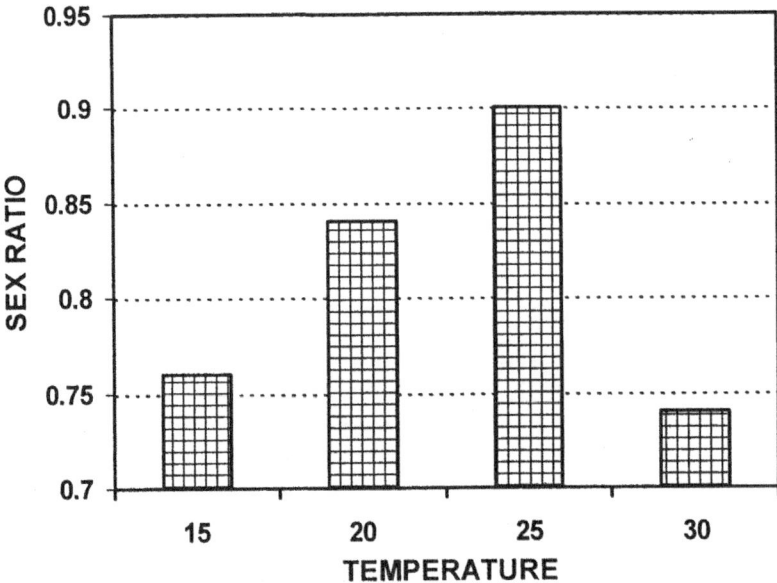

Fig. 40.3 Effect of four temperatures on the sex ratio of *Stratiolaelaps miles* reared on larvae of the sciarid fly *Bradysia*. (Data from Ydergaard *et al.*, 1997.)

noted by Dyer & Swift (1979). At 18°C the SR of *Typhlodromus phialatus* was 0.47 and of *Euseius stipulatus* 0.66, whereas at 32°C it changed to 0.62 and 0.52, respectively (Ferragut *et al.*, 1987). The SR of *S. miles* was affected by the rearing temperatures, being 0.76 and 0.74 at 15 and 30°C, respectively, and rising to 0.9 at 25°C (Fig. 40.3).

Density dependence

Females of *Galendromus occidentalis* kept in isolation (but with abundant prey and access to males) had a significantly higher SR than others held under similar conditions but in groups of ten. Nagelkerke & Sabelis (1991) concluded that these females (and, by extension, all phytoseiids) can make sex-allocation decisions according to group density, and thus regulate the SR of their progeny. Another density-dependent cause for lower SR values in phytoseiid (and other ABA) cultures could be mutual interference. Phytoseiid mating lasts for several hours and interruptions, which are more likely to occur in large cultures, thus curtail the SR (Castagnoli & Liguori, 1991).

In other cases there was a density-dependent, but not linear, effect of crowding on the SR. When three to 40 *Macrocheles muscaedomesticae* females were placed in a container (and provided with abundant food), the descendants' SR was around 0.3. Upon raising their initial number to 160/container, the SR of their offspring came to 0.5, falling again to 0.36 with greater (700/unit) crowding (Geden *et al.*, 1990). As noted, the SR of *H. coccophagus* in the field may be diet (host) related, but it is also strongly density dependent (see below). Cannibalism was ruled out as a factor affecting the SR because it does not occur in *H. coccophagus*.

Disease

Females of *P. persimilis* infected by an intracellular protozoan in the laboratory deposited fewer but more male-biased eggs (0.46 vs 0.67 in healthy mites) (Bjørnson & Keddie, 1999). Infection frequency increases when mites live in close quarters, thus it is also density dependent. In a later survey of disease in *P. persimilis*, Bjørnson *et al.* (2000) obtained females from commercial suppliers that were small, similar to males. The average SR was 0.97, indicating a lack of inseminating males. Infection of *G. occidentalis* with the bacterium *Wolbachia* was correlated with a decrease in the number of progeny and their SR (Johanowicz & Hoy, 1998).

Male inadequacy

Populations of arrhenotokous ABA in which males become rare for various reasons (e.g. sublethal pesticide sprays that kill only the smaller males or reduce their fecundity) may become male biased. This could be the result of an inadequate male/female ratio or impaired potency, or both. The strongly female-biased SR of the progeny of emerging *Pyemotes tritici* (0.91) became progressively more male biased as the males mated consecutively with more than 15 females. Bruce & Wrensch (1990) attributed this result to sperm deprivation, because males recuperated overnight. However, in this species the SR is probably not density dependent, as it was unaffected by crowded conditions.

Evolutionary considerations

Natural selection favours individuals that maximise their contribution of genes to future generations (Luck *et al.*, 1999b). The female's decision whether to invest more in sons or in daughters probably depends on the mating structure and on environmental conditions, which impact on population structure. Factors that affect the SR include the ABA's basic lifestyles (parasitism or predation; breadth of diet; modes of dispersal) and hereditary limits within which the variable environmental components exert their influences.

Population structure, including group size and duration, movement between patches and local density dependence, affect competition between relatives, thus determining sex allocation decisions. Populations may last only for a single generation or for longer, their duration affecting the mating structure. This structure could consist of a single-generation group, where mating occurs locally among the descendants of a single or a few 'founding' females. The males compete for access to females (thus, 'local mate competition' or LMC), a situation that favours a female-biased SR. Only mated females disperse to found new colonies. When such groups last for a few generations (the 'haystack' model), populations expand and then start to disperse; if only mated females disperse, a female-biased SR is likewise selected for. Another population structure (the 'island' model) consists of permanent populations whose size is relatively constant and some of whose inseminated females leave the island every generation. The predicted SR of these panmictic populations would be similar to, but lower than in LMC. Two other models may apply to sex allocation in ABAs. Host quality (HQ) describes adjustments in the primary SR due to one sex being more affected by lack of resources than the other. Operational sex ratio (OSR) applies to situations where imbalances occur in the usual ratio of receptive females to potent males. Scarcity of one sex during a generation may lead to its overproduction in the next. Genetic conflict, an alternative model intended to explain variations in SR, was reviewed by Werren & Beukeboom (1998), but has not yet been investigated in regard to mites.

The great variability in phytoseiid lifestyles (Chapter 26) affects their SR. Type I specialists, such as *P. persimilis*, rapidly find and deplete their prey, usually staying in each patch for no more than one or two generations. Such subdivisions in the haystack were predicted to promote, and shown to cause, a very high SR. Type III and IV phytoseiids, which subsist on various diets and tend to persist in their habitats for many generations, could be considered island species, and have a lower SR. The field data of Dyer & Swift (1979), noted above, support the Nagelkerke & Sabelis (1996) postulate of a haystack to island gradation across species.

Turning to *H. coccophagus*, the size of its female offspring decreased when mothers developed on lower value hosts, and the smaller daughters then produced more sons (HQ effect). OSR was seen when strongly biased female/male ratios in the parents resulted in skewed SR in their progeny. The offspring of young females had a higher SR than that of older mothers, showing LMC. The secondary SR was male biased when mortality during development was due to competition, or female biased when resulting from low HQ. Sex-specific mortality during development can thus engender diverging SRs within the same species, depending on the operative factor. The tertiary SR was also influenced by mother/host density. At high mite density females lived longer and left their natal hosts more than males, causing shifts in the SR. In the field the progeny of young parents mate among themselves; sex allocation at this stage would fit the LMC theory. Over time, as parents age and their offspring attack more and more hosts, the population shifts towards panmixis and factors that had previously operated no longer apply; females then adjust the SR of late progenies to become more male biased. Such ongoing changes in the population structure could bring about the asymmetric, male- or female-biased SRs seen in the field (Izraylevich & Gerson, 1995d, 1996).

There are also special situations, in which the entire population consists only of females (SR = 1.0). Walter & Lindquist (1995) found that although thelytokous Ascidae occurred in many different habitats, they were more characteristic of stable sites. Bisexual members of the family, in contrast, dominated in highly disturbed or ephemeral natural habitats. Manning & Halliday (1994) reported similar findings in regard to beetle- and fly-associated Macrochelidae. As most members of these families are bisexual, it seems that the thelytoky of these ABAs is an adaptation to predictable conditions with an abundant food supply in stable habitats (Kaliszewski & Wrensch, 1993).

Practical aspects

The recommendations formulated by Luck *et al.* (1999b) for culturing parasitic Hymenoptera (which, like many ABAs, are arrhenotokous) were based on knowledge concerning the organisms' biology along with theories of sex allocation and determination. Comparable understanding about ABAs, incomplete as it may be, as well as the theoretical background, could and does lead to more efficient rearing of predatory and parasitic mites.

Although most ABA females that are collected before introductions are likely to be mated, and may thus establish new families in the MRF (as they do in the field), care should be taken to initiate cultures with enough individuals (see Chapter 39). If too few are used, they may be subpopulations with a skewed SR, leading to insufficient inseminations. Such SRs may also indicate diseased mites or hereditary malformations. In the MRF SR could become biased for several reasons, including inbreeding, inappropriate conditions and disease. Inbreeding may not be too harmful to phytoseiids, at least in regard to *G. occidentalis* (Hoy, 1977; see Chapter 39). However, little is known about this problem in most other ABAs. Adequate nutrition, including suitable hosts for parasites, is a prerequisite for obtaining suitable SRs, as are appropriate temperature and humidity conditions. As noted, disease may seriously bias the SR, which would greatly decrease predator growth in the MRF as well as reduce its efficacy in the field. In the same vein, an SR <0.9 in the offspring of *P. tritici* (whose males usually suffice to fertilise all emerging females) indicates less than normal insemination. Owing to the negative correlation between total progeny number and SR in this species, such a decline heralds fewer offspring and, as mating triggers host searching, fewer ABAs to attack target pests (Bruce & Wrensch, 1990).

Arrhenotokous ABAs (e.g. Tarsonemidae and Tetranychidae) can produce thelytokous populations under prolonged stable conditions (Kaliszewski & Wrensch, 1993), a situation that would be unsuitable for subsequent releases. The optimal SR for each ABA should be known, so that the best mating structures can be maintained and females mated as early as possible. Van Lenteren (1993) recommended that in a weekly sample of 500 mites, the SR of commercially produced *P. persimilis* and *N. cucumeris* should be greater than 45%.

Authors have debated the optimal number of natural enemies that should be released for establishment (Chapter 39), but have rarely addressed the all-important issue of their SR. However, practitioners using ABAs require information on their SR and on factors that affect that value. Such data are essential for maintaining optimal SRs in the MRF, and could assist in manipulating the SR of ABAs, for instance reducing *Wolbachia*-induced SR changes by applying heat treatments to infected phytoseiids in the MRF (Johanowicz & Hoy, 1998).

Chapter 41
The effect of host plants and the ground cover on acarine biocontrol agents

Introduction

Plants (called producers or first trophic level in ecological parlance) are the universe in and on which herbivores (primary consumers or second trophic level), their natural enemies (secondary consumers or third trophic level) and various detrivores (tertiary consumers) live. Herbivores exploit plants, using them as sources of nutrients, shelter and substrates. The affected plants have reciprocated by building up an array of defences that, although often insufficient for total pest exclusion and the complete avoidance of damage, allow the plants to survive and reproduce. The defences include the natural enemies of the pests. Although such defences have often been noted, it required the seminal review by Price *et al.* (1980) to initiate a concerted effort that addressed this topic. Price *et al.* (1980) firmly established the tenet that natural enemies are an important component of the plants' array of defences. Nowadays, it is understood that the interactions between natural enemies and their prey cannot be understood (and manipulated) without taking into account the host plant (Sabelis *et al.*, 1999; Cortesero *et al.*, 2000).

Other plant defences include a tough outer layer, diverse hairs and spines that may exude sticky excretions, poor nutrient quality, a vast arsenal of repellent, inhibitory or poisonous secondary plant substances and 'bodyguards'. The latter are predators for which the plant provides shelter and nutrition (e.g. secretions from extrafloral nectaries) before pest attack, or else attracts during herbivore onslaught. The presence of predatory bodyguards is often considered to be a mutualistic relationship between plants and natural enemies; ants are an example (Bentley, 1977). Alterations in plant architecture (e.g. branching pattern, foliage morphology) could affect the time that natural enemies remain on plants, increasing or decreasing accessibility to the herbivores. These factors may also affect the outcome of intraguild predation (IGP; see Chapter 42). Many such modifications of plant morphology were tentatively attributed to selection by natural enemies (top-down effects) (Marquis & Whelan, 1996).

Hereditary or environmental factors that affect the plant will also influence its guests, directly or indirectly, whether plant feeders, predators or detrivores. The influence of these factors on predatory mites is the main topic of this chapter. For convenience plant effects are separated into direct [e.g. influencing the acarine biocontrol agents (ABAs)] and indirect (affecting the ABAs through the prey/host) factors. Direct effects include plant facies, growth pattern and phenology; the influence of plant species and cultivar (cv.); plant nutrients; leaf surface topography (including domatia); 'experience' or 'memory' of plants; leaf age and the influence of agricultural chemicals, e.g. fertilisers and pesticides that are applied to the crop. Indirect effects include most of the above, owing to their effect on the prey, as well as plant resistance to pests and pest-induced volatiles, produced by the plant and attracting predators. The volatiles were labelled information-bearing chemical messengers or infochemicals (Dicke & Sabelis, 1988). These authors defined them as synomones, chemicals that mediate interactions between individuals of different species and are adaptively beneficial to both emitter (e.g. the plant) and receiver (the ABA).

Direct effects

Plant facies and growth pattern

The natural distribution of most ABAs appears to be habitat specific rather than plant specific, although exceptions occur (see below). Species may prefer trees and shrubs, or are found mainly on annual or recumbent plants; others favour the litter layers and the soil. Evidence for such patterns of distribution is often obtained from collections of natural enemies occurring on local floras. A survey of the Japanese Phytoseiidae suggested to Ehara & Amano (1998) that mites in the subfamily Phytoseiinae have a close relationship with stable habitats (e.g. woody plants), members of the tribe Typhlodrominae occurred mainly on trees and bamboo, and many Amblyseiinae were collected from grasses and other low-growing plants. Data on phytoseiids obtained from indigenous plants around Lake Ontario in Canada confirmed these generalities. *Typhlodromus bakeri* (Garman) occurred almost exclusively on shrubs and trees, *Amblyseius meridionalis* (Berlese) was mostly found in the soil and *Amblyseius andersoni* showed no preference for any habitat (Rothman, 1988). Ragusa Di Chiara & Tsolakis (2001) distinguished between phytoseiids on wild plants in natural environments, those on wild plants in agricultural systems and those found on cultivated plants. *Kampimodromus aberrans* occurred mainly on arboreal cultivated plants and rarely on herbs, whereas *Typhlodromus exhilaratus* Ragusa was found mostly on herbs. A third species, *Amblydromella athenas* (Swirski & Ragusa), was found in all habitats, on all plants.

Karg (1989a) suggested that the distribution of phytoseiids (and other mesostigmatid predators) in different habitats is determined mainly by the ambient relative humidity. Although fortified by many experimental data, this hypothesis needs more corroboration, especially from humid subtropical and tropical regions.

Specialised ABAs are often restricted to pest-infested plants, but it is difficult to decide whether the plant determines the specificity, whether it reflects prey presence, or whether both factors play a role. Some Australian *Neoseiulus* spp. were collected only from each of several tree species sampled in southern Queensland, without overlap between these mites, although the trees grew in well-mixed stands (Beard & Walter, 2001).

Such preferences for plants or plant types could affect the establishment rate of ABAs and their success in pest control. *Phytoseiulus persimilis* is usually found on shrubs and recumbent plants (e.g. roses, strawberries and watermelons), seldom in arboreal situations, and provides better pest control on the former group of plants (Chapter 26, *Phytoseiulus persimilis*). Conversely, predators adapted to living on trees do not do well on the understorey vegetation.

Plant architecture and horticultural practices affect ABA activities. ABAs locate prey more quickly on crops of a simple architectural structure; it follows that fewer predators would be needed to control pests on such plants. Branches and foliage that conjoin separated plants or trees (links that are sometimes facilitated by trellising) enable predators to move between pest patches, requiring fewer natural enemies to be placed on each individual plant. Control of the two-spotted spider mite (TSSM) by *P. persimilis* was better on trellised cucumbers whose upper parts were touching than on peppers, which developed individually (Jarosik, 1990). Densely growing rose cultivars provided more refuges for TSSM than sparse varieties, while the profusion of touching foliage on the dense roses enhanced the rapid movement of *P. persimilis* between their leaves. The long-term interactions between TSSM and predator on these roses, which brought about satisfactory pest control, were prolonged and stabilised by the growth habits of the denser varieties (Gough, 1991; see Chapter 26).

In Oregon, USA, hops (*Lupulus lupulus*) grow rapidly and the leaves are attacked by TSSM early in the season. As hops grow upward, the pest moves along to establish new colonies, soon infesting entire plants. *Neoseiulus fallacis*, the local predator, does not keep pace, and about 23–30% of the pest's colonies are thus free of predators. In other words, rapid plant growth provided the pests with refugia (or enemy-free space), which reduced the risk of spider mite (and then predator) extinctions (Strong *et al.*, 1997). The plants' growth pattern thus contributes to the biological control of its spider mite pests.

The plant as shelter

ABA populations on deciduous trees face two hazards during winter. The prey (and alternate food sources) become scarce or disappears, and the predators must find shelter in face of the inclement weather. ABAs that overwinter on trees either locate protected sites on the bark and branches, or hide in the ground. The amount of available shelter often depends on the tree's architecture, number and distribution of cracks, crevices and buds, and on the remnants of various biota (e.g. algae and lichens, abandoned nests, empty shields of scale insects, exuviae of other apple arthropods). The hanging husks on walnut trees offer a unique sheltering site (McMurtry & Flaherty, 1977). Plants that provide more shelter would be able to support more predators, and these would be on hand earlier during the next season. James *et al.* (1992) extracted phytoseiids from the canes and wood of dormant grapevines in amounts that were numerically comparable to values obtained from ten-leaf samples of foliage during the growing season. The apple cultivars Spartan and McIntosh have rough fruit spurs that shelter resident phytoseiids during winter; Delicious, in contrast, has smoother spurs and thus accommodates fewer predators (Downing & Moilliet, 1967). The current horticultural vogue towards densely growing, rapidly maturing fruit trees, and their quick turnover, may not allow the trees to develop crevices, epiphytes and other sheltering sites for predators. Alternate winter refugia should be provided if the use of acarine predators is to be continued there (Gurr *et al.*, 1997).

Phytoseiids hiding in the cavities of the apple calyx initially shelter on the tree. After fruit drop they stay on the ground, remaining within their shelter until spring. Highest numbers (up to six/fruit) were found in the calyx of cv. Red Delicious, fewer in Granny Smith and fewest in Bonza (Gurr *et al.*, 1997). The predators probably entered via a channel that is partially blocked by the remnants of the sepals, a portal that may be harder to transverse in cv. Bonza than in apples of other cultivars. Phytoseiid recolonisation of apples would thus be slower on Bonza, possibly explaining why fewer predators were found on these trees. The survival of predators that overwinter on the orchard floor is mainly determined by the ground cover (discussed below).

Plant species and cultivar

Pests mites prefer certain plant species to others and often discriminate even between cultivars of the same commercial crop. (Plant resistance will be discussed below.) The citrus red mite, *Panonychus citri*, developed more rapidly and raised larger populations on lemon than on orange trees (Fleschner, 1958). Greater numbers of *Panonychus ulmi*, the European red mite (ERM), developed on the apple cv. Oregon Spur than on Delicious in Israel (Palevski *et al.*, 1996). The abundance of predatory mites that track their prey onto commercial varieties thus often, but not invariably, reflects pest density. In Ontario, Canada, the stigmaeid *Zetzellia mali* was most common on the apple cvs Red Delicious and Empire, which were also the preferred hosts of the apple rust mite and of several spider mite pests (Woolhouse & Harmsen, 1984). The resident phytoseiids (*N. fallacis*, *Typhlodromus pyri* and *Typhlodromus caudiglans* Schuster), were inconsistently associated with these cultivars, seeming to be more common on cv. McIntosh. The clear preference of *A. andersoni* for certain peach cultivars in northern Italy appeared to be unrelated to the presence of any prey on these trees (Ivanich Gambaro, 1988).

Varietal factors affect the early establishment of released predators. *Amblyseius andersoni*, mass-released in Italian grape vines, was initially more common than *T. pyri* on cv. Merlot; the converse occurred on cvs Trebbiano and Garganega (Camporese & Duso, 1996). Predator abundance as affected by the glandular trichome secretions is discussed below (see leaf surface topography and glandular hairs). Cassava cultivars differ in the number of shoots that they produce. The leaves at their tips provide shelter and nutrients for *Typhlodromalus aripo*, the predator of cassava green mite (CGM) (see section on feeding on nectar, honeydew and fungi, below). Numbers of *T. aripo* are smaller on shoot-poor cultivars, which thus indirectly reduce its efficacy (Yaninek & Hanna, 1998).

Plant species and cultivars affect the spectrum of synomones that are emitted by spider mite-damaged leaves and that attract predators to the feeding sites (Dicke *et al.*, 1990; Takabayashi *et al.*, 1991); this will be discussed below.

The effect of plant species and cultivars is critical when the specificity of phyto-phagous mites for weed control is being determined. Various strains of the weed *Chondrilla juncea* showed different sensitivities to attack by the Chondrilla gall mite (Carèsche & Wapshere, 1974). Research on the biological control of weeds has revealed that common and widespread wild plants commonly occur as many races or strains.

Predator feeding on plant-provided nutrients: general comments

Many ABAs feed at more than a single trophic level (e.g. omnivores, as defined by Pimm & Lawton, 1978). Pollen is the most common plant product used, but floral nectar and extrafloral exudates, pearl bodies (see below) and even plant sap, are reg-ularly or irregularly taken. Additional non-prey, plant-affected diets include honey-dew (sweetish secretions from phytophagous Hemiptera, e.g. aphids, whiteflies and scale insects) and various fungi. The latter may be growing on the honeydew (e.g. the dark sooty mould fungi), or may be saprophytes or even plant pathogens (e.g. powdery mildews). Pollen as a sole nutrient supports generalist phytoseiids for many generations, whereas other foods serve mainly as alternate diets, sustaining predator populations in the absence of prey. The number of generalist phytoseiids in the field is often determined by the availability of plant-provided nutrients rather than by the presence of prey (McMurtry, 1992).

Feeding on pollen

Many ABAs consume pollen, as alternate or supplementary food, or both. Some generalists (e.g. *Euseius* spp.) attain maturity and reproduce when given only this diet (Zhimo & McMurtry, 1990). Pollen hastens development, increases (or decreases) the rate of predation, enables survival during periods of prey scarcity and even determines whether predators will control pests on certain plants. Diverse pollens affect various predators in different ways. This plethora of interactions suggests that the effects of pollen on predators should be determined individually for every com-bination of mite and pollen. McMurtry & Scriven (1966a) were probably the first to address the dynamic relationship between pollen feeding and predation. *Euseius tularensis* (then called *Amblyseius hibisci*) devoured 60% fewer avocado brown mite, *Oligonychus punicae*, when concurrently offered pollen of the ice plant, *Malephora crocea* (then called *Mesembryanthemum*; see Flechtmann & McMurtry, 1992a). When given only a little pollen, predation rate was reduced only by 26%. Any seemingly deleterious effects of feeding on pollen were offset by the higher fecundity of preda-tors offered enough of that nutrient. This led to the belief that the increased fecun-dity of *E. tularensis*, when enough pollen was available, would enhance pest control, a hypothesis validated in the field by Kennett *et al.* (1979). Changes in the numbers of *E. tularensis*, which feeds on the citrus red mite, were correlated with the seasonal pattern of wind-borne pollen drop during 4 years. High pollen drop in late winter and early spring was followed by elevated predator numbers, whose populations grew more slowly after seasons of low pollen fall. Wei & Walde (1997) reached a generally similar conclusion in regard to an apple system, wherein *T. pyri* feeds on *P. ulmi*, as high pollen levels usually coincided with low pest numbers.

The introduction of *T. pyri* into Australian apple orchards was followed by improved, but not fully successful, pest control, a situation that required applications of supplementary acaricides (Readshaw, 1998). Orchards were then placed under an alternate mowing regimen to enable weeds, especially grasses, to grow and flower. The abundance of *T. pyri* subsequently increased along with reductions in pest numbers. Few predators (one to five/five leaves) occurred on trees that were not surrounded by flowering grasses, and all were pale, without juveniles. Apple trees within a dense understorey growth of flowering grasses carried many *T. pyri* (up to 30/five leaves) with reddish guts (indicating that they had fed on ERM) and with young stages. No other sources of food were at hand, indicating that pollen was the only available nutrient. The spider mite pests were extremely scarce on these trees.

Although pollen from certain plants (e.g. *M. crocea*) is taken by many ABAs, others may be fastidious. The suitability of pollen from 23 plants for *E. tularensis* was compared by Kennett *et al.* (1979). Development and reproduction were normal on the pollen of 11 of these plants, whereas they were altered on pollen from the remaining plants (Table 41.1, supplemented by data about *E. tularensis* from McMurtry & Scriven, 1966a). Avocado (*Persea americana*) pollen was suitable for the reproduction of *E. tularensis*, but Kennett *et al.* (1979) later listed it as inadequate for

Table 41.1 Suitability of various pollens for the reproduction of *Euseius tularensis*

Reproduction	No reproduction
Aesculus californica[†]	*Acacia* sp.[†]
Brassica nigra[†]	*Acer negundo*[†]
Capsicum frutescens[*]	*Amsinickia douglasiana*[†]
Ligustrum japonicum[†]	*Brodiaea laxa*[*]
Mesembryanthemum spp.[*‡]	*Juglans regia*[†]
Nicotiana glauca[†]	*Catalpa bignoniodes*[†]
Persea americana[*]	*Cedrus deodora*[*]
? Pineapple family[†]	*Cupressus* spp.[*]
Prunus amygdalus[†]	*Franseria acanthicarpa*[†]
Quercus agrifolia[*†]	*Gossypium* sp.[†]
Ricinus communis[*]	*Hibiscus* spp.[*]
Rubus sp.[†]	*Melia azedarach*[†]
Salix sp.[†]	*Olea europea*[†]
Typha latifolia[†]	*Persea americana*[†]
Zea mays[*†]	*Phoenix dactylifera*[*]
	Pinus coulteri[†]
	Sambucus coerulea[†]

Based on data in [*]McMurtry & Scriven (1966a), and [†]Kennett *et al.* (1979). [‡]Actually *Malephora crocea*.

this species. Pollen of *Typha orientalis* was very good for *Typhlodromus doreenae* and *Euseius victoriensis* (James, 1993), but no eggs were produced by *Typhlodromus athiasae* when offered only pollen of *Typha australis* (Reuveny *et al.*, 1996). They also noted that apple pollen was unsuitable for *T. athiasae*, a diet which was regarded as optimal for *E. tularensis* (Ouyang *et al.*, 1992). Predators in other families preferred different pollens: the pollen of date palms (*Phoenix dactylifera*), apparently unsuitable for *E. tularensis*, provided better nutrients for the stigmaeid *Agistemus exsertus* than did pollen of corn or castor bean (*Ricinus communis*) (Abo Elghar *et al.*, 1969).

The survival of *E. tularensis* that were fed pollen from eight wild grasses was initially high (63–97% during the first generation), but declined to below 45% in the second generation (Ouyang *et al.*, 1992). Wheat pollen was especially unsuitable for this predator. Data from North America (Coli *et al.*, 1994) and China (Liang & Huang, 1994) indicate that understorey grasses carry fewer ABAs than broadleaved plants. Raworth *et al.* (1994), however, advocated planting barley in apple orchards in southern France because the largest spring colonies of *Neoseiulus californicus* occurred on that plant, owing to its pollen. Smith & Papacek (1991) concurred: by planting Rhodesgrass (*Chloris gayana*) in citrus orchards and mowing it every 7–10 days, a continuous supply of pollen was provided, which resulted in larger populations of *E. victoriensis*. The data of Readshaw (1998) were noted above.

The ability of predators to utilise diverse pollens could be due to their different sizes and the thickness of their walls (exines). Feeding on pine and cotton pollen is probably avoided because of the large size of the grains (75–100 μm). Oat pollen has a thick exine, postulated by Ouyang *et al.* (1992) to account for its unsuitability for *E. tularensis*, whose optimal development on apple pollen was attributed to its thin walls. Van Rijn & Tanigoshi (1999b) noted that phytoseiids often prefer, or reject, pollen from related plants, but there are many exceptions. The nutritional merit of different pollens for ABAs has not been evaluated.

Windbreak trees are often planted around orchards and serve as refugia for pests and predators (Solomon, 1981) and as pollen sources. *Cupressus* spp. and *Pinus radiata*, used as windbreaks in South Africa, produce pollen that was unsuitable for *Euseius addoensis* (Grout & Richards, 1992b). *Casuarina cunninghamiana* pollen, however, provided sufficient nutrients for this predator, and was recommended as the first choice for windbreaks in the eastern Cape Province of South Africa.

In Dutch glasshouses *Neoseiulus cucumeris* provided better control of thrips pests on sweet pepper than on cucumber. van Rijn & van Houten (1991) attributed the predator's unsatisfactory performance on the latter crop to a lack of sufficient pollen. Cucumber is parthenocarpic, providing only scant pollen, whereas peppers blossom continuously, producing abundant pollen. Its availability increased predator activity on sweet peppers before the thrips arrived, as well as after that prey had been exhausted. Differences in leaf setosity between the two crops could be an additional explanation for the differences in the predation rates (see below).

Prolonged or exclusive feeding on pollen may elicit unexpected and sometimes adverse responses from ABAs. Carotenoids are required for diapause induction in *A. andersoni*, being photopigment receptors that are active in photoperiodic light perception by this species (van Zon *et al.*, 1981). Mites that were reared only on broad bean pollen did not enter winter diapause, a result attributed to the low carotenoid content of this pollen (Overmeer & van Zon, 1983). Another adverse response to prolonged feeding on pollen was changed attraction to prey. When *A. andersoni* and *T. pyri* were given only bean pollen during development they responded to kairomones of three prey species (ERM, TSSM and the apple rust mite), but predators reared on TSSM were attracted only to the ERM. The broader range of attraction stimuli shown by pollen-reared predators was engendered by the low carotenoid content of broad bean pollen. Phytoseiids reared on this pollen supplemented with crystalline β-carotene displayed a pattern of attraction similar to that of mites given spider mite prey (Dicke, 1988). The attraction of the stigmaeid *Agistemus exsertus* to spider mites decreased after being reared for ten generations on castor bean pollen, and its sex ratio (SR) shifted from 0.5 when given mite prey to 0.38 on pollen (Rasmy *et al.*, 1996).

The tydeid *Homeopronematus anconai*, an important alternate diet for *Galendromus occidentalis* in Californian vineyards, thrives on weed pollen (Flaherty & Hoy, 1971). The predator is the major natural enemy of grape spider mites, and it was suggested that pollen-bearing plants should be planted in and around vineyards to promote pest control.

Phytoseiids use different methods for feeding on mite prey and on pollen. When attacking mites they wound the cuticle, probably inject proteolytic enzymes and imbibe the partially predigested prey contents. When feeding on pollen, each grain is held between the chelicerae and ground by their rapid alternate movements, the contents are sucked in and the empty exine is removed. The entire process, from taking up individual pollen grains to exine removal, takes 4–14 s (Flechtmann & McMurtry, 1992a). Each of the two species observed [*Euseius stipulatus* (Athias-Henriot) and *Amblyseius similoides* Buchelos and Pritchard] consumed a similar mean of around 100 ice plant pollen grains in one 1 h.

Certain phytoseiids (*Euseius* spp., *I. degenerans*) that feed, develop and reproduce on pollen, have the lobe of the fixed digit of their chelicerae modified into a spoon-like structure on the paraxial (inner) face. Members of this family that feed mostly on spider mites have quite variable cheliceral lobes, and a pollen feeder (*A. similoides*) has a large lobe. Thus, despite some tentative trends, the variability in lobe structure precludes the present use of cheliceral morphology as a guide to the diet preferences of unstudied phytoseiids (Flechtmann & McMurtry, 1992b). The tendency of *I. degenerans* to feed on pepper pollen increases its ability to reduce the numbers of thrips, which often visit these flowers (Chapter 26).

Feeding on the host plant

Many ABAs spend their entire lives on the host plant, which is thus the most easily available food source for non-specific predators and for those with limited dispersal abilities. Various predators require leaf sap or supplement their diets by feeding on that source. Observations that some ABAs, if denied access to plants, cannot be reared for long, support the hypothesis that they require some plant constituent(s) for their development. It follows that such ABAs are affected by their host plants and by the horticultural practices that are applied to them. This direct plant–predator interaction was termed transtrophic, signifying elimination of the intermediate primary consumer (i.e. the herbivore prey) level (Bakker & Klein, 1992).

Larvae of *Z. mali* that had been denied access to apple leaves, even when provided with enough prey eggs, did not moult to the next stage. The females also appear to feed on plants, because they were observed to spend about one-third of their time probing leaf tissues (Santos, 1982). The partiality of this predator for various apple cultivars, noted above, may be due to host–plant nutritional factors, as could be the preference of *A. andersoni* for various peach cultivars (Ivanich Gambaro, 1988). Other stigmaeids, for example *Agistemus fanari* and *A. longisetus*, likewise fed on plant tissues (Dosse, 1967; Gonzalez-Rodriguez, 1965). Partial phytophagy by stigmaeids may be common; members of the genus *Eustigmaeus* (=*Ledermuelleria*) are phytophagous, feeding and reproducing on mosses (Gerson, 1972).

Herbivory by phytoseiid mites was initially demonstrated by Chant (1959). He introduced acid-fuchsin, a vital stain, into apple, blackberry and blackcurrant leaves and confined individuals of *T. pyri* and of *Typhlodromus rhenanus* (Oudemans) on them. The mites' intestines became stained, indicating that they had fed on the leaves. Proof of phytophagy by other phytoseiids was obtained by labelling lemon and avocado leaves with radioactive phosphoric acid (Porres *et al.*, 1975). Females of *E. tularensis* placed on these leaves accumulated significant amounts of radioactivity. No meaningful amounts of such material were detected in the bodies of *E. stipulatus* and *Amblyseius fructicolus* Gonzalez & Schuster, indicating that they did not feed from the leaves.

Fertilisers, applied to plants to promote their growth and health, affect the fecundity of phytophagous predators. Grafton-Cardwell & Ouyang (1996) placed eggs of *E. tularensis* on excised citrus leaves that were supplemented only with ice plant pollen. The leaves were obtained from potted trees given various concentrations of a stock fertiliser [nitrogen–phosphorus–potassium (NPK)]. The concentrations were 1, 1/10, 1/100 and 1/1000 of the stock, applied to the plants weekly for 3 months. The development of predators placed on the various leaves was normal, and there were no differences in either survival frequencies or SRs. However, females that had fed on leaves taken from plants that had received the lowest fertiliser concentration produced the fewest eggs (Fig. 41.1).

Nitrogen and manganese levels in the leaves were significantly correlated with mean number of eggs deposited, suggesting that the increased fecundity of females reared on the leaves was due to the higher concentrations of these minerals (and perhaps also zinc), which were used for progeny production. Populations of *E. tularensis* prefer young citrus foliage and their numbers increase in response to postpruning flushes of growth, even in the absence of prey, suggesting that the predator recognises the presence of suitable nutrients in the leaves. The densities of the predator (and hence its effect on pests) could therefore be manipulated by suitable fertiliser regimes. Since pruning is a normal horticultural practice in citrus groves, it could be another cost-effective method of increasing *E. tularensis* numbers in the orchard (Grafton-Cardwell & Ouyang, 1995).

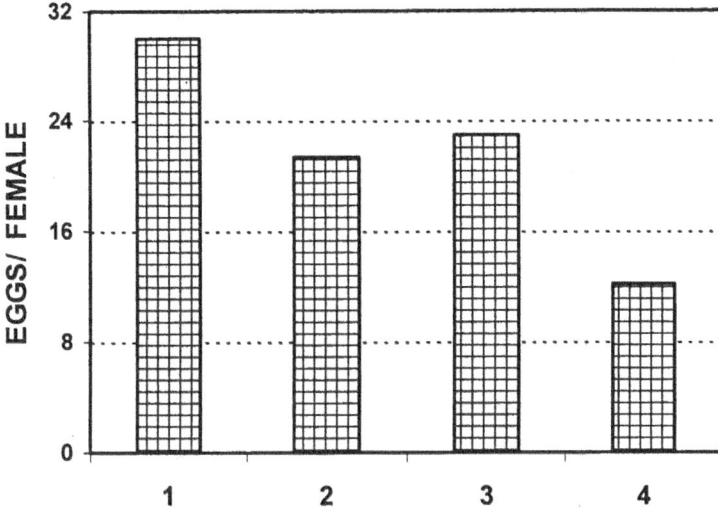

Fig. 41.1 Mean number of eggs deposited by *Euseius tularensis* reared on leaves from citrus trees supplied with (1) 100% of a stock fertiliser solution; (2) 10%; (3) 1% and (4) 0.1% of the solution. (Data from Table 1 in Grafton-Cardwell & Ouyang, 1996.)

Phytophagy by phytoseiids may detract from their efficacy as predators. *Phytoseius plumifer* (Canestrini & Fanzago), kept on leaves of castor bean and supplied with the carmine spider mite as prey, produced very few eggs (Elbadry, 1974). No correlation was obtained between prey consumption and oviposition, a result attributed to prolonged plant feeding.

Pearl bodies are small, round, translucent epidermal outgrowths on the young parts of grapevines (*Vitis vinifera*) and some other plants. Adults of *T. pyri* deposited only 0.3 eggs/day on this diet, but survived for several weeks (Engel & Ohnesorge, 1994).

Do plant-feeding ABAs compete with their own phytophagous prey and thereby affect their populations? No research appears to have been conducted on this issue. Coll & Izraylevich (1997) considered a system in which a predatory bug (Hemiptera) fed on a phytophagous thrips as well as on a common host plant. They developed a heuristic model which predicted that an increase in plant suitability for the predator, and/or increased competition with the thrips, would destabilise the system, leading to extinctions. Conversely, the growth rate of the predator was reduced with less plant suitability, resulting in less competition, leading to a more stable system. The stability of such systems was therefore postulated to be strongly affected by the plants, especially if less suitable for the predators. As phytophages are more efficient plant feeders than ABAs, there would be little competition between them and the system should persist. In any case, predatory mites that feed on plants use this resource mostly to supplement their diets, thus reducing the likelihood of competition with plant specialists. The situation may be different in regard to *E. tularensis*, which can be reared on plant tissues. It would be interesting to determine whether the pest-controlling effect of this predator would be affected by competition with its prey.

Feeding on floral and extrafloral nectar, honeydew and fungi

Although floral nectar is a rich natural food, and phytoseiids often occur within blossoms, little is known about its effect (as separate from that of pollen) on ABAs. Offered only orange nectar in the laboratory, *Amblyseius limonicus* deposited few eggs, but survived for a long time, and 30% were still alive after 60 days at around 22°C (McMurtry & Scriven, 1965a).

Extrafloral nectaries are plant glands that secrete sweet exudates but are not located near flowers (e.g. at the bases of leaves). They entice diverse insects (e.g. ants) as well as mites, and their amino acid content and volume may increase in response to herbivore pressure (Bronstein, 1998). The addition of the extrafloral nectar of castor bean to the pollen slightly improved the reproduction of *I. degenerans* and increased its longevity for several months at 25°C. Predators placed on castor bean bushes could thus be sustained during releases into greenhouses (van Rijn & Tanigoshi, 1999a).

The phytoseiids *T. aripo* and *Typhlodromalus manihoti* (Chapter 26) coexist on cassava plants in Brazil and Colombia, where they feed on CGM (Yaninek & Hanna, 1998). The former is restricted to the shoot tips at the cassava apices, where it lives and shelters during the day. It feeds and its juveniles develop on phloem droplets that are secreted at the bases of the youngest petioles. In the evening this predator moves down into lower plant strata and hunts for CGM, which it requires for reproduction. The preference of *T. aripo* for the upper parts of cassava was shown by its disappearance from plants whose terminal shoots were removed. *Typhlodromalus manihoti*, which lives on cassava leaves, tends to overexploit the CGM, and likewise feeds on other diets. Morphological constraints usually do not allow *T. manihoti* easy access to the upper parts of the plant. When both predators co-occurred with the prey, *T. aripo* was dominant in the upper plant parts. However, in the absence of cassava mites, *T. manihoti* also foraged for the droplets in the upper plant regions. This nutrient thus reduced competition between the two predators, enabling both to survive. The effect of various shoot forms in different cassava cultivars on CGM control was mentioned above. Pemberton (1993) noted that in Korea the anystid *Anystis* sp. fed on the extrafloral nectaries of *Prunus sargentii* and *Populus tementiglandulosa*.

Honeydew contains sugars and other nutrients, including amino acids (Ewart & Metcalfe, 1956), and constitutes an important supplementary diet for generalist predators. Offered as the sole nutrient to *E. tularensis*, honeydew secreted by the mealybug *Planococcus citri* (Risso) allowed only a low rate of reproduction but increased the predator's longevity. Adding honeydew to spider mites or to pollen increased the fecundity *E. tularensis*, enhanced its development and enabled more juveniles to survive (Fig. 41.2) (McMurtry & Scriven, 1966b). Similar results were obtained with other *Euseius* spp. given honeydew from the bean aphid, *Aphis fabae*, and from the woolly whitefly, *Aleurothrixus floccosus* (Maskell) (Zhimo & McMurtry, 1990). The

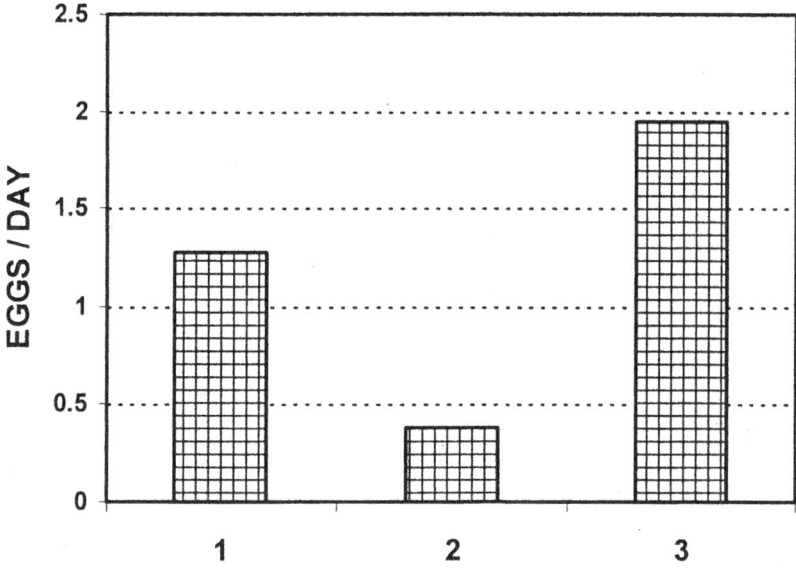

Fig. 41.2   Reproduction of *Euseius tularensis* when given (1) an abundance of the spider mite *Oligonychus punicae*; (2) mealybug honeydew; and (3) an abundance of spider mites and honeydew. (Data from Table 1 in McMurtry & Scriven, 1966b.)

phytoseiid *Typhlodromus phialatus* Athias-Henriot also survived on the honeydew of the woolly whitefly, but deposited no eggs (Ferragut *et al.*, 1987).

Honeydew as a nutrient is usually considered in a generic sense, without distinguishing between the secretions of different hemipteran insects. However, the constituents of honeydew differ from producer to producer (e.g. Ewart & Metcalfe, 1956) and some may even be toxic (Avidov *et al.*, 1970).

Generalist phytoseiids feed on some plant-pathogenic fungi, especially mildews, as an alternate or even as the sole diet. Spores of tobacco mildew (*Erysiphe orontii*) and of strawberry mildew (*Oidium fragariae*) were readily eaten by *T. pyri*, but those of red clover mildew (*Erysiphe polygoni*) were not (Zemek & Prenerová, 1997). The mite developed normally and even reproduced when offered only *E. orontii* spores, but fecundity was very low, at two or three eggs/female. Other phytoseiids that completed their development on mildews included *K. aberrans* and *Typhlodromus umbraticus* Chant, which subsisted on *Podosphaera leucotricha* infecting apple leaves (Chant, 1959). The latter mite also developed on spores of raspberry leaf rust (*Phragmidium rubi-idaei*), feeding on them to the extent that some spore eruptions were destroyed. Many common saprophytic fungi serve as low-value nutrients for various ABAs; only two out of 18 nymphs of *T. caudiglans* kept on fungi growing on decaying leaves reached maturity, and they required twice the time taken by conspecifics that were fed spider mites (Putman, 1962).

Feeding of the tydeid *Tydeus lambi* on powdery mildew reduced the severity of this disease to grapes in eastern USA, and the grazing of another tydeid, *Lorryia formosa*, on sooty mould decreased scale insect damage to citrus in Israel (Chapter 36). The activities of these ABAs were thus indirectly affected by the plants, which were the mildews' and the scales' hosts.

Leaf surface topography and glandular hairs

The major leaf topography features that affect pests as well as their ABAs are veins, non-glandular hairs, glandular trichomes and the domatia (see below).

The veins on the lower side of leaves often serve as preferred resting and oviposition sites for many ABAs. Most *Z. mali* on the apple cv. Red Delicious occurred on the leaf underside, along the mid-rib (Santos, 1976a). The stigmaeid is a sedate predator that seldom wanders away, the lower mid-rib region being its main foraging area. This affects interactions between *Z. mali*, *T. pyri* and *G. occidentalis*. The latter placed more eggs within the mid-rib region than *T. pyri*, leading to more of its eggs being eaten by *Z. mali* (Chapters 32 and 42).

Non-glandular leaf hairs affect ABAs in different ways, often increasing their foraging efficacy. Pubescent leaves reduce the velocity of predator movement and/or induce the predator to change direction, thus increasing its chances of encountering prey. Most predatory mites appear to prefer at least moderately setose leaves. Downing & Moilliet (1967) compared the populations of *T. caudiglans* and of *G. occidentalis* on three apple cultivars, namely Delicious, Spartan and McIntosh, which were infested with ERM. Delicious bore the most pests, suffered the heaviest damage and had the fewest phytoseiids. The predators were more numerous, and control was better, on Spartan and McIntosh. The leaves of the latter two cultivars are hairy and have pronounced veins, whereas those of Delicious are relatively smooth. Given a choice between setose and glabrous leaves of various plants, *A. andersoni* invariably chose the hairy leaves (Overmeer & van Zon, 1984), a selection consistent with its distribution in the field. A similar pattern was noted in the presence of *Z. mali* on grapevine leaves. It preferred woolly (densely covered with hairs) to cobwebbed (loosely covered with fine interwined hairs that resemble spiders' webs) leaves, and cobwebbed to glabrous leaves (Fig. 41.3). Jenser & Koleva (1996) believed that the preference of *Z. mali* for plants with pubescent leaves explained its more frequent occurrence on apples and plums than on sour-cherry trees. As larvae of *Z. mali* require vital substances from the host plant (see above), leaf vesture may not be the full explanation. Species of *Phytoseius* which, as noted, feed on plants, are more common on hairy leaves than on almost smooth leaves (Walter, 1992).

Leaves with more or longer hairs trap more pollen grains. The hairs on the leaves of two fig (*Ficus carica*) cultivars differed in their lengths, the one with longer hairs catching more pollen and carrying more phytoseiids (Barret & Kreiter, 1995) (Table 41.2). Because plants (whether annuals or perennials, cultivated or wild) with hairy leaves carry more phytoseiids, they are a better and more consistent source of predators for crop recolonisation (e.g. Tixier *et al.*, 1998). The surface of glabrous foliage may be too slippery for certain ABAs, thus favouring others. *Anystis* sp., for instance, provided better control of spider mites on smooth than on hairy, foliage (Lange *et al.*, 1974b).

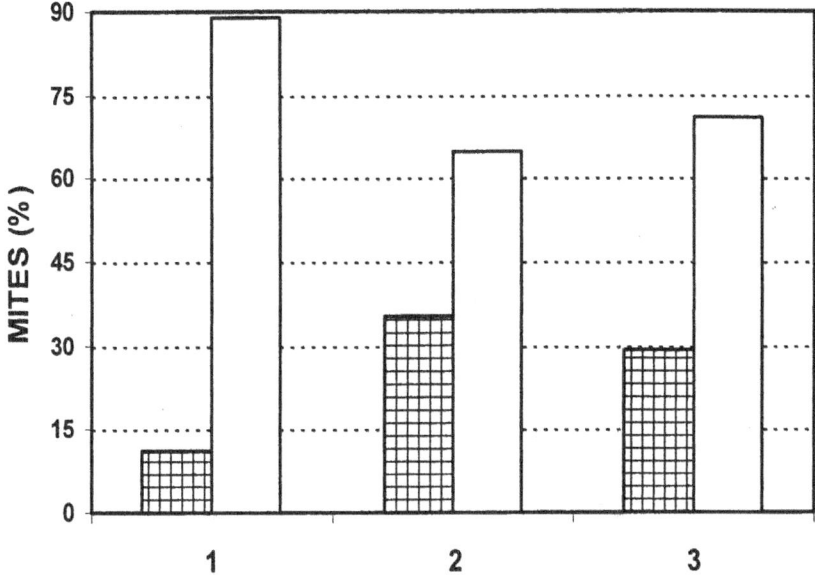

Fig. 41.3 Location of *Zetzellia mali* after 48 h on grape leaves of different hairiness, offered in various combinations. (1) Glabrous (cross-hatched bars) and woolly (white bars); (2) glabrous (cross-hatched bars) and cobwebbed (white bars); (3) cobwebbed (cross-hatched bars) and woolly (white bars). (Based on data in Tables 1 and 2, Jenser & Koleva, 1996.)

Table 41.2 Length of hairs on the leaves of two fig cultivars (A and B), the amount of pollen on them and the number of phytoseiids collected from than

	Cultivar A	Cultivar B
Length of hairs (μm)	252	153
Pollen grains/dm^2	4500	1730
Female phytoseiids/dm^2	19	5

Based on data in Barret & Kreiter (1995).

Predatory mites (e.g. Cunaxidae Phytoseiidae and Stigmaeidae) as well as fungivores (Tydeidae, diverse Cryptostigmata) were much more abundant on setose than on smooth leaves of Australian rainforest trees (Walter & O'Dowd, 1995). The relationship between mites and hairy leaves is thus not restricted to cultivated plants, suggesting that setose foliage confers survival advantages upon many predators, and indicating a prolonged association between hairiness and predatory (and other) mites.

In other cases, hairiness impedes ABAs. As noted, *N. cucumeris* is a better predator of thrips on sweet pepper than on cucumber, requiring less time to handle each individual prey on pepper than on cucumber. Pepper leaf discs (4 cm diameter) carried very few trichomes (about 3.5% of those on cucumber leaf disks, 17 vs 490), leading Shipp & Whitfield (1991) to postulate that trichome density on cucumber might have hindered the predator's mobility, and thus prey finding, on that host plant. Prey seeking by *P. persimilis* is likewise held back on setose leaves, as walking speed and activities were reduced on the foliage of a very setose gerbera cultivar. The predator needed more time to find prey, which detracted from its efficacy on such cultivars (Kleijn *et al.*, 1997). Even more disruptive to *P. persimilis* are the glandular trichomes growing on the leaves and stems of many Solanaceae, especially tomatoes. Some trichomes carry a sticky exudate that repels or entraps small-bodied pests, being a major factor conferring tomato resistance to spider mites (Snyder & Carter, 1984). These pests overcome the sticky trichomes by weaving webs above them and walking along these threads, but *P. persimilis* is often entrapped and killed. A few predators avoid the exudates by walking along spider mite webs, which lead them to new prey patches. Van Haren *et al.* (1987) suggested that the high postrelease mortality of *P. persimilis* on tomato could be reduced if predators were to be released only on plants already infested by spider mites. The number of trichomes per unit of leaf area, and the amount of exudate per trichome, differ among species of *Solanum* and even among tomato cultivars, and are influenced by environmental conditions. Tomatoes form bigger trichome heads when grown in light at 18°C than on plants held in the shade at 24°C, leading to more *P. persimilis* being entrapped on the former plants (Nihoul, 1993b).

A population of *P. persimilis* that had been placed on tomatoes gradually adapted to its new host, suggesting the option of selection for genotypes with better fitness to tomato (Drukker *et al.*, 1997). After several generations this population had a faster rate of increase than another predator population that had more recently been transferred to tomatoes, and provided better spider mite control. This finding suggests that directed selection for tomato-adapted strains of *P. persimilis* could be feasible, and would provide the first case of selecting a predator for suitability to a host plant.

Domatia

Domatia are tiny invaginations, pits or pouches, often accompanied by small tufts of hairs, located in major vein axils on the lower side of tree leaves. They differ from galls (which may be similar) by not being induced by their inhabitants. Domatia have been recorded from plants in many dicotyledonous families and shelter mycophagous and predatory mites, which occur more frequently within them than on other leaf parts (O'Dowd & Willson, 1989, 1997). Most domatia are very small (<2 mm diameter), providing space that is sufficient for mites but too small for most insects; they are seldom occupied by plant-feeding Acari. Leaves of the Australian rainforest tree *Elaeocarpus obovatus* occur with and without domatia; phytoseiids were found on 61% of domatia-bearing leaves, but significantly fewer (only 10%) on leaves that lack these structures (Walter & O'Dowd, 1992). A similar pattern was obtained from leaves of over 30 different tree and shrub species. The predators (and other mites) were postulated to gain several advantages from living within domatia, including not being washed off leaves by rains or blown away by cold or desiccating hot winds, and protection from natural enemies (Norton *et al.*, 2001). The eggs of many ABAs are sensitive to low humidity; placing them within the more humid domatia enhances their survival (Walter & O'Dowd, 1995).

Blocking the entrance to domatia on the leaves of *Elaeocarpus reticulatus* with bitumen paint reduced the number of phytoseiids on these leaves to 24% of untreated control leaves. The experimental excision of domatia from the leaves of *Viburnum tinus* significantly reduced the reproduction of *G. occidentalis* when kept at 30–38% relative humidity (RH), but upon being returned to domatia-bearing leaves they produced their normal egg complement (Grostal & O'Dowd, 1994). Grape leaves with blocked domatia, which harbour *T. lambi*, had much more of their surface area covered by powdery mildew than did plants with intact domatia (Norton *et al.*, 2000). The evidence thus suggests that domatia provide shelter for the plants' resident 'bodyguards'. But are plants that carry domatia always better protected from pests (and plant pathogens)? In other words, is the relationship really mutualistic, the domatia being a constitutive part of the plants' defences? Explorations of this question in regard to mites provided equivocal answers. The density of phytoseiids (mostly *T. caudiglans*) on 20 North American species of *Vitis* was affected neither by the presence of prey nor by plant gender (*Vitis* spp. are dioecious), thereby negating a possible effect of pollen on predator distribution. However, phytoseiid numbers were related to the density of vein hairs and to the presence of domatia (Karban *et al.*, 1995). Grape leaf surface characteristics were seen to be more reliable than any other plant trait for predicting phytoseiid numbers on the leaves of different cultivars, with shelter availability surpassing prey availability in determining the presence of predators on grapevines.

However, the role of domatia in the control of spider mite infestations on avocado appears to be ambiguous. Leaves of some avocado cultivars (e.g. Toro Canyon, used as rootstocks) bear domatia, whereas others (e.g. Hass) do not. Both cultivars are infested by the avocado brown mite, *Oligonychus punicae*, which is attacked by *Galendromus helveolus* (Chant). Domatia were added (by gluing tufts of cotton fibres in the vein axils) to Hass leaves or removed (by blocking with glue) from Toro Canyon leaves, and the effect on both mites was monitored. Predator densities increased fivefold on leaves with artificial domatia, but with little effect on pest numbers, which did not increase on leaves whose domatia had been blocked (Agrawal, 1997). A tally of plant-feeding mites found on smooth and on hairy (or domatia-bearing) leaves of several rainforest trees showed that relatively more phytophages than fungivores and predators occurred on the smooth foliage, whereas ABAs predominated on the hairy and domatia-bearing leaves (Walter & O'Dowd, 1995). However, there was no clear indication that the ABAs reduced the numbers of plant-feeding Acari on the domatia-bearing leaves. O'Dowd & Willson (1989) found up to 70 domatia/leaf on *Hebe townsonii*, but only a few plant-feeding mites. If there are no (or only a few) phytophages on plants with so many domatia, what is the benefit to the trees (especially as these plants are not known to be affected by plant-feeding mites)?

The production of domatia, with their accompanying tufts of hairs, incurs metabolic costs on the plant (Purrington, 2000). Reduced pest damage would probably represent a fair return on such costs, but what is the benefit in having leaves, entire trees, or cultivars of a certain plant species with or without domatia, especially as the pests are likely to attack all of them?

The sources and significance of this variation need to be addressed if one accepts that domatia represent a plant–predator mutualism. First, random intraspecific plant variations could have some effect on maintaining the interaction; an analogy with another mutualism may be invoked here. Bronstein (1998) listed three possible causes for the considerable intraplant variation seen in ant–plant mutualisms. The first is diversity in partner species, meaning that different ant species provide various degrees of protection, and different plants offer dissimilar rewards. The second is variation in mutualist attendance, in that various ants occur in different numbers. The third cause is variation in the need for mutualists, because plants are usually attacked by phytophages that exert different degrees of stress. These three sources of variation may interact and are also affected by environmental factors (e.g. soil fertility, humidity, light). Variability could thus also be expected when applying these hypotheses to the occurrence of domatia, because none of the predatory and fungivorous mites found therein seems to be specific. Karban *et al.* (1999) postulated that plant 'plasticity' (meaning that plant genotypes have the capacity to produce different phenotypes) depends on the cues or information that the plant receives about future herbivore risk. Thus, it is possible that the variability in domatia (and other defences, discussed below) production reflects the uncertainty of future herbivore attack; the plants 'hedge their bets'. This is consistent with the third of Bronstein's suggested causes.

Other data support the mutualism hypothesis. Domatia develop within leaf buds before any association with mites (O'Dowd & Willson, 1989), implying that their occurrence, and thus their role in plant protection, is hereditary. Further, domatia were observed on fossilised plants dating back to the Eocene, indicating an ancient (*c.* 40-million-year-old) interaction between plants and arthropods (O'Dowd *et al.*, 1991).

Plants with domatia thus possess an in-built, constitutive, pre-emptive defence mechanism, based on having bodyguards at hand before the plant is attacked. Other leaves on the same plant, or populations of the same plant species, may invest in emitting synomones (see below) after being infested by herbivores; this is a post-attack, induced defence mechanism. Such polymorphism (whether or not related to the variability noted above) would increase the versatility of the plant's defences by encouraging generalist predators to reside *in situ*, as well as by attracting non-resident, specialist natural enemies. In addition, some individuals in a plant population may lack defences, because natural enemies present on adjacent plants would protect them, and/or because they serve to attract the pests and thus the natural enemies, and/or because they have invested in becoming resistant to the pests. Atsatt & O'Dowd (1976) discussed 'plant defense guilds', meaning that different plants would be better defended against herbivores when growing in groups than when growing alone. The same may be said about different cultivars of the same crop, which exhibit different defence mechanisms against spider mites (Giménez-Ferrer *et al.*, 1994).

Another tentative explanation (not mutually exclusive to the above) for the variability in mite domatia occurrence could be that some domatia have had more survival value in the evolutionary past than they have in the present. Tentative support for this argument comes from the lack of ABA specificity for the various domatia. The postulated 40-million-year-long association would have been expected to select for increased, rather than for decreased specialisation. However, if a selecting factor, such as phytophagous mites, that had caused much damage in the past has since relaxed its pressure, the domatia on certain plants could have been left as a residue of redundant evolution.

Domatia in northern regions are mostly in the form of simple tufts of hairs, often associated with a shallow pit or pocket-like flap of tissue, whereas larger, deeper domatia are common on tree leaves in warm–temperate to tropical forests (Walter, 1996). The two forms of domatia serve the same purpose, but the origin of their differences is unclear. The structure and growth pattern of leaves of deciduous trees in temperate regions could differ in some fundamental manner from those of evergreen, tropical trees, or the variance in structure could be an adaptation to the presence of different predatory mites.

Spider mites are major pests of cotton, and several predatory insects, including bugs and thrips, are among their most important predators (van den Bosh & Hagen, 1966). An attempt to reduce spider mite populations on cotton by fortifying its leaves with artificial domatia (tufts of cotton fibers), was reported by Agrawal & Karban (1997). They obtained significantly larger populations of non-acarine natural enemies on treated plants, whose yields were higher by almost one-third than those of control plants. This demonstration of the benefits of domatia to a commercial crop, although not conducted with ABAs, could lead to enhanced pest control.

Host plant experience

Do ABAs that occur on different plants recognise and 'remember' some as providing more prey? How long does the memory last? Starved *P. persimilis* that had been born and reared on bean leaves infested by spider mites spent more time on bean than on rose leaves with spider mite residues, but the response became insignificant when the predators were well fed (Zhang & Sanderson, 1992). Predators with a similar experience on roses showed no plant preferences, suggesting that bean cues may be stronger in imprinting predators than rose cues. The observed preferences were not observed after 5 days. Thus, it seems that (at least for *P. persimilis*) experience on a certain host plant could affect foraging behaviour for a short while. Further, the postembryonic predation experience of the present generation, not that of parent generations or of the site where the eggs were deposited, may affect foraging. Lastly, this response was influenced by an individual mite's hunger state. These observations relate to biocontrol because *P. persimilis* is usually reared on beans and sold to control spider mites infesting other crops.

Age of leaf or plant

Young plant tissues are soft, succulent and unsclerotised, usually of high nutrient quality (Slansky & Scribner, 1985), and contain few of the protective secondary chemicals (see section on plant resistance, below) that plants later metabolise. They also carry few (or no) ABAs (with the exception of resident bodyguards or predators that had overwintered on the plant). Such tissues (or entire young plants) are readily consumed and utilised by phytophages. Most natural enemies arrive later, often by random drifting on air currents, and remain on infested plants as a result of cues emitted by them. The abundance and activities of predatory mites on a plant are thus often related to its age.

As will be discussed below, plants emit predator-attracting volatiles as a consequence of spider mite feeding. Young leaves attract more predators than mature foliage; old cucumber leaves infested by spider mites emitted more volatiles that appeared to mask the predator-attracting components of the synomones (Takabayashi *et al.*, 1994a).

Fertilisers and plant health

Plant quality and health are improved by adequate mineral and organic fertilisation (the 'bottom–up' effect), engendering larger pest populations which, in turn, increase the number of natural enemies, a topic discussed below. Healthier plants also accommodate more honeydew-producing insects that provide nutrients for predatory mites. The direct effect of various fertiliser regimes on the feeding of a phytoseiid was discussed above (Fig. 41.1).

The European *Aculus hyperici* was released to control St John's wort in Australia (Chapter 15). Its establishment rates were affected by the health of the weed. Establishment rates reached 78% where the weeds were in good health, whereas at locations with plants in poor to fair health the rates were only 40–42% (Jupp & Cullen, 1996).

Effect of pesticides through plants

The concept of physiological selectivity of pesticides was used to describe cases in which the use of a systemic insecticide, the organophosphate (OP) schradan, caused high pest mortality without harming the associated natural enemies (Ripper, 1956). Crops infested by spider mites were drenched with the pesticide, which moved through the plants, reached the pests' feeding sites and killed them without harming the natural enemies. The mode of selectivity of schradan was attributed to its lack of contact toxicity to insects (other systemics kill arthropods by contact). McClanahan (1967) later demonstrated that the application of systemics does not invariably produce the desired selective results. Three systemic OPs were applied as root drenches to cucumbers to determine their effect on *P. persimilis* and TSSM. Phorate and thionazin caused about the same mortality to predator and prey, whereas dimethoate was about three times more toxic to *P. persimilis* than to the pest. This food chain toxicity was further explored by Daneshvar & Rodriguez (1975). Roots of bean plants were immersed in a nutrient solution with labelled phosphorus and their leaves were colonised by starved females of *N. fallacies*. No radioactivity was found in these predators, indicating that they did not feed on the plant. Radioactivity was, however, detected in TSSM individuals that had fed on the labelled leaves. With these findings in mind, spider mites were exposed to sublethal dosages of systemics that were applied through the plant's nutrient solution. TSSM mortality was low but all predators that fed on the spider mites died. These and other experiments (e.g. Lindquist & Wolgamott, 1980) were restricted to exploring the effect of pesticides that reach the predators via the food chain. Another plant-mediated mode by which ABAs become exposed to insecticides is by feeding on crops that have been treated with systemics. Populations of *E. tularensis* suffered heavy mortality when placed for 48 h on grapefruit leaves whose petioles were dipped into dimethoate (Congdon & Tanigoshi, 1983). The direct effect of pesticides on predators is discussed in Chapter 43.

The means of neutralising agricultural toxicants by detoxification can also reach predatory mites through the food chain. To detoxify insecticides mites use several enzymic systems (Chapter 43, Pesticide-resistant ABAs), the activities of which can be induced by plant chemicals (Cohen *et al.*, 1993). The inducing materials, secondary substances that are referable to various chemical groups (see below), may reach ABAs via their phytophagous prey or, to a lesser degree, by feeding on the plants. Levels of mixed-function oxidases, a detoxification mechanism that is affected by plant toxins (Dowd *et al.*, 1983), were lower in *N. fallacis* and its spider mite prey when the latter was reared on cucumbers than on beans (Strickler & Croft, 1985). This suggests that a modicum of predator resistance to pesticides could be induced through the food chain.

Can plants further modify the effect that pesticides have on ABAs? Populations of *T. pyri* on foliage of the apple cv. Belle de Boskoop in The Netherlands recovered from fungicidal treatments better than those on leaves of cv. Golden Delicious. Neither differential leaf vesiture nor initial predator numbers seemed to be the answer; Blommers & Helsen (1986) postulated that variable prey densities could have caused the observed disparity. This would be an indirect effect, but the possibility that the plant had directly affected the outcome cannot be ruled out. The effect of the acaricide cyclopyrate (Zardex®) on spider mite eggs differed according to the host plant; eggs on soya bean leaves were less susceptible than those on peach, orange or apple leaves (Asano & Kamei, 1982). This was attributed to the higher ability of soya beans to alter the pesticide's transport through their foliage. Spider mite eggs deposited on apple leaves were less affected by the carbamate (CRB) hexythiazox than those on bean leaves (Marris & Chapman, 1987). Eggs on leaves that contain or retain higher concentrations of pesticides would have a greater effect on their predators.

Apart from the effect of insecticides and fungicides on prey and predators, pesticides could indirectly influence both groups by affecting plant growth. These chemicals are known to increase or reduce photosynthesis and yield (Jones *et al.*, 1986), hinder or promote vegetative growth (Lloyd & Krieg, 1987) and affect the metabolism of secondary chemicals (Lydon & Duke, 1989).

Indirect effects (through prey)

Agricultural chemicals and plant health

Applying NPK fertilisers is standard horticultural practice. Spider mite increases are often correlated with the amounts of nitrogen and phosphorus at low to moderate levels, whereas high mineral concentrations engender decreases in spider mite numbers (Rodriguez & Rodriguez, 1987; Walde, 1995). The relationships between nitrogen and natural enemies, via their spider mite prey, are not symmetrical. *Typhlodromus pyri*, the resident ERM predator on apple trees in eastern North America, had a delayed density-dependent response to prey numbers on unfertilised trees, a pattern that became variable on fertilised ones. Densities of another resident predator, *Z. mali*, increased along with nitrogen levels, but both predators had small populations (in relation to those of the pest) on fertilised trees, and their impact on ERM was stronger at low nitrogen concentrations. Plant nutrition should thus be considered in biological control efforts.

Failures to control thrips by *Neoseiulus barkeri* in certain Danish glasshouses were attributed, in part, to the pest's improved reproduction on some cucumber cultivars (Hansen, 1989). Such plant effects are seldom incorporated into relevant models (Baumgärtner *et al.*, 1988).

Drought stress in plants has been implicated in spider mite outbreaks, but the effect on predators is not well known. TSSM populations were highest on well-irrigated, heavily stressed bean plants, and lowest on moderately stressed plants. *Phytoseiulus persimilis* reduced pest numbers on the latter but had the least effect on severely stressed beans, possibly because of the low RH that prevailed on such leaves (English-Loeb, 1990).

Plant resistance or direct plant defence

Plant resistance to pests is due to one or more of three mechanisms, namely tolerance, non-preference (also called antixenosis) and antibiosis (Painter, 1951). Tolerance designates the plant's ability to withstand pest attack without serious reductions in yield quantity and quality. Non-preference is the plant's ability to be 'uninteresting' to the pest, and antibiosis refers to adverse plant effects on the pest. These modalities are not mutually exclusive; strawberry cultivars may employ all three (Giménez-Ferrer *et al.*, 1994). Tolerance, which does not affect pest numbers, is the only mechanism that is fully compatible with the use of ABAs. Non-preference, by discouraging pest sojourn on plants, may lead to fewer resident or visiting predators. Antibiosis causes a reduction in herbivore fitness through repellency, diminished (or curtailed) oviposition, aborted progenies and/or early death. Most interactions between resistance and biological control were studied with plants that use antibiosis. Their effects are due mostly to the secondary substances that are produced (or less often, accumulated) by plants in great variety, such as alkaloids, cyanogenic compounds, glycosides, lectins, resins, tannins, terpenoids and many others that are repellent, deterrent and/or poisonous to arthropods. Other modes of direct plant defence include deficient diets and physical and chemical barriers, such as a thick epidermis, spines or glandular hairs (trichomes). Some of these modes of defence (e.g. spines) could also be adaptations to surviving in stressful environments, such as dry or hot climates.

The uneasy relationship between plant resistance (e.g. antibiosis) and biological control has long been recognised (Painter, 1951); resistant plants are seen to have a variable effect, even becoming disruptive to biocontrol (Hare, 1992). Resistant plants decrease and/or moderate the pests' growth rate, which on the one hand reduces the predators' prey (and thus reproduction), while on the other prolonging pest exposure to natural enemies. The passage of secondary plant substances, ingested by herbivores and passed on to their predators, may hinder biological control. *Phytoseiulus persimilis*, fed with TSSM that was reared on nightshade (*Solanum douglasii*), lost much weight (Moraes & McMurtry, 1987). Nightshade is not a commercial plant and the hindering mechanism (whether ingestion of inhibiting factors via TSSM or lower prey nutritional value) is unclear, but the implications for mass-rearing of prey and ABAs (Chapter 39) are clear. Another example of the incompatibility of plant resistance with ABAs is the glandular tomato trichomes that entrap *P. persimilis*, noted above. The indirect effect of plants on ABAs was also seen when *P. persimilis* was given TSSM reared on different hosts (Table 41.3). Predator performance was best when the prey were reared on soya beans and worst when from carnations, the other hosts (roses and chrysanthemums) being intermediate (Popov & Khudyakova, 1989).

Although evidence for the theory that natural enemies were agents of selection (or coevolution) for traits that enhance plant survival is tenuous (Hare, 1992; Sabelis *et al.*, 1999), selection of crop plants for expressing such attributes has been contemplated (Dicke, 1995; Bottrell *et al.*, 1998). Because pests could be attacked by several natural enemies, Bottrell *et al.* (1998) advocated breeding plants for traits that are beneficial to more than a single enemy (e.g. more pollen). A compatible choice would be to breed for genotypes of natural enemies that are better suited to certain cultivars or crop plants, as suggested by Drukker *et al.* (1997), above.

Table 41.3 Performance of *Phytoseiulus persimilis* (measured by various life history parameters) when offered *Tetranychus urticae* reared on soya beans and on carnations

Parameter	Soya bean	Carnation
Oviposition (days)	18.4±1.1	12.5±1.1
Fecundity (eggs/female)	69.1±3.8	32.5±4.4
Juvenile mortality (%)	5.4	14.5
Sex ratio (% F/F+M)	0.65	0.56
Longevity (days)	22.4±1.7	15.8±1.0
R_0	44.13	15.69
r_m	0.334	0.244

Based on data in Popov & Khudyakova (1989).
All differences were statistically significant.

Crops engineered for resistance to pests (e.g. by expressing the endotoxin of *Bacillus thuringiensis*, commonly abbreviated to Bt) could be more suitable for bio-control because they would need fewer pesticide applications. Spider mites attained damaging levels on potatoes in the Pacific north-west of the USA after pesticides eliminated other pests. Bt-engineered plants may encourage ABA activity and reduce the use of pesticides (Hoy *et al.*, 1998).

Indirect plant defence

'Indirect defence' refers to plant effects on ABAs that are mediated through the prey/host. One example, increases in the volume of sweet exudates (which attract generalist enemies) that are produced by extrafloral nectaries in response to herbivore damage, was noted above. However, in order to entice specialists, injured plants need defenders that are attracted by specific volatiles emitted only by pest-damaged tissues, not as a result of random mechanical injury. To ensure that suitable natural enemies would repeatedly arrive at the site where they are needed, the alluring signal(s) must contain unequivocal information about the availability of a reward, e.g. prey. Thus, it is to the plant's advantage that the signals are emitted only as a result of pest feeding, providing 'a chemical burglar alarm' (Dicke, 1995). The plants are spared from the wasteful, random or continuous production of volatiles by the specificity of the biochemical reactions that metabolise the synomones.

The 'chemical burglar alarm' is triggered when spider mites damage plants, which in response emit a blend of around 60 attractant volatiles that entice predators (Dicke *et al.*, 1999). The chemical spectrum of the synomones emitted by a given plant differs according to the herbivore (and plant; see below) and attracts different predators. Females of the spider mite specialists *P. persimilis* and *G. occidentalis* were attracted only to air blown over apple leaves infested by TSSM, but showed little response to air from leaves infested by ERM, which weaves only sparse webbing. The reverse occurred with regard to predators that prefer ERM [*Neoseiulus finlandicus* (Oudemans), formerly placed in *Amblyseius* and *A. andersoni* (Sabelis & van de Baan, 1983)]. Foliage of the apple cultivars Summer Red and Cox Orange Pippin, infested by TSSM, emitted somewhat different blends of mostly the same volatiles (Takabayashi *et al.*, 1994b). As noted earlier, synomones from young cucumber leaves differed from those given off by older leaves.

Plants seem to influence blend composition more than the pests. The difference between volatiles emitted by apple leaves attacked by TSSM and by ERM was smaller than the disparity among attractants produced by the leaves of two cultivars infested by TSSM (Takabayashi *et al.*, 1991). Undamaged bean and cotton plants exposed to volatiles from neighbouring mite-infested plants also became attractive to *P. persimilis* (Dicke *et al.*, 1990), and there is evidence of infochemical exchange between infested and healthy plants through the soil (Dicke & Dijkman, 2001). As to prey location in the field, *P. persimilis* migrated to TSSM-infested plants from a distance of about 0.5 m when given a choice between mite-infested and uninfested plants

(Zemek & Nachman, 1998), arriving at infested plants upwind from a distance of 1.2 m (Sabelis & van der Weel, 1993).

Resistance costs to the plants are reductions in fitness due to redirecting valuable resources away from growth, reproduction and competitive ability, but their dimensions are not clear (Purrington, 2000). Costs would probably depend on the extent of other stress factors, such as soil and climate, nutrient and water availability, other herbivores, pathogens and competitors, on whether the given mode of defence against one pest also provides protection towards others, and on whether the return is high enough. Costs may also differ according to the time lag between signal and expression of resistance. Sabelis & De Jong (1988) pointed out that because of their volatile properties, the benefits of each individual synomone 'spurt' would be of short duration and costs might be high. However, plants would 'save' resources if volatiles were emitted only when needed. Thus, it would be of interest to determine the costs incurred by plants being protected by ABAs via a combination of domatia and synomones (as well as by resistance), and the blends that plants emit when concurrently attacked by different pests. There is probably intraspecific variation in amounts (and composition) of synomone blends produced and in their ratios, and consequently in the resultant tritrophic interactions. Such variation would be more common in the field than in the laboratory. Costs should also include the increased fitness of competing plants (due to the invited bodyguards), which have not invested in emitting the synomone (Sabelis *et al.*, 1999).

At present there is no evidence that genetically modified (GM) crops have directly affected the attractiveness of plants to natural enemies (Hoy *et al.*, 1998), but this has not been studied in regard to ABAs. GM plants may, however, differ from unmodified plants in their nutrient quality (Donegan *et al.*, 1999), thereby affecting pest and predator populations.

Most research efforts were aimed at exploring the effects of attractants emitted by plants in response to spider mite damage. However, predator orientation to prey may require the simultaneous reception of at least two groups of chemical cues, coming from different sources (Hagen, 1987). One group is the plant volatiles, and the other comprises those that originate from the prey or its faeces. Manipulation of ABAs for enhanced mite control through synomone use may therefore require the concurrent application of volatiles emanating from both sources.

Other indirect effects

Webbing is a major defence mechanism that is used by spinning spider mites against their enemies (Gerson, 1985). Specialists (e.g. *P. persimilis*) are guided to their prey by web threads and forage freely within the colony. Generalists, in contrast, are repelled by the webbing or die in it. The quantity of webbing produced by spider mites is affected by their host plants. Carmine spider mites that were reared on leaves of seven cultivated plants produced most webbing on beans, and least on ivy (Gerson & Aronowitz, 1981). This also reflected plant suitability for pest development, as mites were most fertile on beans, and least on ivy. Plants on which mites do not grow well thus offer fewer prey and, because they have less webbing, are less attractive to the predators, and vice versa (e.g. beans).

Plants (mostly through their dietary components) affect the dimensions of attacking phytophages. Differently sized host herbivores provide more, or fewer, nutrients for their natural enemies, affecting the number of ABA progeny, their size and fecundity. The prevalence of *Hemisarcoptes coccophagus* was dependent on the size of its host scale (the chaff scale, *Parlatoria pergandii*) which, in turn, was affected by its own host citrus species. Scales on grapefruit were larger than those on orange and also carried more mites (see Fig. 19.2).

Ground cover, weeds and nearby plants

Ground cover and weeds

Ground cover plants grow between the rows of crops. They may be adventitious (and then sometimes regarded as weeds) or purposely planted (understorey crops for economic reasons, weed strips to promote natural enemies). The cover plants can be annuals, biennials, perennials or mixed, comprising members of one or more botanical families. Understorey crops are managed by various modes (no-tillage, mowed or ploughed under), with diverse effects on associated pests and natural enemies (Bugg & Waddington, 1994). The understorey vegetation (whether crop or weed) ameliorates conditions for resident and visiting arthropods by lowering ground temperatures along with increasing humidity, often serving as hosts for pests and providing them with refuges. Ground cover plants are more important in temperate regions, because they provide overwintering shelter for many pests and natural enemies, including mites of both categories. In the spring this vegetation serves the pests as stepping-stones that facilitate their renewed attack on crops. TSSM infested pears (Gotoh, 1997) and peachs (Meagher & Meyer, 1990) more rapidly, and in larger numbers, in orchards with ground cover than in plantations where chemical weed control was practised. Weeds (e.g. field bindweed, *Convolvulus arvensis*) growing in the field or in glasshouses are often the first plants to be attacked by spider mites (Fig. 41.4), and could be used for monitoring and as release sites of natural enemies (see below).

The ground vegetation, especially when consisting of different species, provides prey, shelter and alternate food (e.g. pollen, honeydew) for ABAs. Most weeds produce pollen that is beneficial to phytoseiids; the pollen of only three out of 18 weeds failed to sustain the development of a generalist predator (Ragusa & Swirski, 1975). The importance of weed pollen for maintaining alternate prey for phytoseiids on grapevines was discussed above.

Plants growing at the bases of apple trees in Michigan, USA, were considered necessary for the successful on-site overwintering of *N. fallacis* (Croft & McGroarty, 1977). The early season upward growth of this vegetation into the apple canopy encouraged predator migration into the trees. Pests (e.g. TSSM) were seldom found on apple trees after they were controlled on the ground cover. If predator-damaging

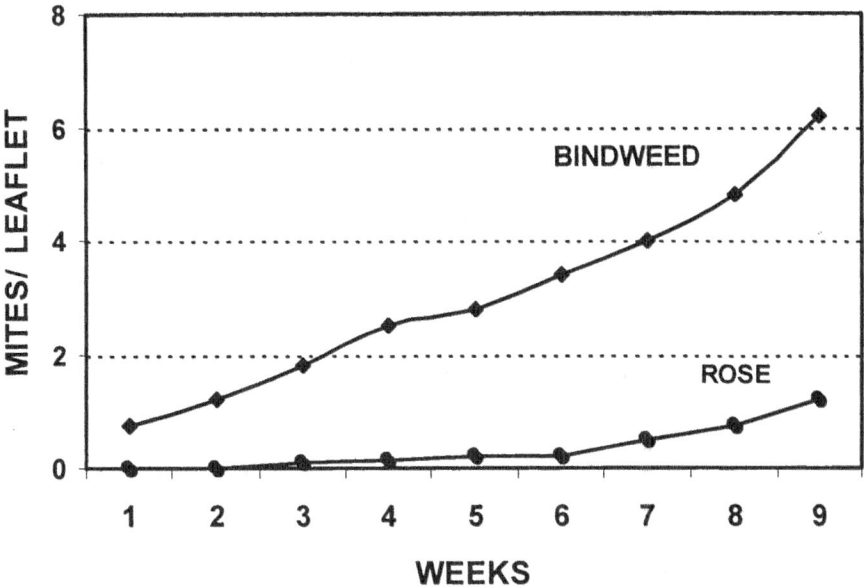

Fig. 41.4 Infestation of the carmine spider mite (*Tetranychus cinnabarinus*) on rose and field bindweed (*Convolvulus arvensis*) leaflets in a rose greenhouse, Israel. (Redrawn from Dubitzki, 1981.)

pesticides were applied, TSSM populations initially became numerous on the weeds and then migrated onto apple trees, causing much damage. The development of *N. fallacis* was little affected by the composition of the ground flora. Nyrop *et al.* (1994) arrived at a different conclusion. They found similar populations of *N. fallacis* on apple trees whose understorey vegetation had been removed and on those with intact ground cover, concluding that these plants did not play a large part in predator dynamics.

In the USA, apple orchards in Utah whose total ground vegetation cover was at least 50% had ABA populations that sufficed to keep pest mites below their damaging levels (Alston, 1994). Broadleaved plants, e.g. *Rubus* sp. and wild strawberry, carried more spider mites and phytoseiids in Massachusetts apple orchards than did grasses (Coli *et al.*, 1994). In China, broadleaved dicotyledonous plants were considered to be more beneficial to predatory mites than many monocotyledons (Liang & Huang, 1994). The weed *Ageratum conyzoides* was a special case. Populations of phytoseiids on citrus trees in China were correlated with the plant's presence and most predators were common to citrus and weed. Orchards in which *A. conyzoides* was growing carried 0.1–0.3 predators/leaf and citrus red mite numbers remained at non-injurious levels. Groves without *A. conyzoides* had few predators but high pest numbers. The plant modified the climate within orchards: summer temperatures were reduced by around 5°C and RH was raised by at least 5% (Huang *et al.*, 1983). In recognition of these beneficial effects *A. conyzoides* was planted or conserved in Chinese citrus groves over an area comprising approximately 135,000 ha (Liang & Huang, 1994). The planting of cucumbers and encouraging the growth of *Ageratum houstonianum* between strawberry rows in eastern Australia provided alternate hosts for TSSM and thereby fostered populations of *P. persimilis* between seasons (Waite, 1988a).

Sown weed strips are 'ecological compensation areas', zones that are maintained between agroecosystems, taking up 5–10% of the agricultural land. They are intended to increase the region's biodiversity by providing hibernation sites, refuge and dispersal foci for beneficial species. Although artificial, such zones help to conserve many taxa. The strips consist of mixtures of annual and perennial weeds, the composition of which is determined by specific environmental conditions (e.g. climate; soil), but no effort is being made to exclude plants on which pests may develop. Owing to normal succession processes in the plant populations, the weeds must be managed; appropriate techniques were described by Nentwig *et al.* (1998, and references therein). Populations of generalist predators (e.g. spiders, predatory beetles) were much larger in and near the weed strips, and fed on pests found there. The fitness of the natural enemies (and thus their fecundity and impact on pests) was enhanced by the availability of more diverse diets (e.g. additional prey, pollen, nectar). This interesting protocol has not yet been evaluated for the maintenance and conservation of ABAs, nor is enough known about its economic feasibility.

Ground cover plants, when infested by pests or by alternate prey, can be used as release sites for ABAs. The failure of *P. persimilis* (which seldom settles on erect perennial plants) to colonise grapevines led Takahashi *et al.* (1998) to release the predator on ground cover plants, resulting in satisfactory spider mite control. Bushes of fox berry (*Solanum nigrum*) grow year round as weeds in apple orchards in Israel, and often serve as a winter refuge for the carmine spider mite. Bushes growing near or in touch with apple tree branches served for early, preblossom releases of *N. californicus* (E. Palevsky, personal communication). Another approach was to grow the perennial leguminous shrub *Psoralea bituminosa* between rows of apple trees, infest the shrubs with spider mites and then inoculate them with the predator. Spider mite-favoured weeds, such as bindweed, may be grown (in separate containers) and used as foci for predator releases among commercial crops. Ramakers & Voet (1996) placed potted castor beans, with large numbers of *I. degenerans*, among sweet peppers in glasshouses. The predators moved onto the peppers, while others continued their development, feeding on the pollen of castor beans. Another use of weeds is as 'sentinel plants', intended to monitor the establishment and spread of colonising predatory mites (Chapter 39).

Phytoseiid abundance and species variety were correlated with ground cover on uncultivated land in Ontario, Canada, possibly through a rise in potential living space (Rothman, 1988). The amount of habitable surface area of the vegetation best explained observed trends in phytoseiid abundance and species richness.

Nearby plants

The vegetation that surrounds crop plots, be it commercial, native (e.g. uncultivated) or weedy, serves as a reservoir and colonising stage for invading or reinvading pests.

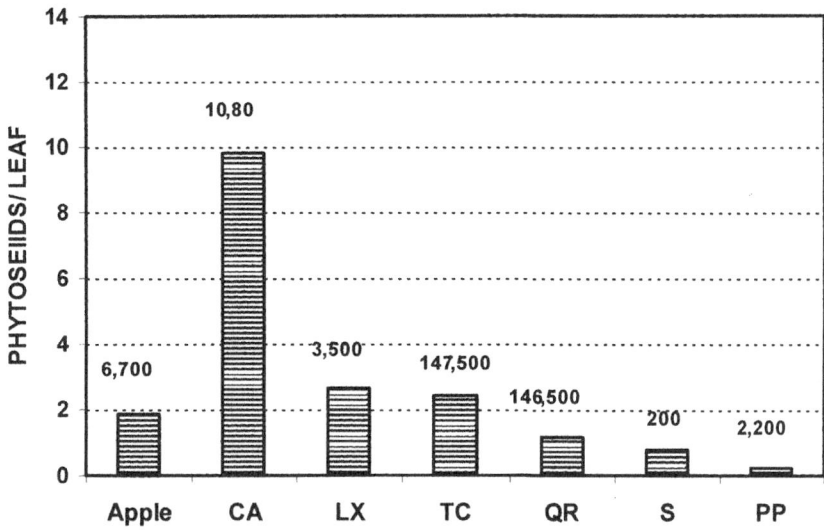

Fig. 41.5 Number of phytoseiids found on apple foliage (eastern aspect) and on leaves of six species of trees growing near an orchard in Finland, and calculated number of predators available thereon for dispersal (values above columns) (CA: *Corylus avellana*; LX: *Lonicera xylosteum*; TC: *Tilia cordata*; QR: *Quercus robur*; S: *Salix* sp.; PP: *Prunus padus*). (Data from Table 5 in Tuovinen, 1994.)

Many natural enemies arrive, their species and variety depending on the trees on which they originated. Tuovinen (1994) collected phytoseiids from trees growing near apple orchards in southern Finland. Hazel (*Corylus avellana*) leaves carried the most phytoseiids, whereas oak (*Quercus robur*) carried very few (Fig. 41.5). However, as oaks are bigger trees, with many more leaves, the total number of predators available on them for dispersal was about 13.5 times larger. Another factor related to the surrounding trees is their size. Tall trees (e.g. oak) growing in the vicinity of orchards would probably be more important than low bushes as sources for the aerial migration of phytoseiids into orchards (Tuovinen, 1994). Border areas of vineyards adjacent to hedges or forests carried large numbers of ABAs in northern Switzerland. Blackberry (*Rubus fruticosus*), with abundant numbers of phytoseiids, was considered to be of special importance in this respect, because predator 'bridgeheads' were found on vines facing blackberry hedges (Boller *et al.*, 1988). Similar results were obtained in southern France, where the diversity of phytoseiids found in vineyards reflected that occurring on the surrounding vegetation. Tall and dense woody vegetation near vineyards ensured the consistent arrival of larger number of predators; narrow border hedges were insufficient (Tixier *et al.*, 1998). Windbreak rows consisting of several tree species would provide a more variable menu of pollens and more species of herbivores and predators (Solomon, 1981).

Chapter 42
Intraguild predation and other interactions among acarine biocontrol agents

Introduction

Intraguild predation (IGP), which takes place when two different natural enemies share a host (or prey), and at least one also feeds on the other (Rosenheim *et al.*, 1995, who described IGP as 'a hybrid of predation and competition') is common among acarine biocontrol agents (ABAs). However, their interactions may also be competitive without predation, or even co-operative. A distinction is made between what is beneficial to a given species and what promotes pest control. A certain ABA may be the winner in the contest, but if it were the less efficient natural enemy, pest control would be reduced (as in the interaction between *Neoseiulus cucumeris* and *Neoseiulus barkeri*, below). This bears on the introduction of natural enemies, as biological control practitioners are divided about whether it would be better to introduce several species of natural enemies or to try only the more promising candidates. Proponents of the former school argue that when several natural enemies co-occur, their IGP would cause the most effective one to dominate and control the pest. The other view is that at least some details of the candidate's biology should be at hand before release, including an evaluation of its relationships with other natural enemies in the field (see Chapter 39).

ABAs often ignore each other; these neutral interactions are probably prevalent, but except when predators act in tandem, they remain little studied. Co-operation takes place when the fitness of the same or different ABAs that attack a pest is increased and adds to its level of control.

This chapter looks first at IGP, then at competition among ABAs without predation, then at co-operation. Cannibalism, which is prevalent among ABAs, will not be discussed separately.

Intraguild predation

Guilds of natural enemies may be natural, restructured or synthetic (Ehler, 1992). Natural guilds consist of species that are associated with the prey (or host) in its native habitat, and may be restructured by the addition of natural enemies that were accidentally or intentionally introduced. Synthetic guilds include exotic natural enemies released against an indigenous or exotic pest. In contrast to the first two, members of the synthetic often consist of species with little or no co-evolutionary history, and their association is believed to be short-lived. Since agricultural systems are not 'natural' systems, most IGP by ABAs takes place within synthetic guilds. It occurs between members of the same ABA family, among members of different families, and between mites and predatory insects. Only a few representative cases from each category will be examined (IGP among nematophagous ABAs is noted in Chapter 38). Although most data are from laboratory studies, and IGP may be less intense in the field (Sabelis & van Rijn, 1997), they suggest what could take place there.

Intraguild predation among phytoseiids

IGP implies that at least one of the predators is not a specialist; much interest has thus been focused on the interactions between the specialist *Phytoseiulus persimilis* and generalist phytoseiids (Schausberger & Walzer, 2001, and earlier papers). *Phytoseiulus persimilis* suffers from attack by other phytoseiids and, like all type I species (Chapter 26), is also more cannibalistic than generalists. In consequence, it may be displaced on annuals or on leaf disks in two-species contests (Pruszynski & Cone, 1972; Yao & Chant, 1989; Kabicek, 1995). However, the final outcome also depends on the crop, its longevity and the prevailing weather conditions (an example of the latter is presented below, under Competition). On strawberry, *Galendromus occidentalis* declined because *P. persimilis* developed larger populations, more rapidly (Laing & Huffaker, 1969). On perennial crops with dense spider mite webbing the phytoseiids may coexist by residing in different sites. The less generalist *Neoseiulus californicus* fed at the fringes of a prey colony, whereas *P. persimilis* foraged in its midst, and spider mite control on such plants could be enhanced by the sequential release of these species (Schausberger & Walzer, 2001).

Neoseiulus cucumeris and *N. barkeri* co-occur in European glasshouses and feed on the same thrips prey: a natural guild coexisting in an artificial environment. The former is considered to be the better predator, but on parthenocarpic cucumbers (which do not produce pollen), as well as in common mass-rearing units, it was displaced by *N. barkeri*. This was attributed to the latter's greater mobility, its lower dependence on pollen and/or its intrinsically superior abilities on some crops (Brødsgaard & Stegaard Hansen, 1992). However, on sweet pepper *N. cucumeris* established better and attained higher numbers than *N. barkeri*, even when the latter was released first or in larger numbers (Ramakers, 1988). In northern Italy *Kampimodromus aberrans* usually displaced *Amblyseius andersoni* and *Typhlodromus pyri* only from grape varieties whose leaves were more pubescent on their undersides, IGP in this case also being mediated by the host plant (Camporese & Duso, 1996).

Croft & Zhang (1999) listed biological traits (e.g. fecundity, cannibalism, interspecific feeding and the use of alternate diets) of juvenile and adult phytoseiids that might affect the outcome of their contests. *Galendromus occidentalis* and *Neoseiulus fallacis* were more fecund than *T. pyri* and *A. andersoni*, but the latter two used more supplementary diets than the former pair and were also better searchers. Nutritional benefits accruing to generalist and specialist phytoseiids were explored by Schausberger & Croft (2000). Generalists gained more nutritional benefit from IGP than from cannibalism, whereas specialists, owing their limited diet spectra, obtained more from cannibalism. Populations of generalists would therefore persist longer than those of specialists when both types of predators occurred in systems with little alternate prey. Such data assist in understanding the results of IGP among phytoseiids in the field.

Intraguild predation between phytoseiids and stigmaeids

Phytoseiidae and Stigmaeidae coexist on several crops and may devour the same prey as well as each other, with different pest control results. The best known case is the interaction between several phytoseiids and *Zetzellia mali* on apple trees (Chapter 32), the outcome of which may enhance (or impede) biological control. *Zetzellia mali* overwinters on the trees and feeds mainly on the apple rust mite, *Aculus schlectendali*, on eggs and quiescent stages of, the European red mite (ERM), as well as on the eggs of other mites. Phytoseiids are aggressive predators that prefer ERM to apple rust mite, and readily feed on *Z. mali*. Because more eggs of *T. pyri* are deposited within the stigmaeid's foraging area than those of *G. occidentali*, the former phytoseiid suffers more from their co-occurrence. This, in turn, impacts on pest control, as detailed in Chapters 26 and 32. Long-term studies (Croft, 1994) indicated that *Z. mali* was capable of displacing *T. pyri* after several years. Antagonism between *Agistemus* sp. and phytoseiids on tea in Taiwan detracts from overall pest control (Lo, 1986). When the stigmaeid occurred in densities similar to those of the phytoseiids, the latter could not exert their maximal effect on the pestiferous mites; the specific effect of the stigmaeids was not examined.

Thus, despite complaints that Stigmaeidae may hinder biological control efforts by Phytoseiidae, the situation appears to be complex. Phytoseiids are often seen as the 'stronger' predators, and are important at high pest densities, but stigmaeids are more efficient at maintaining low prey numbers. Members of the two families could therefore reduce pest populations in combination (or in tandem) over a wider range of prey densities than either group alone (Clements & Harmsen, 1992). Their different patterns of resistance to pesticides would ensure the survival of at least one group after a chemical application.

On apples *Z. mali* co-occurs with *Agistemus fleschneri*, but the two occupy different tree parts. The former lives mostly on fruit-cluster leaves on the periphery of the tree, whereas *A. fleschneri* is more common on inner-tree water sprout leaves (Holdsworth, 1972); it is not known whether they engage in IGP.

Intraguild predation between Cheyletidae and predatory Mesostigmata

The Cheyletidae (Chapter 12), which usually ambush their prey, are capable of devouring and eliminating co-occurring ABAs. *Hemicheyletia bakeri* subdued much larger prey with a toxin injected into the victim's body. It fed on various phytoseiids when they were confined together in small cells, although *P. persimilis* was dominant on strawberries in greenhouse experiments (Laing, 1973). Females of *Euseius hibisci* placed with *Cheletomimus berlesei* on avocado leaves in arenas were killed after 24 h. (McMurtry & Scriven, 1964). *Hemicheyletia mori* Ehara ambushes and feeds on phytoseiids that approach its nests (Mori *et al.*, 1999; see below). In the laboratory *Cheyletus eruditus* may displace the ascid *Blattisocius dentriticus* (see below), but it is not known what takes place in nature. *Cheyletus malaccensis* and *Blattisocius tarsalis* are the two most abundant acarine predators in tropical and subtropical storage (Haines, 1984), indicating that cheyletids and ascids often coexist under natural conditions. This is another case of a natural predator guild surviving in an artificial (synthetic) environment.

Intraguild predation among predators of nuisance flies

Parasitidae (Chapter 25) have a detrimental effect on the beneficial activity of other filth fly predators. Ignatowicz (1979) found that in mixed cultures of the macrochelid *Macrocheles glaber* (Chapter 23) and *Parasitus coleoptratorum* (even at a ratio of 15/1), the former produced no progeny, apparently because of intense egg predation by the latter. The separate and combined effects of *Parasitus* sp. and *Macrocheles muscaedomesticae* on house fly juveniles were studied by Geden *et al.* (1988). The predators were confined with prey (house fly eggs) in small cups in a suitable medium; their effect on the pests was estimated by subsequent fly emergence. Placing the two predators together (four *Parasitus* to one *Macrocheles*) resulted in fewer flies being killed than when either was assayed alone, owing to the parasitid's feeding on the juveniles of *M. muscaedomesticae*. Another parasitid, *Poecilochirus* sp., interfered in the feeding of the beetle *Carcinops pumilio* (Erichson) on house flies by disturbing it during prey handling. These Parasitidae represent another case of less efficient ABAs reducing overall pest control by disturbing more efficient predators.

IGP between *Macrocheles robustulus* and *M. glaber* reduced the latter's populations when both were placed in cow dung. The smaller, but more aggressive, *M. robustulus* attacked and killed *M. glaber*, consuming its young (Wallace & Holm, 1984, see Fig. 23.2).

Intraguild predation between predatory Phytoseiidae and predatory bugs

Several predatory bugs of the genus *Orius* (Anthocoridae) are used to control pest thrips (Thysanoptera) in glasshouses (Chapter 26, Phystoseiids that feed on thrips). *Orius tristicolor* attacked *N. cucumeris* as well as thrips, the presence of the latter reducing bug predation on the mite. The bugs were mobile and moved away after reducing thrips density, leaving the more sedate *N. cucumeris* in place. The phytoseiid may later prevent pest resurgences, leading Gillespie & Quiring (1992) to conclude that the bug and *N. cucumeris* could be used together for thrips control in greenhouses. Wittmann & Leather (1997) doubted whether western flower thrips (WFT) control would be improved if *Orius laevigatus* co-occurred with *N. cucumeris* on the same plants, owing to the bug's feeding on the mite as well as on the pest. The outcome of such encounters was affected by the host plant, as the predators were less likely to meet on pepper than on cucumbers. Another phytoseiid, *Iphiseius degenerans*, suffered little IGP from *O. laevigatus*, indicating that these species could be used together for thrips control.

A third bug, *Orius insidiosus* (Say), had another interaction with *N. cucumeris* on sweet peppers. Both fed on WFT and coexisted from March to October, but *N. cucumeris* populations were only about one-sixth as large in the bug's presence as in its absence (Ramakers, 1993). Competition for prey (contest competition, see below) rather than direct predation was probably the mechanism for this result. The release of *T. pyri* and of *K. aberrans* in Italian vineyards was adversely affected by the presence of *Orius vicinus* (Ribaux); Duso & Pasqualetto (1993) believed that on grape cultivars that are not benign to phytoseiids such reductions could become crucial for their survival.

Competition

Two types of competition are usually recognised: contest (or interference) competition and scramble (or exploitive) competition. The first refers to cases where the winner has the almost exclusive use of the resource that is in short supply and other contestants are displaced. In scramble competition all contestants have equal access to the limiting resource (Price, 1997). As a result, many of the competitors survive but are often weaker and smaller.

Contest competition occurred when the stored product mites *C. eruditus* and *B. dentriticus* were concurrently placed in arenas, with *Acarus siro* as prey. The latter predator disappeared within 6 weeks, possibly owing to specific responses of the two predators to variations in prey density. The oviposition of *B. dentriticus* is dependent on prey availability, thus its numbers declined as soon as *A. siro* became scarce. *Cheyletus eruditus*, however, shows a delayed numerical response to decreasing prey densities, with the result that its numbers did not decline along with reductions in *A. siro* numbers (Burnett, 1977). A similar case is the interaction between *N. cucumeris* and the soil-dwelling laelapid *Stratiolaelaps miles* (Chapter 21), which are used concurrently to control thrips on cut roses. The phytoseiid devoured active pest stages on above-ground plant parts, whereas the laelapid fed on prepupae and pupae in the soil. Pest numbers were significantly reduced only when both predators were released, but the numbers of *S. miles* later declined; this was attributed to indirect food competition (Linnamäki *et al.*, 1998). As pest populations declined through predation by *N. cucumeris*, fewer soil-inhabiting pupae were available to *S. miles*, whose numbers then decreased. Contest competition between *O. insidiosus* and *N. cucumeris* was noted above. Scramble competition may take place when different phytoseiids are concomitantly released against thrips in glasshouses.

The climate and the predators' diets may affect the outcome of competition between ABAs. *Typhlodromus pyri*, being susceptible to low humidities, is often displaced on various crops by more desiccation-tolerant phytoseiids (Duso & Pasqualetto, 1993). One of the latter is *G. occidentalis*, which prefers drier areas in western USA, whereas *T. pyri* favours wetter regions (Croft *et al.*, 1993); in intermediate humidity zones their distributions overlap and they compete. When *P. persimilis* co-occurred with *N. californicus* on spider mite-infested *Ceanothus* plants grown outdoors in the UK, the latter predator became dominant, probably because of its higher tolerance to low temperatures (Buxton, 1999). The larvae of *T. pyri* do not feed, whereas those of *G. occidentalis* devour spider mite eggs; thus, when spider mites are scarce, *T. pyri* becomes dominant because it does not require food for continuing its development. Many similar interactions probably occur in nature.

Co-operation among ABAs

Intraspecific co-operation increases the fitness of individual species; for instance, juveniles of *N. cucumeris* develop more rapidly upon feeding on prey caught by their adults (Chapter 26). Group predation occurs in the subsocial cheyletid *H. mori*, which ambushes prey that passes by the woven nests. The nearest predator grasps the prey and other nest members, including the larvae, join in feeding (Mori *et al.*, 1999). Such co-operation promotes pest control, as seen in the concurrent parasitisation of several trombidiid larvae on aphids, which hastens the death of the host pest (Chapter 35). Interspecific co-operation that enhances pest control takes place when *Typhlodromus doreenae* and *Euseius victoriensis* together provide long-term control of grapevine mites in Australia (James & Whitney, 1991; see Chapter 26). Although polyphagous, each phytoseiid prefers different mite pests (the former tenuipalpids, the latter eriophyids). Both also feed on other diets, which is crucial to their coexistence in the absence of pest prey. The control of two-spotted spider mite on red clover by *P. persimilis* and *Neoseiulus longispinosus* in a greenhouse was better by both predators together than by either alone (Mori & Saito, 1979).

Discussion

According to Rosenheim *et al.* (1995), IGP between predators is more likely to disrupt biological control than IGP by parasitoids. IGP among parasitoids is mediated by a shared host/prey, leading to declines in its numbers. However, encounters between predators, which usually occur while they forage, do not require the death of the common prey. Owing to the consequent removal of one (or more) of the intraguild predators, the number of available natural enemies would decline without bringing about reductions in pest populations, possibly even decreasing the level of their biological control.

While not contesting the general conclusions reached by Rosenheim *et al.* (1995), there should be some reservations. First, the presence of competing ABAs does not necessarily disrupt pest control; second, as IGP by ABAs may be mediated by the host plant or the climate, the outcome could be variable; third, pesticides affect the outcome of contests in unexpected ways; and fourth, the effect of mite toxins on IGP has not been evaluated.

Several cases were noted where the presence of competing ABAs has increased overall pest control. In brief, these included *N. californicus* and *P. persimilis* on perennials, *Z. mali* with apple phytoseiids, *O. tristicolor* and *N. cucumeris* on greenhouse crops, *O. laevigatus* with *I. degenerans* on greenhouse pepper and *N. cucumeris* with *S. miles* on cut roses. Common to these interactions (and probably to others) was that the host plant could change the outcome. Such mediation is not restricted to mites. The severity of bug, coccinellid and predatory thrips attack on ABAs was reduced by leaf domatia (Chapter 41) and by the trichomes of the host plant (Norton *et al.*, 2001; Roda *et al.*, 2000; respectively). To balance the picture, we noted cases of IGP by ABAs that resulted in overall reductions in pest control.

The diverse effects of pesticides on the interactions among *G. occidentalis*, *T. pyri* and *Z. mali* were mentioned above; additional cases are described in Chapters 26 and 43.

It is usually believed that predators (or winners in IGP) are at least as big as their prey (or losers in IPG) (Sabelis & van Rijn, 1997). Pyemotids (Chapter 30), along with anystids (Chapter 6), cheyletids (above and Chapter 12) and erythraeids (Chapter 16), however, represent mites that use toxins to overcome much bigger prey as well as larger guild members. The effect of such toxins in IGP is little known.

The finding that ABAs can avoid conspecifics be detecting volatiles (Janssen *et al.*, 1997) suggests that they may also be able to discern and thus avoid encounters with other intraguild mites. An indication of such perception could be the increased aggregation of ABAs that are susceptible to predation. Slone & Croft (2000) postulated that in the presence of other predators, the more vulnerable phytoseiids would show significant increases in aggregation. *Galendromus occidentalis* and *Z. mali* were the most aggregated, and *T. pyri* became aggregated only in the presence of *Z. mali*. This suggests that ABAs may be able to identify specific odours produced by other predators.

Finally, although agricultural systems are artificial, the year-after-year consistency of the relationships within their guilds of natural enemies, whether restructured or entirely synthetic, suggests that they are undergoing incipient coevolution.

Chapter 43
The effect of agricultural chemicals on acarine biocontrol agents

Introduction

The contribution of acarine biocontrol agents (ABAs) to pest control was often inferred from outbreaks of phytophagous mites after the inadvertent destruction of their acarine natural enemies by plant-protection chemicals (Cutright, 1944; Lord, 1949; Clancy & Pollard, 1952). Such consequences, feared even before World War II (Gilliatt, 1935), became very noticeable after the war. These aspects of pesticide usage were recognised and analysed in an early review (Ripper, 1956), which included five predatory mites (in the families Anystidae, Bdellidae, Hemisarcoptidae, Phytoseiidae and Stigmaeidae) in the list of natural enemies whose beneficial effects were reduced by pesticides. McMurtry *et al.* (1970) later produced a comprehensive review that pertained mostly to the Phytoseiidae (Chapter 26). As noted, the beneficial activities of other acarine families were also curtailed by pesticides, leading to continuing quests for plant-protection chemicals whose selective toxicity would affect pests more than ABAs. Such efforts began with field and laboratory studies on resistance, and were broadened to include attempts to find, select and utilise ABAs resistant to pesticides ('genetic improvement'). They also encompass attempts to understand the different patterns of pesticide resistance in acarine pests and in ABAs, and more recently, efforts to create a genetically engineered pesticide-resistant phytoseiid.

Gilliatt (1935), Lord (1947, 1949) and Lord & MacPhee (1953) were among the first to assay the effect of various spray programmes on ABAs. Their conclusions that common pesticides, such as sulfur and winter oils, were detrimental to phytoseiids and to *Hemisarcoptes malus*, helped to explain outbreaks of the European red mite and the oystershell scale, respectively, on apples in Nova Scotia, eastern Canada, leading the way to an integrated pest management (IPM) approach. Such studies were later extended to other acarine predators of fruit tree pests (MacPhee & Sanford, 1954; Herne & Putman, 1966). Ristich (1956) conducted an early large-scale assay on the effects of 25 commercial pesticides on a single phytoseiid (*Neoseiulus fallacis*, then placed in *Typhlodromus*), an approach expanded and reviewed by Bartlett (1964). The importance of continuously assaying the effects of agricultural chemicals (new and those in use) on natural enemies (including predatory mites) was recognised by many individual scientists (e.g. Oomen *et al.*, 1991; Kreiter *et al,*. 1998b) as well as by international bodies. An example of the latter is the International Organisation for Biological Control (IOBC), which has a working group on Pesticides and Beneficial Organisms. This group periodically publishes reports on the effects of new pesticides (insecticides, fungicides and herbicides) on a selected array of natural enemies, in the laboratory and in the field. Their most recent report (Sterk *et al.*, 1999) includes data on the effect of 20 chemicals on three phytoseiids. Owing to the fact that the tested compounds were developed to kill pests, these results also supply data on the physiological selectivity of the toxicants, including their relative toxicity to pest and natural enemy (Croft, 1990).

Bakker & Jacas (1995) criticised the methods used by that working group. Faults found included exposing predators to tested chemicals on glass arenas or on detached leaves, neither of which ensures maximum coverage, and using unrealistic doses in field assays. In addition, species-specific attributes and indirect effects (some of which are discussed below) might be ignored. A chemical that delays mite reproduction would affect the fast-growing *Phytoseiulus persimilis* more than a slower predator, such as *Typhlodromus pyri*. The latter, because it complements its diet with rust mites or fungi, will be harmed more by fungicides (many of which kill rust mites) than a specialist (e.g. *P. persimilis*) that only feeds on spider mites. For these reasons field trials should be made compulsory in order to demonstrate that agricultural chemicals would not have negative side-effects on the environment. Ruberson *et al.* (1998) added that sex-specific data should also be sought, and that while conducting the all-important field assays, the particular crop system and the size of the experimental plot must be considered. Amano & Haseeb (2001) suggested additional improvements.

'Soft pesticides', such as abamectin, azadirachtin, *Bacillus thuringiensis* (Bt) and insect growth regulators (IGRs) (see below), usually affect arthropods for longer periods, often being sublethal. Their effects are thus more difficult to evaluate by standard laboratory tests. The importance of such effects was discussed by Croft (1990).

Two major sources of information on the effect of pesticides on ABAs (besides the working group reports) are Croft (1990) and the compilation prepared by Kostiainen & Hoy (1996). The former covers the Phytoseiidae and the Stigmaeidae, whereas the latter pertains only to the Phytoseiidae.

Summary of the pesticides

In the following summary the mode of action of pesticides that are mentioned (only by their generic names) will briefly be noted, moving from insecticides through fungicides to herbicides, and then to other, minor groups. Most data, including modes of action, are from Tomlin (1997).

Insecticides and acaricides

These pesticides may be grouped in several ways, including their mode of action. Organophosphates (OPs) and carbamates (CRBs) are neurotoxins that inhibit acetylcholine esterases (AChE), enzymes that mediate the transfer of pulses in the nervous system. OPs include azinphos-methyl, chlorpyrifos, chlorpyrifos-methyl, diazinon, dichlorvos, dimethoate, malathion, methyl-parathion, monocrotophos;

pirimiphos-methyl and terbufos. CRBs include carbaryl, carbofuran, hexythiazox and pirimicarb. Chloronicotinyls (e.g. imidacloprid) have a similar mode of action, blocking acetylcholine receptors, and formamidines (e.g. chlordimeform) are another group of neurotoxins.

Organochlorines (OCLs) disrupt the sodium balance of nerve membranes; they include dimethyldiphenyltrichloroethane (DDT), lindane, dicofol and chlorobenzilate. Pyrethroids (PYRs) (e.g. permethrin) block the passage of impulses through the nervous system. Most PYRs (and DDT) have negative temperature coefficients of toxicity, being more toxic at low than at high temperatures. Newer insecticides, such as cyhexatin, an organotin, or fenazaquin, inhibit electron transport in the cells.

Soft pesticides often act slowly, which renders their effects more difficult to evaluate by standard laboratory tests. All have low to no mammalian toxicity. IGRs either interfere in arthropod endocrine-induced activities, such as moulting (e.g. cyromazine, flubenzimine), or disrupt chitin synthesis. Azadirachtin, extracted from the neem tree (*Azadirachta indica*), also disrupts insect moulting (e.g. acts as a moulting hormone antagonist), as well as being an antifeedant and a repellent; a compound is neem. Abamectin is a fermentation product of the fungus *Streptomyces avermitilis* that inhibits neurotransmission, causing paralysis; a commercial compound is avemectin B. The Gram-positive bacterium *Bacillus thuringiensis* and its subspecies contain crystals that are toxic to insects upon ingestion. The sobriquet Bt is applied to various pesticides (e.g. Dipel) obtained from that bacterium.

Fungicides

Fungicides are regarded either as plant protectants or as plant systemics (capable of moving through the plant), with various modes of action. The protectants include dithiocarbamates (inhibitors of intracellular energy production) such maneb, mancozeb, propineb and zineb. Other protectants are captan, dichlofluanid and dodine, as well as dinocap, which comes in two isomers, one being more acaricidal, the other more fungicidal. The systemics include steroid (ergosterol) inhibitors such as benomyl, bitertanol, cyproconazole, difenoconazole, fenpropimorph, flusilazole, flutriafol, myclobutanil, penconazole and tebuconazole.

Herbicides

Most herbicides affect plants by interfering with photosynthetic processes (amitrole, glufosinate-ammonium; ioxynil, paraquat dichloride, terbacil), disrupting the synthesis of fatty acids (e.g. haloxyfop) or amino acids (dalapon, glyphosate), or by inhibiting mitochondrial activity (e.g. azoxystrobin). An older compound is the systemic 2,4-D, an analogue of plant hormones. The mode of action of herbicides on arthropods is not known; the adverse effect of some herbicides in the field may be indirect, through reducing the plant cover, the universe in which ABAs live.

General poisons

Arsenates, copper and sulfur inhibit various enzymes, especially those that mediate respiration. Cryolite, a mineral compound containing fluor and aluminium, damages the stomach lining of arthropods.

Summary of the effect of agricultural chemicals on ABAs

The following summary (in which the ABA families are presented alphabetically) does not contain information on the Acaridae, Eriophyidae, Tarsonemidae or Tetranychidae, because these families include the main pest species against which acaricides were developed. Chemical groups rather than individual compounds are usually noted, except in cases of special interest.

Data were obtained from four major sources:

- field experiments in chemical pest control, with results showing side-effects on ABAs
- field experiments intended to demonstrate the effects of pesticides on ABAs, or to explore their efficacy (including the insecticide check method)
- laboratory investigations on the effects of pesticides on ABAs
- laboratory experiments on the chemical control of ABAs that become pests when insects are being mass-reared.

Anystidae

Anystis spp. were very sensitive to CRBs, OCLs and OPs, as well as to arsenate and sulfur compounds, but diazinon had little effect. PYRs, when applied at recommended field rates, were toxic; James *et al.* (1995) postulated that some recovery of anystids could take place. Fungicides, such as iron carbamate and copper compounds, were harmless (MacPhee & Sanford, 1954; Herne & Putman, 1966).

Ascidae

Malathion was believed by Graham (1970) to reduce populations of *Blattisocius tarsalis* in bagged maize in Kenya. Chlorpyrifos-methyl, applied in Canadian farm granaries, virtually eliminated *Blattisocius keegani* from treated oats (White & Sinha, 1990).

Bdellidae

The susceptibility of a bdellid to DDT was utilised by Wallace (1954) to demonstrate the mite's efficacy in the field (Chapter 9). James *et al.* (1995) examined the effects of two PYRs, applied at field rates, on *Bdellodes affinis* Atyeo and *Cyta latirostris* (Hermann). The former was very sensitive, whereas the latter predator was more tolerant, indicating within-family differences in susceptibility.

Ceratozetidae, Galumnidae and Scheloribatidae

As relevant data on these families are scarce, they will be discussed together. The Cryptostigmata, to which they belong, seem to be insensitive to most commonly used pesticides (Steven & Lofroth, 1987). None of five OPs assayed had any lasting effects on *Orthogalumna terebrantis* (Ganga Visalakshy, 1992), used in water weed control. Paraquat and glyphosate were almost harmless, but 2,4-D caused significant (24.6%) adult mortality at a 0.6% application rate. Soil populations of *Scheloribates laevigatus* were reduced for several months by OP surface sprays (Al-Assiuty & Khalil, 1995). Applications of lindane to forest soils did not affect numbers of a *Scheloribates* sp., while greatly reducing those of a *Ceratozetes* sp. (J.B. Hoy, 1990). A soil treatment with carbofuran greatly increased the numbers of resident Cryptostigmata (Wright & Coleman, 1988).

Cheyletidae

OPs are often harmful to cheyletids; chlorpyrifos-methyl was very toxic to *Cheyletus eruditus* (White & Sinha, 1990). Zdárková (1994) concurred, adding that chlorpyrifos and pirimiphos-methyl were less harmful. *Cheletogenes ornatus* was relatively insensitive to parathion-methyl and malathion, as well as to other chemicals, such as the fungicides cyhexatin, sulfur and zineb, but chlorobenzilate was very toxic (Avidov *et al.*, 1968). Dicofol, mixed with mancozeb, reduced this mite's populations on citrus in Egypt by around 61% (Rezk & Gadelhak, 1996).

Cunaxidae

The organobrome compound bromopropylate, and fenazaquin were less toxic to a *Cunaxa* spp. on tea in India than to PYRs (Sah *et al.*, 1999).

Erythraeidae

Increases in the numbers of *Balaustium putmani* in sprayed North American orchards (e.g. Childers & Rock, 1981) suggested that this ABA might have an innate tolerance to pesticides. Cadogan & Laing (1981) noted, however, that the predator occurred only in orchards treated with low-volume applications; whether the mite's survival is due to tolerance or to tentative escape mechanisms remain untested possibilities. Sulfur, DDT, parathion, azinphos-methyl and permethrin were lethal to *B. putmani*

(Herne & Putman, 1966; Hagley & Simpson, 1983). In contrast, laboratory bio-assays in Australia suggested that an application of 2.5 g of PYR/ha would permit the survival of 10–30% of local populations of *Balaustium murorum* (James *et al.*, 1995).

Hemisarcoptidae

Hemisarcoptes malus was susceptible to sulfur, dinitro-*o*-cresol and winter oils, as well as to DDT. Sellers & Robinson (1950) used bromopropylate to eliminate *H. malus* when it became a pest in laboratory cultures. *Hemisarcoptes coccophagus* was relatively tolerant to some OPs and PYRs (Keesing, 1990). Copper-based fungicides (e.g. fermate), lead arsenate, synthetic cryolite, nicotine sulfate and summer oils had little effect under field conditions in Canada (Lord, 1947; MacPhee & Sanford, 1954).

Laelapidae

Chlorpyrifos-methyl, used in farm granaries, eliminated *Androlaelaps casalis* from treated oats in Canada (White & Sinha, 1990). Another OP, dimethoate, stimulated the reproduction of *Geolaelaps aculeifer* when assayed at field rates (0.390 mg/kg), but caused a 90% reduction at three times that rate. Pirimicarb and the fungicide fen-propimorph had no effect on mite reproduction at concentrations up to five times their field doses (Krogh, 1995). Two IGRs applied in mushroom cultivation had no effect on *Stratiolaelaps miles* (Ali *et al.*, 1999). Insecticides used to control laelapid pests of bees (e.g. *Varroa*) will not be discussed here.

Macrochelidae

OCLs and carbamates applied onto manure greatly reduced these ABAs, but OPs were less toxic (Rodriguez *et al.*, 1970). Cyromazine, dimethoate (sprayed on wet manure in spots only) and permethrin had insignificant effects on mite numbers when applied onto chicken manure (Wills *et al.*, 1990). *Bacillus thuringiensis* was also safe for the mites (Wicht & Rodriguez, 1970). Anderson (1983) advocated the selection of pesticide-resistant fly predators (such as Macrochelidae) that could be incorporated into integrated fly management programmes.

Parasitidae

Parasitus fimetorum was very sensitive to PYRs when applied at 5 g AI/ha. At half that rate, the recommended field dose in Australia, up to 30% of the predator's populations survived (James *et al.*, 1995).

Phytoseiidae

In keeping with their prominence as ABAs, the reactions of phytoseiids to agricultural chemicals have been studied more often than those of other mites. Extensive lists of the effects of various toxicants (including insecticides, acaricides, IGRs, fungicides, herbicides and thinning agents) to individual species are thus available (e.g. Croft, 1990; Kostiainen & Hoy, 1996; Kreiter *et al.*, 1998b; Sterk *et al.*, 1999). Regarding the results of laboratory tests, the critique of Bakker & Jacas (1995) should be kept in mind.

Although phytoseiids are usually sensitive to OCLs, OPs, CRBs, PYRs and diverse nematicides, there are many exceptions as well as unexpected results, some possibly due to increasing resistance to pesticides. A PYR mixed with Dipel controlled apple pests in eastern Canada without seriously affecting *T. pyri* numbers (Hardman & Gaul, 1990). Neem, applied at recommended dosages onto bean leaves, did not shorten the lives of *Neoseiulus cucumeris* or *P. persimilis*. The fecundity of the former decreased when exposed to the higher dosage (60 ppm), and both predators were repelled by the toxicant (Spolen & Isman, 1996). Earlier reports (Mansour *et al.*, 1987) that neem caused heavy mortality (up to 80%) to *P. persimilis* were probably due to using the chemical at concentrations above the recommended rates. Different strains or formulations of *B. thuringiensis* appear to be either innocuous or harmful to phytoseiids (Anonymous, 1998; Chapman & Hoy, 1991, respectively). Abamectin, applied either at recommended or at sublethal rates, reduced the longevity and fecundity of several phytoseiids (Sanderson & Zhang, 1995; Ibrahim & Yee, 2000), but to a lesser degree than those of spider mites. It may thus be used in IPM programmes, notwithstanding its side-effects on predators, for 3–10 days (Anonymous, 1998). Cyromazine was toxic to test species in the laboratory but not in the field (Hassan *et al.*, 1994), whereas flubenzimine was harmful to citrus phytoseiids (Ragusa & Ciulla, 1988). Mineral oils had only a fleeting, minor effect on these mites (Oomen *et al.*, 1991; Smith & Papacek, 1991).

Fungicides have variable effects. Sulfur was very toxic; two sulfuric compounds tested by Hassan *et al.* (1994) were lethal to *T. persimilis, Amblyseius* spp. and *T. pyri*. The ergosterol inhibitors cyproconazole, difenoconazole and tebuconazole were moderately harmful to *T. pyri*, but harmless to *P. persimilis* (Sterk *et al.*, 1999), suggesting different susceptibilities among phytoseiids. Another fungicide of the same group, penconazol, was harmless to both mites. Oomen *et al.* (1991) and van Zon & Wysoki (1978) explored the impact of fungicides and noted their diversity of effects on phytoseiids. Regarding herbicides, 2,4-D, dalapon, paraquat and terbacil caused considerable mortality to *N. fallacis* in the laboratory (Rock & Yeargan, 1973), and Pfeiffer (1986) obtained similar results with glyphosate and paraquat in orchards. Amitrole and glufosinate-ammonium were very toxic to *P. persimilis, Amblyseius* spp. and *T. pyri* (Hassan *et al.*, 1991; Oomen *et al.*, 1991). Ioxynil was also harmful to these predators; the pesticide-resistant strain of *T. pyri* was much less affected by that herbicide than was its susceptible strain (Hassan *et al.*, 1994). Sterk *et al.* (1999) reported that haloxyfop had a moderately harmful effect on *T. pyri*. Only one of the 12 herbicides tested by Oomen *et al.* (1991) was totally harmless to *P. persimilis*.

Pyemotidae

Sulfur dust eradicated *Pyemotes tritici* that attacked a culture of wood borers in the laboratory (Hanks *et al.*, 1992). A *Pyemotes* sp. was controlled by carbaryl and by the broad-range fumigant methyl bromide (Muttrie & Anderson, 1984).

Stigmaeidae

Older, widely used insecticides (e.g. arsenates, cryolite, DDT) had little effect on stigmaeids (Lord, 1949; Collyer, 1964a), which appear to be tolerant to some OPs (e.g. diazinon and azinphos-methyl) (Nelson *et al.*, 1973; White & Laing, 1977b; Thistlewood, 1991). CRBs (including carbaryl, often used for fruit thinning) were very detrimental (White & Laing, 1977a; Thistlewood, 1991), whereas fenazaquin was relatively benign to *Agistemus* spp. in India (Sah *et al.*, 1999). These mites are usually susceptible to PYRs (Wieres & Smith, 1978), but in eastern Canada *Zetzellia mali* could have become resistant to these toxicants (Villanueva & Harmsen, 1998). Dicofol mixed with mancozeb reduced a population of *Agistemus exsertus* on citrus in Egypt by around 55% (Rezk & Gadelhak, 1996). An IGR increased the feeding of this mite while decreasing its longevity (Farag *et al.*, 1990).

The fungicides sulfur, dodine, dinocap, benomyl, captan, mancozeb and flusilazole were toxic (Nelson *et al.*, 1973; Childers & Enns, 1975a; Hagley & Biggs, 1989); copper did not harm *A. fleschneri*. In later studies (Bostanian & Larocque, 2001) flusilazole and myclobutanil were harmless to that predator. Stigmaeids living on tea in Indonesia were not affected by maneb (Oomen, 1982). A fruit thinner, naphthaleneacetic acid (NAA), reduced the number of mobile stages and eggs of *Z. mali* in east Canadian apple orchards for about 14 days, causing nearly 50% mortality in laboratory tests. Benzyladenine, another thinner, induced some mortality in the laboratory but not in the field (Thistlewood & Elfving, 1992).

This mosaic of susceptibilities, somewhat different from that of other predators, is a useful tool for assaying the potential of Stigmaeidae for pest control (Oomen, 1982; Croft & MacRae, 1993).

Trombidiidae

Allothrombium pulvinum was very susceptible to DDT and to dimethoate (Chen & Zhang, 1991).

Tydeidae

These mites are very sensitive to plant protection chemicals; most pesticides used in California were detrimental to *Homeopronematus anconai* (Knop & Hoy, 1983b), but selective doses of avermectin had little effect on its numbers (Royalty & Perring,

1987). Mancozeb and sulfur were lethal to *Tydeus lambi*, but myclobutanil and azoxy-strobin were almost benign (English-Loeb *et al.*, 1999). Mineral oil, a mild toxicant, killed two *Tydeus* spp. that occurred on persimmons in New Zealand (Tomkins *et al.*, 1997).

Uropodidae

OP pesticides were only moderately toxic to poultry manure uropodids (mostly *Uroobovella marginata*), and a Bt formulation provided effective house fly control without harming the mites (Wicht & Rodriguez, 1970). Dimethoate and permethrin had a transitory effect on *U. marginata* when applied directly onto chicken manure (Wills *et al.*, 1990).

A note on the effect of fungicides and herbicides

The effect of many fungicides and herbicides on ABAs differs from their influence on insects. A few examples were taken from Hassan *et al.* (1991), Hassan *et al.* (1994) and Sterk *et al.* (1999). Propineb and sulfur were very toxic to three phytoseiids, but had little effect on hymenpterous parasitoids and predators belonging to various insect orders. Cyproconazole, difenoconazole and tebuconazole were moderately harmful to *T. pyri*, but not to most tested parasitoids and predators. As for herbicides, protein inhibitors (e.g. dalapon, glyphosate) as well as compounds that interfere in photosynthesis (e.g. amitrole, glufosinate-ammonium, ioxynil, paraquat and terbacil) were toxic to *P. persimilis*, *Amblyseius* spp. and *T. pyri*, but had variable effects on other organisms. The sources of this differential toxicity remain obscure, suggesting that some acarine toxification and detoxification mechanisms are more similar to those of fungi and of plants than to those of insects.

Not many cases of phytoseiid resistance to fungicides are known (Kreiter *et al.*, 1998a), although that could reflect a bias in observing and recording. Apparently harmless fungicides may have an adverse, additive action on ABAs: a single treatment of dichlofluanid did not affect phytoseiids, but when applied three times it hindered the growth of the predators' populations in an orchard (Jermini *et al.* 1989).

Factors that affect the impact of pesticides on ABAs

Formulation, solvents and application mode

Emulsifiable concentrates of permethrin were more toxic to *Neoseiulus womersleyi* than was the wettable powder formulation (Mochizuki, 1997). Neem extracted by methanol, ethanol or acetone was more toxic to *P. persimilis* than were pentane extracts (Mansour *et al.*, 1987). Pesticides that are sprayed in low volume penetrate deeper into plant canopies and would therefore reach ABAs more than those applied in high volume. However, the smaller low volume drops decay more quickly, affecting predators for a shorter time. The latter reason may have allowed the survival of *Z. mali* under low volume applications, whereas high volume sprays eliminated this predator in Canadian orchards (White & Laing, 1977a). The differential impact of pesticide formulations could thus promote ABA conservation. A recent example is the use of a micro-encapsulated PYR on apples in Canada, a formulation that enabled ABA activity (Lester *et al.*, 1999). In addition, predators could be conserved by applying pesticides only on alternate rows or by spraying on the outside or the lower parts of plants (ecological selectivity) (Croft, 1990).

Persistence

Pesticides differ in their persistence and thus in their short- or long-term residual effects on ABAs, with implications for IPM programmes. Disregarding predator resistance to pesticides (discussed below), there are at least four possible outcomes of using chemicals that affect ABAs. Toxic materials whose residues break down rapidly (e.g. certain OPs) will destroy all ABAs, but they can recolonise the crop after pesticide dissipation, whereas other chemicals, whose residues have a long retention time (e.g. OCLs), will render the crop unsuitable for much longer. Mildly toxic compounds with a short retention time will permit ABA activity, but those that do not dissipate quickly would have a long-term deleterious effect.

Toxicants are usually tested first as fresh deposits in the laboratory, then as 'aged' residues, and finally in the field, often with reasonable agreement among the various methods (e.g. Bartlett, 1964; Oomen *et al.*, 1991; Malezieux *et al.*, 1992; for critiques see Bakker & Jacas, 1995; Blümel *et al.*, 2000). A different approach consists of assaying the effect of pesticides after 'weathering' in the field. Bellows *et al.* (1992) determined the survival of *Euseius scutalis* females that were placed on citrus leaves sprayed in the orchard. CRBs and dithiocarbamates were toxic after 30 days and PYRs 60 days post-treatment. Mite death was correlated with the quantities of pesticide residues left on the leaves.

Mode of action

Pesticides with apparently similar modes of action differ in their actual effect on ABAs. Chlorpyrifos-methyl is about three times more poisonous to *Cheyletus eruditus* than are two other OPs, chlorpyrifos and pirimiphos-methyl (Zdárková, 1994), although the former is less toxic to test birds and fish (Tomlin, 1997). Other examples were provided in the section on Pyemotidae, above. Ovicides (toxicants meant to kill eggs) will have different effect from pesticides that affect other stages. In addition, some pesticides repel ABAs (see below), causing apparent changes in their numbers without actually killing them.

Plants

The effect of pesticides on ABAs as affected by plants was discussed in Chapter 41. Suffice to note that phytotoxic pesticides (and herbicides) reduce the plant area available to pest mites, and that even non-phytoxic insecticides can affect herbivore (and indirectly, ABAs) densities by increasing, or decreasing, plant growth. PYRs increased the size of cotton plants, whereas chlordimeform reduced their vegetative growth (Lloyd & Krieg, 1987). High DDT dosages applied to beans through the soil increased the densities of *Tetranychus urticae* on these plants, a change attributed by Rodriguez *et al.* (1960) to higher levels of reducing sugars. Plants differ in their ability to move toxicants through their systems; the same pesticide may thus have a different food-chain effect on the same ABA on various plants. Finally, owing to specific interactions between secondary substances and pesticides, the same toxicant could affect predators (via the food chain) differently on various plants (e.g. Strickler & Croft, 1985).

Biology of ABAs and stage affected

ABAs that search for prey in sheltered or hidden habitats (e.g. within manure pads or under scale insect shields) are more protected and thus less prone to pesticide effect. ABAs that actively hunt for their prey (e.g. phytoseiids) are more likely to contact pesticide residues and would therefore develop more resistance to pesticides than would ambushing predators (e.g. cheyletids). Specialists such as *P. persimilis* will be less affected by fungicides than will generalists (e.g. *T. pyri*) that complement their diets with rust mites or fungi. *Phytoseiulus persimilis*, because of its short life cycle, will be more harmed by pesticides that delay reproduction than will *T. pyri*, which develops more slowly. The food-chain factor was discussed in Chapter 41.

   Juvenile ABAs, weighing less than their adults and with a relatively large surface/volume ratio, succumb to lower pesticide dosages. European guidelines for testing pesticides on *P. persimilis* therefore recommend using the predator's juveniles in routine tests; surviving mites are then exposed to toxicant residues and their reproduction is assayed. This step-wise procedure is termed 'a sequential decision making scheme for classifying the side effects of pesticides' (Oomen *et al.*, 1991). Although compounds are usually more lethal to juveniles, there are exceptions; dichlorvos was toxic to *P. persimilis* adults for 5 days, but only for 1 day to its larvae (Malezioux *et al.*, 1992).

Side-effects of pesticides on ABAs

Increased fecundity

OP systemics (e.g. monocrotophos, terbufos) stimulated the fecundity of *Euseius citri* in the laboratory (Grout *et al.*, 1997); a similar effect was seen when *G. occidentalis* was exposed to the CRB hexythiazox (Hoy & Ouyang, 1986). A spray of imidacloprid to control fruit tree aphids initially reduced and then increased a population of *Euseius victoriensis* to more than twice its initial size, suggesting the option of enhancing predator populations in the field (James, 1997). As noted, an IGR increased the feeding and curtailed the longevity of *A. exsertus* (Farag *et al.*, 1990).

Removal of competitors and alternate prey

IGRs usually affect insects more than mites, thus removing predatory or omnivorous competitors from the habitat. Insecticide-resistant phytoseiids might not be damaged by many toxicants, but their populations could still decline if the alternate prey (e.g. Tydeidae) is destroyed. The same (compounded by the removal of fungi) may happen after a treatment with fungicides that do not affect ABAs. This effect (part of the 'food limitation hypothesis') is discussed below.

Repellence and dispersal

Sublethal residues of many pesticides repel and otherwise change the behaviour of phytoseiids (e.g. Hislop *et al.*, 1981). This would reduce contact between ABAs and toxicants, lessening their impact, as well as distorting results of tests intended to gauge the chemicals' effect (Blümel *et al.*, 2000). Spider mites are strongly repelled by PYRs, thus increasing the number of their colonies and providing more prey patches for predators (reviewed by Gerson & Cohen, 1989). Finally, resistant ABAs (see below) with a high rate of dispersal would convey their resistance to other habitats (Dunley & Croft, 1994).

Pesticide-resistant ABAs

The recognition that ABAs (especially Phytoseiidae) survive pesticide treatments, whether through tolerance or resistance, can be traced back to Huffaker & Kennett (1953). They noted that field populations of *G. occidentalis* on strawberry persisted after being treated with parathion, and that only 37% of the predators died after the toxicant was dusted on the leaves, a result attributed to tolerance. Similar results with other pesticides were ascribed over a number of years to 'tolerance' until Croft & Jeppson (1970) confirmed the presence of OP resistance in some strains of *G. occidentalis*, by conventional toxicological tests. Susceptibility or resistance to a pesticide in arthropods may be established in the laboratory by exposing their populations to increasing doses of the toxicant and comparing the resultant death rates. The graph on which the data are plotted describes the dose–response mortality of the investigated population. Resistance to pesticides is demonstrated when the dose–response fatality (expressed in probit values, obtained by transforming the mortality percentages into probability units) of one arthropod population is appreciably higher (usually >10 in ABAs) than that of another (Fig. 43.1). The resistant

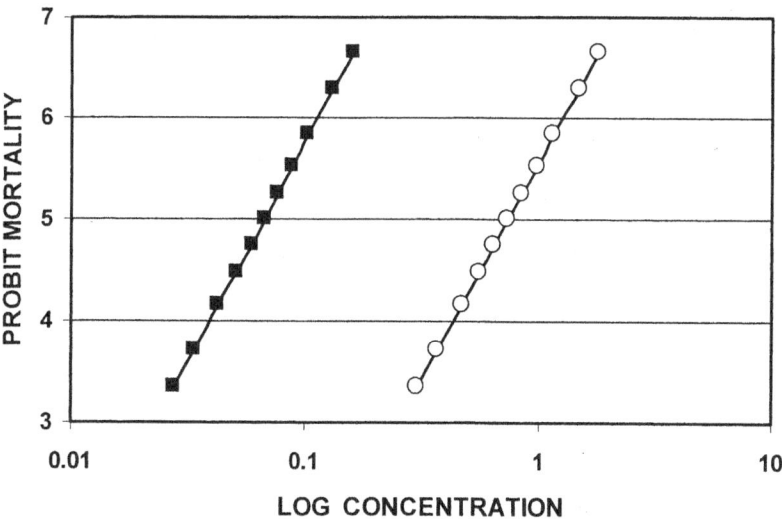

Fig. 43.1 Standard demonstration of the effects of increasing the concentration of a pesticide on the mortality of arthropod populations, measured in probability units (probits). The susceptible population (black squares) suffers more mortality at lower toxicant concentrations than the resistant population (open circles). The values obtained at five probit units indicate the median lethal dosage, or the concentration at which 50% of the population succumbs (thus LD_{50}.)

population, which shows the higher dose response (e.g. succumbing only to higher toxicant concentrations), is termed R, whereas the susceptible population is called S. Strains (or populations) of species may differ in their responses to the same toxicant, usually as a result of past exposure (see below). The use of resistant phytoseiids in IPM programs was discussed in Chapter 26.

Although stigmaeids seem to be acquiring resistance to pesticides (Villanueva & Harmsen, 1998), in ABAs this phenomenon has mostly been documented in the Phytoseiidae, especially in type II, III and IV species (see Chapter 26). The only type I species with high resistance to pesticides (×200–300) is *P. persimilis*, the most often mass-reared ABA. Although the reason(s) underlying its resistance are not clear, one possible cause, the food limitation hypothesis, could explain why strict specialists (*Phytoseiulus* spp.) have developed few R strains in the field. Stated briefly, specialists that had survived a pesticide treatment, owing to incipient resistance, would not have enough food (because their prey were killed by the toxicant) to raise a new generation. In contrast, R generalists, which subsist on alternate diets, would be able to reproduce. Laboratory selections to increase R in *P. persimilis* probably succeeded because of the continuous availability of abundant prey (in other words, unlimited food). Resistance in type I phytoseiids may also be delayed owing to their restricted diets. This is part of the preadaptation hypothesis, formulated to explain the differences in resistance onset between pests and natural enemies. Pests, armed with detoxification enzymes that they had evolved to overcome the plants' poisonous secondary substances, were preadapted to surmount pesticides, an advantage that natural enemies did not have. This theory is supported by the fact that type II and III phytoseiids feed on prey and plant material (including pollen and tissues), thus also ingesting diverse plant toxicants. The number and polymorphism of non-specific esterases (active in OP degradation) increased from the specialist *P. persimilis* to the herbivore *T. urticae* to the omnivore *Rhizoglyphus robini*, consistent with the breadth of their diets (Capua *et al.*, 1990).

Resistance in arthropods is usually attributed to three main mechanisms: slower toxicant penetration, changes in the target (e.g. the AChE system) or rapid pesticide breakdown due to detoxification enzymes, the latter being the prevalent mechanism in ABAs. Resistant phytoseiids employ a variety of enzymic systems to neutralise pesticides, including detoxification by esterases, glutathione S-transferases (GST) and mixed function oxidases (MFO). Phytoseiids may show multiple resistance, for instance being resistant to different toxicant groups owing to more than a single protective mechanism; an example is *N. fallacis*, detailed below (Croft & Meyer, 1973). Mechanisms of pesticide detoxification in phytoseiids were explored by Motoyama *et al.* (1971) and their relative importance was discussed by Croft (1990).

Reported cases of resistance in pests outnumber those recorded in their natural enemies by a factor of greater than 30, a discrepancy addressed by Tabashnik & Johnson (1999). They grouped the hypotheses that were intended to explain this phenomenon into three, not mutually exclusive, general categories: bias in documentation, differential preadaptation to pesticides and variance in population ecology. Bias in documentation, such as more attention being paid to resistance in pests than in natural enemies, could be neither supported nor refuted. Parts of the hypothesis of differential preadaptation (whether due to the presence of more detoxification systems, greater intrinsic tolerance to pesticides with more genetic variation, or lower fitness costs in pests than in natural enemies) were seen as possible but of minor importance (despite the above). Variance in population ecology, especially the food limitation hypothesis, was seen to be important in slowing the onset of pesticide resistance in natural enemies more than in pests.

Previous exposure to pesticides increases the chances of ABAs to become resistant. The intensity of *Euseius tularensis* resistance to OPs, formetanate and especially to carbaryl was probably enhanced by former exposure to these chemicals (Grafton-Cardwell & Ouyang, 1993). The basis for initiating genetic improvement procedures with phytoseiids (see below) was the presence of strains that had acquired pesticide resistance in the field (Hoy, 1985a). The advantage accruing from such strains is that higher levels of resistance (which is often monogenic) among the improved mites were obtained when their source populations had been preselected in the field. In contrast, those with no known field resistance and that were selected only in the laboratory usually show low, mostly polygenic levels of resistance (Croft, 1990).

Genetic improvement

Genetic improvement of a natural enemy is the directed alteration of its genome in order to increase the enemy's efficiency as a biocontrol agent, correcting or improving a trait that is limiting efficacy (Hoy, 1986, 1991). ABA improvement, mostly in regard to pesticide resistance (selection for non-diapause in *Neoseiulus barkeri* was described in Chapter 26), was achieved by hybridisation and/or by artificial selection. Recombinant DNA techniques for this purpose are also available. As improved ABAs are meant for augmentative releases, the candidates should be natural enemies of proven efficacy that can be mass-reared cheaply.

The first phase in genetic improvement is the identification of the trait that needs to be improved (e.g. susceptibility to pesticides) (Hoy, 1991). Natural variability in the candidate species should then be determined, after which selection is applied for the required trait. Resultant strains are evaluated in the laboratory and in the greenhouse for fitness and efficacy, and to ascertain the stability of the acquired trait, and its mode of inheritance is determined. The next phase includes field releases, evaluation of the modified predator's effect, its persistence and rate of autonomous dispersal. Cost–benefit analyses and documentation conclude the project.

Hybridisation was used to obtain improved *N. fallacis*. Populations that survived in heavily sprayed orchards had spontaneously developed resistance to many OP compounds (OP-R), carbaryl and PYRs. By crossing OP-R strains with those resistant to carbaryl, Croft & Meyer (1973) obtained a strain (OP×CL) that maintained its multiple resistances in the laboratory for 25 generations. OP×CL was released in apple orchards and became established (Meyer, 1975). However, pesticides had to be applied to reduce the number of resident, susceptible mites, which could have diluted the OP×CL populations.

Artificial selection was used to obtain improved *G. occidentalis* (Chapter 26). Resistance to OPs (OP-R) and to sulfur was present in field populations, thus providing a platform on which to extend the selections. However, because *G. occidentalis* was susceptible to both permethrin and carbaryl, selections were initiated for resistance to these toxicants (Hoy & Standow, 1982). To increase the probability of obtaining major dominant or semidominant genes for resistance, toxicants were applied with doses intended to kill around 80% of the mites. Resistance to carbaryl, conferred by a major semidominant gene, was thus obtained (Roush & Hoy, 1981a). A permethrin-OP-R strain was concurrently selected, but with polygenic resistance (Hoy & Knop, 1981). Both strains became established in the field, had similar life histories (Roush & Hoy, 1981b), overwintered and survived pesticide applications. The carbaryl-R strain became commercially available, retaining its eminence in the field owing to the dominant resistance gene. But here, too, pesticides had to be applied to reduce the number of resident, susceptible mites, which could have diluted the R populations. Other options were to use R strains that are reproductively isolated from the resident mites, or to release the R predators in such large numbers that they would overwhelm the field population. Headley & Hoy (1987) later calculated that the release of multiple-resistant *G. occidentalis* in almond groves would save US$350/ha during 5 years (Chapter 46).

Galendromus occidentalis was also improved by a recombinant DNA technique. The inserted material (the construct) consisted of a plasmid containing the donor DNA and a gene that expressed an enzyme which mediates a colorimetric reaction, and a heat-shock-regulating promoter. The construct (meant to serve as marker) was inserted into gravid *G. occidentalis* by microinjection. The colour of the mites after they were squashed in a reagent that stained them dark blue (due to the enzyme's activity) confirmed the construct's incorporation into the genome. Several progeny of the females subsequently expressed the colouring enzyme, implying that stable, heritable transformation was obtained. It was also successful when larvae [of *G. occidentalis* and *Neoseiulus finlandicus* (Oudemans)] were injected with the construct (Presnail & Hoy, 1994a, b).

Hoy (1992) listed the criteria for releasing genetically improved natural enemies, using a scenario based on *G. occidentalis*, and concluded that owing to this species' biological attributes the risk was low. However, gaps in relevant knowledge led to risk assessment exercises that should be conducted before any proposed field releases of transgenic arthropods. (A relevant exercise with *G. occidentalis* is detailed in Chapter 45).

Concluding remarks

Pesticides have brought about many changes in the composition of phytoseiid (and other ABA) populations, such as the disappearance of resident susceptible species which have been displaced by their own resistant strains or by resistant exotics. In other cases the greater sensitivity of certain species brought about changes in their relative abundance. Two fungicides applied onto pecan foliage in Florida caused high but differential mortality to three resident phytoseiids; the most abundant was almost eliminated, whereas another became dominant (Ball, 1982). Thus, while resistant phytoseiids have advanced the implementation of many IPM programmes in which toxicants are applied without disrupting ABA activities (Chapter 26), in others they may have encouraged the continued use of pesticides. Later on, reduction or cessation of chemical applications is followed by declines in the resistant exotics, because their survival is dependent on continued pesticides usage. Resistant *G. occidentalis*, *T. pyri* and *P. persimilis* were introduced into Australia, became established, but were displaced by indigenous phytoseiids in orchards under low-pesticide regimens (James, 2001).

Knowledge on the differential susceptibilities of ABAs to pesticides is a powerful research tool when trying to isolate the specific contribution of predators in a guild to overall pest control (Chapter 44).

Most studies on pesticide resistance in ABAs were conducted on phytoseiids. However, the increasing recognition of the role of Stigmaeidae in biocontrol (Chapter 32), and observations that some may be resistant, indicate that these mites are also worthy of similar efforts. Anderson (1983) advocated the selection of pesticide-resistant fly predators (e.g. Macrochelidae, Chapter 23) that could be incorporated into integrated fly management programmes.

Ruberson *et al.* (1998) listed several means that could promote pesticide use along with natural enemies, such as the temporal and spatial separation of the latter from toxicants and the use of selective pesticides. Relevant examples were noted above. Data on pesticide persistence would indicate when predatory mites could be released in a sprayed crop, thus providing temporal separation. Spatial separation was obtained by using resistant phytoseiids or by spraying in alternate rows; the application of information on the effects of current or new plant protection chemicals facilitates the use of selective pesticides.

Pesticides and ABAs can reduce pest populations in tandem. At times pest numbers must be reduced before ABAs can be used. Zdárková (1997) advocate the use of OPs to decrease pest numbers in stored products, before *Cheyletus eruditus* is released there.

Before leaving this chapter, it should be noted that pest increases after a pesticidal treatment, often attributed to the 'elimination' of natural enemies, could be due to more than a single factor. The differential 'kill' of ABAs but not of pests is the simplest. However, the pests could also have been less affected than the predators, giving them a clear edge. Both ABA and pest may be eliminated to the same extent, but the latter, whose only nutritional requirement is the plant, often recolonise it more rapidly. The behaviour of a predator could be affected by sublethal residues, detracting from its efficacy. Another variation could be that even if ABAs arrive on the plant along with the pests (or, being resistant to the chemical, have remained on the plant), there is insufficient prey to allow specialists to survive (the 'food limitation hypothesis', discussed above). In addition, the populations of some pests (e.g. spider mites) may increase greatly owing to the effect of certain pesticides (e.g. PYRs), even in the absence of natural enemies (Gerson & Cohen, 1989).

Chapter 44
Demonstrating the efficacy of acarine biocontrol agents

Introduction

'In biological control today, as in the past, statistical proof of the efficacy of entomophagous insects in the field is all too often lacking'. These words, written over 50 years ago (DeBach, 1946), resound clearly to this day. Although such proof is the key to the further promotion of biological control, the theoretical framework that would explain past successes and predict future outcomes is still not at hand. Models for host–parasite and host–pathogen interactions have been formulated (e.g. May & Hassell, 1988), but represent only a few real-world situations (Neuenschwander, 1996; Worner, 1991). Some models operate on the assumption that pest and predator(s) have attained a stable state condition, in which most other variables [such as natural enemy idiosyncrasies, specific plant characteristics and intraguild predation (IGP)] are also more-or-less stable. This assumption may be more valid in controlled cultivation situations than in the field, which explains why modelling has been more successful in predicting greenhouse or poultry house prey–predator interactions. Evaluation efforts are particularly difficult in regard to mites, because many Acari do not behave like other biological control agents. Some successful acarine biocontrol agents (ABAs) are not very specific to the pest they control, their basic biological attributes [life table data; sex ratios (SRs)] do not seem to be related to pest control in the field, plant traits affect their performance in unexpected ways and they may not show a functional response to their prey. Further, pest populations (and/or their damage) are reduced not only by predation or parasitism, but also through competition or changing host plant attributes, mechanisms that lie outside the framework of biological control models.

Some authors do not clearly define the purpose of their research, which hinders efforts at evaluation. Does 'biological control' mean reductions in pest numbers and/or in the level of damage (e.g. Oatman *et al.*, 1976)? Is the aim to obtain long-term control or to demonstrate the effect of short-term augmentations? How are reductions in the extent of economic damage to the commodity measured? What is the expected or allowed time frame for the project? Has a pre-introduction census of natural enemies been conducted, to ensure that reductions in pest and/or damage levels were exclusively due to the released predator? How are natural enemy numbers evaluated after establishment, and is there any correlation between such values and reductions in pest (or injury) levels? Can a quantitative method be used for monitoring? Have suitable control plots (e.g. without natural enemies) been prepared, with appropriate buffer zones? Correctly addressing these (and many other) issues would

384

help in demonstrating the efficacy of ABAs. Finally, the impact of natural enemies on weeds is difficult to evaluate, as their effect would be seen only over several years.

Luck *et al.* (1988) suggested six general experimental approaches (or techniques) for evaluating the impact of natural enemies on pest populations. These consist of (1) introduction and augmentation, (2) cages and barriers, (3) removal of natural enemies, (4) prey enrichment, (5) direct observations and (6) chemical evidence of natural enemy feeding. In this chapter molecular evidence is added to the last technique, which Sunderland (1996) treated more quantitatively. Indirect (inference) methods are also noted.

Introduction and augmentation

The basic argument underlying this approach is that the introduction (or augmentation) of the natural enemy brings about a significant decline in pest numbers and/or level of damage. Non-quantitative accounts and claims about the success of introduced ABAs, often based on anecdotal field observations, abound in the literature. Early examples are reports that the bdellid *Bdellodes lapidaria* was reducing populations of the lucerne flea as the predator was spreading from and within fields in Western Australia (WA) (Womersley, 1933) (Chapter 9). Turnbull & Chant (1961) scored the introduction of *Hemisarcoptes malus* from eastern to western Canada to control an apple pest as a success, a score based only on anecdotal evidence, without statistical data on efficacy.

An early effort to quantify the effect of an ABA on a pest showed the impact of *B. lapidaria* on the lucerne flea. Currie (1934) collected several samples from a clover field in Waroona, WA, during 1933, to provide a picture of the mite's progress in controlling the pest. Because of the predator's uneven distribution, only a generalised estimate of its population could be provided. The pests, however, live on the crops and could be counted when a glass cylinder covered the plants. A traverse was marked in the field from the point where the lucerne flea appeared to be controlled to another site, 170 yards (155 m) away, at which the pest was still abundant. Estimates were made every 5 yards (4.5 m), at about fortnightly intervals. The resulting diagram (Fig. 44.1), which incorporates pest numbers with predator estimates, shows how the population of the former had been reduced (from 10–15/5 yards to *c.* 5/5 yards) towards the right-hand area of the field, followed by increases in the mite population. 'The way in which the mites had reduced the flea populations is sufficiently striking to require no elaboration' (Currie, 1934).

Proper quantitative evaluations are based on counts (or estimates) of pest numbers made before and after the introduction, along with assessments of damage. Gough (1991) sampled two-spotted spider mite (TSSM) populations on greenhouse roses for about 3 years in south-eastern Queensland, Australia, before and after introducing *Phytoseiulus persimilis*. Pest numbers totalled ten to 20/leaf before the introduction, dropping to less than one/leaf afterwards; on only eight out of 66 sampling dates (12%) did spider mite densities exceed one/leaf. Furthermore, the fluctuations of prey and predator population were similar, and the numerical trends of *P. persimilis*

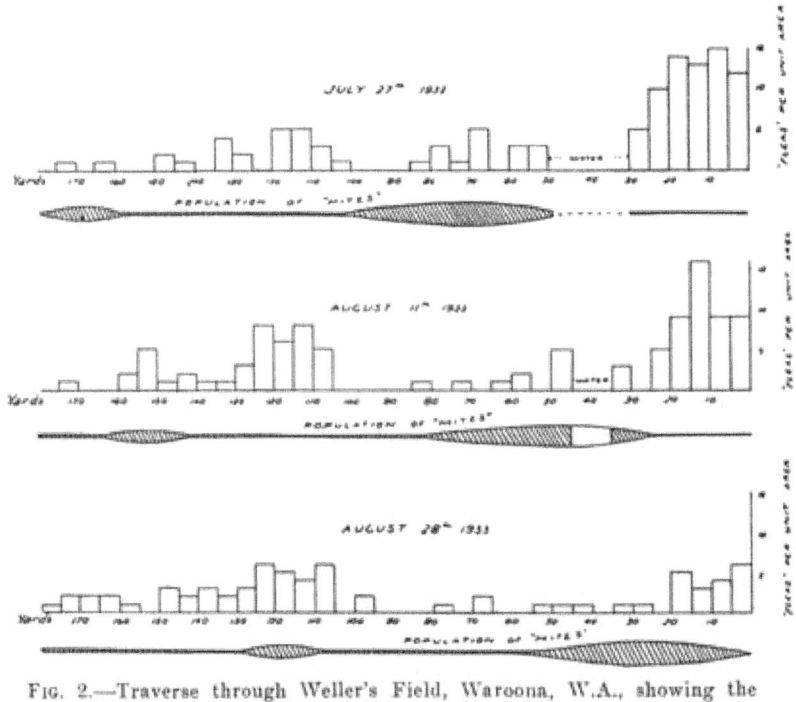

FIG. 2.—Traverse through Weller's Field, Waroona, W.A., showing the
decrease in "flea" population at the different dates with the
advance of the mite.

"Flea" = *Sminthurus viridis*.   "Mite" = *Biscirus lapidarius*.

Fig. 44.1 Field demonstration of the effect of the bdellid *Bdellodes lapidaria* on populations
of the lucerne flea, *Sminthurus viridis*, during one month (27 July to 28 August 28th, 1932).
Bars: number of prey tallied every 10 yards along a transect of 170 yards; variable-thickness
line: representation of predator numbers at the same sampling points. Upper: 27 July, 1932;
middle: 11 August 1932; lower: 28 August 1932. (Figure 2 in Currie, 1934; original legend left
in place.)

were highly correlated with those of TSSM (see Fig. 26.3). An evaluation of the damage, made during a period of above-average pest activity, showed that 70% of the
leaves were totally unmarked, 24% had negligible damage, observable only on close
scrutiny, and the rest (in the lowest part of the stems) had some injury. Satisfactory
spider mite control on those roses was thus demonstrated by continuous monitoring
of pest and predator numbers and by evaluating the extent of postrelease damage.
Other demonstrations of ABA efficacy include *Stratiolaelaps miles* and increases in
mushroom yields (see Fig. 21.2) and *Pyemotes tritici*, which reduced losses of flour
weight due to a moth (see Fig. 30.1). The same quantitative approach can be used to
show that biocontrol with ABAs does not provide suitable returns. Oatman *et al.*
(1976) released large numbers of *P. persimilis* (up to 250,000/0.4 ha) in strawberry
beds infested by TSSM and compared their fruit yield with that from control beds
(i.e. without predators). Pest numbers were significantly reduced (Fig. 44.2), but
neither yield nor mean fruit size differed between the treatment and control plots.
Other quantitative demonstrations of ABA failures pertain to the gall mite *Aceria*

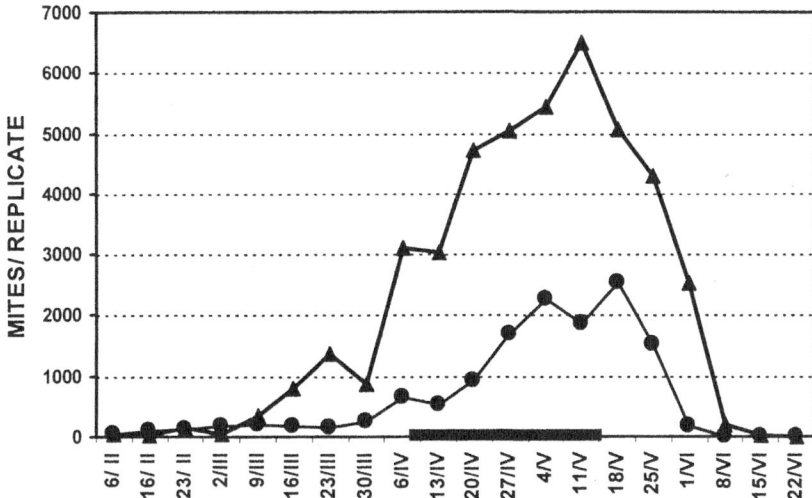

Fig. 44.2   Mean numbers of *Tetranychus urticae* on replicates of strawberry leaflets from plants with (triangles) and without (squares) adding ten *Phytoseiulus persimilis*/plant. The black bar denotes a period of significant differences between mite values in the two treatments. (Data from Table 1 in Oatman *et al.*, 1976.)

condrillae and to *P. tritici*. The former was introduced into California to control a perennial weed (Chapter 15) and became established, but there was no relationship between number of galled plants and changes in weed density (Supkoff *et al.*, 1988). Thorvilson *et al.* (1987) tried to control the fire ant in Texas by introducing around 1.4 million *P. tritici* into nests, but found no change in the number of ant workers (Chapter 30).

Cages and barriers

These techniques are intended to determine the quantitative effect of the natural enemy on prey numbers in the field. They consist of restricting pests to a small area, using barriers such as glues or cages to isolate them from predators. Changes in pest populations are then compared with those that had occurred in the unrestricted controls. However, the practitioner should bear in mind that results obtained by these methods may be biased owing to side-effects. Fleschner (1958) explored the effect of predatory phytoseiids on citrus spider mites by enclosing their colonies within organdy sleeves that enveloped citrus branches (the 'sleeve cage method'). Populations of the enclosed pest, the citrus red mite, became about 12 times larger than the exposed colonies, an increase ascribed to microclimatic changes within the enclosure and to the pest's inability to disperse, rather than to predator exclusion. The use of glues, sticky bands or other mechanical barriers implies that the predators can arrive only by walking, because such obstacles will be ineffective against wind- or insect-borne ABAs.

Removal of natural enemies

Chemical and non-chemical methods were used for ABA removal. The former have served to demonstrate the efficacy of ABAs in three different ways.

First, pesticide applications were often followed by vast increases in target and non-target pest mites, outbreaks usually attributed to the inadvertent elimination of natural enemies. The following citation reflects a widely held opinion: '... data indicating that predators have, on so many occasions, kept mites under control except when pesticides were applied, are collaterally supportive of the predator inhibition hypothesis' (Huffaker *et al.*, 1969). The use of broad-spectrum pesticides thus served as a vast exclusion experiment, which by serendipity brought about a greater appreciation of the role of predatory mites (and other natural enemies) in controlling many pests. (Other postulated causes for spider mite eruptions are discussed below.)

The ability of *Cheyletus eruditus* to control storage mites spontaneously was indirectly demonstrated when food stores were fumigated with methyl bromide. The stores were overrun by huge numbers of acarid pests 2–4 months post-treatment, but only very few *C. eruditus* were found. The inference was that the resident predators had hitherto controlled pest populations, and that the fumigation had disrupted this natural control (Chapter 12). Another case where the significance of the natural enemy was realised only after its unintentional removal by a pesticide was the moth *Ephestia cautella*. Applications of the organophosphate (OP) malathion to stored products in Kenya were followed by moth outbreaks, presumably because the pesticide had eliminated the moth's most important natural enemy, *Blattisocius tarsalis*, without harming the moth (Chapter 8).

Second, the differential toxicity of pesticides to pests and to natural enemies has been utilised to evaluate as well as to demonstrate the beneficial effect of predators and parasites. A suitable chemical that removes the natural enemies but leaves the pest intact is applied to the experimental plants. The size of the pest population that subsequently develops is compared with that found on control plants, where the predators had not been affected. This is the 'pesticide check method' (DeBach, 1946), which provides qualitative as well as quantitative data on the ability of a natural enemy to control a pest (provided the chemical does not increase the pest's population; Gerson & Cohen, 1989). Wallace (1954) used this method to demonstrate control of the lucerne flea by *B. lapidaria* (Chapter 9, Fig. 9.1).

After spraying selected tea bushes with dichlorodiphenyltrichloroethane (DDT) and estimating the numbers of scarlet mites (*Brevipalpus phoenicis*) thereon, Oomen (1982) calculated that a resident phytoseiid species had decreased pest abundance to 83% of the predator-free controls, whereas a guild of stigmaeids reduced it to 27%. These results, along with regression data (see below), led Oomen to state that stigmaeids are more efficient predators of the pest in Indonesia than phytoseiids (Chapter 32).

As acaricides have little effect on plants (but see Chapter 43), they can be used to evaluate the effect of ABAs on weeds. The impact of the eriophyid *Aculus hyperici* on St John's wort was examined by spraying target plants with the acaricide Omite.

Unsprayed plants had smaller root diameters and less root and shoot weight than sprayed plants (Jupp & Cullen, 1996), attesting to the mites' activity.

The third use has been to isolate the specific role of a single predator in a guild whose members control a given pest, by applying a selective pesticide that affects only that predator. Croft & MacRae (1993) demonstrated the effect of the stigmaeid *Zetzellia mali* on two apple pests, the European red mite (ERM) and the apple rust mite, and on their two phytoseiid predators, *Typhlodromus pyri* and *Galendromus occidentalis*, by this method (Chapter 26; see Fig. 32.3). This sophisticated insecticidal check method showed that *Z. mali* was capable of controlling ERM by itself, as well as indicating what occurs when the stigmaeid is present along with either of the two phytoseiids.

In this context, the use of pesticides can engender unexpected results, such as enhanced predator or prey fecundity. The OP dimethoate, when assayed at field rates (0.389 mg/kg), increased the fecundity of the laelapid *Geolaelaps aculeifer* by around 25% (Krogh, 1995; see Chapter 21). The insecticide imidacloprid had a similar effect on the phytoseiid *Euseius victoriensis* (James, 1997). Similar or greater stimulation may follow the application of sublethal pesticide doses. This variable could become important when using the insecticide check method for longer than a single predator generation.

As noted, other mechanisms besides pesticide toxicity were invoked to account for post-treatment outbreaks of pests. They include direct and/or indirect stimulation of pest fecundity, female-based shifts in the SR and greater redistribution of pest populations on the plants caused by certain pesticides. The recognition of more mechanisms of pest outbreaks made it necessary to distinguish between them in order to pinpoint the effect of the natural enemies. Hand removal (see below) was one procedure employed. Huffaker & Kennett (1956) established with the insecticide check method that phytoseiids controlled the cyclamen mite on strawberries. However, to bolster their argument as well as to counter possible criticisms, they also hand-removed the predators from the plants, leaving other leaflets as untouched controls. Pest populations became 20–30 times larger on the 'removal' leaflets and blossoms, than on the control plants, with their predators intact (Fig. 44.3). The compatible results obtained with both methods indicated that 'stimulation' theories should be rejected.

Collyer (1964a) used three methods to demonstrate the effect of *T. pyri* on the ERM. First, plots treated with lime sulfur consistently carried more pests, but fewer predators, than those sprayed with the fungicide captan. Observations that *T. pyri* overwinters and shelters in bands of sacking tied around tree trunks (meant to attach the trees to support stakes) suggested an indirect, non-chemical method of gauging the effect of the predators. Trees are sprayed in the winter with dormant oils, which are toxic to *T. pyri*. By removing the sacking in some plots, trees were obtained with more, or with fewer, predators. Many *T. pyri* and few ERM occurred on trees where the sacking was left in place. Other trees, on which the predator was annually reduced owing to a lack of shelter, had high pest numbers. Finally, the inverse relationships between prey and predator numbers suggested that the presence or absence of *T. pyri* was the major cause of numerical fluctuations in pest populations.

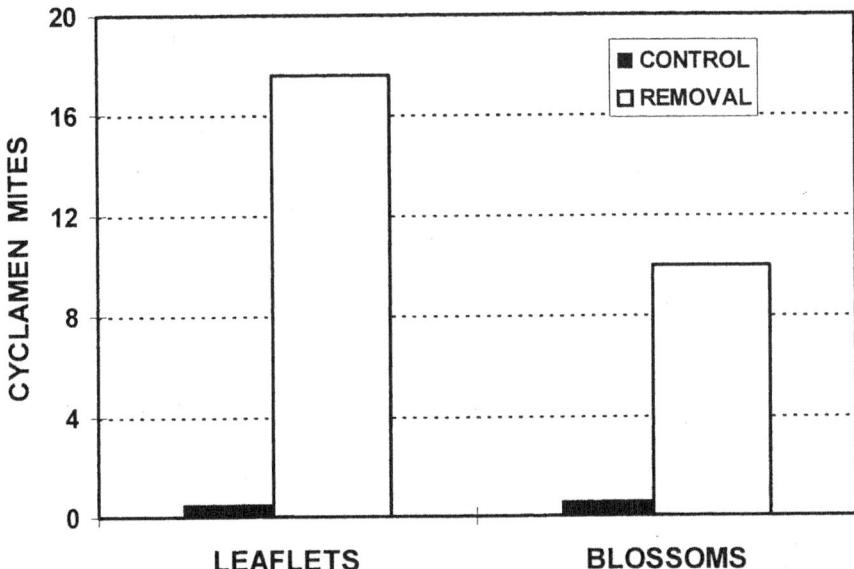

Fig. 44.3 Numbers of cyclamen mite, *Phytonemus pallidus*, counted on strawberry leaflets and blossoms after hand removal ('Removal') of predators, compared with pest numbers on untouched controls ('Control'). (Based on data in Huffaker & Kennett, 1956.)

Moving to other non-chemical modes, Wassenaar (1988) vacuumed a domestic carpet to demonstrate the effect of *Cheyletus* sp. on the house dust mite, *Dermatophagoides pteronyssinus* (Chapter 12). The vacuuming did not affect pest numbers, but reduced those of *Cheyletus*, with a subsequent increase in house dust mite populations.

Fleschner (1958) compared the efficacy of several methods used to estimate the effect of natural enemies on spider mites. Hand removal was the most effective and dependable method and avoided the introduction of new factors into the system; thus it was suitable for evaluating results obtained by other methods. However, hand removal is very tedious and time consuming, demanding full attention for extended periods, 7 days a week.

Prey enrichment

This method enables the researcher to control the number of prey individuals offered, and consumed, within given time frames. The predator is given prey in known numbers within an arena, and the resulting data on its voracity are used to estimate its effect on the pest in the field. This is the common method of obtaining data for calculating the functional response of natural enemies (see below) and to calculate life history parameters required to set up mass-rearing protocols.

Axtell (1963) demonstrated the effect of the macrochelids *Macrocheles muscaedomesticus* and *Glyptholaspis confusa* on house fly populations by placing 20,000 fly eggs in several cages containing fresh cattle manure (Chapter 23). The effect on the target insects was estimated by counting the number of flies that were captured on sticky fly ribbons placed in each cage. Adult pest emergence was reduced by 94%, attesting to the mites' effect (see Fig. 23.1).

The great voracity of several mesostigmatids that feed on nematodes (e.g. see Fig. 38.2) indicated to Sharma (1971) that some of these predators could have the potential to reduce nematode populations in the field or the greenhouse. In contrast, the inability of the cheyletid *Cheletogenes ornatus* to feed on more than two prey individuals/day suggested to Avidov *et al.* (1968) that it was not an important predator of armoured scale insects (Chapter 12).

Rates of feeding obtained in cage experiments should be interpreted with caution, as they may not reflect what happens in the field. The large number of prey within a limited, usually uniform space (the experimental arena), and their inability to utilise possible escape mechanisms and refuges ('enemy-free space') could detract from the validity of the results. Predators of nematodes have different movement and prey-finding patterns in sandy compared with heavy, water-logged soils. Predictions of nematode control extrapolated from arena studies may thus be misleading (Chapter 38). IGP and other interactions between predators (which could be rare or common under field conditions) may also affect the results. Prey enrichment experiments nevertheless provide useful data and can serve as preliminary steps in understanding prey–predator relationships in the field.

Direct observations

Observations reveal useful data about what actually happens when predators encounter prey, how they react to it, what are the grasping mechanisms, whether prey are eaten singly or in groups, which predator stages feed, and so on. Predators can be watched attacking the prey with which they had been collected, on the same material (e.g. leaves), or they may be placed with various food organisms in separate arenas on suitable substrates. The inability of a given ABA to feed, develop and reproduce on an assumed or candidate prey (or weed) is taken as evidence for a lack of dietary association. Rockett (1980) offered nematodes to 14 cryptostigmatids; five fed on the prey and the indifferent others were not considered as nematophages. Direct observations will show whether a tested predator survives on additional foods (e.g. fungi), or prefers them, or may even require these other diets.

Observational methods are used to evaluate the effect of natural enemies on weeds, because their damage is immediately evident. The effect of *Tetranychus lintearius*, introduced into New Zealand to control gorse, was clearly demonstrated because it encased the entire target weeds in its profuse webbing (Hill *et al.*, 1991). Damage caused by eriophyids includes leaf rolling, yellowing, deformation of tissues and formation of galls, reduction of flowering and plant stunting, all easily observable and amenable to quantification.

Direct observations have the advantages listed for hand removal; many of the disadvantages can be surmounted (at least in the laboratory) by using video equipment (Wharton & Arlian, 1972). This also provides data on predator behaviour (albeit under artificial conditions), including modes of searching and alternate prey selection procedures. The behaviour of phytoseiid juveniles was analysed by Palevsky *et al.* (1999) from data obtained by short exposure periods made at 6, 12 and 24 h after mite eclosion with a video camera mounted on a stereomicroscope.

Parasitised hosts can be collected and observed for the effects of their enemies. Lanciani & Boyt (1977) obtained female mosquitoes (*Anopheles crucians*), parasitised or unparasitised by the water mite *Arrenurus pseudotenuicollis*, and kept them in the laboratory. The healthy insects lived significantly longer, and deposited more eggs, than parasitised hosts. Furthermore, engorged mosquitoes carried a lower mite load, a result consistent with the assumption that unengorged, heavily parasitised mosquitoes die before obtaining their first meal. These observations led to the hypothesis that the mites shorten the lives of the mosquitoes and reduce the fecundity of surviving hosts (Chapter 7).

Prey consumed by ABAs may be identified by a microscopic examination of the predator's gut contents or by chromatography. Predatory Astigmata and Cryptostigmata ingest solid food, which facilitates the visual identification of prey remains. The presence of identifiable nematode stylets within mite gut boluses constituted proof that predators had fed on nematodes (Walter, 1988b). Gut content analysis of Cryptostigmata was discussed by Behan-Pelletier & Hill (1983). However, recognisable arthropod remains accumulate in a mite's gut throughout life, so it may be difficult to attribute the remains to any discrete period (unless the predator had been under observation for given durations, or its age is known).

Predatory Prostigmata and Mesostigmata feed by sucking out the body fluids of their prey. These fluids, if pigmented (e.g. the ERM), can be observed within the whitish bodies of some ABAs (e.g. most phytoseiids). To demonstrate quantitative or stage-related predation it is necessary to determine whether all prey stages provide this marker, and how long it remains readable. Jones & Morse (1995) used an observable graded scale of the red coloration and gut distension of *Euseius tularensis* that had fed on the citrus red mite. The coloration persisted for 24–48 h when feeding on prey deutonymphs, but only for 12 h when larvae were consumed. Putman & Herne (1964) used paper chromatography to demonstrate the feeding of phytoseiids on spider mites. *Hemisarcoptes* has a whitish body, which acquires the colour (yellow, red or purple) of the body fluids of the host scale insects on which it feeds (Gerson, 1967b).

Chemical and molecular evidence for natural enemy feeding

A major problem in applying these methods to estimating predation by fluid-feeding mites is that the amount of prey material ingested is often very small. Electrophoretic analysis, immunological techniques and molecular methods are the three main approaches used to overcome this problem and to demonstrate ABA activities.

Electrophoresis of esterases is often preferred over other proteins. The products of esterase activity on a suitable substrate (e.g. 1-naphtyl-acetate) react well with the colouring agent Fast Blue, providing stains with a high extinction coefficient. Polyacrylamide gel electrophoresis (PAGE) focuses the migrating proteins in the slab gels into narrow bands, imparting good resolution. Murray & Solomon (1978) used this method to detect remnants of two prey species (ERM and TSSM) in the bodies of two predators, *T. pyri* and *Neoseiulus fallacis*. A reading taken from a *T. pyri* that had been kept without food after feeding on spider mites showed that the enzymes persisted in the predator's gut for at least 31 h. However, use of the method depends on the prey having at least one esterase that is located in a characteristic position on the gel; otherwise it would be indistinguishable from other predator or prey enzymes. Also, some prey have relatively low esterase activity, which could be masked within the diet of the predator. The effect of the anystid *Anystis baccarum* on cereal aphid populations was similarly clarified by El Banhawy et al. (1993). Solomon *et al.* (1996) advocated using electrophoretic analyses for recognising the remains of small prey species and for assaying the predation spectrum of polyphagous predators, at least as an initial step in narrowing the range of potential prey species before selecting candidates for further work.

PAGE of allozymes (also called isoelectric focusing) was used to follow the spread of newly released, pesticide-resistant strains of *N. fallacis* in Canadian fruit crops (Navajas *et al.*, 2001). This method (along with analysing variations in DNA sequences; see below) can also be applied to assay the efficacy of ABA strains. Results obtained with such electrophoresis led Jones & Morse (1995) to doubt whether *E. tularensis* is a major predator of the citrus thrips in California. Only seven out of the 556 (1.3%) predators assayed (collected from prey-rich trees) had a positive stain reaction to thrips enzymes, indicating very low predation rates on this prey.

The sensitivity of immunological methods, such as enzyme-linked immunosorbent assay (ELISA), makes them highly suitable for detecting minute residues of prey in predator extracts. Application of ELISA involves the preparation of prey anti-serum, in which prey material is injected into (usually) a rabbit, and subsequently obtained from it. This may introduce two factors that could lead to mistakes: the need to have enough unambiguous, clean prey material and variability in the rabbit's immunological responses. Sunderland (1996) updated methods used for the indirect detection of invertebrate predation and Hagler (1998) compared the relative sensitivities of various gut content immunoassays.

The presence of prey within the body of a predator can be demonstrated by various molecular methods, for example random amplification of polymorphic DNA–polymerase chain reaction (RAPD-PCR) assays. DNA primers are prepared from the prey, cloned into a plasmid vector and transformed into the bacterium *Escherichia coli*. Once the mite DNA has been incorporated into the bacterium, it can be sequenced and used to prepare specific primers for detecting prey residues in predators. Greatorex (1996) used primers based on DNA from a tydeid mite to determine the presence of its residues in the gut of *T. pyri*. One primer recognised tydeid DNA that was diluted to 1/100,000 of predator remnants, a level of sensitivity suitable for determining the prey diets of polyphagous predators.

One source of bias in using the presence of prey residues in the bodies of natural enemies (unless the state of the prey is clearly known) is that some predators will feed on cadavers, whether dead by natural causes, killed by others or frozen. The sclerotised body parts, the pigments or the molecular markers will not be seriously affected (at least in the short term) by the death of the prey.

Inference methods

Calculations

Chiang (1970) calculated the effect of manure applied to cornfields on the predation rate of laelapids on rootworms (Chapter 21). He estimated the number of resident acarine predators and how many pests they usually kill, the increase in laelapid numbers after manure applications and the number of rootworms that they had destroyed, and concluded that the manure had caused a 63% reduction in pest numbers.

The effect of water mites on mosquito populations in the field was inferred by collecting mosquitoes of different ages and determining the number of stylostomes found in their bodies (see Fig. 7.1). Average stylostome number was consistently and significantly higher in young (unfed) than in older (fed) insects, suggesting that there was a 42.5% mite-caused loss of heavily parasitised hosts from the former group (Smith & McIver, 1984b).

Correlations

The effectiveness of natural enemies is sometimes demonstrated by calculating correlations between numbers of pests and of natural enemies (which could be separated by a half or a full generation, or 'phase') (e.g. see Fig. 26.3). A positive correlation may suggest that increases in natural enemy populations were affected by those of the prey, but not necessarily provide proof of control. A negative correlation, meaning that reductions in prey numbers occur concurrently (or after a certain period) with increases in predator densities has often been used to argue that the natural enemy has reduced pest populations. Inverse numerical relations that developed on apple trees between *T. pyri* and ERM suggested to Collyer (1964a) that the former was decreasing the latter. Numbers of lucerne fleas and *B. lapidaria* counted during 8 years provided a consistent, significant negative correlation (Wallace, 1967), indicating the predator's role in pest suppression. Oomen (1982) plotted numbers of tea pests and stigmaeids against each other and obtained significant negative correlations in seven out of ten censuses; these predators were therefore considered to contribute significantly to the pests' control. Ireson & Webb (1995) correlated the numbers of the bdellid *Neomolgus capillatus* recorded on certain dates and those of the lucerne flea counted 2 weeks previously, a relationship that explained 80% of the variation in pest numbers. Jupp *et al.* (1997) infested glasshouse-grown St John's wort plants with

Aculus hyperici (Chapter 15). Mite numbers were rated and their effect was determined by weighing the plants 14 weeks later. An inverse relationship was obtained between *A. hyperici* ratings and root, as well as shoot weight, indicating an effect of the mites on the weed. A lack of significant correlations between numbers of skeletonweed plants galled by *A. condrillae* and changes in the weed's density suggested that the ABA had little effect on skeletonweed (Supkoff *et al.*, 1988).

Life table analysis

Laing (1973) studied the effect of the cheyletid *Hemicheyletia bakeri* on TSSM populations and calculated its life history parameters (intrinsic rate of increase, net reproductive rate and generation time). These rates were quite low compared with those of several phytoseiids that fed on the same prey, leading to the conclusion that the cheyletid had little effect on these pests (see next section). Multifactor analyses of life tables were used by Samarasinghe & LeRoux to demonstrate that *H. malus* (and an aphelinid wasp) were the key factors regulating the populations of an apple pest in eastern Canada.

Bellows & van Driesche (1999) discussed the construction and analysis of life tables for evaluating the impact of natural enemies on their hosts and/or prey. Although their emphasis was mostly on insect parasitoids, the reader is referred to that essay for further details.

Functional and numerical responses

The functional responses of ABAs (changes in number of prey killed by individual predators as prey numbers change) and their numerical responses (changes in predator numbers in response to changes in prey numbers) were used to demonstrate predator efficacy (for examples, see Chapter 26). In the laboratory *N. fallacis* showed good responses of both types to increases in TSSM numbers. This indicated to Croft & Blyth (1979) that the predator could rapidly respond to pest outbreaks in the field. The functional response of *Neoseiulus cucumeris* on first instar larvae of the western flower thrips indicated that male predators were ineffective (Shipp & Whitfield, 1991). The finding that *H. bakeri* lacked either of these responses to increases in spider mite densities confirmed Laing's opinion (1973) that the cheyletid has only a minor role in TSSM control. Many acarine predator–prey models (see below) are based on the functional and numerical responses of ABAs. However, relying on functional and numerical response criteria alone could be misleading. Laing & Knop (1983) believed that members of the Anystidae, Cheyletidae, Erythraeidae, Stigmaeidae and Tyeidae are unlikely to control acarine pests because their numerical responses are much lower than those of their prey, and that this deficiency is seldom offset by a very high functional response. Data presented in the various family chapters suggest that such generalisations may have been premature.

Models

The laboratory and field data and their inferred conclusions may be consolidated into models to produce predictions of ABA efficacy in the field, for the use of farmers and policy makers. The successful prediction of control, or model validation, is another demonstration of the efficacy of these natural enemies. Models can also pinpoint environmental factors that affect the association, the importance of the predators for the control of specific pests and their probability of achieving that aim.

A model that predicts ERM numbers and their probability of being controlled by *T. pyri* on apple trees, as affected by initial pest and predator numbers, was prepared by Dover *et al.* (1979), and another by Hardman *et al.* (1999). The latter, which was validated at predator/prey ratios of >1/10, further indicated the value of introducing *T. pyri* early in the season. Berry *et al.* (1991) developed MiteSim, a model intended to study processes leading to outbreaks of the Banks grass mite (*Oligonychus pratensis*). Their simulations showed that under warm and dry conditions the pest raised very large populations that could not be controlled by *N. fallacis*, but that during cool and humid seasons the predator eliminated the pest. MiteSim thus demonstrated the predator's efficacy, emphasising the significance of prevailing climatic conditions for achieving pest control. A simulation model based on functional responses enabled Clements & Harmsen (1992) to show that in Ontario apple orchards *Z. mali* controlled ERM when it occurred at low densities, whereas the phytoseiid *Typhlodromus caudiglans* was the more efficient predator when the pest was abundant.

Sabelis & van Rijn (1997) reviewed models of phytoseiid relationships with spider mites, with a view to understanding these interactions for better pest control.

Attributes of an efficient ABA

Having demonstrated the efficacy of various predatory and parasitic mites, it is time to tackle the oft-asked question (albeit about other natural enemies): what are the attributes of an efficient ABA and how can they be recognised? Formerly the authors responded in detail to this all-important question (Gerson & Smiley, 1990), but with more data at hand, it has become difficult to provide unequivocal answers.

In Chapter 39, the search for ABAs attributes that are desired in effective biocontrol agents was discussed. To recapitulate, the major traits were high specificity and searching capacity, good dispersal, strong voracity, a high power of increase, adaptability to the target pest environment, and survival at prey scarcity. How do ABAs shape up in this context? The answer is that predatory and parasitic mites are too diverse in their habits and in their pest-suppression modes to fit comfortably into any of the above moulds.

The variability within the Phytoseiidae illustrates this statement. Apparent discrepancies between the life history traits of some phytoseiids and their ability to control pests were resolved by placing them into four groups (McMurtry & Croft, 1997). Type I includes *Phytoseiulus* spp., specialised to feed on heavily webbing spider mites; types II and III are progressively less fastidious predators and type IV consists of

Euseius spp., generalists that develop and reproduce best on pollen. However, all four groups include members that are efficient ABAs (Chapter 26). Type I species show all traits attributed by Rosen & Huffaker (1983) and McMurtry (1982) to effective ABAs (except for survival at prey rareness). However, many type III and IV phytoseiids that lack most of these traits also control pests. They include *T. pyri*, which regulates ERM in many parts of the world, *Typhlodromalus aripo*, which controls the cassava green mite, and successful predators of thrips.

Some ABAs, such as type I and II phytoseiids (and eriophyids), are highly specific. Others, however, do not show high specificity to their prey; examples include not only many type III and IV phytoseiids, but also Anystidae, Bdellidae, Cheyletidae, Erythraeidae, Laelapidae and Macrochelidae. Their success in controlling pests is probably due to their preferences for particular prey and to their use in specific situations against specific pests, a point that is continued in Chapter 47.

A high reproductive potential, the ability to increase at least as rapidly as the prey does, is also evident in type I and II phytoseiids, but its real value as a general characteristic of ABAs is uncertain. Because the intrinsic rate of increase (r_m) is intended to represent the maximum rate of population increase, it is usually determined in the laboratory with unlimited prey (so that lack of food does not become a limiting factor) and without competitors and natural enemies. In the field, however, fecundity is affected by the available food. Examples (see Table 21.1 and Fig. 26.7) demonstrate this point. The reproductive potential of *T. pyri* and of *T. aripo* is lower than that of their prey. Anystide and Trombidiidae usually have only one or two annual generations (far fewer than their prey) and still provide satisfactory control.

ABAs are often associated with their target pests in the field, thus being preadapted to the new environments that had been invaded by the pests; the ability to occupy all pest habitats could be a problem only when trying 'new associations' (Chapter 39). Climatically adapted, suitable ABAs may be located by using the CLIMEX programme (Chapter 39).

Searching capacity (often through infochemicals) remains the single most important attribute of an effective ABA, but even that may not be general. Some efficient ABAs ambush their victims, or locate them at random (Chapters 32 and 35), coming upon their prey when living in the same habitat. In the latter category could be Australian species of *Neoseiulus*, which appear to be specific to particular host plants (Beard & Walter, 2001) and thus preadapted to prey that lives there.

Should the efficient ABA be large or small? The size of ABAs in comparison to their prey often determines success in pest control. *Neoseiulus barkeri* killed more small first instar thrips nymphs than the larger second instar individuals. The latter had a stronger, more frequent wagging reaction, which hindered the predators' grip (Bakker & Sabelis, 1989). The erythraeid *Lasioerythraeus johnstoni* attacks the tarnished plant bug (Chapter 16), provided the prey is at most four times the size of the mite. Childers & Rock (1981) noted that erythraeids fed only on small aphids, as bigger hosts twitched vigorously and dislodged the mites. The ability of ABAs to feed on maggots of filth flies in dung and manure depends on prey size; the Macrochelidae and Uropodidae (Chapter 37) consume only first instar larvae, but cannot subdue the larger second instar maggots. Some ABAs (e.g. *Anystis*, Camerobiidae, Cheyletidae,

Piona, Pyemotidae) overcome the limitation of size by using toxins to overcome larger prey or hosts. At the other end of the scale, small ABAs can forage within buds, crevices, galls and leaf folds, wherein some pests cannot be reached by larger predators.

In addition, certain ABAs reduce pest (or disease) injury in unusual ways. A ceratozetid reduced numbers of a forest pest by feeding on the woolly filaments of its ovisacs, causing the eggs to drop off branches and the emerging juveniles to perish (Chapter 11). A tetranychid diminished damage due to another spider mite through plant 'vaccination' (Chapter 34), and a tydeid reduced grape powdery mildew by feeding on its mycelia. These beneficial activities of ABAs were unpredictable and efforts to locate them by applying the 'desired attributes' paradigm would have met with little success. More difficulties arise when trying to formulate the qualities desired of ABAs for nematode or weed control. Jatala (1986) listed the attributes of successful biocontrol agents of nematodes. Aside from a high reproductive rate, strong dispersal abilities and rejecting plants and other animals, the traits included persistence in the soil, rapid germination of infective propagules, the ability to produce antibiotics, compatibility with agricultural chemicals and genetic stability. Non-specificity was listed among the important attributes of these natural enemies. Qualities required from natural enemies for weed control comprise other traits, such as prevention of seed production by weeds that reproduce only by seeds, attacking during the entire growth and/or reproductive season, vectoring virulent pathogens or having a dry weight of over 50 mg (Harris, 1973; Goeden, 1983).

These very different requirements serve to underline the contention that there cannot be one, global, list of desired attributes for all ABAs. The unique features of each group of target organisms lead to different lists of attributes that are required for their control. Therefore, in conclusion, it is conceded that, where no experimental field data are available, one can neither know nor predict whether a given mite will be an efficient ABA.

Chapter 45
Environmental constraints in the use of acarine biocontrol agents

Introduction

Biological control, often seen as the most successful and promising alternative to chemical control (DeBach & Rosen, 1991), has been under environmental onslaught since the 1980s. Conflicts of interest first arose in regard to biological weed control (Bennett, 1985) and were reinforced by data showing the adverse effects of introduced natural enemies on indigenous, non-target organisms (summarised by Hopper, 2001; Lynch & Thomas, 2000). The integrity of the endemic biodiversity due to diluting the indigenous biota by exotic enemies that are introduced for biocontrol is a more recent concern (Emberson, 2000). Criticisms of classical biological control (Howarth, 1991; Simberloff & Stiling, 1996) have generated debates on the risks and the non-target effects of this practice. Gould *et al.* (1996) led a student debate on these issues, and the topic was discussed in several reviews (e.g. Thomas & Willis, 1998; Ewel *et al.*, 1999; Lynch & Thomas, 2000), as well as in two multi-authored books (Follett & Duan, 2000; Wajnberg *et al.*, 2001). Concerns about environmental safety were expressed even about exotic natural enemies of cosmopolitan pests in artificial habitats such as food stores; as a result, their candidate enemies must now be screened for possible impacts on native organisms, according to varying national nature protection rules (Schöller *et al.*, 1997).

Environmental objections are not the only constraints on biological control; others were discussed in Chapter 1. The above-noted sources (and references therein) can provide the reader with arguments presented on either side of the debate; only a few will be repeated, mostly as they pertain to acarine biocontrol agents (ABAs).

Risk assessment

When an invading organism is known to be or becomes a pest, and chemical or other means of containment cannot be used, a biocontrol decision is often made. The options are: (1) to introduce exotic natural enemies (given that they are available), and hopefully control the pest, or (2) not to introduce the natural enemy and face the damage incurred. Adopting the first option may leads to environmental concerns, whereas choosing the second option may lead to economic costs. Central to the environmental concerns is the need to assess the potential risks (direct and indirect) of the introduced natural enemies, and pivotal to that question is the availability of data on the impact of the exotic species on its new environment. For some natural enemies such data could be at hand, but more often they are scarce or even lacking, requiring practitioners to use any available information to assess potential risks.

The assessment of risk (which may be defined as the chances of incurring undesirable results and their consequences) of introducing biocontrol agents was applied to general issues of biological control (Bourchier & McCarty, 1995; van Driesche & Hoddle, 1997), with emphasis on weeds (Londsdale *et al.*, 2001). Risk assessment is part of risk management, which along with policy, leads to effective administration. Policy objectives should be defined before initiating the procedure of risk assessment. Risk (in the present context) is the prospect that environmental damage will be incurred (directly or indirectly; see below). The introduction of exotics can be seen as planned invasions, the choice of the invader being based on the best predictions about the outcome. Such predictions are usually based on available data (discussed in Chapters 38 and 39) and on educated guesses. The data include climate matching (e.g. the CLIMEX system, Chapter 39), the stability of the association between invader and host/prey (whether recent or long-term, e.g. lasting for several million years; McClay, 1996) and the exotics' specificity to the target pest (e.g. host range testing; van Driesche & Hoddle, 1997). Such data, however, are hard to interpret, as it is seldom possible to determine all potential hosts or prey (Hopper, 1995), first, because economic species are more often examined than others, then because target pests may have strains or populations that are not attacked by the exotics (Chapter 15), and finally because the records are not predictive. Such tenuous reliability suggests that field data from the collection site(s), the habitats of the natural enemy, are to be preferred. Despite these drawbacks, in most cases the prediction methodology appears to have been sound, especially in regard to more recent introductions (Charles, 1998; Lynch & Thomas, 2000; Londsdale *et al.*, 2001).

Direct risks to the environment are adverse effects on the indigenous flora and fauna, some of which can be predicted (e.g. by gauging the specificity of the candidate agent) and thus avoided. Indirect impacts are 'cascading' effects, such as competition with native species that may lead to their dilution or displacement, or to secondary pest outbreaks, events that are unforeseeable. As noted in Chapter 39, postcolonisation changes in host (or prey) preferences, although rare, have been recorded but cannot be foreseen

Successful invaders are often R-selected organisms, fecund generalists with short life cycles and good dispersal abilities, which are capable of surviving in diverse habitats (Enserink, 1999). Regarding exotic ABAs as invaders, their known biological traits could provide some predictions about their potential danger to new environments. Of the few ABAs that were intentionally transported across international borders, about half are restricted in their feeding (Marked by + in Table 45.1). Other ABAs, although generalists, are restrained by climate or vectors. *Anystis wallacei*, for instance, is restricted to Mediterranean coastal areas (Chapter 6) and *Macrocheles peregrinus* requires dung beetles for dispersal. Others, such as *Geolaelaps aculeifer, Neoseiulus cucumeris* and *Pyemotes tritici*, have broad prey (or host) ranges. Such data can form the basis for predictions about the risks of introduced predatory and parasitic mites.

ABAs may interfere with each other or with other natural enemies, at times detracting from the overall pest control that could be achieved (Chapter 42).

Table 45.1 Representative acarine biocontrol agents introduced across international borders for pest or weed control, based on data in the relevant family chapters

Family	ABA	Target pest or weed
Anystidae	*Anystis wallacei*	Red-legged earth mite
Bdellidae	*Neomolgus capillatus*	Lucerne flea
Eriophyidae	*Aculus hyperici*[†]	St John's wort*
Galumnidae	*Orthogalumna terebrantis*[†]	Waterhyacinth*
Hemisarcoptidae	*Hemisarcoptes coccophagus*[†]	Armoured scale insects
Laelapidae	*Geolaelaps aculeifer*	Bulb mite
Macrochelidae	*Macrocheles peregrinus*	Filth flies
Phytoseiidae	*Neoseiulus cucumeris* *Phytoseiulus persimilis*[†] *Typhlodromus pyri*	Spider mites and thrips Spider mites European red mite
Podapolipidae	*Coccipolipus epilachnae*[†]	Mexican bean beetle
Pyemotidae	*Pyemotes* sp.	Cotton boll weevil
Tetranychidae	*Tetranychus lintearius*[†]	Gorse*

* Weed.
[†] Host- or family-specific.

Direct environmental risks of ABAs to the indigenous flora and fauna

Direct risks (e.g. physical interactions) due to ABAs may be caused by their attacking indigenous non-target organisms (non-pests or other natural enemies) and becoming pests themselves. An example of the first was the introduction of *Aculus hyperici* into Australia to control St John's wort. The mite reproduced on a native relative, but its effects were believed to be milder than on the weed. Experiments and surveys indicated that it would have only a minor effect on the indigenous plant (Chapter 15).

Many of the data about environmental risks of ABAs to the indigenous biota pertain to the Phytoseiidae, as its members have most often been introduced into new habitats. Australian hesitation in granting import permits for exotic phytoseiids was due in part to concerns that these predators might affect indigenous rainforest trees by feeding on their pollen (Westcott, 1993). The release of exotic, pesticide-resistant phytoseiids was believed to have led to the disappearance of native species, whether by displacement or through direct predation. Three phytoseiids (*Typhlodromus pyri*, *Galendromus occidentalis* and *Neoseiulus fallacis*) introduced into Australia became established and appeared to displace native species. They were rarely found outside crop systems and required pesticides for their maintenance there. When low pesticide regimens were adopted in these systems, the indigenous species returned to displace the introduced ones (Chapters 26 and 43; James, 2001). The potential risk of the exotic *Phytoseiulus persimilis*, which occurs on many introduced commercial plants and weeds, establishing on the indigenous Australian flora was tested experimentally in Queensland. Walter *et al.* (1998) placed spider mite-infested bean plants ('trap plants') in fields and rainforests and collected all associated predators, as well as checking Australian records for plants on which *P. persimilis* had been collected. It was found on the plants placed in the field, very rarely in the rainforest, only once on an indigenous plant growing along a road, and never in any native habitats. By extension, the chances of exotic phytoseiids becoming established on the native flora (and affecting the local fauna) appear to be remote (although exotic predators could feed on indigenous arthropods that had colonised introduced plants).

Direct risk incurred by an ABA through becoming a pest was feared when *Tetranychus lintearius* was considered for introduction into New Zealand to control gorse (*Ulex*) (Chapter 34). That spider mite is closely related to the two-spotted spider mite (TSSM), *Tetranychus urticae*, and could become a problem, and/or the two species might interbreed to form a more dangerous pest. Such fears were allayed by showing that *T. lintearius* would neither form permanent colonies on any plants except *Ulex* spp., nor breed with *T. urticae* (Hill & O'Donnell, 1991a; see Chapter 34). Some generalist phytoseiids and stigmaeids feed on plant tissue (Chapters 26 and 32), but none appears to cause any damage.

Indirect environmental risks of ABAs

Indirect risks may be incurred without actual physical interaction between agent and non-target. Two pests, namely *Halotydeus destructor*, the red-legged earth mite (RLEM), and *Sminthurus viridis*, the lucerne flea, co-occur and compete in Western and South Australian pastures. *Anystis wallacei* was introduced to enhance the biological control of RLEM (Chapter 6), but is not very effective against the lucerne flea. During periods of prey scarcity the exotic predator and the naturalised bdellid *Bdellodes lapidaria*, the major predator of the lucerne flea (Chapter 9), compete for food. The bdellid may then be killed by *A. wallacei*, leading to increases in lucerne flea numbers (Michael *et al.*, 1991).

Environmental risks of genetically modified ABAs

Mackauer (1976) expressed concern that owing to genetic drift and to the take-up of deleterious traits, natural enemy selection under artificial laboratory conditions might result in strains that would perform badly in the field. In rebuttal Hoy (1992) noted that there was no evidence that pesticide-resistant, transgenic phytoseiids have strayed beyond the agricultural systems into which they had been released, that none was known to be a pest, and that they are not considered to be risky by the US Department of Agriculture.

Hoy (1992, 2000) further addressed risk assessments that should be undertaken before releasing transgenic ('engineered') phytoseiids, with special reference to *G. occidentalis* (Chapter 26). Risks engendered were attributed to various components that required different degrees of scrutiny. The components were the wild-type organism, the transferred genetic material, the wild-type compared with the engineered organism, and the environment into which the latter was to be released. The degree of scrutiny needed was qualitative, expressed as being placed along a continuum between 'less' and 'more'. The wild-type organism had two elements that required more scrutiny, which were frequent gene exchange in nature and self-propagation. It also had three that suggested less scrutiny (the wild-type could be controlled, it was a free-living species and its relatives were not pests), and three elements, namely survival, geographical range and being indigenous, that were intermediate. The transferred genetic material needed less scrutiny because: (1) it was fully characterised (only a single gene was added); (2) the source of the gene was an unrelated species; (3) insertion of exogenic material was by microinjection (e.g. without a vector); and (4) it was regarded as stable. The other attributes (the engineered vs the wild-type organism and the target environment) also had elements that required some scrutiny, or were intermediate.

One difficulty with this assessment was that no rankings of risk were associated with the degrees of the required scrutiny. In a differently constructed risk assessment exercise, the fact that the mite was not a pest indicated minimal concern. Fears that the transgene would interbreed with a local population might engender some worries over safety, which could be allayed by the fact that the engineered trait, resistance to organophosphate (OP) pesticides, is common in nature. The wind-borne dispersal of the mite also caused some concern. Taking all of the information into account Hoy (2000) suggested that the risk of releasing the transgenic *G. occidentalis* was low. Notwithstanding these results, assessments should be made separately for each engineered phytoseiid, in accordance with its biological traits, the mode of transformation and the release site.

After *G. occidentalis* (a dry weather species) had been 'engineered' in the laboratory, a preliminary step in gauging potential risks was to assess its ability to survive the humid Florida summer. Normal mites were released on the campus of the University of Florida but could not survive the summer (McDermott & Hoy, 1997). Two field releases, undertaken after consultations with various biosafety and regulatory authorities and with their approval, were made in an experimental plot on the same campus. The site was fenced and contained trap plants, and sticky boards were placed around the plot to monitor the predator's dispersal. Access to the site was limited and laboratory clothing that was worn there was sterilised. The first field experiment, in which mites from a single line were used, was terminated after a few weeks (a period in which the mite could raise three generations), mostly because of unsuitable weather. Only a few of the surviving predators contained the transgene (the relevant marker used for its recognition was described in Chapter 43), indicating its instability in the field. Similar results (no transgene survival) were obtained from the second release, in which mites from six lines were assayed. These trials showed that if all contingencies are properly evaluated and prudently acted upon, the risks of releasing transformed phytoseiids into the environment would be less than feared. No specific guidelines for the containment of transgenic arthropods in the laboratory or for their release in the field are yet available. Various policy issues must be resolved before these arthropods are deployed in pest management programmes.

Based on her experience, Hoy (2000) argued that transgenic natural enemies would most readily be used in relatively small areas, such as temporary cropping systems or individual orchards, as well as in geographical regions where the wild strain does not occur. She also advocated that vigorous precautions be taken when rearing transgenic arthropods. These safety measures are to include an insect-proof and self-contained laboratory with filters to avoid accidental escapes, facilities to destroy the transgenic material, suitably trained personnel and minimal movement of equipment. Public concern about transgenic arthropods requires that the highest standards be observed until all issues about transgene biosafety are resolved.

Concluding remarks

By invoking a few cases of damage to indigenous, non-target biota, Howarth (1991), Simberloff & Stiling (1996) and others advocated much stricter prerelease studies of candidate biocontrol organisms (to paraphrase the latter authors, an introduction will be guilty until proven innocent). The lack of monitoring of non-target organisms leaves any deleterious effects of natural enemies, and especially generalists, unstudied and thus unknown. Reductions of endemic floras and faunas were the fears most often expressed, a danger with an applied aspect. Organisms that control minor local pests may be among the endangered biota and their destruction would lead to pest outbreaks. In addition, it was argued that the effect of pesticides is limited in time, but biocontrol agents, once established, might remain 'forever'. It is in pristine areas that environmental fears are greatest, because once the 'wrong' generalist becomes established it would be impossible to remove and it might be of great danger to intricate food webs that are based on native organisms, which are the 'riskiest of targets' (Strong & Pemberton, 2001).

How real are such dangers? Lynch & Thomas (2000) devised a quantitative categorisation of non-target effects that could be attributed to classical and inundative biological control. Their 'severity index' ran from level 0 (no evidence of adverse effects) through 1 (<5% mortality, no population consequences) and 5 (>40% long-term suppression of a local non-target population), culminating at level 9, certified extinctions in an area of 100×100 km or more. Data indicating non-target effects were found in 1.7% of the 5279 recorded cases of tallied classical introductions, with a somewhat higher proportion in inundative biocontrol. Safety with the latter mode of control was justified by the transience of the effects, along with the apparent lack of natural enemy persistence. Thus, while deleterious effects are known, their overall frequency is low. Although ABAs were not separately discussed, there is no indication that overall conclusions about their safety would be different.

In his 'historical guide to the future' on the control of fruit tree pests in New Zealand, Charles (1998) thought that risks posed by classical biological control cannot be viewed in isolation from the continuous accidental arrival of exotic pests and natural enemies (including spider mites and their phytoseiid predators). An informed pest control programme is better than doing nothing, hoping for the lucky arrival of suitable natural enemies. Emberson (2000), discussing the risk of 'diluting' native arthropod faunas by introducing natural enemies of invasive weeds, concluded that dense infestations of such weeds would probably have a worse effect on indigenous insect biodiversity. Nor should the socioeconomic dimension be ignored. Neuenschwander & Markham (2001), reviewing the effect of ABAs on the cassava green mite in Africa, wondered whether that hugely successful project (Chapter 26) would have been allowed to proceed nowadays, because the best predator (out of around 50 assayed) is not a strict carnivore. Without these exotic predators farmers would have had to meet the lower market value for mite-damaged cassava by additional forest clearing, with a devastating impact on the local biodiversity. The other (albeit more expensive, and thus less likely) option for the farmers would have been to use more chemicals, leading to increases in health hazards and ecological damage.

There is some scope for biocontrol practitioners to improve risk assessment procedures of planned introductions (e.g. van Driesche & Hoddle, 1997). However, they must strongly reject the argument that owing to the impossibility of predicting how 'exotics' would behave in a new environment, no natural enemy should ever be introduced. In this context the scientific challenge would be to understand the ecological factors underlying successful biological control projects and to improve the predictability of the new food webs that may arise when an exotic establishes in an ecosystem. However, regulatory frameworks should not become so prescriptive as to negate or impede all biocontrol projects. If this form of control ceases to be an option, it would be replaced by pest control alternatives that are far more destructive. Imposing a set of 'rich country' environmental values on malnourished people would lead to rather different impacts on biodiversity than those intended.

Notwithstanding the above, the controversy between 'classical' biocontrol practitioners and environmentalists seems to be moving towards a consensus. This is seen in a document issued by the Convention on Biological Diversity (Glowka *et al.*, 1994), which makes biological control a priority while emphasising the obligation of all countries to protect their indigenous biodiversity. Ewel *et al.* (1999) formulated a series of research needs that should be implemented in order to monitor deliberate introductions of all organisms (including the development of specificity tests for biocontrol agents). Strong & Pemberton (2001), who advocated reform in biological control, argued that not every invasive species has to be a target for biocontrol and not every available natural enemy promises relief. Such 'reform' should include more studies on indigenous pests (including weeds) and their endemic relatives, and the suggestion that exotic natural enemies of indigenous pests are to be introduced only under extreme circumstances. Vigorous prerelease as well as postrelease investigations should be undertaken, especially in regard to specificity. Finally, more public transparency and debate are needed.

Guidelines, procedures and codes of conduct for introducing exotic natural enemies are available (Greathead, 1997; Knutson & Coulson, 1997 and others), and should strictly be adhered to. This chapter concludes by citing Strong & Pemberton (2001) that without a 'reform' in biocontrol practices, the support of the public and the regulatory bodies for this mode of pest control, still the best hope for combating some of the most damaging insect and weed pests, might not be maintained.

Chapter 46
Commercial and economic aspects in the use of acarine biocontrol agents

Definition

The term 'biopesticide' (an abbreviation of 'biological pesticides') is applied to natural (biological), living or non-living entities offered in commerce to provide non-chemical solutions to pest problems, and indicating a means of control analogous to 'pesticides'. Since different meanings have been bestowed on this term by various authors, 'biopesticides' will first be defined. Waage (1997) characterised biopesticides as 'any mass produced and marketed natural enemies, including predators, parasitoids, nematodes and microbial agents'. Copping (1998) and Copping & Menn (2000) applied a very broad definition of biopesticides. They included natural products (plant-derived pesticides, or 'botanicals'), pheromones, living organisms (viruses, bacteria, protozoa, fungi, nematodes, insects and mites), as well as genes used to transform crops to express resistance to insect, fungal and viral attack or to render them tolerant to herbicides. Others (Menn & Hall, 1999) considered only viruses, bacteria, nematodes, plant-derived pesticides and insect pheromones in this category, and Hokkanen & Menzler-Hokkanen (2000) restricted the term to viruses, bacteria, fungi and nematodes. A resolution of sorts was adopted by Newton *et al.* (1996), who separated biopesticides into 'macro-biologicals', living and self-renewing natural enemies, and 'microbial pesticides', which are formulated and applied like chemicals. However, as *Phytoseiulus persimilis* (included in the 'macro-biologicals') could be used as a once-only, non-renewable natural enemy, the semantic ambiguity persists. It was compounded by the US Environmental Protection Agency (EPA), which officially defined biopesticides as belonging to one of three groups (Whalon & Norris, 1999): biochemical pesticides [e.g. pheromones, insect growth regulators (IGRs)], microbial pesticides [fungi, *Bacillus thuringiensis* (Bt) products] and transgenic plant pesticides (plants engineered to express insect toxins, especially Bt). Natural enemies were not included in this definition.

Another tier of perplexity revolves about the use of these organisms. Crump *et al.* (1999), who limited their discussion to biological weed control, defined 'biopesticides' as living organisms that are applied in an inundative mode which results in target death. However, even when restricted to weed control, the eriophyid *Aceria malherbae*, a biopesticide sold to control field bindweed, does not destroy its host (i.e. it does not cause 'target death'). This ongoing ambiguity has led to ignoring the use of potentially very useful biopesticides (Waage, 1996). Herein, acarine biopesticides are defined as acarine biocontrol agents (ABAs) that are mass-reared and offered in commerce to control pests and/or their level of damage. Commercial ABA production will be discussed first, then relevant commercial aspects, and finally the future and constraints.

Trade in ABAs

The trade in ABAs expanded greatly during the 1990s, a notable development when seen against the background of the entire biopesticide industry, which seems to have 'stumbled, badly' (Gaugler, 1997) and of increasing regulation. Using as one yardstick the directories of the least toxic pest control products (Anonymous, 1991, 1993, 1997, 2000; Zdárková, 1991), only 30 suppliers offered a total of 13 ABAs (in three families) in 1991, a value that rose to over 50 companies selling 18 species (in six families) by 1993. Towards the year 2001 the number of companies (those that produce and sell as well as those that proffer only repackaged ABAs) remained about the same, but they were selling 28 species, in seven families (Table 46.1). Pride of place was taken by *P. persimilis*, offered by 50 companies (52 in 1993), and second place by *Neoseiulus cucumeris*, probably because it is being advertised to control spider mites as well as thrips. At the other end of the scale, about half of the ABAs listed in Table 46.1 was offered only by a single company, suggesting either very narrow 'niche' markets or problems in production. ABAs are mostly sold for use on high-value crops, on which chemicals are not applied because of market demands (the 'biomarkets'), strict regulation or pest resistance.

Although biopesticides account for only about 1% of all pesticide sales (Newton *et al.*, 1996; Anonymous, 1999; Copping & Menn, 2000), and that percentile is dominated by Bt products, the demand for ABAs seems to entice both large and small companies (which also sell other natural enemies). The manufacturers of biocontrol agents have organised themselves in the International Biocontrol Manufacturers Association (IBMA), which co-ordinates regional as well as regulatory committees (Anonymous, 1999). IBMA is in contact with various governments, especially in the European Union (EU), and promotes the implementation of biological control, including that with ABAs. A noteworthy regional group is the Australian Association of Beneficial Arthropod Producers (ABC), which published a list of beneficials available in that region, with user-friendly instructions (Broadley & Thomas, 1995).

Producers and suppliers 'create' eye-catching trade names for their products. For instance, *P. persimilis* is sold by Novartis BCM as 'Phyto-line p' and by Koppert as 'Spidex' and 'Spidex Plus'. *Neoseiulus cucumeris* is advertised by Novartis BCM as 'Ambly-line cu 25000', and by Koppert as 'Thripex-C' and 'Thripex Plus' (Copping, 1998).

Economics and pricing

Biocontrol research and application appear to pay off, despite problems with mass-production and distribution (see below). The benefit/cost ratio for the first 10 years of several classical biological control projects was estimated by Gutierrez *et al.*

Table 46.1   Acarine biocontrol agents offered in international trade

Family	Species	Number of suppliers	Target pests or weeds
Cheyletidae	*Cheyletus eruditus*	1	Grain mite
Cunaxidae	*Cunaxoides* spp.	1	Scale insects
Eriophyidae	*Aceria malherbae*	1	Bindweed
Laelapidae	*Hypoaspis aculeifer*	1	Bulb mite
		15	Fungus gnats
	Hypoaspis miles	5	House flies
		9	Thrips
	Geolaelaps spp.	2	Fungus gnats
Macrochelidae	*Glyptholaspis confusa*	1	House flies
	Macrocheles spp.	1	House flies
Phytoseiidae	*Amblyseius barkeri*	6	Thrips
	Amblyseius californicus	23	Spider mites
	Amblyseius cucumeris	29	Thrips
		20	Spider mites
	Amblyseius degenerans	10	Thrips
		3	Spider mites
	Amblyseius fallacis	6	Spider mites
	Euseius rubini	1	Whiteflies
	Euseius sp.	1	Scale insects, Whiteflies
	Galendromus annectens	3	Spider mites
	Galendromus helveolus	5	Spider mites
	Galendromus occidentalis	22	Spider mites
	Mesoseiulus longipes	11	Spider mites
	Neoseiulus scyphus	1	Spider mites
	Neoseiulus setulus	1	Cyclamen mite
	Phytoseiulus macropilis	3	Spider mites
	Phytoseiulus persimilis	49	Spider mites
	Typhlodromus athiasae	1	Spider mites
	Typhlodromus mcgregori	1	Spider mites
	Typhlodromus pyri	1	Spider mites
	Typhlodromus rickeri	1	Spider mites
Pyemotidae	*Pyemotes tritici*	3	Ants

Based on The *IBM Practitioner* (Anonymous, 2000) and Zdárková (1991).
Mite names are as originally listed; for current names see relevant family chapters.

(1999a) to be high (up to >US$12,000 in a few cases), not taking into account unmeasured values due to lessening health hazards and reduced environment contamination. It is relatively simple to assign costs in biocontrol projects, because expenditures can be measured and they are incurred in a concentrated manner (within a given time frame). Benefits (including health gains), however, occur over a prolonged period, can be manifested over large areas, are spread out among many beneficiaries and are difficult to quantify. Little is known about the benefits of using ABAs, as few relevant benefit/cost analyses have been undertaken. Comparisons of projects conducted in the field with those done under glass (where releases may be repetitive) are especially needed. Public (and institutional) uncertainty about biocontrol might be mitigated if costs could be evaluated and benefits predicted (McEvoy, 1996).

Headley & Hoy (1987) calculated that implementing an integrated mite management programme on almonds in California, with the pesticide-resistant *Galendromus occidentalis*, could save growers US$60–110/ha per year. On the expense side this included salaries, supplies, travel, equipment, cost of mites and monitoring (with interest compounded at 12%), whereas the main savings were in acaricide costs. The benefit/cost ratios for various control strategies were from 14/1 to 34/1. Hoy (2000) later estimated that the annual economic benefits from this project were greater than US$20 million. Decou (1994) calculated that releasing *P. persimilis* in strawberry fields (against spider mites) in Florida cost US$250–375/ha, whereas acaricide cost came to US$925/ha, even without other expenses. Simberloff & Stiling (1996) criticised such 'one-sided' cost/benefit analyses, because loss of indigenous taxa, ecosystem disruption, and so on, are usually not considered (Chapter 45).

At times there are no economic benefits to using biopesticides. Oatman *et al.* (1976) released large numbers (125,000–250,000/0.4 ha) of *P. persimilis* in strawberry beds infested by spider mites. Although the predators reduced pest numbers (see Fig. 44.2), there were no differences in mean fruit yield or size between release and control plots.

The increasing cost of developing chemical pesticides, their relatively short shelf-life owing to increasing resistance, and environmental concerns, have turned the attention of some firms (e.g. Ciba; see Newton *et al.*, 1996) to biopesticides. Another major incentive towards this business is the need for corporations to enhance their environmentally friendly images; a chemical company that can boast of its 'green' semblance may score public-relations points. Management decisions to enter the biopesticide business depend mostly on the size and extent of the perceived demand, on reasonably assured consumer satisfaction and on low production costs. The growing reluctance of consumers to buy pesticide-contaminated foods (with their concomitant readiness to pay more for 'organics') has increased demand. On the side of the manufacturers, the damaging presence of the two-spotted spider mite (TSSM) and thrips on high-input crops such as glass-covered floriculture and vegetables during all seasons ensures a continuous demand for ABAs throughout the year. The very encouraging results obtained with certain pesticide-resistant ABAs (e.g. *P. persimilis*) have reduced fears of disappointing results, and the ease of rearing that species and others (such as *N. cucumeris*) has concurrently decreased costs.

ABAs are sold in batches of several thousand: *N. cucumeris* in units of 5000–50,000, most others in units of 1000, and a few, such as *Iphiseius degenerans* (labelled *Amblyseius* in Tables 46.1 and 46.2), in even smaller groups (Cranshaw *et al.*, 1996; van Lenteren *et al.*, 1997). Prices range from US$0.00097 to 0.1013 per individual predator (Table 46.2), but some suppliers are willing to give large discounts when predators are bought in bulk. Differences between US and European suppliers (US$0.01924 for a single *P. persimilis* in the USA vs US$0.1013 in Europe) probably reflect production costs. The entrance of small producers ('cottage industries') into

Table 46.2   Pricing of single acarine biocontrol agents ABAs in the USA and in Europe

ABA	Commercial unit	Average price/mite (US$)	
		USA*	Europe[†]
Amblyseius cucumeris	Mixed life stages	0.00190	0.0094
Amblyseius degenerans	Mixed life stages	Not listed	0.100
Amblyseius fallacis	Mixed life stages	0.0095	Not listed
Galendromus occidentalis	Mixed life stages	0.0244	Not listed
Hypoaspis miles	Mixed life stages	Not listed	0.00097
Mesoseiulus longipes	Mixed life stages	0.02435	Not listed
Neoseiulus californicus	Mixed life stages	0.02186	0.031
Phytoseiulus persimilis	Mixed life stages	0.01924	0.1013

*Based on Cranshaw *et al.* (1996).
[†]Based on van Lenteren *et al.* (1997).
Mite names are as originally listed; for current names see relevant family chapters.

the market is encouraged by freely available knowledge about rearing methods, by specialising in a few species, by the option of reducing costs by using family members and by ABAs' being in the public domain (i.e. they cannot be patented). Large companies can rapidly change production schedules according to fluctuating demand, using their staff and equipment more efficiently than small producers. However, as the former maintain more staff they must earn more to return investments. Innovative rearing, harvesting, preserving and packaging methods might be patented or at least kept trade secrets, enabling some producers to reduce costs and prices. The quoted prices probably reflect these factors, as well as changing demand. Improved production of natural enemies enabled Australian producers to raise prices by only 14% in the 10 years prior to 1991 (Llewellyn, 1991).

Constraints and future

A central point made in an overview of the problems facing the biopesticide industry (Anonymous, 1999) was that the development of its markets cannot be similar to those of the chemical pesticide industry. Waage (1996) even queried whether industry-developed biopesticides were selected with insecticides as models for product development, created in the same image. The main obstacles that hinder growth of the ABA biopesticide industry include non-competitive production costs, an uneven distribution system, high specificity, limited persistence, inconsistent field results and flawed public (farmer and consumer) awareness. Whatever the reasons for the small market share, biopesticide products (including ABAs) are not realising their mission of reducing the use of chemical pesticides (Gaugler, 1997).

Wada (1998), expressing the views of a large corporation entering the biopesticide business, noted that natural enemies must be produced in competitive cost frames in

order to be accepted by growers. The lower the price of a biopesticide and its particular application costs, the more readily will it be bought by farmers. The need for
reducing production costs of natural enemies by several orders of magnitude to solve
pest problems in major crops was noted by Newton *et al.* (1996).

A major component lacking in the biopesticide industry is a dedicated distribution
system, incorporating well-trained staff who would be able to provide all complementary services to their customers (Anonymous, 1999). The application of one
chemical is similar to that of another, but the use of a new ABA requires new expertise. The delivery and demonstration of integrated pest management (IPM) packages
to farmers, the end ABA consumers, accompanied by full instruction and technical
support, could go a long way towards reassuring farmers (e.g. Newton *et al.*, 1996).

Farmers might thus be reassured that it would be to their advantage to use biopesticides, although these products could not guarantee a totally clean crop nor complete pest removal. Furthermore, if farmers were taught more about specific
implementation issues they could participate more actively in control procedures
(Whalon & Norris, 1999). Differences from pesticides in the modes of action and the
time span required should be explained. One source of confusion comes from adver-
tisements that promote biopesticides as 'green pesticides', emphasising their safety
while de-emphasising their increasing role in overall control (Waage, 1996). This
leads to ignoring the fact that IPM programmes are often based on biopesticides,
whose pace of development is bridled to the pace of IPM implementation (Newton
et al., 1996). Although increasing numbers of consumers are willing to pay more for
'green' produce, they should also be educated to prefer slightly blemished but pesticide-free products over those that are clean and shiny but contaminated.

High specificity, although limiting ABA use to only a few targets (or a single target), which might be seen as an apparent drawback, could become an advantage
because the specificity would reduce danger to other, non-target arthropods. Low or
no field persistence may especially be important when releasing exotic ABAs, which
would control the target pest and then disappear. The problems of short shelf-life and
uneven field results are being tackled; the former by cooling procedures to preserve
ABAs (e.g. Zdárková, 1991), the latter by more vigorous experimentation (e.g.
Nihoul, 1993a).

The uncertain future of biopesticides (Copping & Menn, 2000) will hopefully be
resolved by growing consumer preference for 'green' products along with increasing
pressure by environmentalists to reduce chemical usage. Anonymous (1999) estimated that organic farming (which supplies products for the 'biomarket') in Europe
is increasing at an annual rate of 10%, because two-thirds of the EU population are
currently buying 'bioproducts'. Last but not least, many governments encourage the
use of non-chemical plant protection products, whether by direct support or by
appropriate legislation. An example is *N. cucumeris*, recently advertised in Japan (as
Cucumeris®) for the control of pest thrips (Wada, 1998).

Neale (2000) believed that the new and demanding rules that govern the registration and release of biopesticides are major constraints on their future. In regard to
ABAs, such constraints may be non-existent when using indigenous mites, including
the local phytoseiids found to feed on thrips in The Netherlands (Ramakers, 1978),

the cheyletid that controls stored product mites in the Czech Republic (Zdárková, 1991) and many others. As indigenous natural enemies may be inefficient (or 'they would have done the job and there would not have been a problem'), their efficacy should clearly be demonstrated.

Chapter 47
Recommendations for future work

By way of summation, recommendations for future work are presented. Hoy *et al.* (1983a) recommended further exploration for predatory mites and basic studies on their systematics, biology and mass-rearing methods, as well as conducting field experiments with suitable controls, devising better evaluation procedures and developing means to incorporate promising acarine biocontrol agents (ABAs) into existing integrated pest management (IPM) programmes. While agreeing with these proposals, we expand some and add others.

1. The need for the further exploration for ABAs should be re-emphasised. In the Introduction it was stated that only families with one or more members that were known or were postulated to reduce pest populations and/or to decrease their damage were included. Five families of predatory or parasitic mites were added to the roster of ABA families that was included in the authors' former essay, suggesting that more acarine natural enemies of pests can and should be found. This suggestion is bolstered by noting that perhaps only about 10% of the world's mite fauna is known (Halliday *et al.*, 1999).

   In addition, qualitative rather than quantitative observations (hence not included in the family chapters) are noted here, which were published in recent years on the antagonistic effect of mites on pests. Relevant families include the Caligonellidae and Sperchontidae mentioned in Chapter 1, as well as other water mites (Hygrobatidae and Lebertiidae), whose adults fed on blackfly (Simuliidae) and chironomid larvae (Ten Winkel *et al.*, 1989; Mwango *et al.*, 1995). A member of the water mite family Unionicolidae was a voracious predator of small mosquito larvae in the laboratory (Rajendran & Prasad, 1989), but its effect in the field is not known. Parasitengone families are postulated to affect pest populations (Welbourn, 1983). Larvae of the Microtrombidiidae and Johnstonianidae parasitise grasshoppers (Belovsky *et al.*, 2001), Ceratopogonidae (biting midges) (Baker, 1999b) and members of other brachyceran fly families (Metz & Irwin, 2001). Among the Mesostigmata, the Rhodacaridae were often listed as important predators of nematodes (Karg, 1983; see Chapter 38), and deserve further study. Gomaa *et al.* (1989) found that adults of a member of the family Pachylaelapidae consumed several hundred housefly eggs in the laboratory.

2. Failures in introductions of ABAs might have been caused by releasing them in unsuitable amounts, indicating that there is a need to determine minimal numbers that would ensure establishment. Because of the Allee effect, it is also necessary to study the dispersal and mate-finding patterns of these mites.

3. Some failures in biocontrol with mites may be due to prey avoidance of predator-inhabited patches or to anti-predator behaviour (not only due to differences

in size). The two-spotted spider mite avoided leaf discs with *Phytoseiulus persimilis* residues in the laboratory (Grostal & Dicke, 1999), and in some cases the hosts of water mites overcome their parasites (Lanciani, 1988; Forbes *et al.*, 1999). The role of intraguild predation (IGP) during introductions, including that of indigenous natural enemies, should also be studied.

4. Non-specific ABAs are often deemed unsuitable for introductions into many regions, although providing satisfactory pest control in some cases. This success is probably due to their use in specific situations against particular pests and to their ability to survive on diverse diets while still preferring certain prey. There is a need for 'low-tech' life history studies on such mites, including the possibility of selecting them for higher specificity and thus increasing their usefulness. This fits in well with the current eco-political climate that tends both to encourage the use of biopesticides (including ABAs) at the cost of chemical pesticides, and to discourage the introduction of exotic natural enemies. The case for assaying more indigenous ABAs has been argued by Gerson & Vacante (1993), Steiner & Goodwin (2001) and Tuovinen (2000).

5. Some ABAs could affect pest populations by transmitting diseases. Soil acarids (Chapter 4) dispersed spores of entomopathogenic fungi horizontally as well as vertically without being affected (Samšiňáková & Samšiňák, 1970), and the myco-parasite *Coniothyrium minitans* was carried to a minimum distance of 55 mm in the soil (Williams *et al.*, 1998). Macrochelids (Chapter 23) transmitted an insect disease to forestry pests but were also susceptible, their own cadavers adding to inoculum build-up (Schabel, 1982). The blood-sucking pterygosomatids (Chapter 29) may transmit pathogens (e.g. viruses or bacteria) that could affect the host bugs (Newell & Ryckman, 1966). In addition, some mites could vector diseases of weeds and nematodes, topics that need more study. An extension of this recommendation is to apply ABAs together with pathogens that affect pests. Jacobson *et al.* (2001) released *Neoseiulus cucumeris* along with the fungus *Beauveria bassiana* in order to control the western flower thrips.

6. The considerable specificity of infochemicals emitted by commercial plants that are attacked by different spider mites (Chapter 43) raises three questions. First, how have the crop plants, cultivars that had been under domestication for many generations, managed to conserve their array of defensive volatiles? Second, why are ABAs of different origins sensitive to infochemicals with which they might not have come in contact before? Third, the considerable specificity to volatiles shown by the ABAs suggests that they must possess a great variety of receptors, which so far remain unstudied.

For these reasons interactions between ABAs and their prey/hosts should be studied on the plants (wild or cultivated) on which they occur in nature. Neither of these arthropod groups has evolved on agricultural plants and their current interactions (regarding infochemicals) may reflect patterns of behaviour and dispersal that were evolved on other hosts. Alternately, the prolonged consistency of many pest–ABA associations on crops suggests that they may be undergoing incipient coevolution in the artificial agricultural systems. Data on such interactions would be useful in predicting the performance of exotic ABAs.

7. There is a need to develop means of incorporating promising ABAs into existing IPM programmes. This necessitates close co-operation with extension officers.

8. Despite current difficulties in working with transgenic ABAs, studies on these engineered entities could provide new methods of studying their biology and lead to strain improvement and better pest control. Transformation has the advantage over artificial selection in that it allows access to specified genes and allows the student to follow the dispersal, persistence and activities of the engineered organism in a native population (e.g. Hoy, 2000).

9. Although the effects of genetically modified (GM) plants on ABAs could be wide-ranging owing to changes in nutrient contents, the present comments are restricted to insect-resistant GM plants. All GM plants currently in commerce express toxin genes derived from the bacterium *Bacillus thuringiensis*, to which (and to whose products) the sobriquet 'Bt' is commonly applied. In the laboratory such plants decreased or increased the pressure of natural enemies on pests, but it was unclear whether the negative effects were due to changes in the plants, reduced prey quality, or both (Schuler *et al.*, 1999a). Nothing is known about the effects of GM on ABAs and the scarce published data suggest that the topic requires further research. One immediate result could be a decline in prey density, followed by that of the predators. Another outcome could be changes in host quality. Donegan *et al.* (1999) found no differences in the levels of soil micro-arthropods in soil planted to GM alfalfa; there were, however, changes in host quality, such as shoot weight, nutrient content and enzyme activity. As these factors affect pests and ABAs (Chapter 41), their influence on such organisms (and on non-target species that often serve as alternate prey) should be studied. There is no current evidence that GM plants change their attractiveness for natural enemies (Hoy *et al.*, 1998), but Schuler *et al.* (1999b) speculated that GM plants might have different profiles of volatiles, including those used by ABAs to locate prey. A separate task is to incorporate transformed ABAs as well as GM plants into IPM programmes.

10. More studies on the economics of using ABAs (Headley & Hoy, 1987; Whalon & Norris, 1999) need to be conducted. The best way of sustaining public support in biocontrol is by demonstrating its financial benefits (along with being 'environmentally friendly'), which would differ in the field and in the greenhouse. By their very nature biological control projects in the field require more than the usual 3 years to show clear results (Ireson *et al.*, 2001), whereas in glasshouses such demonstrations may require only a few weeks.

11. Yaninek *et al.* (1989) produced a training manual intended for personnel involved in cassava green mite research. The guide included the pest's biological background, described procedures for fieldwork and noted its natural enemies. A unique feature of this manual was that it was printed on paper specially treated to be water and tear resistant. This made the guide highly suitable for consultation under field conditions. It is recommended that similar manuals should be prepared for other systems.

12. A central archive for ABAs should be established in order to document all experiments conducted in this area, including unsuccessful efforts. Information on the

systematics and biology of these mites and their reactions to pesticides should also be stored there. Too many relevant data are often 'submerged' and even lost through ignorance about systematics and the specific interests of acarologists.

Glossary

Acetabula: Rounded sclerites located on the genital plates of water mites.

Acetylcholine esterases (AChEs): A group of enzymes that mediate pulse transfer in the nervous system, and are inhibited by organophosphate (OP) and carbamate (CRB) pesticides.

Alate (plural: alatae): Winged insect (mostly used in regard to aphids).

Allee effect: The increase in the fitness of a species as its numbers increase.

Allelochemicals: Chemicals of no nutritional value that are produced by certain organisms and affect the biology of others. Allelochemicals may be stimulatory at low doses, repellent or inhibitory at higher concentrations.

Allelopathic: The effect (usually negative) of allelochemicals.

Allomone: A volatile chemical emitted by one species that modifies the behaviour of another species towards the emitter.

Allopatric: Two or more populations of the same species that live in different areas (see Sympatric).

Allozymes: Slightly different enzymes that have the same function.

Ambulacrum (plural: ambulacra): The membranous portion of the pretarsus.

Anal suckers: Rounded, paired sclerites located alongside the acarine anus.

Anhydrobiosis: A state of quiescence, induced by dehydration, usually terminated upon the resumption of humid conditions.

Anopheline mosquitoes: Mosquitoes belonging to the subfamily Anophelinae, known to transmit *Plasmodium*, the protozoan that causes malaria, as well as various filarial worms.

Apodemes: Internal sclerotised projections of the exoskeleton, often in the shape of a blade or a ridge, serving to anchor the muscles.

Apomictic: Reproducing asexually.

Apophysis (plural: apophyses): Strong, recurved spines usually located on the palpi or legs of mites.

Apotele: The reduced, distal (seventh) segment of acarine legs and palpi: the leg ambulacra or a tined structure on the palptarsi of the Mesostigmata.

Arolium: A soft lobe between the tarsal claws, which may bear the empodium.

Arrhenotoky (Arrhenotokous) (or haplo-diploidy): A sex-determining system in which females develop only from fertilised eggs and are always diploid (2n chromosomes), whereas the haploid (n chromosomes) males are formed from unfertilised eggs by parthenogenesis (see Thelytoky).

Basifemur: The basic part of the femur, if subdivided (see Telofemur).

Brooming: The formation of foreshortened terminal branches on a plant, with many growing points and forks, resulting in broom-like formations.

Capitate raylets: Hairs that emerge from (or are part of) the empodium and have broader ends.

Capitulum: Gnathosoma, the anterior part of the mite body, consisting of the mouth parts, e.g. chelicerae and palpi.

Carapace: An outer, hard cover over the body of many aquatic arthropods.

Carbamates (CRBs): Insecticides based on esters of carbamic acid. Many are systemic, moving through the host plant and then being ingested by feeding pests. CRBs inhibit the cholinesterase system.

Carotenoids: Reddish pigments of plant origin that may be precursors for vitamin A.

Chagas' disease: Virulent disease of humans (and many mammals) caused by the protozoan *Trypanosoma cruzi* Chagas. Transmitted by Hemiptera (bugs) of the reduviid subfamily Triatominae. Prevalent in Meso- and South America.

Chephalothorax: The fused head and thorax characteristic of the Arachnida (except the Acari).

Chorion: A membrane covering the eggs of arthropods.

Chromatography: A process of separating chemical substances by their movement within a carrier (paper, gas), according to their molecular weights.

Cladocera: Small crustaceans with a bivalved carapace and a posterior jumping organ. A common genus is *Daphnia*.

Claparède's organ (or Urstigma): A pore-like sensory organ located between coxae I and II of larval water mites and some Cryptostigmata.

Clawlets: Claw-like subdivisions of the claw.

Cleisthotecia: Closed fruiting bodies that are critical to powdery mildew survival in temperate climates.

Cohort: An undefined, usually large group of organisms.

COI (cytochrome oxidase subunit I): Highly conserved regions occurring in the cells of all aerobic organisms. Mutational studies of COI are used to estimate rate of evolutionary changes.

Conidium (plural: conidia): Asexual spore of terrestrial fungi.

Construct: A plasmid engineered in order to include one or more DNA sequences.

Copepods: Minute crustaceans with an elongated, segmented body. A common genus is *Cyclops*.

Coprophagus: Animals that feed on dung (see Fimicolous).

Costula: Rib-like protuberances on the prodorsum of some Cryprostigmata.

Coxa: The basal, immovable segment of the acarine leg and palpus; in the latter it is fused to the base of the gnathosoma.

Crista metopica: A dorsal, elongate, sclerotised plate that bears sensilli; a characteristic of the Erythraeidae and Trombidiidae.

Crustacea: Aquatic arthropods that breathe through gills and possess forked (biramous) appendages, and whose heads may be coalesced with their thoraces.

Cryptogenic: A species that is neither demonstrably native (indigenous) nor introduced (exotic).

Cultivar (cv.): A plant variety that was developed under cultivation.

Cuspis: The apex of a lamella on the prodorsum of cryptostigmatid mites.

Cysts (of nematodes): Rounded structures consisting of an outer membrane and the inner protected animal, which is in a reduced metabolic state.

Dentate: With teeth.

Deuterotoky or amphterotoky: Parthenogenesis in which an unferitilised egg may become either male or female.

Deutogynes: Females of eriophyoid females that reproduce in the year after they were produced (see Protogynes). They may differ morphologically from the protogynes and have no male equivalent.

Deutonymph: The second nymphal stage in the development of a mite. The hypopus is the deutonymph of mites in the suborder Astigmata.

Diapause: A state of suspended development, independent of temperature, expressed in females by lack of reproduction, in eggs by a suspension of metabolic activities. Diapause is usually induced by environmental factors, e.g. photoperiod temperature and diet.

Dioecious: Plants whose male and female sexual organs occur on different individuals.

Diplo-diploidy: A mode of reproduction in which both sons and daughters are produced from fertilised eggs and have two sets (2n) of chromosomes.

Diploid: With the full, double complement of chromosomes.

DNA (deoxyribonucleic acid): The primary genetic material in all organisms, a molecule made up of two complementary strands that are wound around each other in a double helix formation.

Domatium (plural: domatia): A cavity (usually on the leaves) in which mites (and sometimes small insects) live and/or hide.

Dorsum: The dorsal surface of the body.

Doubling time: The rate at which a population will double itself under a given set of environmental conditions.

Downy mildew: A plant disease seen as whitish patches on the underside of leaves, caused by members of the fungal family Peronosporaceae.

Duplex setae: A pair of adjacent dorsal setae that often emerge from a common base. They consist of an elongate, tapering solenidion and a shorter tactile seta, and are borne on legs I and II of most spider mites (Tetranychidae).

Electrophoresis: A process that separates proteins (and other large molecules) on the basis of their size, being driven between a positive and a negative charge through a matrix with a constant pH (see Isoelectric focusing electrophoresis).

ELISA (enzyme-linked immunosorbent assay): A method for estimating minute amounts of materials in a solution. An antibody is coupled to an enzyme as a means of detecting an antigenic match.

Elytron (plural: elytra): The anterior, sclerotised pair of wings of beetles (Coleoptera).

Empodium (plural: empodia): The (usually) pad-like structure that lies between the tarsal claws. May be modified into a claw (with or without tenent hairs) or a cup-like sucker, or be absent.

Emulsifiable concentrate: The liquid formulation of a pesticide that is applied diluted in water.

Entomopathogenic: Fungi or bacteria that cause insect diseases.

Epicuticle: The outermost layer of the epidermis, consisting of highly resistant polymerised non-living secretions.

Epidermis: The outermost layer(s) of cells in the integument.

Epigynial (or genital) plate: A sclerotised plate (or shield) around the opening of the female acarine reproductive system.

Epizootics: Large-scale (plague-like) outbreaks of diseases.

Ereynetal organ: A structure located on tibiae I of all Ereynetidae, consisting of an internal canal opening near a sensory seta, the tibial famulus.

Ergosterol: A sterol of plant or yeast origin.

Esterases: A group of enzymes that hydrolyse a variety of esters; some play major roles in the breakdown of pesticides and secondary metabolites.

Euedaphic: Organisms that are true soil dwellers.

Eupathidium (plural: eupathidia): Spinose sensory setae that are located on the palpus and the anterior legs.

Extrafloral nectary: A nectar-producing organ located outside flowers (e.g. at the base of leaves).

Featherclaw: The feather-like empodium of the Eriophyoidea.

Femur: The third segment of the leg. It may be subdivided to a basal basifemur and a distal telofemur; in the palpus it is usually the first free segment.

Fimicolous: An organism that lives in dung (see Coprophagous).

Functional response: The change in prey number killed, per individual predator per unit time, as a function of changes in prey density (see Numerical response).

Genital suckers: Rounded, paired sclerites located alongside the reproductive organs.

Genu: The fourth segment of the leg and of the palpus.

Geotactic: Responding to the Earth's gravity, thus moving downwards.

Glabrous: A body or organ (e.g. leaf) without hairs.

Glutathione S-transferase (GST): A group of enzymes present in most organisms that mediates in the detoxification of pesticides, especially OPs.

Gnathosoma: The anterior part of the acarine body that carries the palpi and chelicerae. It is usually conjoined ventrally to the anterior end of the idiosoma (see Capitulum).

Gonophore: The external opening of the genital tract.

Guild: A group of different organisms that live on the same resource (e.g. plant) and exploit it in a similar manner.

Haller's organ: A dorsal pit on tarsus I of ticks that contains specialised setae with olfactory and humidity-sensing functions.

Harvestmen: Members of the arthropod group Opiliones.

Hemiedaphic: An organism that lives only part of its life in the soil.

Heterologous: Genetic material from a foreign organism (e.g. not from self).

Heteromorphic male (or female): An organism that has a different morphological aspect from its female (or male) counterpart.

High volume (in sprays): The use of large amounts of water (usually >500 l/ha) when applying a pesticide (see Low volume).

Homomorphic male (or female): An organism that has the same morphological characters as its female (or male) counterpart.

Honeydew: The sticky, often sweetish excretion of sucking insects (e.g. aphids and scale insects), on which dark sooty mould fungi settle.

Hyperplastic: Abnormal growth increases, such as in galls.

Hypopus (plural: hypopodes): The second nymphal stage (deutonymph) of the Astigmata, whose dark aspect is different from that of the preceding (protonymphal) and subsequent (tritonymphal) stages. Its main function is dispersal by clinging or adhering to vertebrates or invertebrates.

Hypostome: The ventral, median wall of the gnathosoma, which is greatly developed in ticks and carries recurving teeth.

Hysteronotum: The dorsum of the posterior part (hysterosoma) of a mite's body.

Idiosoma: The main body of the mite, which carries the legs.

IMC (integrated mite control): A mite control strategy, analogous to IPM, which incorporates different techniques in a compatible manner in order to reduce and maintain pest mite numbers and/or their injury below the economic damage threshold levels.

Infochemical: A chemical that conveys information from one organism to another.

Inoculative releases: The initial releases of natural enemies meant to inoculate the habitat with an exotic natural enemy; also called seeding.

Integument: The external tissue that covers the body, consisting of live cells (see Epidermis) and non-living layers (see Epicuticle and Procuticle).

Intrinsic rate of increase (usually denoted as r_m): The rate of increase of a population under optimal conditions, including an unlimited food supply, no competition or natural enemies. Calculated from age-specific survival and fecundity rates, and used in life-table analyses (see Net reproductive rate).

IPM (integrated pest management): A pest control strategy that incorporates various techniques in a compatible manner in order to reduce and maintain pest numbers and/or their injury below the economic damage threshold levels.

Isoelectric focusing electrophoresis: A process that separates proteins (and other large molecules) on the basis of their pH, being driven from a positive to a negative charge through a pH gradient matrix. Also called equilibrium electrophoresis (see Electrophoresis).

Juvenoids: Synthetic insect growth regulators that mimic the effects of arthropod juvenile hormones; used as pesticides.

Kairomone: A chemical emitted outwardly by an organism that promotes the survival of another but not that of the emitting species (see Pheromone, Allomone).

Karyotype: The chromosome complement (number, size, shape) of a species.

K-selected organisms: Organisms that have evolved to survive in stable, predictable habitats. Usually these are indigenous, resident species that produce relatively few progeny which are efficient competitors (see r-selected organisms).

Lamellae: Blade-like ridges on the prodorsum of the Cryptostigmata.

Larva: The first active stage in the development of a mite; it bears only three pairs of legs.

LC_{50} (lethal dose$_{50}$): The dosage of a pesticide that kills 50% of a random sample of an animal population, expressed as mg toxicant/kg target body weight.

LC_{50} ratio: The ratio between doses of pesticides causing the death of 50% of a resistant population of a given pest and the doses killing the same rate of a susceptible population.

Leg grooves: Ventrolateral concavities that are usually adjacent to coxae III and IV of the Uropodidae, into which the legs can be withdrawn.

Low volume (in sprays): The use of small amounts of water (usually 50–500 l/ha) during pesticide application (see High volume).

Mean intensity (as used in parasitology): Mean number of active parasites per infested host (see Prevalence).

Metapodosoma: The posterior part of the body that bears the third and fourth pairs of legs.

Metasternal shields: Sclerotised, ventral shields that are usually placed laterally to the sternal shield in the Mesostigmata.

Microsporidia: Parasitic one-celled organisms that live within cells; *Nosema* is a common genus.

Mixed function oxidase (MFO): A group of metabolic enzymes that break down many compounds, including pesticides, formed by oxidation.

Monodactyl: Claw made up of a single part, a modification of the empodium (see Tridactyl).

Monogenic: A trait that is due to a single gene.

Multivoltine: Animals that raise several generations each year.

Mutualism: An interaction between two (or more) organisms in which both partners are fitter when interacting than when living separately.

Mycoparasite: Fungi or bacteria that parasitise fungi.

Mycopathogenic: Fungi or bacteria that cause diseases of fungi.

Mycophagous: An organism that feeds on fungi.

Nematocera: A suborder of the Diptera, which includes the Ceratopogonidae (biting midges), Chironomidae (midges) and Culicidae (mosquitoes).

Net reproductive rate (usually denoted as R_0): An estimate of the number of times a female will reproduce itself (produce females) during a generation under optimal environmental conditions, including an unlimited food supply, no competition or natural enemies. Used in life-table analyses (see Intrinsic rate of increase).

Notogaster: The posterior dorsal area of the body.

Nulliparous: Insects (e.g. blood-sucking Diptera, like mosquitoes) in whose bodies there is no egg development, indicating that they have not yet taken their first blood meal (see Parous).

Numerical response: The change in predator numbers in response to changes in prey numbers (see Functional response).

Ocellum (plural: ocelli): Simple eyes.

Olfactometer: An instrument to measure the attractiveness of various volatiles.

Omnivores: Animals that feed at more than a single trophic level (e.g. feed on plants and animals).

Opisthosoma: The posterior region of the mite's idiosoma, beyond its legs.

Organochlorines (OCLs): Insecticides that enter the target's body through the skin (contact poisons), impeding the activity of sodium channels in the nervous system. They are not water soluble and persist (e.g. in the soil) for long periods. They have little effect on phytophagous mites.

Organophosphates (OPs): Insecticides that enter the target's body through the skin (contact poisons), inhibiting its AChE system. Some are systemic, moving through the plant (or the host animal, e.g. livestock) to affect the feeding pests. OPs are water soluble and not persistent. They differ in their toxicity to mammals; some (e.g. parathion) are very harmful, whereas others (e.g. malathion) are only moderately toxic.

Ostracoda: Small Crustacea whose unsegmented body is enclosed within a carapace.

Palpal-claw complex: See Thumb-claw process.

Palpus (plural: palpi): Free limb-like appendage that is articulated to the palpcoxal region of the gnathosoma and usually overreaches the anterior margin of the capitulum. The palpi are held forward and bear sense organs, and are often used to catch and kill prey (e.g. Bdellidae; Cheyletidae).

Panmixis: Interbreeding in a population, a situation wherein each member may mate with any other member.

Papillate integument: Integument showing small protuberances.

Parahaploidy (or pseudo-arrhenotoky): A sex-determining system in which male and female progeny are formed from fertilised eggs, but only the females are diploid (2n chromosomes), whereas the males are haploid (with n chromosomes). The male egg loses its paternal genome during early stages of embryo development (see Arrhenotoky; Thelytoky).

Parasitism: The development of an organism (the parasite) at the expense of another (the host). The host suffers but does not die, and the parasite's development is faster than that of the host. A host may support more than a single parasite.

Parasitoidism: The development of an organism (the parasitoid) at the expense of another (the host), which usually dies as the parasitoid completes its development. Each host seldom supports more than a single parasitoid.

Parous: Insects (especially mosquitoes) with developing eggs, indicating that they have taken at least one blood meal (see Nulliparous).

Parthenocarpic: Fruit developing without prior fertilisation, thus usually without seeds.

Parthenogenesis: The production of progeny without mating due to the development of unfertilised eggs (see Thelytoky).

Patch: A colony or concentration of organisms that live and feed on the same resource.

PCR (Polymerase Chain Reaction): A technique used to amplify the number of copies of a specific region of DNA, in order to produce enough DNA for assaying.

Pearl bodies: Nutrient-rich outgrowths on many parts of plants (e.g. grapes).

Pelagic: Organisms living in lakes.

Peritreme: Tubes that begin at the stigmata and carry air to the mite's organs (analogous to tracheae in insects).

Pesticide check method (or pesticide exclusion): A method used to demonstrate the effect of a given natural enemy. A pesticide that affects only the natural enemy but not the pest is applied; subsequent changes in the latter's numbers may thus be attributed to the activity, or lack of activity, of the natural enemy.

Phenology: The study of the effect of chronology and climate on the life cycle of organisms.

Pheromone: A chemical, usually volatile, that is secreted outwardly by an organism and affects another member (of the same or the other gender) of the same species (e.g. sex pheromone).

Phoretic: Being carried about by another animal (the host or phorobiont), usually without feeding on the latter.

Phototactic: Responding to light, thus moving towards its source (usually upwards).

Physogastric: The much-enlarged body (usually consisting of the posterior part of the hysterosoma) of heterostigmatic female mites, wherein the juveniles complete their entire development (e.g. Pyemotidae).

Phytophagy: Feeding on plants.

Pilosity: Covering of fine hairs (see Setose; Glabrous).

Plasmid: Molecules of double-stranded DNA of bacterial origin used in cloning.

Polygenic: A trait that is due to more than a single gene.

Porous areas (or sacculi): Depressions on the dorsal surface of the acarine body, consisting of minute open pores.

Powdery mildew: A common plant disease seen as white–grey patches on leaf surfaces, caused by members of the fungal family Erysiphaceae.

Predator: An animal that has to consume more than a single individual food organism (usually called prey) in order to attain maturity.

Pretarsus: The distal part of the tarsus that connects the tarsus with the claws.

Prevalence (as used in parasitology): Number of parasitised hosts per total number of potential hosts (see also Mean intensity).

Primer: A short DNA fragment designed to be complementary for a specific sequence on a relevant DNA.

Probit (Probability unit): Percentages expressed after transformation to probability units, used to evaluate the susceptibility or resistance of an arthropod to pesticides (Fig. 43.1).

Procuticle: A relatively thick tensile layer located on top of the epidermis.

Prodorsum: The dorsal part of the propodosoma.

Propagule: The smallest unit of a living pathogen (or pest) available for infection or infestation.

Propodosoma: The anterior part of the idiosoma, which bears the two anterior legs.

Protandry: The emergence of adult males before females.

Protogynes: Females of eriophyoid mites that reproduce in the same year in which they were born and are similar to the males (see Deutogynes).

Protonymph: The first nymphal instar of a mite, usually with four pairs of legs.

Pseudostigmatic organ (sometimes called sensillum): Sensory setae, which may be bulb-like, feather-like or needle-like, situated on the prodorsum of the Cryptostigmata, Bdellidae, Cunaxidae, Tydeidae and most Heterostigmata.

Pteromorphae: A pair of thin, lateral roof-like processes that overhang and flank the legs of some cryptsostigmatid mites. They may be hinged or wing-like.

Pulvillus: An elongated extension of the pretarsus that is not flanked by the true claws. May be membranaceous, claw-like or sucker-like.

Pygidium: The posterior, often fused segments of the bodies of certain insects (e.g. Hemiptera: Diaspididae).

Pyrethroids (PYRs): Insecticides that act on the insect nervous system and paralyse flying insects ('knockdown effect'). PYRs are more potent at low temperatures (thus 'negative temperature dependent'), are toxic to most predatory Acari but usually have little effect on phytophagous mites.

RAPD (Random amplification of polymorphic DNA): A method of creating genomic 'fingerprints' from species of which little is known, by amplifying their DNA ('DNA fingerprinting'). Because strain-specific arrays of DNA fingerprints are generated by PCR, may also be called PCR/RAPD.

Receptacle (of flowers): The part of the flower bearing its reproductive organs.

Reflexive bleeder: An animal that exudes fluids from certain body parts upon being touched (e.g. some beetles); possibly a defence mechanism.

Resistance level (to pesticides): The value obtained by dividing the LC_{50} level acquired by an organism after exposure to a toxicant, by that obtained before its initial exposure.

RFLP (Restriction fragment length polymorphism analysis): A technique by which organisms may be separated by analysing the patterns obtained by cleaving their DNA.

Rhizosphere: The area around the plant root.

Rosette: Group of leaves growing from the same stem, sometimes an anomaly that is due to the feeding of gall mites.

r-Selected organisms: Organisms that have evolved to survive in unpredictable, ephemeral habitats. They are good dispersers but poor competitors, thus leaving the resource as it becomes depleted or overcrowded. They produce relatively many progeny (see K-selection).

Sacculi: Small sacs with minute openings lying below the cuticle of the notogaster of Cryptostigmatid mites.

Saprophage: An animal that obtains its nutrients from non-living organic sources.

Scale insects: Members of the Hemipteran superfamily Coccoidea, whose bodies are covered by adhering or removable shields ('scales'). It includes margarodids (Margarodidae), soft scales (Coccidae), mealybugs (Pseudococcidae) and armoured scale insects (Diaspididae).

Scanning electron microscope (SEM): A microscope that uses electrons which move along the contours of a specimen's body, resulting in a three-dimensional view.

Sclerotia: A compact, often rounded mass of fungal threads (hyphae) that serves to preserve the fungus for long periods.

Sclerotised: Thickened cuticle.

Scrapie: A fatal, degenerative, contagious disease of sheep and goats caused by modified proteins or prions (proteinaceous infectious particles).

Secondary substances (or chemicals or metabolites): Substances specific to various plants, usually produced in small quantities and conferring special attributes to the organisms that produce them (e.g. colour, taste). Believed to be a major mode of plant defence against herbivores, some of which use the chemicals to locate their producers.

Sejugal furrow: A line delineating the propodosoma of a mite from its hysterosoma.

Sensillum (plural: sensilli): A sensory seta that occurs in many forms (e.g. solenidium). Sometimes called trichobothria or pseudostigmatic organ.

Setose: Covered with setae (see Glabrous).

Sex ratio: The ratio between females and males in a population (herein denoted as females/females + males).

Soil-borne pathogens: Pathogens that spend part of or their entire lives in the soil.

Solenidium (plural: solenidia): A hollow chemosensory seta located on the anterior segments of the legs in some mite families (e.g. Cheyletidae).

Sooty mould: Dark fungi that develop on the honeydew of aphids and scale insects.

Spermadactyl: a modification of the male's chelicerae in the Mesostigmata, which serves to transport spermatophores to the female.

Spermatheca: The female's sexual receptacle, into which the male deposits its spermatophore and where it is stored.

Spermatophore: A sac-like envelope containing the male's spermatozoa, which is either placed into the female's spermatheca (e.g. Phytoseiidae) or deposited on substrates and taken up by the female (e.g. Eriophyidae).

Stigmata: Paired openings on the mite's body through which oxygen enters (analogous to the insects' spiracle).

Stylostome: A feeding tube that is formed in the bodies of vertebrate or invertebrate hosts, being a reaction between the secretions of a parasitic mite and the host's body fluids. The parasites use the stylostome to suck the host's body fluids. The stylostomes (or wounds denoting their past location) remain in the host's body after the mite drops off.

Supracoxal seta: A large, usually pectinated seta borne above coxa I in several Astigmata.

Symbiotes: Microorganisms (usually bacteria or fungi, including yeasts) that live in close association within the bodies of their hosts and provide them with essential nutrients (e.g. vitamins) that are not otherwise available.

Sympatric: Two or more populations of the same species that live in the same area (see Allopatric).

Synanthropic: Living in association with humans.

Systemics: Pesticides that are transported within the host to different parts of plants or animals, thus affecting pests at a distance from the site of application.

Tarsus: The sixth, terminal segment of the leg and of the palpus. On the leg it usually carries an empodium and the claws, which may be modified.

Taxon: An undefined taxonomic unit.

Tectum: The anterior part of the dorsal plate that covers the chelicerae.

Telofemur: The distal part of the femur, if subdivided (see Basifemur).

Tenent hairs: Capitate setae that are borne on the empodium, often located between the claws.

Thelytoky: A sex-determining system, in which the females are always diploid (2n), formed by parthenogenesis; males are absent or very rare (see Arrhenotoky).

Thumb-claw process: Structure consisting of the palptibial claw and the palptarsus (the 'thumb'), which together form a pincer-like structure. Present in many families of the suborder Prostigmata.

Tibia: The fifth segment of the leg and of the palpus.

Tolerance (to a pesticide): A natural lack of susceptibility to a tested toxicant.

Tospovirus: Virus causing tomato-spotted disease, vectored by thrips.

Trade-off: An evolutionary term signifying that increases in a certain trait of fitness (e.g. fecundity) would be at the expense of another trait (e.g. flight).

Transgene (Transgenic organism): An organism that contains genetic material from another organism, inserted by molecular manipulation.

Translamella: A thickened bridge between the lamellae on the prodorsum of cryptostigmatid mites.

Trichobothria: Sensory, usually dorsal modified setae (see Sensillum).

Tridactyl: Claw with three parts, made up of the true claws and the claw-like empodium (see Monodactyl).

Tritonymph: Third nymphal stage in the development of a mite.

Tritrophic: The situation in which an effect of the primary producer, the plant, is transferred through an herbivore, the primary consumer, to a natural enemy of the herbivore, the secondary consumer.

Trochanter: The second segment of the leg and of the palpus; in the leg it is usually movable, whereas in the palpus it is fused to the gnathosoma.

Uncate palp: A palpus whose fifth segment can fold against the enlarged fourth segment. Present in some water mites.

Univoltine: Animal that raises only a single annual generation.

Urstigma: See Claparède's organ.

Vescicular–arbuscular mycorrhizal fungi (VAM): Internal symbiotic fungi that live in the roots of most plants.

Wettable powder: The formulation of a pesticide that is used as powder suspended in water.

Wolbachia: Gram-negative bacteria that occur in many insects and mites and may affect their sex ratio.

Zooplankton: The floating or weakly swimming animals in water bodies.

References

Abdel-Samad, M.A., El-Halawany, M.E. & El-Saied, K.M. (1996) Utilizing *Euseius scutalis* (Athias-Henriot) to control *Eutetranychus orientalis* (Klein) on citrus trees. *Egypt J. Agric. Res.*, **74**, 671–84.

Abo Elghar, M.R., Elbadry, E.A., Hassan, S.M. & Kilany, S.M. (1969) Studies on the feeding, reproduction and development of *Agistemus exsertus* on various pollen species (Acarina: Stigmaeidae). *Z. Angew. Entomol.*, **63**, 282–4.

Abou-Awad, B.A. & Reda, A.S. (1992) Studies on copulation, egg production and sex-ratio of the predacious mite *Agistemus exsertus* Gonzalez (Acari, Stigmaeidae). *J. Appl. Entomol.*, **113**, 472–5.

Abou-Awad, B.A., Reda, A.S. & Elsawi, S.A. (1992) Effects of artificial and natural diets on the development and reproduction of two phytoseiid mites *Amblyseius gossipi* and *Amblyseius swirskii* (Acari: Phytosaeiidae). *Insect Sci. Appl.*, **13**, 441–5.

Abou-Awad, B.A., El-Sawaf, B.M. & Abdel Kader, A.A. (1999) Life history and life table of *Pronematus ubiquitus* (McGregor) as a predator of eriophyoid mites in Egypt (Acari: Tydeidae). *Acarologia*, **40**, 29–32.

Abou-Setta, M.M. & Childers, C.C. (1987) A modified leaf arena technique for rearing phytoseiid or tetranychid mites for biological studies. *Fl. Entomol.*, **70**, 245–8.

Åbro, A. (1982) The effects of parasitic water mite larvae (*Arrenurus* spp.) on zygopteran imagoes (Odonata). *J. Invert. Path*, **39**, 371–81.

Åbro, A. (1988) The mode of attachment of mite larvae (*Leptus* spp.) to harvestmen (Opilines). *J. Nat. Hist.*, **22**, 123–30.

Addison, J.A., Hardman, J.M. & Walde, S.J. (2000) Pollen availability for predaceous mites on apple: spatial and temporal heterogeneity. *Exp. Appl. Acarol.*, **24**, 1–18.

Adler, C.R.L. & Browning, M.W. (1986) Attachment and dispersal of *Callidosoma metzi* (Acari: Erythraeidae) parasitizing *Platynota idaeusalis* (Lepidoptera: Tortricidae). *Ann. Entomol. Soc. Am.*, **79**, 56–9.

Aeschlimann, J.P. & Vitou, J. (1986) Observations on the association of *Allothrombium* sp. (Acari: Trombidiidae) mites with lucerne aphid populations in the Mediterranean region. In: *Ecology of Aphidophaga* (Ed. by I. Hodek), pp. 405–10. Academia, Prague.

Agrawal, A.A. (1997) Do leaf domatia mediate a plant–mite mutualism? An experimental test of the effects on predators and herbivores. *Ecol. Entomol.*, **22**, 371–6.

Agrawal, A.A. & Karban, R. (1997) Domatia mediate plant–arthropod mutualism. *Nature*, **387**, 562–3.

Ahlstrom, K.R. & Rock, G.C. (1973) Comparative studies on *Neoseiulus fallacis* and *Metaseiulus occidentalis* for azinphosmethyl toxicity and effects of prey and pollen on growth. *Ann. Entomol. Soc. Am.*, **66**, 1109–13.

Ahmad, R., Muzaffar, N. & Munawar, M.S. (1985) Studies on biological control of hornet predators of honeybees in Pakistan. *Proc. 30th Int. Congr. Apiculture*, Nagoya, Japan, pp. 403–4.

Al-Amidi, A.H.K., Dunne, R. & Downes, M.J. (1991) *Parasitus bituberosus* (Acari: Parasitidae): an agent for control of *Lycoriella solani* (Diptera: Sciaridae) in mushroom crops. *Exp. Appl. Acarol.*, **11**, 159–66.

Al-Amidi, A.H.K. & Downes, M. J. (1990) *Parasitus bituberosus* Acari, Parasitidae) a possible agent for biological control of *Heteropeza pygmaea* (Diptera, Cecidomyiidae) in mushroom compost. *Exp. Appl. Acarol.*, **8**, 13–25.

Al-Assiuty, A.I.M. & Khalil, M.A. (1995) The influence of insecticide–pheromone substitution on the abundance and distributional pattern of soil oribatid mites. *Exp. Appl. Acarol.*, **19**, 399–410.

Albert, R. (1999) Integrated pest management in *Dendranthema indicum. IOLB/WPRS Bull.*, **22**(1), 1–4.

Alberti, G. & Coons, L.B. (1999) Acari: mites. In: *Microscopic Anatomy of Invertebrates* (Ed. by F.W. Harrison & R.F. Foelix), pp. 515–1265. Wiley-Liss, New York.

Alberti, G. & Ehrnsberger, R. (1977). Rasterelektronenmikroskopische untersuchungen zum Spinnvermögen der Bdelliden und Cunaxiden (Acari, Prostigmata). *Acarologia*, **19**, 55-61.

Alberti, G., Fernandez, N.A. & Kümmel, G. (1991) Spermatophores and spermatozoa of oribatid mites (Acari: Oribatida). Part II: Functional and systematical considerations. *Acarologia*, **32**, 435–49.

Ali, O. & Brennan, P. (1997) Development, feeding and reproduction of the predatory mite, *Hypoaspis miles* (Acari: Mesostigmara: Laelapidae) on different types of prey. *Syst. Appl. Acarol.*, **2**, 81–8.

Ali, O. & Brennan, P. (2000) Observations on the feeding behaviour of *Hypoaspis miles* (Mesostigmata: Laelapidae). *Syst. Appl. Acarol.*, **5**, 41–5.

Ali, O., Dunne, R. & Brennan, P. (1997) Biological control of the sciarid fly, *Lycoriella solani* by the predatory mite, *Hypoaspis miles* (Acari: Laelapidae) in mushroom crops. *Syst. Appl. Acarol.*, **2**, 71–80.

Ali, O., Dunne, R. & Brennan, P. (1999) Effectiveness of the predatory mite *Hypoaspis miles* (Acari: Mesostigmata: Laelapidae) in conjunction with pesticides for control of the mushroom fly *Lycoriella solani* (Diptera: Sciaridae). *Exp. Appl. Acarol.*, **23**, 65–77.

Alston, D.G. (1994) Effect of apple floor vegetation on density and dispersal of phytophagous and predaceous mites in *Utah. Agric. Ecosyst. Environ.*, **50**, 73–84.

Amano, H. & Chant, D.A. (1990) Species diversity and seasonal dynamics of Acari on abandoned apple trees in southern Ontario, Canada. *Exp. Appl. Acarol.*, **8**, 71–96.

Amano, H. & Haseeb, M. (2001) Recently-proposed methods and concepts of testing the effects of pesticides on the beneficial mite and insect species: study limitations and implications in IPM. *Appl. Entomol. Zool.*, **36**, 1–11.

Amin, A.W., Mowafe, M.H. & Fatma, S.A. (1999) Effect of predaceous mesostigmatid mites in the control of *Meloidogyne javanica* root-knot nematode on kidney bean. *Pakistan J. Nematol.*, **17**, 91–6.

Amrine, J.W., Jr (1996a) *Keys to the World Genera of the Eriophyoidea (Acari: Prostigmata).* Indira, West Bloomfield, MI.

Amrine, J.W., Jr (1996b) *Phyllocoptes fructiphilus* and biological control of multiflora rose. In: *Eriophyoid Mites, Their Biology, Natural Enemies and Control* (Ed. by E.E. Lindquist, M.W. Sabelis & J. Bruin), pp. 741–9. Elsevier, Amsterdam.

Amrine, J.W., Jr & Stasny, T.A. (1994) *Catalog of the Eriophyoidea (Acarina: Prostigmata) of the World*. Indira, West Bloomfield, MI.

Anderson, J.R. (1983) Mites as biological control agents of dung-breeding pests: practical considerations and selection for pesticide resistance. In: *Biological Control of Pests by Mites* (Ed. by M.A. Hoy, G.L. Cunningham & L. Knutson), pp. 99–102. University of California, Special Publ., No. 3304, Berkeley, CA.

Anderson, R.C. (1968a) Ecological observations on three species of *Pimeliaphilus* parasites of Triatominae in the United States (Acarina: Pterygosomidae) (Hemiptera: Reduviidae). *J. Med. Entomol.*, **5**, 459–64.

Anderson, R.C. (1968b) The biology of the conenose bug parasite, *Pimeliaphilus plumifer* Newell and Ryckman (Acarina: Pterygosomidae) (Hemiptera: Reduviidae). *J. Med. Entomol.*, 5, 473–7.

André, H.M. (1979) A generic revision of the family Tydeidae (Acari: Actinedia). I. Introduction, paradigms and general classification. *Ann. Soc. R. Zool. Belgique*, **108**, 189–208.

André, H.M. (1980) A generic revision of the family Tydeidae (Acari: Actinedia). IV. Generic descriptions, keys and conclusions. *Bull. Ann. Soc. R. Belgique Entomol.*, **116**, 103–30, 139–68.

André, H.M. & Fain, A. (2000) Phylogeny, ontogeny and adaptive radiation in the superfamily Tydeoidea (Acari: Actinedida), with a reappraisal of morphological characters. *Zool. J. Linn. Soc.*, **139**, 405–48.

Anonymous (1985) Biological control of *Solenopsis invicta*. *Trop. Pest Manage.*, **31**, 238.

Anonymous (1991) Directory of producers of natural enemies of common pests. *IPM Practitioner*, **13**(4), 15–9.

Anonymous (1993) 1994 directory of least-toxic pest control products. *IPM Practitioner*, **15**(11/12), 1–40.

Anonymous (1997) 1998 directory of least-toxic pest control products. *IPM Practitioner*, **19**(11/12), 1–52.

Anonymous (1998) Effects of active substances of plant protection products on biological control agents used in glasshouses. *EPPO Bull.*, **28**, 425–31.

Anonymous (1999) IBMA Bio news: International Biocontrol Manufacturers Association. *Integr. Pest Manage. Rev.*, **4**, 167–73.

Anonymous (2000) 2001 directory of least-toxic pest control products. *IPM Practitioner*, **21**(11/12), 1–48.

Aoki, J.-I. (1971) Soil mites (oribatids) climbing trees. In *Proc. 3rd Int. Congr. Acarology* (Ed. by M. Daniel & B. Rosický), pp. 59–65. Junk, The Hague.

Argov, Y., Amitai, S., Beattie, G.A.C. & Gerson, U. (2002) Rearing, release and establishment of imported predatory mites to control citrus rust mite in Israel. *BioControl*, **47**, 399–409.

Arlian, G.L., Morgan, M.S. & Houck, M.A. (1999) Allergenicity of the mite *Hemisarcoptes cooremani*. *Ann. Aller. Asth. Immunol.*, **83**, 529–32.

Asano, S. & Kamei, M. (1982) Ovicidal activity of cycloprate for several phytophagous mite species and its relationship with the test host plant. *Appl. Entomol. Zool.*, **17**, 67–74.

Ashihara, W. (1995) Studies on the occurrence of the Kanzawa spider mite, *Tetranychus kanzawai* Kishida (Acarina: Tetranychidae) and its biological control by *Phytoseiulus persimilis* Athias-Henriot (Acarina: Phytoseiidae) on grapevine in greenhouses. *Bull. Fruit Tree Res. Stn*, **E6**, 55–151 (in Japanese with English summary).

Astarita, C., Franzese, A., Scala, G. *et al.* (1994) Farm workers' occupational allergy to *Tetranychus urticae*: clinical and immunological aspects. *Allergy*, **49**, 466–71.

Ateyo, W.T. (1960) A revision of the mite family Bdellidae in North and Central America (Acarina, Prostigmata). *Kansas Univ. Sci. Bull.*, **40**, 345–499.

Atsatt, P.R. & O'Dowd, D.J. (1976) Plant defense guilds. *Science*, **193**, 24–9.

Avidov, Z., Blumberg, D. & Gerson, U. (1968) *Cheletogenes ornatus* (Acarina: Cheyletidae), a predator of the chaff scale on citrus in Israel. *Israel J. Entomol.*, **3**, 77–93.

Avidov, Z., Balshin, M. & Gerson, U. (1970) Studies on *Aphytis coheni*, a parasite of the California red scale, *Aonidiella aurantii*, in Israel. *Entomophaga*, **15**, 191–207.

Avilla, J., Bosch, D., Sarasúa, M.J. & Costa-Comelles, J. (1993) Biological control of *Panonychus ulmi* in apple orchards in Lleida (NE of Spain). *Acta Hort.*, **347**, 267–72.

Axelsen, J.A., Holst, N., Hamers, T. & Krogh, P.H. (1997) Simulations of the predator–prey interactions in a two species ecotoxicological test system. *Ecol. Model.*, **101**, 15–25.

Axtell, R.C. (1963) Effect of Macrochelidae (Acarina: Mesostigmata) on house fly production from dairy cattle manure. *J. Econ. Entomol.*, **56**, 317–21.

Axtell, R.C. (1968) Integrated house fly control: populations of fly larvae and predaceous mites, *Macrocheles muscaedomesticae*, in poultry manure after larvicide treatment. *J. Econ. Entomol.*, **61**, 245–9.

Axtell, R.C. (1970) Integrated fly-control program for caged-poultry houses. *J. Econ. Entomol.*, **63**, 400–5.

Axtell, R.C. (1986) Fly management in poultry production; cultural, biological and chemical. *Poult. Sci.*, **65**, 657–67.

Axtell, R.C. (1991) Role of mesostigmatid mites in integrated fly control. In: *Modern Acarology* (Ed. by F. Dusbábek & V. Bukva), Vol. 2, pp. 639–46. SPB, The Hague.

Babikir, E.-T. A. (1978) Factors affecting biological control of the red spider mite in glasshouses. The effect of fungicides and light intensity on the population dynamics of *Tetranychus urticae* and *Phytoseiulus persimilis* with reference to the efficiency of biological control in glasshouses. DPhil Thesis, University of Bradford, UK.

Baccetti, B. (1960) Le cocciniglie Italiane delle Cupressacee. *Redia*, **45**, 23–111.

Bader, C. (1980) Some biological and ecological data on water mites, mainly some significant data on the life-duration. *Int. J. Acarol.*, **6**, 239–43.

Badii, M.H. & McMurtry, J.A. (1983) Effect of different foods on develoment, reproduction and survival of *Phytoseiulus longipes* (Acari: Phytoseiidae). *Entomophaga*, **28**, 161–6.

Badii, M.H. & McMurtry, J.A. (1984) Feeding behavior of some phytoseiid predators on the broad mite, *Polyphagotarsonemus latus* (Acari: Phytoseiidae, Tarsonemidae). *Entomophaga*, **29**, 49–53.

Badii, M.H., McMurtry, J. A. & Flores, A.E. (1999) Rates of development, survival and predation of immature stages of *Phytoseiulus longipes* (Acari: Mesostigmata: Phytoseiidae). *Exp. Appl. Acarol.*, **23**, 611–21.

Baillod, M., Erard, F., Antonin, Ph. & Stäubli, A. (1992) Distribution, méthodes de controle, estimation du risque pour *Panonychus ulmi* (Koch) et *Tetranychus urticae* Koch en vergers de poiriers et incidence de deux prédateurs *Anthocoris nemoralis* F. et *Typhlodromus pyri* Scheuten. *Rev. Suisse Vitic. Arbor. Hort.*, **24**, 179–87.

Baker, A.S. (1999a) *Mites and Ticks of Domestic Animals. An Identification Guide and Information Source.* Natural History Museum, Grimsby.

Baker, A.S. (1999b) Two new species of larval mites (Acari: Trombidioidea: Microtrombidiidae and Johnstonianidae) parasitising *Culicoides impunctatus*, the highland midge (Insecta: Ceratopogonidae), in Scotland. *Syst. Parasitol.*, **44**, 37–47.

Baker, B.A. (1970) The food of *Riccardoella limacum* (Schrank) (Acari: Trombidiformes) and its relationship with pulmonate molluscs. *J. Nat. Hist.*, **4**, 521–30.

Baker, E.W. (1965) A review of the genera of the family Tydeidae (Acarina). In: *Advances in Acarology* (Ed. by J.A. Naegele), Vol. 2, pp. 95–133. Cornell University Press, Ithaca, NY.

Baker, E.W. (1968) The genus *Pronematus* Canestrini. *Ann. Entomol. Soc. Amer.*, **61**, 1091–7.

Baker, E.W. & Tuttle, D.M. (1994) *A Guide to the Spider Mites (Tetranychidae) of the United States.* Indira, West Bloomfield, MI.

Baker, E.W. & Wharton, G.W. (1952) *An Introduction to Acarology.* Macmillan, New York.

Baker, E.W., Kono, T., Amrine, J.W., Jr. *et al.* (1996) *Eriophyoid Mites of the United States.* Indira, West Bloomfield, MI.

Baker, G.T. (1996) Chemoreception in four species of water mites (Acari: Hydrachnida): behavioural and morphological evidence. *Exp. Appl. Acarol.*, **20**, 203–13.

Baker, T.C. & Eickwort, G.C. (1975) Development and bionomics of *Chrysomelobia labidomerae* (Acari: Tarsonemina; Podapolipidae), a parasite of the milkweed leaf beetle (Coleoptera: Chrysomelidae). *Can. Entomol.*, **107**, 627–38.

Baker, W.V. (1967) Some observations on predation in an anystid mite. *Entomol. Mon. Mag.*, **103**, 58–9.

Bakker, F.M. (1993) *Selecting phytoseiid predators for biological control, with emphasis on the significance of tri-trophic interactions*. DPhil Thesis, University of Amsterdam.

Bakker, F.M. & Jacas, J.A. (1995) Pesticides and phytoseiid mites: strategies for risk assessment. *Ecotox. Environ. Safe.*, **32**, 58–76.

Bakker, F. & Klein, M.E. (1992) Transtrophic interactions in cassava. *Exp. Appl. Acarol.*, **14**, 293–311.

Bakker, F.M. & Klein, M.E. (1999) Extrafoliar domatia and extrafoliar nectar. In *Acarology IX, Symposia* (Ed. by G.R. Needham, R. Mitchell, D.J. Horn & W.C. Wellbourn), Vol. 2, pp. 221–6. Ohio Biological Survey, Columbus, OH.

Bakker, F.M. & Sabelis, M.W. (1989) How larvae of *Thrips tabaci* reduce the attack success of phytoseiid predators. *Entomol. Exp. Appl.*, **50**, 47–51.

Bakker, F.M., Klein, M.E., Mesa, N.C. & Braun, A.R. (1993) Saturation deficit tolerance spectra of phytophagous mites and their phytoseiid predators on cassava. *Exp. Appl. Acarol.*, **17**, 97–113.

Ball, J.C. (1980) Development, fecundity, and prey consumption of four species of predaceous mites (Phytoseiidae) at two constant temperatures. *Environ. Entomol.*, **9**, 298–303.

Ball, J.C. (1982) Impact of fungicides and miticides on predatory and phytophagous mites associated with pecan foliage. *Environ. Entomol.*, 11, 1001–4.

Balogh, J. & Balogh, P. (1990) Identification key to the genera of the Galumnidae Jacot, 1925 (Acari: Oribatei). *Acta Zool. Hung.*, **36**, 1–24.

Balseiro, E.G. (1992) The role of pelagic water mites in the control of cladoceran population in a temperate lake of the southern Andes. *J. Plankt. Res.*, **14**, 1267–77.

Banerjee, P. & Datta, S. (1980) Biological control of red cotton bug, *Dysdercus koenigii* Fabricius by mite, *Hemipteroseius indicus* (Krantz and Khot). *Indian J. Entomol.* **42**, 265–7.

Barbosa, P. & Segarra-Carmona, A. (1993) Criteria for the selection of pest arthropod species as candidates for biological control. In: *Steps In Classical Arthropod Biological Control* (Ed. by R.G. van Driesche & T.S. Bellows Jr), pp. 5–23. Entomological Society America, Lanham, MD.

Barker, P.S. (1967) Bionomics of *Blattisocius keegani* (Fox) (Acarina: Ascidae), a predator on eggs of pests of stored grains. *Can. J. Zool.*, **45**, 1093–9.

Barker, P.S. (1968a) Notes on the bionomics of *Haemogamasus pontiger* (Berlese) (Acarina: Mesostigmata), a predator on *Glycyphagus domesticus* (De Geer). *Manitoba Entomol.*, **2**, 85–7.

Barker, P.S. (1968b) Bionomics of *Androlaelaps casalis* (Berlese) (Acarina: Laelapidae) a predator of mite pests of stored cereals. *Can. J. Zool.*, **46**, 1099–102.

Barker, P.S. (1991) Bionomics of *Cheyletus eruditus* (Schrank) (Acarina: Cheyletidae), a predator of *Lepidoglyphus destructor* (Schrank) (Acarina: Glycyphagidae), at three constant temperatures. *Can. J. Zool.*, **69**, 2321–5.

Barker, P.S. (1992) Bionomics of *Nodele calamondin* Muma (Acarina: Cheyletidae) fed on *Lepidoglyphus destructor* (Schrank) (Acarina: Glycyphagidae), at two constant temperatures. *Can. J. Zool.*, **70**, 2333–7.

Barr, D. (1973) *Methods for the Collection, Preservation and Study of Water Mites (Acari: Parasitengona)*. Life Sciences Misc. Publ., R. Ontario Museum.

Barret, D. & Kreiter, S. (1995) Morphometrics of some phytoseiid predatory mites and characteristics of their habitat: consequences for biological control. In: *The Acari, Physiological and Ecological Aspects of Acari–Host Relationships* (Ed. by D. Kropczyńska, J. Boczek & A. Tomczyk), pp. 461–73. Oficyna DABOR, Warsaw.

Bartlett, B.R. (1964) The toxicity of some pesticide residues to adult *Amblyseius hibisci*, with a compilation of the effects of pesticides upon phytoseiid mites. *J. Econ. Entomol.*, **57**, 559–63.

Bartlett, B. & DeBach, P. (1952) New natural enemies of avocado pests. *Citrus Leaves*, **32**(10), 16–7.

Bartlett, B.R. & van den Bosch, R. (1964) Foreign exploration for beneficial organisms. In: *Biological Control of Insect Pests and Weeds* (Ed. by P. DeBach), pp. 283–304. Chapman & Hall, London.

Baumgärtner, J., Gutierrez, A.P. & Klay, A. (1988) Elements for modelling the dynamics of tritrophic population interactions. *Exp. Appl. Acarol.*, **5**, 243–63.

Bautista, R.C., Mau, R.F.L., Cho, J.J. & Custer, D.M. (1995) Potential of tomato spotted wilt tospovirus plant hosts in Hawaii as virus reservoirs for transmission by *Frankliniella occidenatlis* (Thysanoptera: Thripidae). *Phytopathology*, **85**, 953–8.

Beard, J.J. (1999) Taxonomy and biological control: *Neoseiulus cucumeris* (Acari: Phytoseiidae), a case study. *Aust. J. Entomol.*, **38**, 51–9.

Beard, J.J. (2001) A review of Australian *Neoseiulus* Hughes and *Typhlodromips* de Leon (Acari: Phytoseiidae: Amblyseiinae). *Invert. Taxon.*, **15**, 73–158.

Beard, J.J. & Walter, G.H. (2001) Host plant specificity in several species of generalist mite predators. *Ecol. Entomol.*, **26**, 562–70.

Bedford, E.C.G. (1949) Report of the plant pathologist. In: *Report of the Department of Agriculture* 1949, pp. 11–19, Bermuda Board of Agriculture.

Beerling, E.A.M. & van der Geest, L.P.S. (1991) A microsporidium (Microspora: Pleistophoridae) in mass-rearings of the predatory mites *Neoseiulus cucumeris* and *A. barkeri* (Acarina: Phytoseiidae): analysis of a problem. *IOBC/WPRS Bull.*, **14**(7), 5–8.

Beerling, E.A.M., van der Voort, J.N.A.M.R. & Kwakman, P. (1993) Microsporidiosis in mass-rearings of the predatory mites: development of a detection method. *Proc. Exp. Appl. Entomol., N.E.V. Amsterdam*, **4**, 199–204.

Behan-Pelletier, V.M. (1985) Ceratozetidae of the western North American Arctic. *Can. Entomol.*, **117**, 1287–366.

Behan-Pelletier, V.M. (1986) Ceratozetidae (Acari: Oribatei) of the western North American subarctic. *Can. Entomol.*, **118**, 991–1057.

Behan-Pelletier, V.M. (2000) Ceratozetidae (Acari: Oribatida) of arboreal habitats. *Can. Entomol.*, **132**, 153–82.

Behan-Pelletier, V.M. & Hill, S.B. (1983) Feeding habits of sixteen species of Oribatei (Acari) from an acid peat bog, Glenamoy, Ireland. *Rev. Ecol. Biol. Sol.*, **20**, 221–67.

Beirne, B.P. (1975) Biological control attempts by introductions against insect pests in the field in Canada. *Can. Entomol.*, **107**, 225–36.

Bellotti, A.C., Smith, L. & Lapointe, S.L. (1999) Recent advances in cassava pest management. *Annu. Rev. Entomol.*, **44**, 343–70.

Bellows, T.S. & van Driesche, R.G. (1999) Life table construction and analysis for evaluating biological control agents. In: *Handbook of Biological Control, Principles and Applications of Biological Control* (Ed. by T.S. Bellows & T.W. Fisher), pp. 199–223. Academic Press, San Diego, CA.

Bellows, T.S., Jr, Morse, J.G. & Gaston, L.K. (1992) Residual toxicity of pesticides used for control of lepidopteran insects in citrus to the predaceous mite *Euseius stipulatus* Athias-Henriot (Acarina, Phytoseiidae). *J. Appl. Entomol.*, **113**, 493–501.

Belovsky, G.E., Branson, D., Chase, J. *et al.* (2001) Mites and nematode parasites of grasshoppers. In: *Grasshoppers: Their Biology, Identification and Management* (Ed. by D. Branson, & B. Redlin), Section 9.1. US Department of Agriculture, ARS (at website http://www.sidney.ars.usda.gov/grasshopper/index.htm).

Bengtsson, G., Erlandsson, A. & Rundgren, S. (1988) Fungal odour attracts soil Collembola. *Soil Biol. Biochem.*, **20**, 25–30.

Bennett, F.D. (1965) Observations on the natural enemies of *Gynaikothrips ficorum* Marchal in Brazil. *Com. Inst. Biol. Contr., Tech. Bull.*, **5**, 116–25.

Bennett, F.D. (1985) Conflicts of interest in CIBC biological control of weeds programs. In: *Proc. VI Int. Symp. on Biological Control of Weeds* (Ed. by E.S. Delfosse), pp. 241–7. Agriculture Canada, Ottawa.

Bennison, J.A. & Jacobson, R. (1991) Integrated control of *Frankliniella occidenatlis* (Pergande) in UK cucumber crops – evaluation of a controlled release system of introducing *Amblyseius cucumeris*. *Med. Fac. Landbouww. Rijkuniv. Gent*, **56**, 251–5.

Bentley, B.L. (1977) Extrafloral nectaries and protection by pugnacious bodyguards. *Annu. Rev. Ecol. Syst.*, **8**, 407–27.

van den Berg, M., Classen, V.E., Maritz, M. & Ueckermann, E.A. (1995) *Anystis baccarum* (Acarina: Anystidae) – a newly recorded predator of the citrus psylla, *Trioza erytreae* (Hem. Triozidae) and the yellow pecan aphid, *Monelliopsis pecanis* (He., Aphididae). *Bull. Inst. Trop. Subtrop. Gwasse*, **275**, 44–7.

Berkenkamp, S.D. & Landers, E.J. (1983) Observations on the scorpion parasite *Pimeliaphilus joshuae* Newell and Ryckman, 1966 (Acarina: Pterygosomidae). *J. Arizona–Nevada Acad. Sci.*, **18**, 27–31.

Berkett, L.P. & Forsythe, H.Y., Jr (1980) Predaceous mites (Acari) associated with apple foliage in Maine. *Can. Entomol.*, **112**, 497–502.

Bernstein, C. (1983) Some aspects of *Phytoseiulus persimilis* (Acarina: Phytoseiidae) dispersal behaviour. *Entomophaga*, **28**, 185–98.

Berreen, J.M. (1984) The functional response of *Cheyletus eruditus* Schrank to changes in the density of its prey *Acarus siro* L. In: *Acarology VI* (Ed. by D.A. Griffiths & C.E. Bowman), Vol. 2, pp. 980–6. Ellis Horwood, Chichester.

Berry, J.S., Holtzer, T.O. & Norman, J.M. (1991) Experiments using a simulation model of the Banks grass mite (Acari: Tetranychidae) and the predatory mite *Neoseiulus fallacis* (Acari: Phytoseiidae) in a corn microenvironment. *Environ. Entomol.*, **20**, 1074–8.

Berry, R.E. (1973) Biology of the predaceous mite *Pergamasus quisquiliarum*, on the garden symphylan, *Scutigerella immaculata*, in the laboratory. *Ann. Entomol. Soc. Am.*, **66**, 1354–6.

Berry, R.W. (1995) Controlling the common furniture beetle. *Pest Outlook*, **6**, 26–30.

Bhattacharyya, D.R., Dutta, P., Mahanta, J. & Srivastava, V.K. (1994) Acarine parasites of mosquitoes in Upper Assam. *Indian J. Malariol.*, **18**, 173–6.

Bilgrami, A.L. (1994) Pradatory behaviour of a nematode feeding mite *Tyrophagus putrescentiae* (Sarcoptiformes: Acaridae). *Fundam. Appl. Nematol.*, **17**, 293–7.

Bilgrami, A.L. (1997) Evaluation of the predation abilities of the mite *Hypoaspis calcuttaensis*, predaceous on plant and soil nematodes. *Fundam. Appl. Nematol.*, **20**, 96–8.

Binns, E.S. (1973) Predatory mites – neglected allies? *Mushroom J.*, **12**, 540–4.

Binns, E.S. (1974) Notes on the biology of *Arctoseius cetratus* (Sellnick) (Mesostigmata: Ascidae). *Acarologia*, **16**, 557–82.

Bjørnson, S. & Keddie, B.A. (1999) Effect of *Microsporidium phytoseiuli* (Microsporidia) on the performance of the predatory mite, *Phytoseiulus persimilis* (Acari: Phytoseiidae). *Biol. Contr.*, **15**, 153–61.

Bjørnson, S., Steiner, M.Y. & Keddie, B.A. (1996) Ultrastructure and pathology of *Microsporidium phytoseiuli* n.sp. infecting the predatory mite, *Phytoseiulus persimilis* Athias-Henriot (Acari: Phytoseiidae). *J. Invert. Pathol.*, **68**, 223–30.

Bjørnson, S., Steiner, M.Y. & Keddie, B.A. (1997) Birefringent crystals and abdominal discoloration in the predatory mite *Phytoseiulus persimilis* (Acari: Phytoseiidae). *J. Invert. Pathol.*, **69**, 85–91.

Bjørnson, S., Rawort, D.A. & Bédard, C. (2000) Abdominal discoloration and the predatory mite, *Phytoseiulus persimilis* Athias-Henriot: prevalence of symptoms and their correlation with short-term performance. *Biol. Cont.*, **17**, 17–27.

Blahutiak, A. & Alayo Soto, R. (1982) Función de *Pyemotes boylei* Krczal (Acarina: Pyemotidae) en la dinámica poblacional de *Diadiplosis cocci* Felt (Diptera: Cecidomyiidae) en Cuba. *Acad. Cien. Cuba, Informe Cientifico*, **194**, 1–8.

Blommers, L. (1976) Some Phytoseiidae (Acarina: Mesostigmata) from Madagascar, with descriptions of eight new species and notes on their biology. *Bijdr. Dierkunde*, **46**, 80–106.

Blommers, L.H.M. (1994) Integrated pest management in European apple orchards. *Annu. Rev. Entomol.*, **39**, 213–41.

Blommers, L. & Helsen, H. (1986) Host plant influence on the effect of pesticides on the predacious mite *Typhlodromus pyri*. *IOBC/WPRS Bull.*, **9**(3), 55–9.

Blum, M.S., Foottit, R. & Fales, H.M. (1992) Defensive chemistry and function of the anal exudate of the thrips *Haplothrips leucanthemi*. *Comp. Biochem. Physiol.*, **102C**, 209–11.

Blümel, S., Pertl, C. & Bakker, F.M. (2000) Comparative trials on the effects of two fungicides on a predatory mite in the laboratory and in the field. *Entomol. Exp. Appl.*, **97**, 321–30.

Boczek, J. & Petanović, R. (1996) Eriophyid mites as agents for the biological control of weeds. In: *Proc. IX Int. Congr. Biological Control of Weeds* (Ed. by V.C. Moran & J.H. Hoffmann), pp. 127–31. University of Cape Town.

Boldt, P.E. & Sobhian, R. (1993) Release and establishment of *Aceria malherbae* (Acari: Eriophyidae) for control of field bindweed in Texas. *Environ. Entomol.*, **22**, 234–7.

Bolland, H.R. (1983) A description of *Neophyllobius aesculi* and its developmental stages (Acari: Camerobiidae). *Entomol. Berl.*, **43**, 42–7.

Bolland, H.R. (1986) Review of the systematics of the family Camerobiidae (Acari: Raphignathoidea). 1. The genera *Camerobia, Decaphyllobius, Tillandsobius* and *Tycherobius*. *Tijd. Entomol.*, **129**, 191–215.

Bolland, H.R. & Magowski, W.L. (1990) *Neophyllobius succineus* n. sp. from Baltic amber (Acari: Raphignathoidea: Camerobiidae). *Entomol. Berl.*, **50**, 17–21.

Bolland, H.R., Gutierrez, J. & Flechtmann, C.W.H. (1998) *World Catalogue of the Spider Mite Family (Acari: Tetranychidae)*. Brill, Leiden.

Bollen, G.J., Middlekoop, J. & Hofman, T.W. (1991) Effects of soil fauna on infection of potato sprouts by *Rhizoctonia solani*. In: *Biotic Interactions and Soilborne Diseases* (Ed. by A.B.R. Beemster, G.J. Bollen, M. Gerlach *et al.*), pp. 27–34. Elsevier, Amsterdam.

Boller, E.F., Remund, U. & Candolfi, M.P. (1988) Hedges as potential sources of *Typhlodromus pyri*, the most important predatory mite in vineyards of northern Switzerland. *Entomophaga*, **33**, 249–55.

Bonde, J. (1989) Biological studies including population growth parameters of the predatory mite *Amblyseius barkeri* (Acarina: Phytoseiidae) at 25°C in the laboratory. *Entomophaga*, **34**, 275–87.

Booth, J.P. & Learner, M.A. (1978) The parasitization of chironomid midges (Diptera) by water mite larvae (Hydracarina: Acari) in a eutrophic reservoir in south Wales. *Arch. Hydrobiol.*, **84**, 1–28.

Borden, E.E.R. (1989) The phoretic behavior and olfactory preference of *Macrocheles muscaedomesticae* (Scopoli) (Acarina: Macrochelidae) in its relationship with *Fannia canicularis* (L.) (Diptera: Muscidae). *Pan-Pacific Entomol.* **65**, 89–96.

Bornemissza, G.E. (1976) The Australian dung beetle project 1965–1975. *Austral. Meat Res. Comm.*, Sydney, No. 30.

Borthakur, M. (1981) Biological notes on a mite predator of the scarlet mite *Brevipalpus phoenicis* (Geijskes). *Two and a Bud*, **28**, 18–19.

Borthakur, M., Handique, R. & Banerjee, B. (1998) Effect of pollen feeding on development and reproduction of *Agistemus hystrix* Gupta (Acarina: Stigmaeidae). *Two and a Bud*, **44**, 18–20.

Boshko, G.V. & Skylar, V.E. (1981) Parasitization of ixodid ticks on horse flies. *Meditsin. Parazitol.*, **50**, 80–1 (in Russian).

Bostanian, N.J. & Coulombe, L.J. (1986) An integrated pest management program for apple orchards in southwestern Quebec. *Can. Entomol.*, **118**, 1131–42.

Bostanian, N.J. & Larocque, N. (2001) Laboratory tests to determine the intrinsic toxicity of four fungicides and two insecticides to the predacious mite *Agistemus fleschneri*. *Phytoparasitica*, **29**, 215–22.

Botha, J.H. & Pringle, K.L. (1995) The predatory mite, *Phytoseiulus persimilis* Athias-Henriot in deciduous fruit orchards of western Cape Province. *J. South Afr. Soc. Hort. Sci.*, **5**, 51–4.

Bottrell, D.G., Barbosa, P. & Gould, F. (1998) Manipulating natural enemies by plant variety selection and modification: a realistic strategy? *Annu. Rev. Entomol.*, **43**, 347–67.

Bounfour, M. & McMurtry, J.A. (1987) Biology and ecology of *Euseius scutalis* (Athias-Henriot) (Acarina: Phytoseiidae). *Hilgardia*, **55**(5), 1–23.

Bourchier, R.S. & McCarty, L.S. (1995) Risk assessment of biological control (predators and parasitoids). *Bull. Entomol. Soc. Can.*, **27**, 126–38.

Bower, C.C. & Thwaite, W.G. (1982) Development and implementation of integrated control of orchard mites in New South Wales. In: *Proc. Australasian Workshop Development and Implementation of IPM* (Ed. by P.J. Cameron, C.H. Wearing & W.M. Kain), pp. 177–90. Government Printer, Auckland.

Bowie, M.H., Worner, S.D. & Chapman, R.B. (1999) The use of image analysis to study the effects of residues of esfenvalerate on the locomotory behaviour of *Panonychus ulmi* and *Typhlodromus pyri* (Acari: Tetranychidae: Phytoseiidae). *Exp. Appl. Acarol.*, **23**, 1–9.

Bowman, C.E. (1984) Some aspects of feeding and digestion in the soil predatory mite *Pergamasus longicornis* (Berlese) (Mesostigmata: Parasitidae). In: *Acarology VI* (Ed. by D.A. Griffiths & C.E. Bowman), Vol. I, pp. 316–33. Ellis Horwood, Chichester.

Boyne, J.V. & Hain, F.P. (1983) Responses of *Neoseiulus fallacis* (Acarina: Phytoseiidae) to different prey densities of *Oligonychus ununguis* (Acarina: Tetranychidae) and to different relative humidity regimes. *Can. Entomol.*, **115**, 1607–14.

Braun, A.R., Bellotti, A.C., Guerrero, J.M. & Wilson, L.T. (1989) Effect of predator exclusion on cassava infested with tetranychid mites (Acar: Tetranychidae). *Environ. Entomol.*, **18**, 711–14.

Bravenboer, L. & Dosse, G. (1962) *Phytoseiulus riegeli* Dosse als Prädator einiger Schadmilben aus der *Tetranychus urticae*-gruppe. *Entomol. Exp. Appl.*, **5**, 291–304.

Breeuwer, J.A.J. & Jacobs, G. (1996) *Wolbachia*: intracellular manipulators of mite reproduction. *Exp. Appl. Acarol.*, **20**, 421–34.

Brickhill, C.D. (1958) Biological studies of two species of tydeid mites from California. *Hilgardia*, **27**, 601–20.

Bridges, J.R. & Moser, J.C. (1986) Relationship of phoretic mites (Acari: Tarsonemidae) to the bluestaining fungus, *Ceratocystis minor*, in trees infested by southern pine beetle (Coleoptera: Scolytidae). *Environ. Entomol.*, **15**, 951–3.

Briese, D.T. (1997) Biological control of St. John's wort: past, present and future. *Pl. Protect. Q.*, **12**, 73–80.

Briese, D.T. & Cullen, J.M. (2001) The use and usefulness of mites in weed biological control. In: *Acarology: Proc. 10th Int. Congr.* (Ed. by R.B. Halliday, D.E. Walter, H.C. Proctor *et al.*), pp. 453–63. CSIRO, Melbourne.

Broadley, R. & Thomas, M. (1995) *The Good Bug Book*. Australian Biological Control, Richmond, NSW.

Brodeur, J. & Cloutier, C. (1992) A modified leaf disk method for rearing predaceous mites (Acarina: Phytoseiidae). *Phytoprotection*, **73**, 69–72.

Brodeur, J., Bouchard, A. & Turcotte, G. (1997) Potential of four species of predatory mites as biological control agents of the tomato russet mite, *Aculops lycopersici* (Massee) (Eriophyidae). *Can. Entomol.*, **129**, 1–6.

Brødsgaard, H.F. & Stengaard Hansen, L. (1992) Effect of *Amblyseius cucumeris* and *Amblyseius barkeri* as biological control agents of *Thrips tabaci* on glasshouse cucumbers. *Biocont. Sci. Tech.*, **2**, 215–23.

Bronstein, J.L. (1998) The contribution of ant-plant protection studies to our understanding of mutualism. *Biotropica*, **30**, 150–61.

van Bronswijk, J.E.M.H., Schooonen, J.M.C.P., Berlie, M.A.F. & Lukoschus, F.S. (1971) On the abundance of *Dermatophagoides pteronyssinus* (Trouessart), 1897 (Pyroglyphidae: Acarina) in house dust. *Res. Popul. Ecol.*, **8**, 67–79.

Brower, J.H., Smith, L., Vail, P.V. & Flinn, P.W. (1996) Biological control. In: *Integrated Management of Insects in Stored Products* (Ed. by B. Subramanyam & D.W. Hagstrum), pp. 223–86. Marcel Dekker, New York.

Bruce, W.A. (1971) Perception of infrared radiation by the spiny rat mite *Laelaps echidnina* (Acari: Laelapidae). *Ann. Entomol. Soc. Am.*, **64**, 925–31.

Bruce, W.A. (1983) Mites as biological control agents of stored product pests. In: *Biological Control of Pests by Mites* (Ed. by M.A. Hoy, G.L. Cunningham & L. Knutson), pp. 74–8. University of California, Special Publ., No. 3304, Berkeley, CA.

Bruce, W.A. (1984) Temperature and humidity: effects on survival and fecundity of *Pyemotes tritici* (Acari: Pyemotidae). *Int. J. Acarol.*, **10**, 135–8.

Bruce, W.A. (1989) Artificial diets for the parasitic mite *Pyemotes tritici* (Acari: Pyemotidae). *Exp. Appl. Acarol.*, **6**, 11–18.

Bruce, W.A. & LeCato, G.L. (1980) *Pyemotes tritici*: a potential new agent for biological control of the red imported fire ant, *Solenopsis invicta* (Acari: Pyemotidae). *Int. J. Acarol.*, **6**, 271–4.

Bruce, W.A. & Wrensch, D.L. (1990) Reproductive potential, sex ratio and mating efficiency of the straw itch mite (Acari: Pyemotidae). *J. Econ. Entomol.*, **83**, 384–91.

Brust, G.E. & House, G.J. (1988) A study of *Tyrophagus putrescentiae* (Acari: Acaridae) as a facultative predator of southern corn rootworm eggs. *Exp. Appl. Acarol.*, **4**, 335–44.

Buchanan, G.A., Bengston, M. & Exley, E.M. (1980) Population growth of *Brevipalpus lewisi* McGregor (Acarina: Tenuipalpidae) on grapevines. *Austr. J. Agric. Res.*, **31**, 957–65.

Budge, S.P., Mcquilken, M.P., Fenlon, J.S. & Whipps, J.M. (1995) Use of *Coniothyrium minitans* and *Gliocladium virens* for biological control of *Sclerotina sclerotiorum* in glasshouse lettuce. *Biol. Cont.*, **5**, 513–22.

Bugg, R.L. & Waddington, C. (1994) Using cover crops to manage arthropod pests of orchards: a review. *Agric. Ecosyst. Environ.*, **50**, 11–28.

van de Bund, C.F. (1972) Some observations on predatory action of mites on nematodes. *Zesz. Prob. Postepow. Nauk Rolniczych*, **129**, 103–10.

Burden, J.P., Hails, R.S., Windass, J.D. *et al.* (2000) Infectivity, speed of kill and productivity of a baculovirus expressing the itch mite toxin Tpx-1 in second and fourth instar larvae of *Trichoplusia ni. J. Invert. Pathol.*, **75**, 226–36.

Burnett, T. (1977) Biological models of two acarine predators of the grain mite, *Acarus siro* L. *Can. J. Zool.*, **55**, 1312–23.

Burrell, R.W. & McCormick, W.J. (1964) *Typhlodromus* and *Amblyseius* (Acarina: Phytoseiidae) as predators on orchard mites. *Ann. Entomol. Soc. Am.*, **57**, 483–7.

Butler, M.I. & Burns, C.W. (1991) Prey selectivity of *Piona exigua*, a planktonic water mite. *Oecologia*, **86**, 210–22.

Buxton, J. (1999) Biological control of the two-spotted spider mite, *Tetranychus urticae*, on hardy nursery stock. *WPRS/OILB Bull.*, **22**(1), 25–7.

Buxton, J.H. & Finlay, R. (1993) Integrated pest management in AYR chrysanthemums. *IOBC/WPRS Bull.*, **16**(8), 33–41.

Cadogan, B.L. & Laing, J.E. (1977) A technique for rearing the predaceous mite *Balaustium putmani* (Acarina: Erythraeidae), with notes on its biology and life history. *Can. Entomol.*, **109**, 1535–44.

Cadogan, B.L. & Laing, J.E. (1981) A study of *Balaustium putmani* (Acarina: Erythraeidae) in apple orchards in southern Ontario. *Proc. Entomol. Soc. Ontario*, **112**, 13–22.

Calderon-Arguedas, O. (1998) Distribucion topografica de *Pimeliaphilus zeledoni* (Acari: Pterygosomatidae) en su hospedador *Triatoma dimidiata* (Latreille, 1811) (Hemiptera: Reduviidae). *Parasitol. Dia*, **22**, 85–9.

Calis, J.N.M., Overmeer, W.P.J. & van der Geest, L.P.S. (1988) Tydeids as alternative prey for phytoseiid mites in apple orchards. *Med. Fac. Landbouww. Rijksuniv. Gent*, **53**, 793–8.

Callaway, R.M., DeLuca, T.H. & Belliveau, W.M. (1999) Biologically-controlled herbivores may increase competitive ability of the noxious weed *Centaurea maculosa*. *Ecology*, **80**, 1196–201.

Caltagirone, L.E. (1970) Overwintering sites for *Metaseiulus occidentalis* in peach orchards. *J. Econ. Entomol.*, **63**, 340–1.

Camporese, P. & Duso, C. (1995) Life history and life table parameters of the predatory mite *Typhlodromus talbii*. *Entomol. Exp. Appl.*, **77**, 149–57.

Camporese, P. & Duso, C. (1996) Different colonization patterns of phytophagous and predatory mites (Acari: Tetranychidae, Phytoseiidae) on three grape varieties: a case study. *Exp. Appl. Acarol.*, **20**, 1–22.

Cantwell, G.E., Cantelo, W.W. & Cantwell, M.A. (1985) Effect of a parasitic mite, *Coccipolipus epilachnae*, on fecundity, food consumption and longevity of the Mexican bean beetle. *J. Entomol. Sci.*, **20**, 199–203.

Capua, S., Cohen, E. & Gerson, U. (1990) Non-specific esterases in mites – a comparative study. *Comp. Biochem. Physiol.*, **96C**, 125–30.

Carèsche, L.A. & Wapshere, A.J. (1974) Biology and host specificity of the Chondrilla gall mite, *Aceria chondrillae* (G. Can.) (Acarina: Eriophyoidea). *Bull. Entomol. Res.*, **64**, 183–92.

Carey, J.R. (1982) Demography of the twospotted spider mite, *Tetranychus urticae* Koch. *Oecologia*, **52**, 389–95.

Carlton, J.T. (1996) Biological invasions and cryptogenic species. *Ecology*, **77**, 1653–5.

Casanueva, M.E. (1993) Phylogenetic studies of the free-living and arthropod associated Laelapidae (Acari: Mesostigmata). *Gayana Zool.*, **57**, 21–46.

Cassagne-Méjean, F. (1966) Contribution a l'étude des Arrenuridae (Acari, Hydrachnellae) de France. *Acarologia*, **8** (Fasc. Suppl.), 1–186.

Castagnoli, M. & Falchini, L. (1993) Suitability of *Polyphagotarsonemous latus* (Banks) (Acari Tarsonemidae) as prey for *Amblyseius californicus* (McGregor) (Acari Phytoseiidae). *Redia*, **76**, 273–9.

Castagnoli, M. & Liguori, M. (1991) Laboratory observations on duration of copulation and egg production of three phytoseiid species fed on pollen. In: *The Acari: Reproduction, Development and Life History Strategies* (Ed. by P.W. Murphy & R. Schuster), pp. 231–39. Ellis Horwood, Chichester.

Castagnoli, M. & Oldfield, G.N. (1996) Other fruit trees and nut trees. In: *Eriophyoid Mites, Their Biology, Natural Enemies and Control* (Ed. by E.E. Lindquist, M.W. Sabelis & J. Bruin), pp. 543–59. Elsevier, Amsterdam.

Castagnoli, M. & Simoni, S. (1991) Influence of various kinds of food on some biological parameters of *Amblyseius cucumeris* (Oud.) (Acarina: Phytoseiidae). In *Modern Acarology* (Ed. by F. Dusbábek & V. Bukva), Vol. 2, pp. 469–73. SPB, The Hague.

Castagnoli, M. & Sobhian, R. (1991) Taxonomy and biology of *Aceria centaureae* (Nal.) and *A. thessalonicae* n. sp. (Acari: Eriophyoidea) associated with *Centaurea diffusa* Lam. in Greece. *Redia*, **74**, 509–24.

Castagnoli, M., Del Bene, G., Gargani, E. & Simoni, S. (1990) Possibilita di controllo di *Thrips tabaci* Lind. e *Frankliniella occidentalis* (Pergande) (Thys. Thripidae) con *Amblyseius cucumeris* (Oud.). *Redia*, **73**, 53–61.

Castineiras, A., Baranowski, R.M. & Glenn, H. (1997) Distribution of *Neoseiulus cucumeris* (Acarina: Phytoseiidae) and its prey, *Thrips palmi* (Thysanoptera: Thripidae) within eggplants in south Florida. *Fl. Entomol.*, **80**, 211–17.

Čatská, V. & Smrž, J. (1989) Relationships between soil mites and microorganisms in apple seedling rhizosphere. In: *Interrelationships Between Microorganisms and Plants in the Soil* (Ed. by V. Vancura & F. Kunc), pp. 377–82. Academia, Prague.

Chambers, R.J., Wright, E.M. & Lind, R.J. (1993) Biological control of glasshouse sciarid flies (*Bradysia* spp.) with the predatory mite, *Hypoaspis miles*, on cyclamen and poinsettia. *Biocont. Sci. Tech.*, **3**, 285–93.

Chant, D.A. (1959) Phytoseiid mites (Acarina: Phytoseiidae). Part I. Bionomics of seven species in southeastern England. Part II. A taxonomic review of the family Phytoseiidae, with descriptions of 38 new species. *Can. Entomol.*, **91**, (Suppl. 12), 1–166.

Chant, D.A. (1961) An experiment in biological control of *Tetranychus telarius* (L.) (Acarina: Tetranychidae) in a greenhouse using the predacious mite *Phytoseiulus persimilis* Athias-Henriot (Phytoseiidae). *Can. Entomol.*, **93**, 437–43.

Chant, D.A. (1992) Trends in the discovery of new species and adult setal patterns in the family Phytoseiidae (Acari: Gamasina), 1839–1989. *Int. J. Acarol.*, **18**, 323–62.

Chant, D.A. & McMurtry, J.A. (1994) A review of the subfamilies Phytoseiinae and Typhlodrominae (Acari: Phytoseiidae). *Int. J. Acarol.*, **20**, 223–310.

Chapman, M.H. & Hoy, M.A. (1991) Relative toxicity of *Bacillus thuringiensis* var. *tenebrionis* to the two spotted spider mite (*Tetranychus urticae* Koch) and its predator *Metaseiulus occidentalis* (Nesbitt) (Acari, Tetranychidae and Phytoseiidae). *J. Appl. Entomol.*, **111**, 147–54.

Charles, J.G. (1998) The settlement of fruit crop arthropod pests and their natural enemies in New Zealand: an historical guide for the future. *Biocont. News Inform.*, **19**, 47–58.

Charles, J.G. & White, V. (1988) Airborne dispersal of *Phytoseiulus persimilis* (Acarina: Phytoseiidae) from a raspberry garden in New Zealand. *Exp. Appl. Acarol.*, **5**, 47–54.

Charles, J.G., Hill, M.G. & Allan, D.J. (1995a) Persistence of the predatory mite, *Hemisarcoptes coccophagus* Meyer (Hemisarcoptidae) on low populations of *Hemiberlesia lataniae* (Signoret) (Diaspididae) in New Zealand. *Israel J. Entomol.*, **29**, 297–300.

Charles, J.G., Hill, M.G. & Allan, D.J. (1995b) Releases and recoveries of *Chilocorus* spp. (Coleoptera: Coccinellidae) and *Hemisarcoptes* spp. (Acari: Hemisarcoptidae) in kiwifruit orchards: 1987–93. *N.Z. J. Zool.*, **22**, 319–24.

Charles, J.G., Allan, D.J., Wearing, C.H. *et al.* (1998) Releases of *Hemisarcoptes coccophagus* Meyer (Acari: Hemisarcoptidae), a predator of armoured scale insects, in the South Island. *N.Z. Entomol.*, **21**, 93–8.

Charudatan, R. (1986) Integrated control of waterhyacinth (*Eichhornia crassipes*) with a pathogen, insects and herbicides. *Weed Sci.*, **34** (Suppl. 1), 26–30.

Chen, H., Xu, Y., Fei, D. & Deng, G. (1988) A preliminary study on the predaceous mite *Tydeus* sp. of *Phyllocoptruta oleivora* Ashmead. *Chin. J. Biol. Cont.*, **4**, 181–2 (in Chinese).

Chen, P.-R., & Zhang, Z.-Q. (1991) Biology of *Allothrombium pulvinum* Ewing (Acari, Trombidiidae) and its impact on twospotted spider mite (Acari, Tetranychidae) in cotton fields. *J. Appl. Entomol.*, **112**, 31–7.

Chen, P.-R., Zhang, Z.-Q., Wang, K. *et al.* (1994) *Allothrombium pulvinum* Ewing (Acari, Trombidiidae), an important early-season natural enemy of *Aphis gossypii* Glover (Hom., Aphididae) in cotton. *J. Appl. Entomol.*, **117**, 113–21.

Cheng, H.-Y., Ming, W.-J. & Ge, Q.-J. (1983) Experiments on integrated pest control of *Matsucoccus matsumurae* Kuwana. *J. Nanjing Tech. Coll. Forest Prod.*, **1**, 11–30 (in Chinese with English abstract).

Chiang, H.C. (1970) Effects of manure applications and mite predation on corn rootworm populations in Minnesota. *J. Econ. Entomol.*, **63**, 934–6.

Childers, C.C. (1994) Biological control of phytophagous mites on Florida citrus utilizing predatory arthropods. In: *Pest Management in the Subtropics; Biological Control – A Florida Perspective* (Ed. by D. Rosen, F.D. Bennett & J.L. Capinera), pp. 255–88. Intercept, Andover.

Childers, C.C. & Enns, W.R. (1975a) Field evaluation of early season fungicide substitutions on tetranychid mites and the predators *Neoseiulus fallacis* and *Agistemus fleschneri* in two Missouri apple orchards. *J. Econ. Entomol.*, **68**, 719–24.

Childers, C.C. & Enns, W.R. (1975b) Predaceous arthropods associated with spider mites in Missouri apple orchards. *J. Kansas Entomol. Soc.*, **48**, 453–68.

Childers, C.C. & Rock, G.C. (1981) Observations on the occurrence and feeding habits of *Balaustium putmani* (Acari: Erythraeidae) in North Carolina apple orchards. *Int. J. Acarol.*, **7**, 63–8.

Chow, Y.S. (1996) Biology and life stages of *Charletonia taiwanensis* Tsai & Chow (Erythraeidae) in Taiwan. In: *Acarology: Proc. 10th Int. Congr.* (Ed. by R. Mitchell, D.J. Horn, G.R. Needham & W.C. Wellbourn), pp. 359–61. Ohio Biological Survey, Columbus, OH.

Cicolani, B. (1979) The intrinsic rate of natural increase in dung macrochelid mites, predators of *Musca domestica* eggs. *Boll. Zool.*, **46**, 171–8.

Clancy, D.W. & Pollard, H.N. (1952) The effect of DDT on mite and predator populations in apple orchards. *J. Econ. Entomol.*, **49**, 109–14.

Clarke, A.R. & Walter, G.H. (1995) 'Strains' and the classical biological control of insect pests. *Can. J. Zool.*, **73**, 1777–90.

Clarke, G.M. & McKenzie, L.J. (1992) Fluctuating asymmetry as a quality control indicator for insect mass rearing processes. *J. Econ. Entomol.*, **85**, 2045–50.

Clements, D.R. & Harmsen, R. (1990) Predatory behavior and prey-stage preference of stigmaeid and phytoseiid mites and their potential compatibility in biological control. *Can. Entomol.*, **122**, 321–8.

Clements, D.R. & Harmsen, R. (1992) Stigmaeid–phytoseiid interactions and the impact of natural enemy complexes on plant-inhabiting mites. *Exp. Appl. Acarol.*, **14**, 327–41.

Clements, D.R. & Harmsen, R. (1993) Prey preferences of adult and immature *Zetzellia mali* Ewing (Acari: Stigmaeidae) and *Typhlodromus caudiglans* Schuster (Acari: Phytoseiidae). *Can. Entomol.*, **125**, 967–9.

Cloutier, C. & Johnson, S.G. (1993) Interaction between life stages in a phytoseiid predator: western flower thrips prey killed by adults as food for protonymphs of *Amblyseius cucumeris*. *Entomol. Exp. Appl.*, **17**, 441–9.

Cohen, A.C., Nordlund, D.A. & Smith, R.A. (1999) Mass rearing of entomophagous insects and predaceous mites: are the bottlenecks biological, engineering, economic or cultural? *Biocont. News Inform.*, **20**(3), 85–90.

Cohen, E., Gerson, U. & Mozes-Koch, R. (1993) Epoxide hydrolase in the bulb mite *Rhizoglyphus robini*: properties and induction. *Exp. Appl. Acarol.*, **17**, 381–92.

Coli, W.M., Ciurlino, R.A. & Hosmer, T. (1994) Effect of understory and border vegetation composition on phytophagous and predatory mites in Massachusetts commercial apple orchards. *Agric. Ecosyst. Environ.* **50**, 49–60.

Colkesen, T. & Sekeroglu, E. (2000) The effect of *Amblyseius longispinosus* Evans (Acarina: Phytoseiidae) on *Tetranychus cinnabarinus* Boisd. (Acarina: Tetranychidae) on different cucumber cultivars. *IOBC/WPRS Bull.*, **23**(1), 187–93.

Coll, M. & Izraylevich, S. (1997) When predators also feed on plants: effects of competition and plant quality on omnivore–prey population dynamics. *Ann. Entomol. Soc. Am.*, **90**, 155–61.

Collyer, E. (1958) Some insectary experiments with predaceous mites to determine their effect on the development of *Metatetranychus ulmi* (Koch) populations. *Entomol. Exp. Appl.*, **1**, 138–46.

Collyer, E. (1964a) A summary of experiments to demonstrate the role of *Typhlodromus pyri* Scheut. in the control of *Panonychus ulmi* (Koch) in England. *Acarologia*, **6** (fasc. hors ser.), 363–71.

Collyer, E. (1964b) Phytophagous mites and their predators in New Zealand orchards. *N.Z. J. Agric. Res.*, **7**, 551–68.

Collyer, E. (1980) Integrated control of apple pests in New Zealand. 16. Progress with integrated control of European red mite. *N.Z. J. Zool.*, **7**, 271–9.

Colwell, R.K. & Naeem, S. (1994) Life-history patterns of hummingbird flower mites in relation to host phenology and morphology. In: *Mites, Ecological and Evolutionary Analyses of Life-history Patterns* (Ed. by M.A. Houck), pp. 23–44. Chapman & Hall, New York.

Congdon, B.D. & McMurtry, J.A. (1985) Biosystematics of *Euseius* on California citrus and avocado with the description of a new species (Acari: Phytoseiidae). *Int. J. Acarol.*, **11**, 23–30.

Congdon, B.D. & McMurtry, J.A. (1988) Prey selectivity in *Euseius tularensis* (Acari: Phytoseiidae). *Entomophaga*, **33**, 281–7.

Congdon, B.D. & Tanigoshi, L.K. (1983) Indirect toxicity of dimethoate to the predaceous mite *Euseius hibisci* (Chant) (Acari: Phytoseiidae). *Environ. Entomol.*, **12**, 933–5.

Conijn, C.G.M. & Groen, N.P.A. (1990) Physical, biological and chemical control of *Taeniothrips simplex* in gladiolus. *Acta Hort.*, **266**, 539–46.

Conijn, C.G.M., van Aartrijk, J. & Lesna, I. (1996) Flower bulbs. In: *Eriophyoid Mites, Their Biology, Natural Enemies and Control* (Ed. by E.E. Lindquist, M.W. Sabelis & J. Bruin), pp. 651–9. Elsevier, Amsterdam.

Conijn, C.G.M., Lesna, I. & Altena, K. (1997) Biological control of the bulb mite *Rhizoglyphus robini* by the predatory mite *Hypoaspis aculeifer* on lilies: implementation in practice. *Acta Hort.*, **430**, 619–24.

Cook, D.R. (1974) Water mite genera and subgenera. *Mem. Am. Entomol. Inst.*, **21**, 1–860.

Cook, R.J. (1981) Biological control of plant pathogens: overview. In: *Biological Control in Crop Production* (Ed. by G.C. Papavizas), pp. 23–44. Allanheld, Osmun, Totowa.

Cook, R.J. & Baker, K.F. (1983) *The Nature and Practice of Biological Control of Plant Pathogens*. American Phytopathology Society, St. Paul, MN.

Cook, W.J., Smith, B.P. & Brooks, R.J. (1989) Allocation of reproductive effort in female *Arrenurus* spp. water mites (Acari: Hydrachnidia; Arrenuridae). *Oecologia*, **79**, 184–8.

Copping, L.G. (Ed.) (1998) *The BioPesticide Manual*, 1st Edn. British Crop Protection Council, Farnham.

Copping, L.G. & Menn, J.J. (2000) Biopesticides: a review of their action, applications and efficacy. *Pest. Manage. Sci.*, **56**, 651–76.

Cordo, H.A. & DeLoach, C.J. (1976) Biology of the waterhyacinth mite in Argentina. *Weed Sci.*, **24**, 245–9.

Corpuz-Raros, L.A. (1996) Philippine predatory mites of the family Cunaxidae (Acari). 7. Genus *Pulaeus* Den Heyer with records of two species from central Kalimantan, Borneo and Java, Indonesia. *Philipp. Entomol.*, **10**, 119–38.

Corpuz-Raros, L.A. (2000) Guide to Philippine predatory mites: family Cheyletidae Leach. *University of the Philippines, Museum Pub. Natural History*, No. 2. Los Baños.

Corpuz-Raros, L.A. & Garcia, R.C. (1995) Philippine predatory mites of the family Cunaxidae (Acari). 1. Genus *Cunaxa* Von Heyden. *Philipp. Entomol.*, **9**, 605–24.

Cortesero, A.M., Stapel, J.O. & Lewis, W.J. (2000) Understanding and manipulating plant attributes to enhance biological control. *Biol. Cont.*, **17**, 35–49.

Costa, M. (1969) The associations between mesostigmatic mites and coprid beetles. *Acarologia*, **11**, 411–28.

Costa, M. (1971) Mites of the genus *Hypoaspis* Canestrini, 1884 s. *str.* and related forms (Acari: Mesostigmata) associated with beetles. *Bull. Br. Mus. (Nat. Hist.) (Zool.)*, **21**, 67–98.

Cox, P.D. & Wilkin, D.R. (1998) A review of the options for biological control against invertebrate pests of stored grain in the UK. *IOBC Bull.*, **21**(3), 27–32.

Craemer, C. (1995) Host specificity, and release in South Africa, of *Aceria malherbae* Nuzzaci (Acari: Eriophyoidea), a natural enemy of *Convolvulus arvensis* L. (Convolvulaceae). *Afr. Entomol.*, **3**, 213–15.

Cranshaw, W., Casey Sclar, D. & Cooper, D. (1996) A review of 1994 pricing and marketing policy by suppliers of organisms for biological control of arthropods in the United States. *Biol. Cont.*, **6**, 291–6.

Croft, B.A. (1990) *Arthropod Biological Control Agents and Pesticides*. Wiley, New York.

Croft, B.A. (1994) Biological control of apple mites by a phytoseiid mite complex and *Zetzellia mali* (Acari: Stigmaeidae): long-term effects and impact of azinphosmethyl on colonization by *Amblyseius andersoni* (Acari: Phytoseiidae). *Environ. Entomol.*, **23**, 1317–25.

Croft, B.A. & Barnes, M.M. (1971) Comparative studies on four strains of *Typhlodromus occidentalis*. III. Evaluations of releases of insecticide resistant strains into an apple orchard ecosystem. *J. Econ. Entomol.*, **64**, 845–50.

Croft, B.A. & Blyth, E.J. (1979) Aspects of the functional, ovipositional and starvation responses of *Amblyseius fallacis* to prey density. In: *Recent Advances in Acarology* (Ed. by J.G. Rodriguez), Vol. 1, pp. 41–7. Academic Press, New York.

Croft, B.A. & Brown, A.W.A. (1975) Responses of arthropods natural enemies to insecticides. *Annu. Rev. Entomol.*, **20**, 285–335.

Croft, B.A. & Croft, M.B. (1996) Intra- and interspecific predation among adult female phytoseiid mites (Acari: Phytoseiidae): effect on survival and reproduction. *Environ. Entomol.*, **25**, 853–8.

Croft, B.A. & Hoying, S.A. (1977) Competitive displacement of *Panonychus ulmi* (Acarina: Tetranychidae) by *Aculus schlechtendali* (Acarina: Eriophyidae) in apple orchards. *Can. Entomol.*, **109**, 1025–34.

Croft, B. A. & Hoyt, S.C. (Eds) (1983) *Integrated Management of Insect Pests of Pome and Stone Fruits*. John Wiley, New York.

Croft, B.A. & Jeppson, L.R. (1970) Comparative studies on four strains of *Typhlodromus occidentalis*. II. Laboratory toxicity of ten compounds common to apple pest control. *J. Econ. Entomol.*, **63**, 1528–31.

Croft, B.A. & McGroarty, D.L. (1977) The role of *Amblyseius fallacis* (Acarina: Phytoseiidae) in Michigan apple orchards. *Farm Sci.*, **333**, 2–22.

Croft, B.A. & McMurtry, J.A. (1972) Minimum releases of *Typhlodromus occidentalis* to control *Tetranychus mcdanieli* on apple. *J. Econ. Entomol.*, **65**, 188–91.

Croft, B.A. & MacRae, I.V. (1992) Persistence of *Typhlodromus pyri* and *Metaseiulus occidentalis* (Acari: Phytoseiidae) on apple after inoculative release and competition with *Zetzellia mali* (Acari: Stigmaeidae). *Environ. Entomol.*, **21**, 1168–77.

Croft, B.A. & MacRae, I.V. (1993) Biological control of apple mites: impact of *Zetzellia mali* (Acari: Stigmaeidae) on *Typhlodromus pyri* and *Metaseiulus occidentalis* (Acari: Phytoseiidae). *Environ. Entomol.*, **22**, 865–73.

Croft, B.A. & Meyer, R.H. (1973) Carbamate and organophosphate resistance patterns in populations of *Amblyseius fallacis*. *Environ. Entomol.*, **2**, 691–5.

Croft, B.A. & Nelson, E.E. (1972) An index to predict efficient interactions of *Typhlodromus occidentalis* in control of *Tetranychus mcdanieli* in southern California apple trees. *J. Econ. Entomol.*, **65**, 310–12.

Croft, B.A. & Zhang, Z.-Q. (1999) Assessing the roles of species in a predator mite complex using life history traits of female adult and immature mites. In: *Acarology IX, Symposia* (Ed. by G.R. Needham, R. Mitchell, D.J. Horn & W.C. Welbourn), pp. 197–204. Ohio Biological Survey, Columbus, OH.

Croft, B.A., Messing, R.H., Dunley, J.E. & Strong, W.B. (1993) Effects of humidity on eggs and immatures of *Neoseiulus fallacis, Amblyseius andersoni, Metaseiulus occidentalis* and *Typhlodromus pyri* (Phytoseiidae): implications for biological control on apple, caneberry, strawberry and hop. *Exp. Appl. Acarol.*, **17**, 451–9.

Croft, B.A., Kim, S.S. & Kim, D.I. (1995a) Absorption and cannibalism: do adult female phytoseiids conserve egg resources when prey densities decline rapidly? *Exp. Appl. Acarol.*, **19**, 347–56.

Croft, B.A., Kim, S.S. & Kim, D.I. (1995b) Leaf residency and interleaf movement of four phytoseiid mites (Acari: Phytoseiidae) on apple. *Environ. Entomol.*, **24**, 1344–51.

Croft, B.A., Monetti, L.N. & Pratt, P.D. (1998a) Comparative life histories and predation types: are *Neoseiulus californicus* and *N. fallacis* (Acari: Phytoseiidae) similar Type II selective predators of spider mites? *Environ. Entomol.*, **27**, 531–8.

Croft, B.A., McMurtry, J.A. & Luh, H.-K. (1998b) Do literature records of predation reflect food specialization and predation types among phytoseiid mites (Acari: Phytoseiidae)? *Exp. Appl. Acarol.*, **22**, 467–80.

Croft, B.A., Pratt, P.D., Koskela, G. & Kaufman, D. (1998c) Predation, reproduction, and impact of phytoseiid mites (Acari: Phytoseiidae) on cyclamen mite (Acari: Tarsonemidae) on strawberry. *J. Econ. Entomol.*, **91**, 1307–14.

Cromroy, H.L. (1979) Eriophyoidea in biological control of weeds. In: *Recent Advances in Acarology* (Ed. by J.G. Rodriguez), Vol. 1, pp. 473–5. Academic Press, New York.

Cromroy, H.L. (1983) Potential use of mites in biological control of terrestrial and aquatic weeds. In: *Recent Advances in Knowledge of the Phytoseiidae* (Ed. by M.A. Hoy, G.L. Cunningham & L. Knutson), pp. 61–6. University of California, Div. Agric. Special Pub., No. 3304, Berkeley, CA.

Cross, E.A. (1965) The generic relationships of the family Pyemotidae (Acarina: Trombidiformes). *Kansas Univ. Sci. Bull.*, **45**, 29–275.

Cross, E.A. & Krantz, G.W. (1964) Two new species of the genus *Acarophenax* Newstead and Duvall 1918 (Acarina: Pyemotidae). *Acarologia*, **6**, 287–95.

Cross, J.V., Burgess, C.M. & Hanks, G.R. (1996) Integrating insecticide use with biological control of two spotted spider mite (*Tetranychus urticae*) by *Phytoseiulus persimilis* on strawberry in the UK. In: *1996 Brighton Crop Prot. Conf.*, Vol. 3, pp. 899–906. Farnham.

Cross, W.H., McGovern, W.L. & Cross, E.A. (1975) Insect hosts of the parasitic mites called *Pyemotes ventricosus* (Newport). *J. Geor. Entomol. Soc.*, **10**, 1–8.

Crump, N.S., Cother, E.J. & Ash, G.J. (1999) Clarifying the nomenclature in microbial weed control. *Biocont. Sci. Tech.*, **9**, 89–97.

Cullen, J.M., Groves, R.H. & Alex, J.F. (1982) The influence of *Aceria chondrillae* on the growth and reproductive capacity of *Chondrilla juncea*. *J. Appl. Ecol.*, **19**, 529–37.

Cullen, J.M. & Moore, A.D. (1983) The influence of three populations of *Aceria chondrillae* on three forms of *Chondrilla juncea*. *J. Appl. Ecol.*, **20**, 235–43.

Cunliffe, F. (1952) Biology of the cockroach parasite, *Pimeliaphilus podapolipophagus* Tragardh, with a discussion of the genera *Pimeliaphilus* and *Hirstiella* (Acarina, Pterygosomidae). *Proc. Entomol. Soc. Wash.*, **54**, 153–69.

Currie, G.A. (1934) The bdellid mite *Biscirus lapidaria* Kramer, predatory on the lucerne flea *Sminthurus viridis* L. in Western Australia. *J. Aust. Counc. Sci. Ind. Res.*, **7**, 9–20.

Cutright, C.R. (1944) Populations of the European red mite as affected by spray schedules. *J. Econ. Entomol.*, **37**, 499–502.

Daneshvar, H. & Rodriguez, J.G. (1975) Toxicity of organophosphorus systemic pesticides to predator mites and prey. *Entomol. Exp. Appl.*, **18**, 297–301.

Daniel, M. (1981) Bionomics of the predaceous mite *Amblyseius channabasavanni* (Acari: Phytoseiidae), predaceous on the palm mite. In: *Contributions to Acarology in India* (Ed. by G.P. ChannaBasavanna), pp. 167–73. Acarological Society of India, Bangalore.

Davidson, J.A. (1958) A new species of lizard mite and a generic key to the family Pterygosomidae (Acarina, Anystoidea). *Proc. Entomol. Soc. Wash.*, **60**, 75–9.

Davies, D.M. (1959) The parasitism of black flies (Diptera, Simuliidae) by larval water mites mainly of the genus *Sperchon*. *Can. J. Zool.*, **37**, 353–69.

Davis, R., Flechtmann, C.H.W., Boczek, J. & Barké, H.E. (1982) *Catalog of Eriophyid Mites (Acari: Eriophyoidea)*. Warsaw Agricultural University Press, Warsaw.

DeBach, P. (1946) An insecticidal check method for measuring the efficacy of entomophagous insects. *J. Econ. Entomol.*, **39**, 695–7.

DeBach, P. (Ed.) (1964) *Biological Control of Insect Pests and Weeds*. Chapman and Hall, London.

DeBach, P. & Rosen, D. (1991) *Biological Control by Natural Enemies*, 2nd Edn. Cambridge University Press, Cambridge.

Decou, G.C. (1994) Biological control of the two-spotted spider mite (Acarina: Tetranychidae) on commercial strawberries in Florida with *Phytoseiulus persimilis* (Acarina: Phytoseiidae). *Fl. Entomol.*, **77**, 33–41.

De Courcy Williams, M.E. (2001) Biological control of thrips on ornamental crops: interactions between the predatory mite *Neoseiulus cucumeris* (Acari: Phytoseiidae) and western flower thrips, *Frankliniella occidenatlis* (Thysanoptera: Thripidae) on cyclamen. *Biocont. Sci. Tech.*, **11**, 41–55.

De Jesus, L.A.R. & Rueda, L.M. (1992) Seasonal abundance, life history and predatory activity of common mites (Acarina) attacking house fly (*Musca domestica* Linn.) and other filth flies (Diptera) associated with poultry and livestock manure. *Philipp. Entomol.*, **8**, 1213–27.

De Jong, J.H., Lobbes, P.V. & Bolland, H.R. (1981) Karyotypes and sex determination in two species of laelapid mites (Acari: Gamasida). *Genetica*, **55**, 187–90.

De Klerk, M.-L. & Ramakers, P.M.J. (1986) Monitoring population densities of the phytoseiid predator *Amblyseius cucumeris* and its prey after large scale introductions to control *Thrips tabaci* on sweet pepper. *Med. Fac. Landbouww. Rijksuniv. Gent*, **51/3a**, 1045–8.

Delattre, P. (1974) Étude de l'efficacité prédatrice de *Zetzellia mali* (Acarina, Stigmaeidae) vis-à-vis du tétranyque du pommier, *Panonychus ulmi* (Acarina, Tetranychidae). *Entomophaga*, **19**, 13–31.

Delbac, L., Lecharpentier, P., Fos, A. & Stockel, J. (1996) La confusion sexuelle contre l'Eudémis vers un équilibre biologique de l'acarofaune du vignoble. *Phytoma*, **484**, 43–7.

Del Fosse, E.S. (1977a) Temperature optima for development of *Neochetina eichhorniae* and *Orthogalumna terebrantis*. *Fl. Entomol.*, **60**, 109–13.

Del Fosse, E.S. (1977b) Effect of *Orthogalumna terebrantis* (Acari: Galumnidae) on *Neochetina eichhorniae* (Col., Curculionidae) eggs and oviposition. *Entomophaga*, **22**, 359–63.

Del Fosse, E.S. (1978) Interaction between the mottled waterhyacinth weevil, *Neochetina eichhorniae* Warner and the waterhyacinth mite, *Orthogalumna terebrantis* Wallwork. In: *Proc. IV Int. Symp. Biological Control of Weeds* (Ed. by T.E. Freeman), pp. 93–7. University of Florida, Gainesville, FL.

De Leon, D. (1967) *Some Mites of the Caribbean Area*. Allen Press, Lawrence, KS.

De Lillo, E. & Porcelli, F. (1993) *Pyemotes herfsi* (Oud.) (Acari, Pyemotidae) antagonista di *Melanaspis inopinatus* (Leon.) (Coccoidea, Diaspididae) in Puglia. *Entomologia*, **27**, 117–24.

Delucchi, V., Aeschlimann, J.P. & Graf, E. (1975) The regulating action of egg predators on the populations of *Zeiraphera diniana* Guénée (Lep. Tortricidae). *Mitt. Schw. Entomol. Gesell.*, **48**, 37–45.

Denegri, G.M. (1993) Review of oribatid mites as intermediate hosts of tapeworms of the Anoplocephalidae. *Exp. Appl. Acarol.*, **17**, 567–80.

Denmark, H.A. & Schicha, E. (1983) Revision of the genus *Phytoseiulus* Evans (Acarina: Phytoseiidae). *Int. J. Acarol.*, **9**, 27–35.

Denmark, H.A., Evans, G.A., Aguilar, H. *et al.* (1999) *Phytoseiidae of Central America (Acari: Mesostigmata)*. Indira, West Bloomfield, MI.

Dennill, G.B. & Moran, V.C. (1989) On insect–plant associations in agriculture and the selection of agents for weed biocontrol. *Ann. Appl. Biol.*, **114**, 157–66.

Dicke, M. (1988) Prey preference of the phytoseiid mite *Typhlodromus pyri*. 1. Response to volatile kairomones. *Exp. Appl. Acarol.*, **4**, 1–13.

Dicke, M. (1995) Plant characteristics influence biological control agents: implications for breeding for host plant resistance. *IOBC/WPRS Bull.*, **19**(5), 72–80.

Dicke, M. & Dijkman, H. (2001) Within-plant circulation of systemic elicitor of induced defence and release from roots of elicitor that affects neighbouring plants. *Biochem. Syst. Ecol.*, **29**, 1075–87.

Dicke, M. & Sabelis, M.W. (1988) Infochemical terminology: based on cost–benefit analysis rather than origin of compounds? *Func. Ecol.*, **2**, 131–9.

Dicke, M., De Jong, M., Alers, M.P.T. *et al.* (1989) Quality control of mass-reared arthropods: nutritional effects on performance of predatory mites. *J. Appl. Entomol.*, **108**, 462–75.

Dicke, M., Sabelis, M.W., Takabayashi, J. *et al.* (1990) Plant strategies of manipulating predator–prey interactions through allelochemicals: prospects for application in pest control. *J. Chem. Ecol.*, **16**, 3091–118.

Dicke, M., Dijkman, H. & Wunderink, R. (1991) Response to synomones as a parameter in quality control of predatory mites. In: *Proc. 5th Workshop IOBC Working Group, 'Quality Control of Mass Reared Arthropods'*, Wageningen, pp. 56–65.

Dicke, M., Gols, R., Ludeking, D. & Posthumus, M.A. (1999) Jasmonic acid and herbivory differentially induce carnivore-attracting plant volatiles in lima bean plants. *J. Chem. Ecol.*, **25**, 1907–22.

Dicke, M., Schütte, C. & Dijkman, H.M. (2000) Change in behavioral response to herbivore-induced plant volatiles in a predatory mite population. *J. Chem. Ecol.*, **26**, 1497–514.

Di Sabatino, A., Gerecke, R. & Martin, P. (2000) The biology and ecology of lotic water mites (Hydrachnidia). *Freshwater Biol.*, **44**, 47–62.

Dmoch, J. (1995) *Arctoseius semiscissus* (Berlese 1892) (Acarina: Ascidae) phoretic on mushroom sciarid flies in Poland – a possible agent for biological control of sciarids in mushroom houses. In: *Science and Cultivation of Edible Fungi* (Ed. by T.J. Elliot), pp. 533–7. Balkema, Rotterdam.

Donegan, K.K., Seidler, R.J., Doyle, J.D. *et al.* (1999) A field study with genetically engineered alfalfa inoculated with recombinant *Sinorhizobium meliloti*: effect on the soil ecosystem. *J. Appl. Ecol.*, **36**, 920–36.

Dong, Y.-C., Ran, R.-B. & Xiang, J.-Y. (1996) Biology of *Allothrombium ovatum* (Acari: Trombidiidae) and its controlling effect on *Aphis gossypii* (Homoptera: Aphididae). *Syst. Appl. Acarol.*, **1**, 35–40.

Donia, A.R.A., Helal, E.M., El-Hamid, M.M.A. & Zakzouk, E.A. (1995) Mass rearing and field evaluation of released predaceous mite, *Euseius scutalis* (Athias-Henriot) on the citrus whitefly, *Aleurotrachelus citri* (Priesner & Hosny). *Alexandria J. Agric. Res.*, **40**, 209–19.

Dosse, G. (1967) Schädmilben des Libanons und ihre Prädatoren. *Z. Angew. Entomol.*, **59**, 16–48.

Dover, M.J., Croft, B.A., Welch, S.M. & Tummalo, R.L. (1979) Biological control of *Panonychus ulmi* (Acarina: Tetranychidae) by *Amblyseius fallacis* (Acarina: Phytoseiidae) on apple: a prey–predator model. *Environ. Entomol.*, **8**, 282–92.

Dowd, P.F., Smith, C.M. & Sparks, T.C. (1983) Detoxification of plant toxins by insects. *Insect Biochem.*, **13**, 453–68.

Downing, R.S. & Moilliet, T.K. (1967) Relative densities of predacious and phytophagous mites on three varieties of apple tree. *Can. Entomol.*, **99**, 738–41.

van Driesche, R.G. & Hoddle, M. (1997) Should arthropod parasitoids and predators be subject to host range testing when used as biological control agents? *Agric. Human Values*, **14**, 211–26.

Drukker, B., Yaninek, J.S. & Herren, H.R. (1993) A packaging and delivery system for aerial release of phytoseiids for biological control. *Exp. Appl. Acarol.*, **17**, 129–43.

Drukker, B., Janssen, A., Ravensberg, W. & Sabelis, M.W. (1997) Improved control capacity of the mite predator *Phytoseiulus persimilis* (Acari: Phytoseiidae) on tomato. *Exp. Appl. Acarol.*, **21**, 507–18.

Drummond, F.A., Logan, P.A., Casagrande, R.A. & Gregson, F.A. (1985) Host specificity tests of *Chrysomelobia labidomerae*, a mite parasitic on the Colorado potato beetle. *Int. J. Acarol.*, **11**, 169–72.

Drummond, F.A. & Casagrande, R.A. (1989) Effect of the straw itch mite on larvae and adults of the Colorado potato beetle. *Am. Potato J.*, **66**, 161–3.

Drummond, F.A., Casagrande, R.A. & Logan, P.A. (1989) Population dynamics of *Chrysomelobia labidomerae* Eickwort, a parasite of the Colorado potato beetle. *Int. J. Acarol.*, **15**, 31–45.

Drummond, F.A., Casagrande, R.A. & Logan, P.A. (1992) Impact of the parasite, *Chrysomelobia labidomerae* Eickwort, on the Colorado potato beetle. *Int. J. Acarol.*, **18**, 107–15.

Dubitzki, E. (1981) Spider mites (Acari: Tetranychidae) infesting wild plants in Israel. DPhil Thesis, Hebrew University of Jerusalem (in Hebrew, with English abstract).

Dunley, J.E. & Croft, B.A. (1994) Gene flow measured by allozyme analysis in pesticide resistant *Typhlodromus pyri* occurring within and near apple orchards. *Exp. Appl. Acarol.*, **18**, 201–11.

Dunley, J.E. & Croft, B.A. (1996) Eriophyoids as competitors of other phytophagous mites. In: *Eriophyoid Mites, Their Biology, Natural Enemies and Control* (Ed. by E.E. Lindquist, M.W. Sabelis & J. Bruin), pp. 751–5. Elsevier, Amsterdam.

Duso, C. (1989) Role of predatory mites *Amblyseius aberrans* (Oud.), *Typhlodromus pyri* Scheuten and *Amblyseius andersoni* (Chant) (Acari: Phytoseiidae) in vineyards. 1. The effects of single or mixed phytoseiid population releases on spider mite densities (Acari: Tetranychidae). *J. Appl. Entomol.*, **107**, 474–92.

Duso, C. (1992) Role of *Amblyseius aberrans* (Oud.), *Typhlodromus pyri* Scheuten and *Amblyseius andersoni* (Chant) (Acari, Phytoseiidae) in vineyards. *J. Appl. Entomol.*, **114**, 455–62.

Duso, C. & Camporese, P. (1991) Development times and oviposition rates of predatory mites *Typhlodromus pyri* and *Amblyseius andersoni* (Acari: Phytoseiidae) reared on different foods. *Exp. Appl. Acarol.*, **13**, 117–28.

Duso, C. & Pasqualetto, C. (1993) Factors affecting the potential of phytoseiid mites (Acari: Phytoseiidae) as biocontrol agents in North-Italian vineyards. *Exp. Appl. Acarol.*, **17**, 241–58.

Dyer, J.G. & Swift, F.C. (1979) Sex ratio in field populations of phytoseiid mites (Acarina: Phytoseiidae). *Ann. Entomol. Soc. Am.*, **72**, 149–54.

Dyer, L.A. & Gentry, G. (1999) Predicting natural-enemy responses to herbivores in natural and managed systems. *Ecol. Appl.*, **9**, 402–8.

Dyne, D., Campion, K. & Griffin, P. (1996) Occupional allergy among workers producing arthropods for organic pest control purposes. *Ann. Agric. Environ. Med.*, **3**, 33–6.

Easterbrook, M.A. (1992) The possibilities for control of two-spotted spider mite *Tetranychus urticae* on field-grown strawberries in the UK by predatory mites. *Biocont. Sci. Tech.*, **2**, 235–45.

Ebeling, W. (1948) Effect of citrus red mites on a California red scale population. *J. Econ. Entomol.*, **41**, 109.

Edwards, O.R., Melo, E.L., Smith, L. & Hoy, M.A. (1998) Discrimination of three *Typhlodromalus* species (Acari: Phytoseiidae) using random amplified polymorphic DNA markers. *Exp. Appl. Acarol.*, **22**, 101–9.

Egan, M.E. & Moss, W.W. (1969) The life cycle and behavior of a cockroach mite, *Proctolaelaps nauphoetae* (Acari: Mesostigmata: Ascidae). *Notulae Nat, Acad. Nat. Sci. Philadelphia*, **420**, 1–9.

Ehara, S. (1962) Notes on some predatory mites (Phytoseiidae and Stigmaeidae). *Jpn. J. Appl. Entomol. Zool.*, **6**, 53–60.

Ehara, S. & Amano, H. (1998) A revision of the mite family Phytoseiidae in Japan (Acari, Gamasina), with remarks on its biology. *Species Diversity*, **3**, 25–73.

Ehara, S. & Oomen-Kalsbeek, F. (1983) Stigmaeid mites associated with tea plants in Indonesia (Prostigmata: Stigmaeidae). *Int. J. Acarol.*, **9**, 19–26.

Ehler, L.E. (1990) Introduction strategies in biological control of insects. In: *Critical Issues in Biological Control* (Ed. by M. Mackauer, L.E. Ehler & J. Roland), pp. 111–34. Intercept, Andover.

Ehler, L.E. (1992) Guild analysis in biological control. *Environ. Entomol.*, **21**, 26–40.

El Badry, E. (1967) Three new species of phytoseiid mites preying on the cotton white fly, *Bemisia tabaci* in the Sudan (Acarina: Phytoseiidae). *Entomologist*, **100**, 106–11.

Elbadry, E.A. (1968) Biological studies on *Amblyseius aleyrodis*, a predator of the cotton white-fly (Acarina, Phytoseiidae). *Entomophaga*, **13**, 323–9.

Elbadry, E. (1974) Egg laying magnitude as influenced by rate of prey consumption in two predaceous mites. In: *Proc. 4th Int. Congr. Acarology*, (Ed. by E. Piffl), pp. 653–6. Akadémiai Kiado, Budapest.

Elbadry, E.A. (1979) Management of mite pests of cotton in Egypt. In: *Recent Advances in Acarology* (Ed. by J.G. Rodriguez), Vol. I, pp. 49–57. Academic Press, New York.

Elbadry, E.A. & Tawfik, M.S.F. (1970) Life cycle of the mite *Adactylidium* sp. (Acarina: Pyemotidae), a predator of thrips eggs in the United Arab Republic. *Ann. Entomol. Soc. Am.*, **59**, 458–61.

Elbadry, E.A., Abo Elghar, M.R., Hassan, S.M. & Kilany, S.M. (1969a) Life history studies on the predatory mite *Agistemus exsertus*. *Ann. Entomol Soc. Am.*, **62**, 649–51.

Elbadry, E.A., Abo Elghar, M.R., Hassan, S.M. & Kilany, S.M. (1969b) *Agistemus exsertus* as a predator of two tetranychid mites. *Ann. Entomol Soc. Am.*, **62**, 660–1.

El Banhawy, E.M., Carter, N. & Wynne, I.R. (1993) Preliminary observations on the population development of anystid and free-living mesostigmatic mites in a cereal field in southern England. *Exp. Appl. Acarol.*, **17**, 541–9.

El-Halawany, M.E., Abdel-Samed, M.A., Ibrahim, G.A. *et al.* (1993) Evaluating the efficiency of different levels of *Euseius scutalis* (Athias-Henrtiot) released for the control of *Brevipalpus californicus* (Banks) on Baladi oranges. *Menofiya J. Agric. Res.*, **18**, 2697–707.

Elkins, N.Z. & Whitford, W.G. (1982) The role of microarthropods and nematodes in decomposition in semi-arid ecosystems. *Oecologia*, **55**, 303–10.

Ellis-Adam, A.C. & Davids, C. (1970) Oviposition and post-embryonic development of the watermite *Piona alpicola* (Neuman, 1880). *Neth. J. Zool.*, **20**, 122–37.

Elzinga, R.J & Broce, A.B. (1988) Hypopi (Acari: Histiostomatidae) on house flies (Diptera: Muscidae): a case of detrimental phoresy. *J. Kansas Entomol. Soc.*, **61**, 208–13.

Emberson, R.M. (2000) Endemic biodiversity, natural enemies, and the future of biological control. In: *Proc. X Int. Symp. Biological Control of Weeds* (Ed. by N.R. Spence), pp. 875–80. Montana State University, Bozeman, MT.

Enami, Y. & Nakamura, Y. (1996) Influence of *Scheloribates azumaensis* (Acari: Oribatida) on *Rhizoctonia solani*, the cause of radish root rot. *Pedobiologia*, **40**, 251–4.

Engel, R. & Ohnesorge, B. (1994) Die Rolle von Ersatznahrung und Mikroklima im System *Typhlodromus pyri* Scheuten (Acari: Phytoseiidae) – *Panonychus ulmi* Koch (Acari, Tetranychidae) auf Weinreben. I. Untersuchungen im Labor. *J. Appl. Entomol.*, **118**, 129–50.

English-Loeb, G.M. (1990) Plant drought stress and outbreaks of spider mites: a field test. *Ecology*, **71**, 1401–11.

English-Loeb, G.M. & Karban, R. (1988) Negative interactions between Willamette mites and Pacific mites: possible management strategies for grapes. *Entomol. Exp. Appl.* **48**, 269–74.

English-Loeb, G., Norton, A.P., Gadoury, D.M. *et al.* (1999) Control of powdery mildew in wild and cultivated grapes by a tydeid mite. *Biol. Cont.*, **14**, 97–103.

Enkegaard, A., Saradar, M.A. & Brødsgaard, H.F. (1997) The predatory mite *Hypoaspis miles*: biological and demographic characteristics on two prey species, the mushroom sciarid fly, *Lycoriella solani*, and the mould mite, *Tyrophagus putrescentiae*. *Entomol. Exp. Appl.*, **82**, 135–46.

Enserink, M. (1999) Biological invaders sweep in. *Science*, **285**, 1834–6.

Epsky, N.D., Walter, D.E. & Capinera, J.L. (1988) Potential role of nematophagous arthropods as biotic mortality factors of entomogenous nematodes (Rhabditida; Steinernematidae, Heterorhabditidae). *J. Econ. Entomol.*, **81**, 821–5.

Epstein, A.H. & Hall, J.H. (1999) Status of rose rosette disease as a biological control for multiflora rose. *Pl. Dis.*, **83**, 92–101.

Eraky, S.A. (1995) Some biological aspects of *Tyrophagus putrescentiae* (Schrank) (Acari: Acaridae). In: *The Acari, Physiological and Ecological Aspects of Acari–Host Relationships* (Ed. by D. Kropczyńska, J. Boczek & A. Tomczyk), pp. 197–204. Oficyna DABOR, Warsaw.

Evans, G.O. (1992) *Principles of Acarology*. CABI, Wallingford.

Evans, G.O. & Till, W.M. (1979) Mesostigmatic mites of Britain and Ireland (Chelicerata: Acari-Parasitiformes): an introduction to their external morphology and classification. *Trans. Zool. Soc. Lond.*, **35**, 139–270.

Evans, G.O., Sheals, J.G. & Macfarlane, D. (1961) *The Terrestrial Acari of the British Isles*. Trustees of the British Museum, Dorking.

Ewart, W.H. & Metcalfe, R.L. (1956) Preliminary studies of sugars and amino acids in the honeydew of five species of coccids feeding on citrus in California. *Ann. Entomol. Soc. Am.*, **49**, 441–7.

Ewel, J.J., O'Dowd, D.J., Bergelson, J. *et al.* (1999) Deliberate introductions of species: research needs. Benefits can be reaped, but risks are high. *BioScience*, **49**, 619–30.

Ewing, H.E. (1912) The life history and habits of *Cheyletus seminivorous* Packard. *J. Econ. Entomol.*, **5**, 416–20.

Ewing, H.E. (1914) *The common red spider or spider mite*. Oregon Agric. Exp. St., Bull. No. 121.

van Eyndhoven, G.L. (1967) *Tetranychus lintearius* Dufour, 1832, is a valid species (Acari). Notulae ad Tetranychidas 11. *Entomol. Berl.*, **2**, 90–100.

Faasch, H. (1967) Beitrag zur Biologie der einheimischen Uropodiden *Uroobovella marginata* (C.L. Koch 1839) und *Uropoda orbicularis* (O.F. Müller 1776) und experimentelle Analyse ihres Phoresieverhaltens. *Zool. Jb. Syst.*, **94**, 521–608.

Fain, A. (1957) Sur la position systematique de *Riccardoella eweri* Lawrence 1952 et de *Boydaia angelae* Womersley 1953. *Rev. Zool. Bot. Afrique*, **55**, 249–52.

Fain, A. (1980) A method of mounting old preparations of acarines without raising or displacing the cover slip. *Int. J. Acarol.*, **6**, 169–70.

Fain, A. (1964) Chaetotaxy and specialized sensory organs of the Ereynetidae. *Acarologia*, **6** (fasc. hors ser.), 224–7.

Fain, A. & Elsen, P. (1972) Notes sur les acariens parasites ou commensaux des mouches tse-tse. II. Famille Erythraeidae (Trombidiformes). *Acta Zool. Path. Antever.*, **56**, 61–72.

Fain, A. & van Goethem, J.L. (1986) Les acariens du genre *Riccardoella* Berlese, 1923, parasites du poumon de mollusques Gastéropodes pulmonés terrestres. *Acarologia*, **27**, 125–40.

Fain, A. & Krantz, G.W. (1990) Notes on the genus *Asperoseius* Chant, 1957 (Acari, Phytoseiidae), with descriptions of two new species. *J. Afr. Zool.*, **104**, 213–20.

Fan, Y. & Petitt, F.L. (1994) Biological control of broad mite, *Polyphagotarsonemus latus* (Banks), by *Neoseiulus barkeri* Hughes on pepper. *Biol. Cont.*, **4**, 390–5.

Farag, A.I., Abo El-Ghar, G.E.S., Zohdi, G.I. & Sand, A.E. (1990) Predatory efficiency, development and reproduction of *Agistemus exsertus* on juvenoid-treated scale insects (Acarina: Stigmaeidae – Homopt: Coccoidea). In: *Proc. 6th Int. Symp. Scale Insect Studies* (Ed. by J. Koteja), Part II, pp. 153–6. Agricultural University Press, Cracow.

Faroni, L.R. D'A., Guedes, R.N.C. & Matioli, A.L. (2000) Potential of *Acarophenax lacunatus* (Prostigmata: Acarophenacidae) as a biological control agent of *Rhyzopertha dominica* (Coleoptera: Bostrichidae). *J. Stored Prod. Res.*, **36**, 55–63.

Fauvel, G. & Gendrier, J.P. (1992) Problems met in the establishment of phytoseiids in apple orchards of southeastern France. *Acta Phytol. Entomol. Hung.*, **27**, 223–32.

Fenley, W.R. (1966) Field observations of *Aedes ventrovittis* Dyar (Diptera: Culicidae) parasitized by *Panisopsis* sp. (Acarina: Thyasidae). *Mosq. News*, **26**, 583–4.

Fenton, B., Malloch, G., Jones, A.T. *et al.* (1995) Species identification of *Cecidophyopsis* mites (Acari: Eriophyidae) from different *Ribes* species and countries using molecular genetics. *Molec. Ecol.*, **4**, 383–7.

Fernández, M., Val, I.D., Proenza, M.A., Mesa, D. & Burgos, T. (1996) Algunas de las causas de muerta de *Pinnaspis strachani* Cooley (Homoptera: Diaspididae) en el cultivo de toronja en Isla de la Juventud. Parte I. *Rev. Prot. Veg.*, **11**, 91–4.

Ferragut, F., Garcia-Mari, F., Costa-Comelles, J. & Laborda, R. (1987) Influence of food and temperature on development and oviposition of *Euseius stipulatus and Typhlodromus phialatus* (Acari: Phytoseiidae). *Exp. Appl. Acarol.*, **3**, 317–29.

Field, G., Savage, L.B. & Duplessis, R.J. (1966) Note on the cockroach mite, *Pimeliaphilus cunliffei* (Acarina: Pterygosomidae) infesting oriental, German and American cockroaches. *J. Econ. Entomol.*, **59**, 1532.

Field, R.P., Webster, W.J. & Morris, D.S. (1979) Mass rearing *Typhlodromus occidentalis* Nesbitt (Acarina: Phytoseiidae) for release in orchards. *J. Aust. Entomol. Soc.*, **18**, 213–15.

Fitzgerald, J.D. & Solomon, M.G. (1991) Diapause induction and duration in the phytoseiid mite *Typhlodromus pyri. Exp. Appl. Acarol.*, **12**, 135–45.

Fitzgerald, J. & Solomon, M. (1999) The predatory mite *Typhlodromus pyri*; biological characteristics and resistance to insecticides in different mite strains. *IOBC/WPRS Bull.*, **22**(7), 161–7.

Flaherty, D.L. & Hoy, M.A. (1971) Biological control of Pacific mites and Willamette mites in San Joaquin Valley vineyards: Part III. Role of tydeid mites. *Res. Popul. Ecol.*, **13**, 80–96.

Flechtmann, C.H.W. & McMurtry, J.A. (1992a) Studies on how phytoseiid mites feed on spider mites and pollen. *Int. J. Acarol.*, **18**, 157–62.

Flechtmann, C.H.W. & McMurtry, J.A. (1992b) Studies on cheliceral and deutosternal morphology of some Phytoseiidae (Acari: Mesostigmata) by scanning electron microscope. *Int. J. Acarol.*, **18**, 163–9.

Fleschner, C.A. (1958) Field approach to population studies of tetranychid mites on citrus and avocado in California. In: *Proc. 10th Int. Congr. Entomology, Montreal* (Ed. by E.C. Becker), Vol. 2, pp. 669–74. Mortimer, Ottawa.

Follett, P.A. & Duan, J.J. (Eds) (2000) *Nontarget Effects of Biological Control*. Kluwer, Boston, MA.

Fontanillas-Perez, J.C. (1988) La acariosis. Una grave enfermedad de los caracoles. *Hojas Divulgados, Minist. Agric. Pesca y Alimen.*, **16**, 2–15.

Forbes, M.R.L. & Baker, R.L. (1991) Condition and fecundity of the damselfly, *Enallagma ebrium* (Hagen): the importance of ectoparasites. *Oecologia*, **86**, 335–41.

Forbes, M.R., Muma, K.E. & Smith, B.P. (1999) Parasitism of *Sympetrum* dragonflies by *Arrenurus planus* mites: maintenance of resistance particular to one species. *Int. J. Parasitol.*, **29**, 991–9.

Forneck, A., Merkt, N. & Blaich, R. (1998) A tripartite aseptic culture system for grape (*Vitis* spp.) phylloxera (*Daktulosphaera vitifoliae*) and mites (*Tarsonemus* sp.). *Vitis*, **37**, 95–6.

Fournier, D., Millot, P. & Pralavorio, M. (1985) Rearing and mass production of the predatory mite *Phytoseiulus persimilis*. *Entomol. Exp. Appl.*, **38**, 97–100.

Frazer, B.D. & Nelson, C. (1981) Note on the occurrence of predatory *Anystis* mites (Acari: Anystidae) in SW British Columbia. *J. Entomol. Soc. Br. Columbia*, **78**, 46.

Friese, D.D. & Gilstrap, F.E. (1982) Influence of prey availability on reproduction and prey consumption of *Phytoseiulus persimilis, Amblyseius californicus* and *Metaseiulus occidentalis* (Acarina: Phytoseiidae). *Int. J. Acarol.*, **8**, 85–9.

Friese, D.D., Mégevand, B. & Yaninek, J.S. (1987) Culture maintenance and mass production of exotic phytoseiids. *Insect Sci. Appl.*, **8**, 875–8.

Gabryś, G. (1991) Intergeneric relationships within Erythraeidae (Acari, Actinedida). In: *Modern Acarology* (Ed. by F. Dusbabek & V. Bukva), Vol. 2, pp. 207–12. SPB, The Hague.

Gahukar, R.T. (1994) Storage of food grains and insect control in developing countries. *Insect Sci. Appl.*, **15**, 383–400.

Galazzi, D. & Nicoli, G. (1996a) Comparative study of strains of *Phytoseiulus persimilis* Athias-Henriot (Acarina Phytoseiidae). I. Development and adult life. *Boll. Ist. Entomol. 'G. Grandi' Univ. Bologna*, **50**, 215–31.

Galazzi, D. & Nicoli, G. (1996b) Comparative study of strains of *Phytoseiulus persimilis* Athias-Henriot (Acarina Phytoseiidae). II. Influence of mass-rearing on population growth. *Boll. Ist. Entomol. 'G. Grandi' Univ. Bologna*, **50**, 243–52.

Ganga Visalakshy, P.N. (1992) Effect of commonly used weedicides and mosquito larvicides on the exotic waterhyacinth mite, *Orthogalumna terebrantis* Wallwork (Acarina: Galumnidae). *J. Biol. Cont.*, **6**, 35–7.

Ganga-Visalakshy, P.N. & Jayanth, K.P. (1991) Studies on the life history and development of *Orthogalumna terebrantis* Wallwork (Acarina: Galumnidae), an exotic oribatid mite of *Eichhornia crassipes*. *Entomon*, **16**, 53–8.

Gauchat, C.A. (1972) A note on *Podapolipoides grassi* Berlese (Acarina: Podapolipidae), a parasite of *Chortoicetes terminifera* Walker, the Australian plague locust. *J. Aust. Entomol. Soc.*, **11**, 259.

Gaugler, R. (1997) Alternative paradigms for commercializing biopesticides. *Phytoparasitica*, **25**, 172–83.

Gause, G.F., Smaragdova, N.P. & Witt, A.A. (1936) Further studies of interaction between predator and prey. *J. Anim. Ecol.*, **5**, 1–18.

Geden, C.J. & Axtell, R.C. (1988) Predation by *Carcinops pumilio* (Coleoptera: Histeridae) and *Macrocheles muscaedomesticae* (Acarina: Macrochelidae) on the house fly (Diptera: Muscidae): functional response, effects of temperature, and availability of alternative prey. *Environ. Entomol.*, **17**, 739–44.

Geden, C.J. & Stoffolano, J.G., Jr (1988) Dispersion patterns of arthropods associated with poultry manure in enclosed houses in Massachusetts: spatial distribution and effects of manure moisture and accumulation time. *J. Entomol. Sci.*, **23**, 136–48.

Geden, C.J., Stinner, R.E. & Axtell, R.C. (1988) Predation by predators of the house fly in poultry manure: effects of predator density, feeding history, interspecific interference, and field conditions. *Environ. Entomol.*, **17**, 320–9.

Geden, C.J., Stinner, R.E., Kramer, D.A. & Axtell, R.C. (1990) MACMOD: a simulation model for *Macrocheles muscaedomesticae* (Acari: Macrochelidae) population dynamics and rates of predation on immature house flies (Diptera: Muscidae). *Environ. Entomol.*, **19**, 578–86.

van der Geest, L.P.S., Elliot, S.L., Breeuwer, J.A.J. & Beerling, E.A.M. (2000) Diseases of mites. *Exp. Appl. Acarol.*, **24**, 467–560.

Gerecke, R. (1996) Untersuchungen Über Wasserrmilben der Familie Hydryphantidae (Acari, Actinedia) in der Westpalaearktis. II. Die Wasserrmilben der Familie Hydryphantidae Piersig, 1896, in den Mittelmeerlaendern. *Arch. Hydrobiol.*, Suppl. **77**, 337–513.

Gerson, U. (1967a) The natural enemies of the chaff scale, *Parlatoria pergandii* Comstock, in Israel. *Entomophaga*, **12**, 97–109.

Gerson, U. (1967b) Observations on *Hemisarcoptes coccophagus* Meyer (Astigmata: Hemisarcoptidae), with a new synonym. *Acarologia*, **9**, 632–8.

Gerson, U. (1972) Mites of the genus *Ledermuelleria* (Prostigmata: Stigmaeidae) associated with mosses in Canada. *Acarologia*, **13**, 319–43.

Gerson, U. (1985) Webbing. In: *Spider Mites, Their Biology, Natural Enemies and Control* (Ed. by W. Helle & M. W. Sabelis), Vol. 1A, pp. 223–32. Elsevier, Amsterdam.

Gerson, U. (1992) Biology and control of the broad mite, *Polyphagotarsonemus latus* (Banks) (Acari: Tarsonemidae). *Exp. Appl. Acarol.*, **13**, 163–78.

Gerson, U. (1994a) The Australian Eupalopsellidae (Acari: Prostigmata). *Invert. Taxon.*, **8**, 63–73.

Gerson, U. (1994b) First record of the genus *Hemisarcoptes* Lignières (Acari: Astigmata: Hemisarcoptidae) in Australia. *Aust. Entomol.*, **21**, 71–4.

Gerson, U. (1994c) The Australian Cheyletidae (Acari: Prostigmata). *Invert. Taxon.*, **8**, 435–47.

Gerson, U. (2001) Trends in research on acarine biocontrol agents. In: *Proc. 10th Int. Congr. Acarology* (Ed. by R.B. Halliday, D.E. Walter, H.C. Proctor *et al.*), pp. 457–9. CSIRO, Melbourne.

Gerson, U. & Aronowitz, A. (1981) Spider mite webbing. V. The effect of various host plants. *Acarologia*, **22**, 277–81.

Gerson, U. & Blumberg, D. (1969) Biological notes on the mite *Saniosulus nudus*. *J. Econ. Entomol.*, **62**, 729–30.

Gerson, U. & Cohen, E. (1989) Resurgences of spider mites (Acari: Tetranychidae) induced by synthetic pyrethroids. *Exp. Appl. Acarol.*, **6**, 29–46.

Gerson, U. & Izraylevich, S. (1997) A review of host utilization by *Hemisarcoptes* (Acari: Hemisarcoptidae) parasitic on scale insects. *Syst. Appl. Acarol.*, **2**, 33–42.

Gerson, U. & Schneider, R. (1981) Laboratory and field studies on the mite *Hemisarcoptes coccophagus* Meyer (Astigmata: Hemisarcoptidae), a natural enemy of armored scale insects. *Acarologia*, **22**, 199–208.

Gerson, U. & Schneider, R. (1982) The hypopus of *Hemisarcoptes coccophagus* Meyer (Acari: Astigmata: Hemisarcoptidae). *Acarologia*, **23**, 171–6.

Gerson, U. & Smiley, R.L. (1990) *Acarine Biocontrol Agents, An Illustrated Key and Manual.* Chapman and Hall, London.

Gerson, U. & Vacante, V. (1993) The use of indigenous acarine predators to control citrus mite pests. *OILB/SROP Bull.*, **16**, 115–19.

Gerson, U. & van de Vrie, M. (1979) The potential of mites in the biological control of mite and insect hosts. In: *Proc. 4th Int. Congr. Acarology* (Ed. by E. Piffl), pp. 629–35. Akadémiai Kiadó, Budapest.

Gerson, U. & Walter, D.E. (1998) Transfer of *Mecognatha* Wood from Stigmaeidae to Mecognathidae, fam. nov., a new synonymy and a key to families of Raphignathoidea (Acari: Prostigmata). *Syst. Appl. Acarol.*, **3**, 145–7.

Gerson, U., OConnor, B.A. & Houck, M.A. (1990) Acari. In: *Armored Scale Insects: Their Biology, Natural Enemies and Control* (Ed. by D. Rosen), Vol. 4B, pp. 77–97. Elsevier, Amsterdam.

Gerson, U., Cohen, E. & Capua, S. (1991) The mite *Rhizoglyphus robini* (Astigmata: Acaridae) as an experimental animal. *Exp. Appl. Acarol.*, **12**, 103–10.

Gerson, U., Fain, A. & Smiley, R. L. (1999) Further observations on the Cheyletidae (Acari), with a key to the genera of the Cheyletinae and a list of all known species in the family. *Bull. Inst. R. Sci. Nat. Belgique*, **69**, 35–86.

Giles, D.K., Gardner, H.E. & Studer, H.E. (1995) Mechanical release of predacious mites for biological pest control in strawberries. *Amer. Soc. Agric. Engin.*, **38**, 1289–96.

Gillespie, D.R. (1989) Biological control of thrips (Thysanoptera Thripidae) of greenhouse cucumbers by *Amblyseius cucumeris*. *Entomophaga*, **34**, 185–92.

Gillespie, D.R. & Quiring, D.M.J. (1990) Biological control of fungus gnats, *Bradysia* spp. (Diptera: Sciaridae), and western flower thrips, *Frankliniella occidentalis* (Pergande) (Thysanoptera: Thripidae), in greenhouses using a soil-dwelling predatory mite, *Geolaelaps* sp. nr. *aculeifer* (Canestrini) (Acari: Laelapidae). *Can. Entomol.*, **122**, 975–83.

Gillespie, D.R. & Quiring, D.J.M. (1992) Competition between *Orius tristicolor* (White) (Hemiptera: Anthocoridae) and *Amblyseius cucumeris* (Oudemans) (Acari, Phytoseiidae) feeding on *Frankliniella occidentalis* (Pergande) (Thysanoptera: Thripidae). *Can. Entomol.*, **124**, 1123–8.

Gillespie, D.R. & Ramey, C.A. (1988) Life history and cold storage of *Amblyseius cucumeris* (Acarina: Phytoseiidae). *J. Entomol. Soc. Br. Columbia*, **85**, 71–6.

Gillespie, D.R., Quiring, D.J.M. & Greenwood, M. (1997) Collection and selection of natural enemies of twospotted spider mites for biological control. *J. Entomol. Soc. Br. Columbia*, **94**, 7–11.

Gilliatt, F.C. (1935) Some predators of the European red mite, *Paratetranychus pillosus* C. & F., in Nova Scotia. *Can. J. Res.*, **13D**, 19–38.

Giménez-Ferrer, R.M., Erb, W.A., Bishop, B.L. & Scheerens, J.C. (1994) Host–pest relationships between the twospotted spider mite (Acari: Tetranychidae) and strawberry cultivars with differing levels of resistance. *J. Econ. Entomol.*, **87**, 168–75.

Girish, G.K., Goyal, R.K. & Krishnamurthy, K. (1973) Studies on Indian mites. Part II. Association of *Cheyletus malaccensis* Oudemans (Cheyletidae: Prostigmata) with *Trogoderma granarium* Everts, *Bruchus chinensis* Linn., and *Tribolium castaneum* Hbst. *Bull. Grain Tech.*, **11**, 118–20.

Glen, D.M., Spaull, A.M., Mowat, D.J. *et al.* (1993) Crop monitoring to assess the risk of slug damage to winter wheat in the United Kingdom. *Ann. Appl. Biol.*, **122**, 161–72.

Glendenning, R. (1931) The progress of parasite introduction in British Columbia. *Proc. Entomol. Soc. Br. Columbia*, **28**, 29–32.

Glockemann, B. (1992) Biological control of *Frankliniella occidenatlis* on ornamental plants using predatory mites. *EPPO Bull.*, **22**, 397–404.

Glowka, L., Burhenne-Guilman, F., Synge, H. *et al.* (1994) *A Guide to the Convention on Biological Diversity*. IUNC, The World Conservation Union, Gland.

Goeden, R.D. (1983) Critique and revision of Harris' scoring system for selection of insects agents in biological control of weeds. *Prot. Ecol.*, **5**, 287–301.

Goeden, R.D. (1988) A capsule history of biological control of weeds. *Biocont. News Inform.*, **9**, 55–61.

Goeden, R.D., Kovalev, O.V. & Ricker, D.W. (1974) Arthropods exported from California to the U.S.S.R. for ragweed control. *Weed Sci.*, **22**, 156–8.

Goh, K.S. & Lange, W.H. (1989) Microarthropods associated with insecticide-treated and untreated artichoke fields in California. *J. Econ. Entomol.*, **82**, 621–5.

Goldarazena, A., Jordana, R. & Zhang, Z.-Q. (1999) Notes on *Abrolophus neobrevicollis* Z. and G. and *Grandjeanella multisetosa* Z. and G. (Acari: Erythraeidae) with the seasonal abundance in Mediterranean Spain. *Int. J. Acarol.*, **25**, 29–36.

Goldarazena, A., Ochoa, R., Jordana, R. & OConnor, B.M. (2001) Revision of the genus *Adactylidium* Cross (Acari: Heterostigmata: Acarophenacidae), mites associated with thrips (Thysanoptera). *Proc. Entomol. Soc. Wash.*, **103**, 473–516.

Gomaa, E.A., Abou-Awad, B.A., Nasr, A.K. & Abou-Elela, M.M. (1989) Life-history studies and feeding behaviour of the predatory mite, *Pachylaelaps aegyptiacus*, with description of immature stages (Acari: Pachylaelapidae). *Insect Sci. Appl.*, **10**, 691–8.

Gomi, K., Gotoh, T. & Noda, H. (1997) *Wolbachia* having no effect on reproductive incompatibility in *Tetranychus kanzawai* Kishida (Acari: Tetranychidae). *Appl. Entomol. Zool.*, **32**, 485–90.

Gonzalez-Rodriguez, R.H. (1965) A taxonomic study of the *genera Mediolata, Zetzellia* and *Agistemus* (Acarina: Stigmaeidae). *Univ. Calif. Pub. Entomol.*, **41**, 1–64.

Goodwin, S. & Steiner, M.Y. (1996) Survey of Australian native natural enemies for control of thrips. *IOBC/WPRS Bull.*, **19**, 47–50.

Gotoh, T. (1997) Annual life cycles of populations of the two-spotted spider mite, *Tetranychus urticae* Koch (Acari: Tetranychidae) in four Japanese pear orchards. *Appl. Entomol. Zool.*, **32**, 207–16.

Gotoh, T., Gomi, K. & Nagata, T. (1999a) Incompatibility and host plant differences among populations of *Tetranychus kanzawai* Kishida (Acari: Tetranychidae). *Appl. Entomol. Zool.*, **34**, 551–61.

Gotoh, T., Sugasawa, J. & Nagata, T. (1999b) Reproductive incompatibility of the two-spotted spider mite (*Tetranychus urticae*) infested with *Wolbachia. Entomol. Sci.*, **2**, 289–95.

Gough, N. (1991) Long-term stability in the interaction between *Tetranychus urticae* and *Phytoseiulus persimilis* producing successful integrated control on roses in southeast Queensland. *Exp. Appl. Acarol.*, **12**, 83–101.

Gould, F., Kennedy, G. & Kopanic, R. (1996) Environmental issues associated with enhancing the impact of biological control agents. *Am. Entomol.*, **42**, 160–73.

Grafton-Cardwell, E.E. & Ouyang, Y. (1993) Toxicity of four insecticides to various populations of the predaceous mite, *Euseius tularensis* Congdon (Acarina: Phytoseiidae) from San Joaquin Valley California citrus. *J. Agric. Entomol.*, **10**, 21–9.

Grafton-Cardwell, E.E. & Ouyang, Y. (1995) Augmentation of *Euseius tularensis* (Acari: Phytoseiidae) in citrus. *Environ. Entomol.*, **24**, 738–47.

Grafton-Cardwell, E.E. & Ouyang, Y. (1996) Influence of citrus leaf nutrition on survivorship, sex ratio, and reproduction of *Euseius tularensis* (Acari: Phytoseiidae). *Environ. Entomol.*, **25**, 1020–5.

Grafton-Cardwell, E.E., Ouyang, Y. & Striggow, R.A. (1997) Predaceous mites (Acari: Phytoseiidae) for control of spider mites (Acari: Tetranychidae) in nursery citrus. *Environ. Entomol.*, **26**, 121–30.

Grafton-Cardwell, E.E., Ouyang, Y. & Striggow, R.A. (1999a) Predacious mites for control of citrus thrips, *Scirtothrips citri* (Thysanoptera: Thripidae) in nursery citrus. *Biol. Cont.*, **14**, 29–36.

Grafton-Cardwell, E.E., Ouyang, Y. & Bugg, R.L. (1999b) Leguminous cover crops to enhance population development of *Euseius tularensis* (Acari: Phytoseiidae) in citrus. *Biol. Cont.*, **16**, 73–80.

Graham, F.J., Runham, N.W. & Ford, J.B. (1996) Long-term effects of *Riccardoella limacum* living in the lung of *Helix aspersa*. In: *Slug and Snail Pests in Agriculture* (Ed. by I.F. Henderson), pp. 359–64. British Crop Protection Council, Farnham.

Graham, W.M. (1970) Warehouse ecology studies of bagged maize in Kenya – II. Ecological observations on an infestation by *Ephestia (Cadra) cautella* (Walker) (Lepidoptera, Phycitidae). *J. Stored Prod. Res.*, **6**, 157–67.

Grandjean, F. (1954) Essai de classification des Oribates (Acariens). *Bull. Soc. Zool. France*, **78**, 421–46.

Grandjean, F. (1963) Concernant *Sphaerobates gratus*, les Mochlozetidae et les Ceratozetidae (Oribates). *Acarologia*, **5**, 284–305.

Gravena, S., Benetoli, I. Moreira, P.H.R. & Yamamoto, P.T. (1994) *Euseius citrifolius* Denmark & Muma predation on citrus leprosis mite *Brevipalpus phoenicis* (Geijskes) (Acari: Phytoseiidae: Tenuipalpidae). *Anais Soc. Entomol. Brasil*, **23**, 209–18.

Greathead, D.J. (1997) An introduction to the FAO code of conduct for the import and release of exotic biological control agents. *Biocont. News Inform.*, **18**, 119–24.

Greatorex, E.C. (1996) A molecular technique for examining the gut contents of predatory mites. In: *Brighton Crop Protection Conference, Pests and Diseases*, Vol. 1, pp. 437–8, Farnham.

Greco, N.M., Liljesthröm, G.G. & Sánchez, N.E. (1999) Spatial distribution and coincidence of *Neoseiulus californicus* and *Tetranychus urticae* (Acari: Phytoseiidae, Tetranychidae) on strawberry. *Exp. Appl. Acarol.*, **23**, 567–80.

Green, W.F. & Woolcock, A.J. (1978) *Tyrophagus putrescentiae*: an allegenically important mite. *Clin. Allergy*, **8**, 135–44.

Grevstad, F.S. (1999) Factors influencing the chance of population establishment: implications for release strategies in biocontrol. *Ecol. Appl.*, **9**, 1437–47.

Grobler, J.H. (1962) The life history and ecology of the woolly pine needle aphid, *Schizolachnus pini-radiatae* (Davidson) (Homoptera: Aphididae). *Can. Entomol.*, **94**, 35–45.

Grostal, P. & Dicke, M. (1999) Direct and indirect cues of predation risk influence behavior and reproduction of prey: a case for acarine interactions. *Behav. Ecol.*, **10**, 422–7.

Grostal, P. & O'Dowd, D.J. (1994) Plants, mites and mutualism: leaf domatia and the abundance and reproduction of mites on *Viburnum tinus* (Caprifoliaceae). *Oecologia*, **97**, 308–15.

Grout, T.G. (1994) The distribution and abundance of phytoseiid mites (Acari: Phytsoeiidae) on citrus in southern Africa and their possible value as predators of citrus thrips (Thysanoptera: Thripidae). *Exp. Appl. Acarol.*, **18**, 61–71.

Grout, T.G. (1998) Predacious mites as biocontrol agents on citrus in Southern Africa. In: *Citrus Pests in the Republic of South Africa*, 2nd Edn (Ed. by E.C.G. Bedford, M.A. van den Berg & E.A. de Villiers), pp. 29–33. Institute for Tropical and Subtropical Crops, Pretoria.

Grout, T.G. & Richards, G.I. (1992a) *Euseius addoensis addoensis*, an effective predator of citrus thrips, *Scirtothrips aurantii*, in the eastern Cape Province of South Africa. *Exp. Appl. Acarol.* **15**, 1–13.

Grout, T.G. & Richards, R.I. (1992b) The dietary effect of windbreak pollens on longevity and fecundity of a predacious mite *Euseius addoensis addoensis* (Acari: Phytoseiidae) found in citrus orchards in South Africa. *Bull. Entomol. Res.*, **82**, 317–20.

Grout, T.G. & Stephen, P.R. (1993) Predation of alate citrus thrips, *Scirtothrips aurantii* Faure (Thysanoptera: Thripidae), by *Euseius citri* (Van de Merwe & Ryke) (Acari: Phytoseiidae) at low temperatures. *Afr. Entomol.*, **1**, 264–5.

Grout, T.G. & Stephen, P.R. (1994) Importation of *Neoseiulus cucumeris*: how will it affect existing thrips control by *Euseius addoensis* on citrus? *Citrus J.*, **4**(2), 22–4.

Grout, T.G., Richards, G.I. & Stephen, P.R. (1997) Further non-target effects of citrus pesticides on *Euseius addoensis* and *Euseius citri* (Acari: Phytoseiidae). *Exp. Appl. Acarol.*, **21**, 171–7.

Guretzky, J.A. & Louda, S.M. (1997) Evidence for natural biological control: insects decrease survival and growth of a native thistle. *Ecol. Appl.*, **7**, 1330–40.

Gurr, G.M., Thwaite, W.G., Valentine, B.J. & Nicol, H.I. (1997) Factors affecting the presence of *Typhlodromus* spp. (Acarina: Phytoseiidae) in the calyx cavities of apple fruits and implications for integrated pest management. *Exp. Appl. Acarol.*, **21**, 357–64.

Gutierrez, A.P., Caltagirone, L.E. & Meikle, W. (1999a) Evaluations of results; economics of biological control. In: *Handbook of Biological Control* (Ed. By T.S. Bellows & T.W. Fisher), pp. 243–52. Academic Press, San Diego, CA.

Gutierrez, A.P., Yaninek, J.S., Neuenschwander, P. & Ellis, C.K. (1999b) A physiologically-based tritrophic metapopulation model of the African cassava food web. *Ecol. Mod.*, **123**, 225–42.

Hagen, K.S. (1987) Nutritional ecology of terrestrial insect predators. In: *Nutritional Ecology of Insects, Mites, Spiders and Related Invertebrates* (Ed. by J. Slansky Jr & J.G. Rodriguez), pp. 533–77. John Wiley, New York.

Hagler, J.R. (1998) Variation in the efficacy of several predator gut content immunoassays. *Biol. Cont.*, **12**, 25–32.

Hagley, E.A.C. & Biggs, A.R. (1989) Effects of three fungicides on populations of a phytophagous and several predacious mites (Acarina) on apple. *Exp. Appl. Acarol.*, **6**, 253–6.

Hagley, E. & Simpson, C.M. (1983) Effects of insecticides on predators of the pear psylla, *Psylla pyricola* (Hemiptera: Psyllidae), in Ontario. *Can. Entomol.*, **115**, 1409–14.

Haines, C.P. (1981) Laboratory studies on the role of an egg predator, *Blattisocius tarsalis* (Berlese) (Acari: Ascidae), in relation to the natural control of *Ephestia cautella* (Walker) (Lepidoptera: Pyralidae) in warehouses. *Bull. Entomol. Res.*, **71**, 555–74.

Haines, C.P. (1984) Biological methods for integrated control of insects and mites in tropical stored products. III: The use of predators and parasites. *Trop. Stored Prod. Inform.*, **48**, 17–25.

Haines, C.P. (1999) Arthropod natural enemies in stored products – overlooked and underexploited. In: *Proc. 7th Int. Work. Congr. Stored-product Prot.* (Ed. by Z. Jin., Q. Liang, Y. Liang *et al.*) pp. 1205–26. Sichuan, Chengdu, PR China.

Hall, R.W. & Ehler, L.E. (1979) Rate of establishment of natural enemies in classical biological control. *Bull. Entomol. Soc. Am.*, **25**, 280–2.

Halliday, R.B. (2000) The Australian species of *Macrocheles* (Acarina: Macrochelidae). *Invert. Taxon.*, **14**, 273–326.

Halliday, R.B. & Holm, E. (1987) Mites of the family Macrochelidae as predators of two species of dung-breeding pest flies. *Entomophaga*, **32**, 333–8.

Halliday, R.B., Walter, D.E. & Lindquist, E.E. (1998) Revision of the Australian Ascidae (Acarina: Mesostigmata). *Invert. Taxon.*, **12**, 1–54.

Halliday, R.B., OConnor, B.M. & Baker, A.S. (1999) Global diversity of mites. In: *Nature and Human Society: The Quest for a Sustainable World* (Ed. by P.H. Raven), pp. 192–203. National Academy Press, Washington, DC.

Hamlen, R.A. & Poole, R.T. (1980) Effects of a predaceous mite on spider mite populations of *Dieffenbachia* under greenhouse and interior environments. *HortScience*, **15**, 611–12.

Hamstead, E.O. (1970) Greenhouse integrated control studies of the two-spotted spider mite on lima beans with a predaceous mite, *Typhlodromus fallacis*, and insecticides. *J. Econ. Entomol.*, **63**, 1027–8.

Hanks, L.M., McCelfresh, J.S., Millar, J.G. & Paine, T.D. (1992) Control of the straw itch mite (Acari: Pyemotidae) with sulfur in an insect rearing facility. *J. Econ. Entomol.*, **85**, 683–6.

Hanna, R. & Wilson, T. (1991) Prey preference by *Metaseiulus occidentalis* (Acari: Phytoseiidae) and the role of prey aggregation. *Biol. Cont.*, **1**, 51–8.

Hansen, L.S. (1988) Control of *Thrips tabaci* (Thysanoptera: Thripidae) on glasshouse cucumber using large introductions of predatory mites *Amblyseius barkeri* (Acarina: Phytoseiidae). *Entomophaga*, **33**, 33–42.

Hansen, L.S. (1989) The effect of initial thrips density (*Thrips tabaci* Lind. [Thysanoptera, Thripidae]) on the control exerted by *Amblyseius barkeri* (Hughes) (Acarina, Phytoseiidae) on glasshouse cucumber. *J. Appl. Entomol.*, **107**, 130–5.

Hansen, L.S. (1998) Prospects for developing strategies for biological control of the Mediterranean flour moth *Ephestia kuehniella* in flour mills. *IOBC/WPRS Bull.* **21**(3), 33–7.

Hardman, J.M. & Gaul, S.O. (1990) Mixtures of *Bacillus thuringiensis* and pyrethroids control winter moth (Lepidoptera: Geometridae) in orchards without causing outbreaks of mites. *J. Econ. Entomol.*, **83**, 920–36.

Hardman, J.M., Rogers, M.L., Gaul, S.O. & Bent, E.D. (1997) Insectary rearing and initial testing in Canada of an organophosphate/pyrethroid-resistant strain of the predator mite *Typhlodromus pyri* (Acari: Phytoseiidae) from New Zealand. *Environ. Entomol.*, **26**, 1424–36.

Hardman, J.M., van der Werf, W. & Nyrop, J.P. (1999) Modelling mite dynamics on apple trees in eastern North America. *Acta Hort.*, **499**, 201–9.

Hardman, J.M., Moreau, D.L., Snyder, M. *et al.* (2000) Performance of a pyrethroid-resistant strain of the predator mite *Typhlodromus pyri* (Acari: Phytoseiidae) under different insecticide regimes. *J. Econ. Entomol.*, **93**, 590–604.

Hare, J.D. (1992) Effects of plant variation on herbivore–natural enemy interactions. In: *Plant Resistance to Herbivores and Pathogens. Ecology, Evolution and Genetics* (Ed. by R.S. Fritz & E.L. Simms), pp. 278–98. University Chicago Press, Chicago, IL.

Hare, J.D., Yu, D.S. & Luck, R.F. (1990) Variation in life history parameters of California red scale on different citrus cultivars. *Ecology* **71**, 1451–60.

van Haren, R.J.F., Steenhuis, M.M., Sabelis, M.W. & De Ponti, O.M.B. (1987) Tomato stem trichomes and dispersal success of *Phytoseiulus persimilis* relative to its prey *Tetranychus urticae*. *Exp. Appl. Acarol.*, **3**, 115–21.

Hariyappa, A.S. & Kulkarni, K.A. (1988) Biology and feeding efficiency of the predatory mite *Amblyseius longispinosus* (Evans) on chilli mite, *Polyphagotarsonemus latus* (Banks). *J. Biol. Cont.*, **2**, 131–2.

Harley, K.L.S. (1990) The role of biological control in the management of water hyacinth, *Eichhornia crassipes*. *Biocont. News Inform.*, **11**, 11–22.

Harris, J.R.W. & Usher, M.B. (1978) Laboratory studies of predation by the grassland mite *Pergamasus longicornis* Berlese and their possible implications for the dynamics of populations of Collembola. *Sci. Proc. Dublin R. Soc., Ser. A*, **6**, 143–53.

Harris, P. (1973) The selection of effective agents for the biological control of weeds. *Can. Entomol.*, **105**, 1495–503.

Harvey, M.S. (1996) A review of the water mite family Pionidae in Australia (Acarina: Hygrobatoidea). *Rec. West. Aust. Mus.*, **17**, 361–93.

Harvey, M.S. (1998) *The Australian Water Mites: A Guide to Families and Genera*. CSIRO, Collingwood, Victoria.

Hassan, S.A. (1982) Relative tolerance of three different strains of the predatory mite *Phytoseiulus persimilis* A.-H. (Acari, Phytoseiidae) to 11 pesticides used on glasshouse crops. *Zeit. Angew. Entomol.*, **93**, 55–63.

Hassan, S.A., Bigler, F., Bogenschütz, H. *et al.* (1991) Results of the fifth joint pesticide testing programme carried out by the IOBC/WPRS-Working Group 'Pesticides and Beneficial Organisms'. *Entomophaga*, **36**, 55–67.

Hassan, S.A., Bigler, F., Bogenschütz, H. *et al.* (1994) Results of the sixth joint pesticide testing programme of the IOBC/WPRS-Working Group 'Pesticides and Beneficial Organisms'. *Entomophaga*, **39**, 107–19.

Headley, J.C. & Hoy, M.A. (1987) Benefit/cost analysis of an integrated mite management program for almonds. *J. Econ. Entomol.*, **80**, 555–9.

Heimpel, G.E. & Lundgren, J.G. (2000) Sex ratios of commercially reared biological control agents. *Biol. Cont.*, **19**, 77–93.

Helle, W. & Sabelis, M.W. (1985) *Spider Mites, Their Biology, Natural Enemies and Control*, Vols I & II. Elsevier, Amsterdam.

Henderson, C.F. & McBurnie, H.V. (1943) Sampling techniques for determining populations of the citrus red mite and its predators. *USDA Agric. Circ.* 671.

Herbert, H.J. (1962) Influence of *Typhlodromus (T.) pyri* Scheuten on the development of *Bryobia arborea* M. & A. populations in the greenhouse. *Can. Entomol.*, **94**, 870–3.

Herbert, H.J. & Butler, K.P. (1973) Distribution of phytophagous and predaceous mites on apple trees in Nova Scotia. *Can. Entomol.*, **105**, 271–6.

Herne, D.H.C. & Putman, W.L. (1966) Toxicity of some pesticides to predaceous arthropods in Ontario peach orchards. *Can. Entomol.*, **98**, 936–42.

Herren, H.R., Bird, T.J. & Nadel, D.J. (1987) Technology for automated aerial release of natural enemies of the cassava mealybug and cassava green mite. *Insect Sci. Appl.*, **8**, 883–5.

Hess, R.T. & Hoy, M.A. (1982) Microorganisms associated with the spider mite predator *Metaseiulus (=Typhlodromus) occidentali*s: electron microscope observations. *J. Invert. Pathol.*, **40**, 98–106.

Hessein, N.A. & Parrella, M.P. (1990) Predatory mites help control thrips on floriculture crops. *Calif. Agric.*, **44**(6), 19–21.

Hessein, N.A. & Perring, T.M. (1986) Feeding habits of the Tydeidae with evidence of *Homeopronematus anconai* (Acari: Tydeidae) predation of *Aculops lycopersici* (Acari: Eriophyidae). *Int. J. Acarol.*, **12**, 215–21.

Hessein, N.A. & Perring, T.M. (1988a) *Homeopronematus anconai* (Baker) (Acari: Tydeidae) predation on citrus flat mite, *Brevipalpus lewisi* McGregor (Acari: Tenuipalpidae). *Int. J. Acarol.*, **14**, 89–90.

Hessein, N.A. & Perring, T.M. (1988b) The importance of alternate foods for the mite *Homeopronematus anconai* (Acari: Tydeidae). *Ann. Entomol. Soc. Am.*, **81**, 488–92.

Hill, M. (1997) Water hyacinth in Zambia: restoring the balance on the Kafue river. *Pl. Protect. News*, **47**, 11–13.

Hill, R.L. & O'Donnell, D.J. (1991a) Reproductive isolation between *Tetranychus lintearius* and two related species, *T. urticae* and *T. turkestani* (Acarina: Tetranychidae). *Exp. Appl. Acarol.*, **11**, 241–51.

Hill, R.L. & O'Donnell, D.J. (1991b) The host range of *Tetranychus lintearius* (Acarina: Tetranychidae). *Exp. Appl. Acarol.*, **11**, 253–69.

Hill, R.L. & Stone, C. (1985) Spider mites as control agents for weeds. In: *Spider Mites: Their Biology, Natural Enemies and Control* (Ed. by W. Helle & M.W. Sabelis), Vol. 1B, pp. 443–8. Elsevier, Amsterdam.

Hill, R.L., Gourlay, A.H. & Wigley, P.J. (1989) The introduction of gorse spider mite, *Tetranychus lintearius*, for biological control of gorse. *Proc. 42nd NZ Weed Pest Control Conf.*, New Plymouth (Ed. by A.J. Popay), pp. 137–9. Swiftprint Centre, Palmerston North.

Hill, R.L., Grindell, J.M., Winks, C.J. *et al.* (1991) Establishment of gorse spider mite as a control agent for gorse. *Proc. 44th NZ Weed Pest Control Conf.*, New Plymouth (Ed. by A.J. Popay), pp. 31–4. Swiftprint Centre, Palmerston North.

Hill, M.G., Allan, D.J., Henderson, R.C. & Charles, J.C. (1993) Introduction of armored scale predators and establishment of the predatory mite *Hemisarcoptes coccophagus* (Acari:

Hemisarcoptidae) on latania scale, *Hemiberlesia latania* (Homoptera: Diaspididae) in kiwifruit shelter trees in New Zealand. *Bull. Entomol. Res.*, **83**, 369–76.

Hirschmann, W. & Wisniewski, J. (1992) Gangsystematik der Parasitiformes. Teil 543. Die Teilgange (Larve, Protonymphe, Deutonymphe) von *Polyaspinus schweizeri* und *Polyaspinus nicolae* (Uropodini, Uropodinae). *Acarologie*, **39**, 161–8.

Hislop, R.G. & Prokopy, R.J. (1981) Mite predator responses to prey and predator-emitted stimuli. *J. Chem. Ecol.*, **7**, 895–904.

Hislop, R.G., Auditore, P.J., Weeks, B.L. & Prokopy, R.J. (1981) Repellency of pesticides to the mite predator *Amblyseius fallacis. Prot. Ecol.*, **3**, 253–7.

Hluchy, M. (1993) Zur biologischen Bekämpfung der Kräuselmilbe *Calepitrimerus vitis* Nalepa (Acari, Eriophyidae) auf der Weinrebe durch die Raubmilbe *Typhlodromus pyri* Scheuten (Acari, Phytoseiidae). *J. Appl. Entomol.*, **116**, 449–58.

Ho, C.-C. & Chen, W.-H. (1992) Control of phytoseiids in a spider mite mass-rearing system (Acari: Phytoseriidae, Tetranychidae). *Exp. Appl. Acarol.*, **13**, 287–93.

Ho, C.-c., Lo, K.-c. & Chen, W.-a. (1995) Comparative biology, reproductive compatibility, and geographical distribution of *Amblyseius longispinosus* and *A. womersleyi* (Acari: Phytoseiidae). *Environ. Entomol.*, **24**, 601–7.

Ho, T.M. & Auemetua, T.V. (1990) Effects of age and sex of mites, and temperature, on the predation rates of *Macrocheles merdarius* on *Musca domestica* eggs. *Trop. Biomed.*, **7**, 97–101.

Hochmuth, R.C., Hellman, J.L., Dively, G. & Schroder, R.F.W. (1987) Effect of the ectoparasitic mite *Coccipolipus epilachnae* (Acari: Podapolipidae) on feeding, fecundity, and longevity of soybean-fed adult Mexican bean beetles (Coleoptera: Coccinellidae) at different temperatures. *J. Econ. Entomol.*, **80**, 612–16.

Hoffmann, A. (1998) Ácaros queilétidos de México (Prostigmata: Cheyletidae). *Ann. Esc. Nac. Cienc. Biol., México*, **44**, 19–26.

Hokkanen, H.M.T. (1989) Choosing an effective biocontrol agent – an evolutionary perspective. *Acta Entomol. Fenn.*, **53**, 19–24.

Hokkanen, H.M.T. & Menzler-Hokkanen, I. (2000) Field performance of biopesticides. In: *Predicting Field Performance in Crop Protection* (Chaired by L.G. Copping), pp. 77–85. British Crop Protection Council, Farnham.

Hokkanen, H.M.T. & Pimentel, D. (1989) New associations in biological control: theory and practice. *Can. Entomol.*, **121**, 829–40.

Holdsworth, R.P. (1972) *Zetzellia mali* and *Agistemus fleschneri*: difference in spatial distribution. *Environ. Entomol.*, **1**, 532–3.

Holm, E. & Wallace, M.M.H. (1989) Distribution of some anystid mites (Acari: Anystidae) in Australia and Indonesia and their role as possible predators of the cattle tick, *Boophilus microplus* (Acari: Ixodidae). *Exp. Appl. Acarol.*, **6**, 77–83.

Hong, X.Y. & Zhang, Z.-Q. (1996) *The Eriophyoid Mites of China: An Illustrated Catalog and Identification Keys (Acari: Prostigmata: Eriophyoidea)*. Associated Publishers, Florida.

Hopper, K.R. (1995) Potential impacts on threatened and endangered species in the continental United States from introductions of parasitic Hymenoptera for the control of insect pests. In: *Biological Control: Benefits and Risks* (Ed. by H. Hokkanen & J.M. Lynch), pp. 64–74. Cambridge University Press, London.

Hopper, K.R. (2001) Research needs concerning non-target impacts of biological control introductions. In: *Evaluating Indirect Ecological Effects of Biological Control* (Ed. by E. Wajnberg, J.K. Scott & P.C. Quimby), pp. 39–56. CABI Publishing, Wallingford.

Hopper, K.R. & Roush, R.T. (1993) Mate finding, dispersal, number released, and the success of biological control introductions. *Ecol. Entomol.*, **18**, 321–31.

Hoschele, W. & Tanigoshi, L.K. (1993) *Pyemotes tritici* (Acari: Pyemotidae), a potential biological control agent of *Anagasta kuehniella* (Lepidoptera: Pyralidae). *Exp. Appl. Acarol.*, **17**, 781–92.

Houck, M.A. (1989) Isozyme analysis of *Hemisarcoptes* and its beetle associate *Chilocorus*. *Entomol. Exp. Appl.*, **52**, 167–72.

Houck, M.A. (1994) Adaptation and transition into parasitism from commensalism: a phoretic model. In: *Mites, Ecological and Evolutionary Analyses of Life-history Patterns* (Ed. by M.A. Houck), pp. 252–81. Chapman & Hall, New York.

Houck, M.A. (1999) Phoresy by *Hemisarcoptes* (Acari: Hemisarcoptidae) on *Chilocorus* (Coleoptera: Coccinellidae): influence of subelytral ultrastructure. *Exp. Appl. Acarol.*, **23**, 97–118.

Houck, M.A. & Cohen, A.C. (1995) Potential role of phoresy in the evolution of parasitism: radiolabelling (tritium) evidence from an astigmatid mite. *Exp. Appl. Acarol.*, **19**, 677–94.

Houck, M.A. & OConnor, B.M. (1990) Ontogeny and life history of *Hemisarcoptes cooremani* (Acari: Hemisarcoptidae). *Ann. Entomol. Soc. Am.*, **83**, 869–86.

Houck, M.A. & OConnor, B.M. (1991) Ecological and evolutionary significance of phoresy in the Astigmata. *Annu. Rev. Entomol.*, **36**, 611–36.

Houck, M.A. & OConnor, B.M. (1996) Temperature and host effects on key morphological characters of *Hemisarcoptes cooremani* and *Hemisarcoptes malus* (Acari: Hemisarcoptidae). *Exp. Appl. Acarol.*, **20**, 667–82.

Hougen-Eitzman, D. & Karban, R. (1995) Mechanisms of interspecific competition that result in successful control of Pacific mites following inoculations of Willamette mites on grapevines. *Oecologia*, **103**, 157–61.

Houseweart, M.W., Jennings, D.T., Berkett, L.P. & Brann, T.B. (1980) Parasitic mites (Acari: Erythraeidae) on spruce budworm moths (Lepidoptera: Tortricidae). *Can. Entomol.*, **112**, 193–7.

van Houten, Y.M. (1996) Biological control of western flower thrips on cucumber using the predatory mites *Amblyseius cucumeris* and *A. limonicus*. *IOBC/WPRS Bull.*, **19**(1), 59–62.

van Houten, Y.M. & van Stratum, P. (1993) Biological control of western flower thrips in greenhouse sweet peppers using non-diapausing predatory mites. *IOBC/WPRS Bull.*, **16**(2), 77–80.

van Houten, Y.M., van Rijn, P.C.J., Tanigoshi, L.K. & van Stratum, P. (1995a) Preselection of predatory mites to improve year-round biological control of western flower thrips in greenhouse crops. *Entomol. Exp. Appl.*, **74**, 225–34.

van Houten, Y.M., van Stratum, P., Bruin, J. & Veerman, A. (1995b) Selection for non-diapause in *Amblyseius cucumeris* and *Amblyseius barkeri* and exploration of the effectiveness of selected strains for thrips control. *Entomol. Exp. Appl.*, **77**, 289–95.

Howard, C.W. (1918) A preliminary report on the Trombidiidae of Minnesota. *Minnesota St. Entomol. Rep.*, **17**, 111–44.

Howard, D.F., Blum, M.S. & Fales, H.M. (1983) Defense in thrips: forbidding fruitiness of a lactone. *Science*, **220**, 335–6.

Howard, L.O. (1930) *A History of Applied Entomology (Somewhat Anecdotal)*. Smithsonian Misc. Coll., Publ. No. 3065.

Howard, L.O., Dyar, H.G. & Knab, F. (1912) *The Mosquitoes of North and Central America and the West Indies (General Considerations of Mosquitoes, Their Habits and Their Relations to the Human Species*, Vol. 1). Carnegie Institute, Washington, DC.

Howarth, F.G. (1991) Environmental impacts of classical biological control. *Annu. Rev. Entomol.* **36**, 485–509.

Hoy, C.W. & Glenister, C.S. (1991) Releasing *Amblyseius* spp. (Acarina: Phytoseiidae) to control *Thrips tabaci* (Thysanoptera: Thripidae) on cabbage. *Entomophaga*, **36**, 561–73.

Hoy, C.W., Feldman, J., Gould, F. *et al.* (1998) Naturally occurring biological controls in genetically engineered crops. In: *Conservation Biological Control* (Ed. by P. Barbosa), pp. 185–205. Academic Press, San Diego, CA.

Hoy, J.B. (1990) Long-term effects of lindane on the oribatid community in a pine plantation. *Environ. Entomol.*, **19**, 898–900.

Hoy, M.A. (1975) Effect of temperature and photoperiod on the induction of diapause in the mite *Metaseiulus occidentalis. J. Insect Physiol.*, **21**, 605–11.

Hoy, M.A. (1977) Inbreeding in the arrhenotokous predator *Metaseiulus occidentalis* (Nesbitt) (Acari: Phytoseriidae). *Int. J. Acarol.* **3**, 117–21.

Hoy, M.A. (1982) Aerial dispersal and field efficacy of a genetically improved strain of the spider mite predator *Metaseiulus occidentalis. Entomol. Exp. Appl.*, **32**, 205–12.

Hoy, M.A. (1985a) Recent advances in genetics and genetic improvement of the Phytoseiidae. *Annu. Rev. Entomol.*, **30**, 345–70.

Hoy, M.A. (1985b) Almonds (California). In: *Spider Mites, Their Biology, Natural Enemies and Control* (Ed. by W. Helle & M.W. Sabelis), Vol. 1B, pp. 299–310. Elsevier, Amsterdam.

Hoy, M.A. (1986) Use of genetic improvement in biological control. *Agric. Ecosyst. Environ.*, **15**, 109–19.

Hoy, M.A. (1991) Genetic improvement of Phytoseiids: in theory and practice. In: *Modern Acarology* (Ed. by F. Dusbábek & V. Bukva), Vol. 1, pp. 175–84. SPB, The Hague.

Hoy, M.A. (1992) Criteria for release of genetically-improved phytoseiids: an examination of the risks associated with release of biological control agents. *Exp. Appl. Acarol.*, **14**, 393–416.

Hoy, M.A. (2000) Transgenic arthropods for pest management: risks and realities. *Exp. Appl. Acarol.*, **24**, 463–95.

Hoy, M.A. & Cave, F.E. (1988) Premating and postmating isolation among populations of *Metaseiulus occidentalis* (Nesbitt) (Acarina: Phytoseiidae). *Hilgardia*, **56**(6), 1–20.

Hoy, M.A. & Knop, N.F. (1981) Selection for and genetic analysis of permethrin resistance in *Metaseiulus occidentalis*: genetic improvement of a biological control agent. *Entomol. Exp. Appl.*, **30**, 10–18.

Hoy, M.A. & Ouyang, Y.-L. (1986) Selectivity of the acaricides clofentezine and hexythiazox to the predator *Metaseiulus occidentalis* (Acari: Phytoseiidae). *J. Econ. Entomol.*, **79**, 1377–80.

Hoy, M. A. & Smilanick, J.M. (1979) A sex pheromone produced by immature and adult females of the predatory mite, *Metaseiulus occidentalis* (Acarina: Phytoseiidae). *Entomol. Exp. Appl.*, **26**, 291–300.

Hoy, M.A. & Smilanick, J.M. (1981) Non-random prey location by the phytoseiid predator *Metaseiulus occidentalis*: differential responses to several spider mite species. *Entomol. Exp. Appl.*, **29**, 241–53.

Hoy, M.A. & Standow, K.A. (1982) Inheritance of resistance to sulfur in the spider mite predator *Metaseiulus occidentalis. Entomol. Exp. Appl.*, **31**, 316–23.

Hoy, M.A., Castro, D. & Cahn, D. (1982) Two methods for large scale production of pesticide-resistant strains of the spider mite predator *Metaseiulus occidentalis* (Nesbitt) (Acarina: Phytoseiidae). *Z. Angew. Entomol.*, **94**, 1–9.

Hoy, M.A., Cunningham, G.L. & Knutson, L. (1983a) *Biological Control of Pests by Mites*. University of California, Special Publ., No. 3304, Berkeley, CA.

Hoy, M.A., Westigard, P.H. and Hoyt, S.C. (1983b) Release and evaluation of a laboratory-selected pyrethroid-resistant strain of the predaceous mite *Metaseiulus occidentalis* (Acari: Phytoseiidae) in two southern Oregon pear orchards and a Washington apple orchard. *J. Econ. Entomol.*, **76**, 383–8.

Hoyt, S.C. (1969a) Population studies on five mite species on apple in Washington. In: *Proc. 2nd Int. Congr. Acarology* (Ed. by G.O. Evans), pp. 117–33. Akadémiai Kiadó, Budapest.

Hoyt, S.C. (1969b) Integrated chemical control of insects and biological control of mites on apple in Washington. *J. Econ. Entomol.*, **62**, 74–86.

Hoyt, S. C. (1972) Resistance to azinphosmethyl of *Typhlodromus pyri* (Acari: Phytoseiidae) from New Zealand. *N.Z. J. Sci.*, **15**, 16–21.

Hoyt, S.C., Tanigoshi, L.K. & Browne, R.W. (1979) Economy injury level studies in relation to mites on apple. In: *Recent Advances in Acarology* (Ed. by J.G. Rodriguez), Vol. 1, pp. 3–12. Academic Press, New York.

Huang, M.-D., Mai, S.-W., Li, S.-X. & Situ, J. (1983) Biological control of citrus red mite, *Panonychus citri* (McG.) in Guangdong Province. In: *Proc. 1981 Int. Soc. Citriculture* (Ed. by K. Matsumoto), pp. 643–6. Aiko, Tokyo.

Huber, J.T. (1998) The importance of voucher specimens, with practical guidelines for preserving specimens of the major invertebrate phyla for identification. *J. Nat. Hist.*, **32**, 367–85.

Huffaker, C.B. & Flaherty, D.L. (1966) Potential of biological control of two-spotted spider mites on strawberries in California. *J. Econ. Entomol.*, **59**, 786–92.

Huffaker, C.B. & Kennett, C.E. (1953) Differential tolerance to parathion in two *Typhlodromus* predatory on cyclamen mite. *J. Econ. Entomol.*, **46**, 707–8.

Huffaker, C.B. & Kennett, C.E. (1956) Experimental studies on predation: predation and cyclamen-mite populations on strawberries in California. *Hilgardia*, **26**, 191–222.

Huffaker, C.B., van de Vrie, M. & McMurtry, J.A. (1969) The ecology of tetranychid mites and their natural control. *Annu. Rev. Entomol.*, **14**, 125–74.

Huffaker, C.B., van de Vrie, M. & McMurtry, J.A. (1970) Ecology of tetranychid mites and their natural enemies: a review. II. Tetranychid populations and their possible control by predators: an evaluation. *Hilgardia*, **40**, 391–458.

Hughes, A.M. (1976) *The Mites of Stored Food and Houses*. HMSO, London.

Hull, V.J. & Groves, R.H. (1973) Variation in *Chondrilla juncea* L. in Australia. *Proc. Ecol. Soc. Aust.*, **10**, 113–35.

Hunter, P.E. (1960) Plastic paint as a marker for mites. *Ann. Entomol. Soc. Am.*, **53**, 698.

Hunter, P.E. & Rosario, R.M.T. (1988) Associations of Mesostigmata with other organisms. *Annu. Rev. Entomol.*, **33**, 393–417.

Hunter, W.D. & Hinds, W.E. (1904) The Mexican cotton boll weevil. *U.S. Dept Agric. Bull.* 45.

Hurst, G.D.D., Sharpe, R.G., Broomfield, A.H. *et al.* (1995) Sexually transmitted disease in a promiscuous insect, *Adalia bipunctata*. *Ecol. Entomol.*, **20**, 230–6.

Husband, R.W. (1984) *Dilopolipus, Panesthipolipus, Peripolipus* and *Stenopolipus*, new genera of Podapolipidae (Acarina) from the Indo-Australian region. *Int. J. Acarol.*, **10**, 250–69.

Husband, R.W. (2001) A new species of *Eutarsopolipus* (Acari: Podapolipidae) from *Scarites subterraneus* (Coleoptera: Carabidae) from Louisiana, U.S.A. *Int. J. Acarol.*, **27**, 113–17.

Husband, R.W. & Husband, P.S. (1996) Studies of *Locustacarus* spp. (Podapolipidae) tracheal parasites of grasshoppers and bumblebees. In: *Proc. Acarology IX* (Ed. by R. Mitchell, D.J. Horn, G.R. Needham & W.C. Wellbourn), Vol. I, pp. 335–8. Ohio Biological Survey, Columbus, OH.

Ibrahim, G.A., Afifi, A.M. & Abdel-Halim, S.M. (1989) Laboratory observations on the biology of *Proctolaelaps pygmaeus* (Muller) and *Protogamasellus mica* (Athias-Henriot) (Acari: Gamasida: Ascidae). *Bull. Soc. Entomol. Egypte*, **68**, 43–7.

Ibrahim, Y.B. & Yee, T.S. (2000) Influence of sublethal exposure to abamectin on the biological performance of *Neoseiulus longispinosus* (Acari: Phytoseiidae). *J. Econ. Entomol.*, **93**, 1085–9.

Ignatowicz, S. (1979) Inter- and intraspecific competition in mites: *Parasitus coleoptratorum* (L.) and *Macrocheles glaber* (Müller). In: *Proc. 4th Int. Cong. Acarol.* (Ed. by E. Piffl), pp. 623–5. Akadémiai Kiadó, Budapest.

Imbriani, J.L. & Mankau, R. (1983) Studies on *Lasioseius scapulatus*, a mesostigmatid mite predaceous on nematodes. *J. Nematol.*, **15**, 523–8.

Inoue, K. & Tanaka, M. (1983) Biological characteristics of *Agistemus terminalis* (Quayle) (Acarina: Stigmaeidae) as a predator of citrus red mite, *Panonychus citri* (McGregor). *Jpn. J. Appl. Entomol. Zool.*, **27**, 280–8 (in Japanese with English abstract).

Inserra, R. (1970) Observationi morfologische ed appunti de biologia su *Zetzellia graeciana* Gonzalez (Acarina: Stigmaeidae). *Boll. Zool. Agar. Bachia.*, **10**, 85–119.

International Institute of Tropical Agriculture (1999) *Annual Report*, Ibadan, Nigeria.

Ireson, J.E. (1982) A re-examination of the distribution of the pasture snout mite, *Bdellodes lapidaria* (Kramer) (Acari: Bdellidae) and the lucerne flea, *Sminthurus viridis* (L.) (Collembola: Sminthuridae) in Tasmania. *J. Aust. Entomol. Soc.*, **21**, 251–5.

Ireson, J.E. (1984) The effectiveness of *Bdellodes lapidaria* (Kramer) (Acari: Bdellidae) as a predator of *Sminthurus viridis* (L.) (Collembola: Sminthuridae) in north west Tasmania. *J. Aust. Entomol. Soc.*, **23**, 185–91.

Ireson, J.E. & Paterson, S.C. (1991) Progress on the biological control of lucerne flea (*Sminthurus viridis* (L.)) with the spiny snout mite (*Neomolgus capillatus* (Kramer)) in Tasmania. In: *Proc. Natio. Workshop on Redlegged Earth Mite, Lucerne Flea and Blue Oat Mite* (Ed. by J. Ridsdill-Smith), pp. 98–103. Dept of Agriculture, Perth, WA.

Ireson, J.E. & Webb, W.R. (1995) Effectiveness of *Neomolgus capillatus* (Kramer) (Acarina: Bdellidae) as a predator of *Sminthurus viridis* (L.) (Collembola: Sminthuridae) in north-western Tasmania. *J. Aust. Entomol. Soc.*, **34**, 237–40.

Ireson, J.E. & Webb, W.R. (1996) Redistribution and establishment of *Neomolgus capillatus* (Kramer) (Acarina: Bdellidae) for the biological control of *Sminthurus viridis* (L.) (Collembola: Sminthuridae) in Tasmania. *J. Aust. Entomol. Soc.*, **35**, 243–6.

Ireson, J.E., Holloway, R.J. & Chatterton, W.S. (2001) An overview of investigations into the use of predatory mites to control the lucerne flea, *Sminthurus viridis* (L.) (Collembola: Sminthuridae), in Tasmanian pastures. In: *Acarology: Proc. 10th Int. Congr.* (Ed. by R.B. Halliday, D.E. Walter, H.C. Proctor, R.A. *et al.*), pp. 444–52. CSIRO, Melbourne.

Itagaki, N. & Koyama, K. (1986) Rearing of the predacious mite, *Amblyseius eharai* Amitai and Swirski (Acarina: Phytoseiidae) on a synthetic diet. *Jpn. J. Appl. Entomol. Zool.*, **30**, 298–300.

Ivanich Gambaro, P. (1988) Natural alternative food for *Amblyseius andersoni* Chant (Acarina: Phytoseiidae) on plants without prey. Long-term research in orchards with a prey–predator equilibrium. *Redia*, **71**, 161–72.

Izraylevich, S. & Gerson, U. (1993a) Mite parasitization on armored scale insects: host suitability. *Exp. Appl. Acarol.*, **17**, 861–75.

Izraylevich, S. & Gerson, U. (1993b) Population dynamics of *Hemisarcoptes coccophagus* Meyer (Astigmata: Hemisarcoptidae) attacking three species of armored scale insects (Homoptera: Diaspididae). *Exp. Appl. Acarol.*, **17**, 877–88.

Izraylevich, S. & Gerson, U. (1995a) Spatial patterns of the parasitic mite *Hemisarcoptes coccophagus* Meyer (Astigmata: Hemisarcoptidae): host effect, density-dependence of aggregation, and implication for biological control. *Bull. Entomol. Res.*, **85**, 235–40.

Izraylevich, S. & Gerson, U. (1995b) Host scale effects on the parasitic mite *Hemisarcoptes coccophagus* Meyer and their implications for the biological control of diaspidid pests. *Israel J. Entomol.*, **24**, 291–6.

Izraylevich, S. & Gerson, U. (1995c) The hypopus of *Hemisarcoptes coccophagus* Meyer: distribution and apolysis. *Acarologia*, **36**, 333–9.

Izraylevich, S. & Gerson, U. (1995d) Sex ratio of *Hemisarcoptes coccophagus*, a mite parasitic on insects: density-dependent processes. *Oikos*, **74**, 439–46.

Izraylevich, S. & Gerson, U. (1996) Sex allocation by a mite parasitic on insects: local mate competition, host quality and operational sex ratio. *Oecologia*, **108**, 676–82.

Izraylevich, S., Gerson, U. & Wysoki, M. (1995) Karyotype and sex determining mechanism of the mite *Hemisarcoptes coccophagus* Meyer (Acariformes: Astigmata: Hemisarcoptidae). *Int. J. Acarol.*, **21**, 229–32.

Izraylevich, S., Gerson, U. & Hasson, O. (1996) Numerical response of a parasitic mite: host effect and mechanism. *Environ. Entomol.*, **15**, 390–5.

Jacobson, R.J. (1993) Integrated pest management in spring bedding plants: a successful package for commercial crops. *IOBC/WPRS Bull.*, **16**(8), 105–12.

Jacobson, R.J., Chandler, D., Fenlon, J. & Russell, K.M. (2001) Compatibility of *Beauveria bassiana* (Balsamo) Vuillemin with *Amblyseius cucumeris* Oudemans (Acarina: Phytoseiidae) to control *Frankliniella occidentalis* Pergande (Thysanoptera: Thripidae) on cucumber plants. *Biocont. Sci. Tech.*, **11**, 391–400.

Jaffe, H., Huettel, R.N., Demilo, A.B. *et al.* (1989) Isolation and identification of a compound from soybean cyst nematode, *Heterodera glycines*, with sex pheromone activity. *J. Chem. Ecol.*, **15**, 2031–43.

Jalil, M. & Rodriguez, J.G. (1970a) Studies of behavior of *Macrocheles muscaedomesticae* (Acarina: Macrochelidae) with emphasis on its attraction to the house fly. *Ann. Entomol. Soc. Am.*, **63**, 738–44.

Jalil, M. & Rodriguez, J.G. (1970b) Biology of and odor perception by *Fuscuropoda vegetans* (Acarina: Uropodidae), a predator of the house fly. *Ann. Entomol. Soc. Am.*, **63**, 935–8.

James, D.G. (1988) Reproductive diapause in *Typhlodromus occidentalis* (Acarina: Phytoseiidae) from southern New South Wales. *J. Aust. Entomol. Soc.*, **27**, 55–9.

James, D.G. (1989) Influence of diet on development, survival and oviposition in an Australian phytoseiid, *Amblyseius victoriensis* (Acari: Phytoseiidae). *Exp. Appl. Acarol.*, **6**, 1–10.

James, D.G. (1990) Biological control of *Tetranychus urticae* Koch (Acari: Tetranychidae) in southern New South Wales peach orchards: the role of *Amblyseius victoriensis* (Acarina: Phytoseiidae). *Aust. J. Zool.*, **37**, 645–55.

James, D.G. (1993) Pollen, mould mites and fungi: improvements to mass rearing of *Typhlodromus doreenae* and *Amblyseius victoriensis. Exp. Appl. Acarol.*, **17**, 271–6.

James, D.G. (1995) Biological control of earth mites in pasture using endemic natural enemies. *P. Protect. Q.*, **10**, 58–9.

James, D.G. (1997) Imidacloprid increases egg production in *Amblyseius victoriensis* (Acari: Phytoseiidae). *Exp. Appl. Acarol.*, **21**, 75–82.

James, D.G. (2000) Reproductive diapause in *Typhlodromus doreenae* Schicha (Acarina: Phytoseiidae). *Int. J. Acarol.*, **26**, 101–3.

James, D.G. (2001) History and perspectives of biological mite control in Australian horticulture using exotic and native phytoseiids. In: *Acarology: Proc. 10th Int. Congr.* (Ed. by R.B. Halliday, D.E. Walter, H.C. Proctor *et al.*), pp. 436–43. CSIRO, Melbourne.

James, D.G. & Whitney, J. (1991) Biological control of grapevine mites in inland south-eastern Australia. *Aust. N.Z. Wine Indust. J.*, **6**, 210–14.

James, D.G. & Whitney, J. (1993a) Mite populations on grapevines in south-eastern Australia: implications for biological control of grapevine mites (Acarina: Tenuipalpidae, Eriophyidae). *Exp. Appl. Acarol.*, **17**, 259–70.

James, D.G. & Whitney, J. (1993b) Cumbungi pollen as a laboratory diet for *Amblyseius victoriensis* (Womersley) and *Typhlodromus doreenae* Schicha (Acari: Phytoseiidae). *J. Aust. Entomol. Soc.*, **32**, 5–6.

James, D.G. & Whitney, J. (1996) Biological control of mites (Eriophyidae, Tenuipalpidae) in Australian viticulture. In: *Proc. Acarology IX* (Ed. by R. Mitchell, D.J. Horn, G.R. Needham & W.C. Wellbourn), pp. 207–9. Ohio Biological Survey, Columbus, OH.

James, D.G., Warren, G.N. & Whitney, J. (1992) Phytoseiid mite populations on dormant grapevines: extraction using a microwave oven. *Exp. Appl. Acarol.*, **14**, 175–8.

James, D.G., O'Malley, K. & Rayner, M. (1995) Effect of alphacypermethrin and bifenthrin on the survival of five acarine predators of *Halotydeus destructor* (Acari: Penthaleidae). *Exp. Appl. Acarol.*, **19**, 647–54.

James, D.G., Gerson, U. & Heffer, R. (1997) *Eupalopsis jamesi* Gerson (Eupalopsellidae): a new predator of citrus red scale in Australia. In: *Proc. Acarology IX* (Ed. by R. Mitchell,

D.J. Horn, G.R. Needham & W.C. Welbourn), pp. 225–7. Ohio Biological Survey, Columbus, OH.

Janssen, A., Bruin, J., Jacobs, G. *et al.* (1997) Predators use volatiles to avoid prey patches with conspecifics. *J. Anim. Ecol.*, **66**, 223–32.

Janssen, A., Hofker, C.D., Braun, A.R. *et al.* (1990) Preselecting predatory mites for biological control: the use of an olfactometer. *Bull. Entomol. Res.*, **80**, 177–81.

Janssen, A., Pallini, A., Venzon, M. & Sabelis, M.W. (1998) Behaviour and indirect interactions in food webs of plant-inhabiting arthropods. *Exp. Appl. Acarol.*, **20**, 497–521.

Jarosik, V. (1990) *Phytoseiulus persimilis* and its prey *Tetranychus urticae* on glasshouse cucumbers and peppers: key factors related to biocontrol efficiency. *Acta Entomol. Bohem.*, **87**, 414–30.

Jarosík, V. & Plíva, J. (1995) Assessment of *Amblyseius barkeri* (Acarina: Phytoseiidae) as a control agent for thrips on greenhouse cucumbers. *Acta Soc. Zool. Bohem.*, **59**: 177–86.

Jatala, P. (1986) Biological control of plant-parasitic nematodes. *Annu. Rev. Phytopathol.*, **24**, 453–89.

Jayanth, K.P. & Ganga Visalakshy, P.N. (1989) Establishment of the exotic mite *Orthogalumna terebrantis* Wallwork on water hyacinth in Bangalore, India. *J. Biol. Cont.*, **3**, 75–6.

Jedlickova, J. (1992) Increasing the effectiveness of rearing and release of the predatory mite *Phytoseiulus persimilis*. *EPPO Bull.*, **22**, 479–82.

Jenkins, C.F.H. (1935) The bdellid mite *Biscirus lapidarius* Kramer. *J. Agric. West. Aust.*, **12**, 342–7.

Jenser, G. & Koleva, R. (1996) Physical surface features of the grapevine leaf affecting the abundance of *Zetzellia mali* (Ewing) (Acari: Stigmaeidae). *Acta Phytol. Entomol. Hung.*, **31**, 75–81.

Jeppson, L.R., Keifer, H.H. & Baker, E.W. (1975) *Mites Injurious to Economic Plants*. University of California Press, Berkeley, CA.

Jermini, M., Baillod, M. & Stäubli, A. (1989) Side effects of apple fungicides on beneficial organisms: results in Switzerland and practical aspects. *IOBC/WPRS Bull.*, **12**, 308–15.

Jess, S. & Kilpatrick, M. (2000) An integrated approach to the control of *Lycoriella solani* (Diptera: Sciaridae) during production of the cultivated mushroom (*Agaricus bisporus*). *Pest Manage. Sci.*, **56**, 477–85.

Ji, L., Gerson, U. & Izraylevich, S. (1994) The mite *Hemisarcoptes* sp. (Astigmata: Hemisarcoptidae) parasitizing willow oyster scale (Homoptera: Diaspididae) on poplars in northern China. *Exp. Appl. Acarol.*, **18**, 623–7.

Ji, L., Izraylevich, S., Gazit, S. & Gerson, U. (1996) A sex-specific tri-trophic-level effect in a phoretic association. *Exp. Appl. Acarol.*, **20**, 503–9.

Jin, D.C. (1997) *Hydrachnellae – Morphology, Systematics and a Primary Study of Chinese Watermites*. Guizhou Science & Technology Publishing House, Guiyang (in Chinese).

Johanowicz, D.L. & Hoy, M.A. (1998) Experimental induction and termination of non-reciprocal reproductive incompatibilities in a parahaploid mite. *Entomol. Exp. Appl.*, **87**, 51–8.

Johanowicz, D.L. & Hoy, M.A. (1999) *Wolbachia* infection dynamics in experimental laboratory populations of *Metaseiulus occidentalis*. *Entomol. Exp. Appl.*, **93**, 259–68.

Johnson, D.T. & Croft, B.A. (1981) Dispersal of *Amblyseius fallacis* (Acarina: Phytoseiidae) in an apple ecosystem. *Environ. Entomol.*, **10**, 313–19.

Johnson, H.B. (1985) Consequences of species introductions and removals on ecosystem function – implications for applied ecology. In: *Proc. VI Int. Symp. Biological Control of Weeds* (Ed. by E.S. Delfosse), pp. 27–56. Canada Agriculture, Ottawa.

Jones, S.A. & Morse, J.G. (1995) Use of isoelectric focusing electrophoresis to evaluate citrus thrips (Thysanoptera: Thripidae) predation by *Euseius tularensis* (Acari: Phytoseiidae). *Environ. Entomol.*, **24**, 1040–51.

Jones, V.P., Toscano, N.C., Johnson, M.W. *et al.* (1986) Pesticide effects on plant physiology: integration into a pest management program. *Bull. Entomol. Soc. Am.*, **32**, 103–9.

Jupp, P.W. (1996) The establishment of a distribution network for the mite *Aculus hyperici* to control St John's wort (*Hypericum perforatum*) in Australia. In: *Proc. IX Int. Congr. Biological Control of Weeds* (Ed. by V.C. Moran & J.H. Hoffmann), pp. 451–3. University of Cape Town.

Jupp, P.W. & Cullen, J.M. (1996) Expected and observed effects of the mite *Aculus hyperici* on St John's wort, *Hypericum perforatum*, in Australia. In: *Proc. IX Int. Congr. Biological Control of Weeds* (Ed. by V.C. Moran & J.H. Hoffmann), pp. 365–70. University of Cape Town.

Jupp, P.W., Briese, D.T. & Cullen, J.M. (1997) Evidence for resistance in *Hypericum perforatum* to a biological control agent, the eriophyid mite *Aculus hyperici*. *Pl. Protect Q.*, 12, 67–70.

Kabicek, J. (1995) Development and predation by *Amblyseius barkeri* (Acarina: Phytoseiidae) on the mite *Phytoseiulus persimilis* (Acarina: Phytoseiidae). *Ochrana Rostlin.* 31, 57–62.

Kajita, H. (1986) Predation by *Amblyseius* spp. (Acarina: Phytoseiidae), and *Orius* sp. (Hemiptera: Anthocoridae) on *Thrips palmi* Karny (Thysanoptera: Thripidae). *Appl. Entomol. Zool.*, 21, 482–4.

Kaliszewski, M. & Wrensch, D.L. (1993) Evolution of sex determination and sex ratio within the mite cohort Tarsonemina (Acari: Heterostigmata). In: *Evolution and Diversity of Sex Ratio in Insects and Mites* (Ed. by D. Wrensch & M. Ebbert), pp. 192–213. Chapman & Hall, London.

Kaliszewski, M.J., Tobolewski, J., Seyoum, S. *et al.* (1992) The polymerase chain reaction and sequencing of mite DNA. *Int. J. Acarol.*, 18, 231–9.

Kaliszewski, M., Athias-Binche, F. & Lindquist, E.E. (1995) Parasitism and parasitoidism in Tarsonemina (Acari: Heterostigmata) and evolutionary considerations. *Adv. Parasitol.*, 35, 335–67.

Kamath, M.K. (1979) A review of biological control of insect pests and noxious weeds in Fiji (1969–1978). *Fiji Agric. J.*, 41, 55–72.

Kamburov, S.S. (1971) Feeding, development and reproduction of *Amblyseius largoensis* on various food substances. *J. Econ. Entomol.*, 64, 643–8.

Kanavel, R.F. & Selhime, A.G. (1967) Biological studies on *Paracheyletia bakeri* (Acarina: Cheyletidae). *Fl. Entomol.*, 50, 107–13.

Kapetanakis, E.G. & Cranham, J.E. (1983) Laboratory evaluation of resistance to pesticides in the phytoseiid predator *Typhlodromus pyri* from English apple orchards. *Ann. Appl. Biol.*, 103, 389–400.

Kapil, R.P. & Bhanot, J.P. (1973) Feeding behaviour of the predatory mite *Acaropsis docta* (Berlese). *J. Stored Prod. Res.*, 9, 1–6.

Karban, R. (1986) Induced resistance against spider mites in cotton: field verification. *Entomol. Exp. Appl.*, 42, 239–42.

Karban, R. & English-Loeb, G.M. (1990) A 'vaccination' of Willamette spider mites (Acari: Tetranychidae) to prevent large populations of Pacific spider mites on grapevines. *J. Econ. Entomol.*, 83, 2252–7.

Karban, R. & Zalom, F. (1998) Success of mite-fighting tactics evaluated. *Calif. Agric.*, 52(6), 21–4.

Karban, R., Adamchak, R. & Schnathorst, W.C. (1987) Induced resistance and interspecific competition between spider mites and a vascular wilt fungus. *Science*, 235, 678–80.

Karban, R., English-Loeb, G. & Hougen-Eitzman, D. (1997) Mite vaccinations for sustainable management of spider mites in vineyards. *Ecol. Appl.*, 7, 183–93.

Karban, R., English-Loeb, G., Walker, M.A. & Thaler, J. (1995) Abundance of phytoseiid mites on *Vitis* species: effects of leaf hairs, domatia, prey abundance and plant phylogeny. *Exp. Appl. Acarol.*, 19, 189–97.

Karban, R., Agrawal, A.A., Thaler, J.S. & Adler, L. S. (1999) Induced plant responses and information content about risk of herbivory. *Trends Ecol. Evol. (TREE)*, 14, 443–7.

Karg, W. (1961) Ökologische untersuchungen von edaphischen Gamasiden (Acarina, Parasitiformes). *Pedobiologia*, **1**, 77–98.

Karg, W. (1983) Verbreitung und Bedeutung von Raubmilben der Cohors Gamasina als Antagonisten von Nematoden. *Pedobiologia*, **25**, 419–32.

Karg, W. (1986) Vorkommen und Ernührung der Milbencohorts Uropodina (Schildkröten-milben) sowie ihre Eignung als Indikatoren in Agroökosystemen. *Pedobiologia*, **29**, 285–95.

Karg, W. (1989a) Die ökologische Differenzierung der Raubmilbenarten der Überfamilie Phytoseioidea Karg (Acarina, Parasitiformes). *Zool. Jb. Syst.*, **116**, 31–46.

Karg, W. (1989b) Acari, *Milben – Unterordnung Parasitiformes (Anactinochaeta). Uropodina Kramer, Schildkrotenmilben*. Tierwelt Deutschland, Vol. 67. Gustav Fischer, Jena.

Karg, W. (1992) The importance of so-called indifferent mite species for the equilibrium between spider mites and their antagonists. *Acta Phytopathol. Entomol. Hung.*, **37**, 333–42.

Karg, W. (1993) *Acari (Acarina), Milben. Parasitiformes (Anactinochaeta), Cohorts Gamasina Leach Raubmilben*. Die Tierwelt Deutschlands, Teil 59. Gustav Fischer, Jena.

Karg, W. (1994) *Raubmilben, nützliche Regulatoren im Naturhaushalt*. Die Neue Brehm-Bücherei, Bd. 624, Westarp Wissenschaften, Magdeburg.

Karg, W. & Mack, S. (1986) Bedeutung und Nutzung oligophager Raubmilben der Cohorts Gamasina Leach. *Arch. Phytopathol. Pflanzen.*, **22**, 107–18.

Karg, W., Mack, S. & Baier, B. (1987) Advantages of oligophagous predatory mites for biological control. *SROP Bull.*, **10**(2), 66–73.

Kaufmann, T. (1977) *Hemisarcoptes* sp. and biological control of the date palm scale, *Parlatoria blanchardi* Targioni, in the Sahel region of Niger. *Environ. Entomol.*, **6**, 882–4.

Kaźmierski, A. (1998a) A review of the genus *Proctotydaeus* (Actinedida: Tydeidae: Pronematinae). *Acarologia*, **39**, 33–47.

Kaźmierski, A. (1998b) Tydeinae of the world: generic relationships, new and redescribed taxa and keys to all species. A revision of the subfamilies Pretydeinae and Tydeinae (Acari: Actinedida: Tydeidae) – part IV. *Acta Zool. Cracov.*, **41**, 283–455.

Keesing, V.F. (1990) The toxicity of four insecticides to the mite *Hemisarcoptes coccophagus* and its host scale *Hemiberlesia lataniae*. In: *Proc. 43th N.Z. Weed Pest Control Conf.* (Ed. by A.J. Popay), pp. 247–51. Swiftprint Centre, Palmerston North.

Keetch, D.P. (1972) Ecology of the citrus red mite, *Panonychus citri* (McGregor), (Acarina: Tetranychidae) in South Africa. 3. The influence of the predacious mite, *Amblyseius (Typhlodromus) addoensis* van der Merwe & Ryke. *J. Entomol. Soc. South. Afr.*, **35**, 69–79.

Keilbach, R. (1978) Zusammenbruch einer Plage der Klienen Stubenfliege (*Fannia canicularis*) durch den phoretischen Parasiten *Macrocheles muscaedomesticae* im Kleintiertstall. *Angew. Parasitol.*, **19**, 221–3.

Kennett, C.E. (1970) Resistance to parathion in the phytoseiid mite *Amblyseius hibisci*. *J. Econ. Entomol.*, **63**, 1999–2000.

Kennett, C.E. & Hamai, J. (1980) Oviposition and development in predaceous mites fed with artificial and natural diets (Acari: Phytoseiidae). *Entomol. Exp. Appl.*, **28**, 116–22.

Kennett, C.E., Flaherty, D.L. & Hoffmann, R.W. (1979) Effect of wind-borne pollens on the population dynamics of *Amblyseius hibisci* (Acarina: Phytoseiidae). *Entomophaga*, **24**, 83–98.

Kinn, D.N. (1966) Predation by the mite, *Macrocheles muscaedomesticae* (Acarina: Macrochelidae), on three species of flies. *J. Med. Entomol.*, **3**, 155–8.

Kinn, D.N. (1980) Mutualism between *Dendrolaelaps neodisetus* and *Dendroctonus frontalis*. *Environ. Entomol.*, **9**, 756–8.

Kinn, D.N. (1983a) Mites as biological control agents of bark and sawyer beetles. In: *Biological Control of Pests by Mites* (Ed. by M.A. Hoy, G.L. Cunningham & L. Knutson), pp. 67–73. University of California, Special Publ., No. 3304, Berkeley, CA.

Kinn, D.N. (1983b) The life cycle of *Proctolaelaps dendroctoni* Lindquist and Hunter (Acari: Ascidae): a mite associated with pine bark beetles. *Int. J. Acarol.*, **9**, 205–10.

Kinn, D.N. & Doutt, R.L. (1972) Natural control of spider mites on wine grape varieties in northern California. *Environ. Entomol.*, **1**, 513–18.

Kishimoto, H. & Takafuji, A. (1997) Variations in the life-history parameters among populations of *Amblyseius womersleyi* Schicha with different diapause characteristics (Acari: Phytoseiidae). *Appl. Entomol. Zool.*, **32**, 395–401.

Kleijn, P.W., Krips, O.E., Willems, P.E.L. & Dicke, M. (1997) The influence of leaf hairs of *Gerbera jamesoni* on the searching behaviour of the predatory mite *Phytoseiulus persimilis. Proc. Exp. Appl. Entomol., N.E.V. Amsterdam*, **8**, 171–6.

Kluge, R.L. & Caldwell, P.M. (1992) Microsporidian diseases and biological weed control: to release or not to release? *Biocont. News Inform.*, **13**, 43–7N.

Knop, N.F. (1985) Mating behavior in the tydeid mite *Homeopronematus anconai* (Acari: Tydeidae). *Exp. Appl. Acarol.*, **1**, 115–25.

Knop, N.F. & Hoy, M.A. (1983a) Biology of a tydeid mite, *Homeopronematus anconai* (n. comb.) (Acari: Tydeidae), important in San Joaquin vineyards. *Hilgardia*, **51**(5), 1–30.

Knop, N.F. & Hoy, M.A. (1983b) Factors limiting the utility of *Homeopronematus anconai* (Acari: Tydeidae) in integrated pest management in San Joaquin vineyards. *J. Econ. Entomol.*, **76**, 1181–6.

Knutson, L. & Coulson, J.R. (1997) Procedures and policies in the USA regarding precautions in the introduction of classical biological control agents. *EPPO Bull.*, **27**, 133–42.

Kogan, M. (1998) Integrated pest management: historical perspectives and contemporary development. *Annu. Rev. Entomol.*, **43**, 243–70.

Kolar, C.S. & Lodge, D.M. (2001) Progress in invasion biology: predicting invaders. *Trends Ecol. Evol. (TREE)*, **16**, 199–204.

Komlovszky, S. & Jenser, G. (1992) Little known predatory mite species of Hungary (Acari: Stigmaeidae). *Acta Phytopathol. Entomol. Hung.*, **37**, 361–3.

Kostiainen, T. & Hoy, A.M. (1994) Egg-harvesting allows large scale rearing of *Amblyseius finlandicus* (Acari: Phytoseiidae) in the laboratory. *Exp. Appl. Acarol.*, **18**, 155–65.

Kostiainen, T.S. & Hoy, M.A. (1996) *The Phytoseiidae as Biological Control Agents of Pest Mites and Insects: A Bibliography (1960–1994)*. University of Florida Publications, Monogr. 17. Gainesville, FL. (also available at website http://gnv.ifas.ufl.edu/~predbibhtm/predbib.htm).

Kovalev, O.V. (1973) Modern outlooks of biological control of weed plants in the U.S.S.R. and the international phytophagous exchange. In: *Proc. 2nd Int. Symp. Biological Control of Weeds* (Ed. by P.H. Dunn), pp. 166–72. Commonwealth Agricultural Bureaux, Slough.

Kovalev, O.V., Schevchenko, V.G. & Danilov, L.G. (1974) *Aceria acroptiloni*, sp. n. (Acarina, Tetrapodili), a promising phytophage for the biological control of Russian knapweed [*Acroptilon repens* (L.) DC.]. *Entomol. Rev.*, **53**(2), 25–34.

Krantz, G.W. (1962) A review of the genera of the family Macrochelidae Vitzthum 1930 (Acarina: Macrochelidae). *Acarologia*, **4**, 143–73.

Krantz, G.W. (1973) Dissemination of *Kampimodromus aberrans* by the filbert mite. *J. Econ. Entomol.*, **66**, 575–6.

Krantz, G.W. (1978) *A Manual of Acarology*, 2nd Edn. Oregon State University Book Store, Corvallis, OR.

Krantz, G.W. (1983) Mites as biological control agents for dung-breeding flies, with special reference to the Macrochelidae. In: *Biological Control of Pests by Mites* (Ed. by M.A. Hoy, G.L. Cunningham & L. Knutson), pp. 91–98. University of California, Special Publ., No. 3304, Berkeley, CA.

Krantz, G.W. (1998) Reflections on the biology, morphology and ecology of the Macrochelidae. *Exp. Appl. Acarol.*, **22**, 125–37.

Krantz, G.W. & Ainscough, B.D. (1990) Acarina: Mesostigmata (Gamasidae). In: *Soil Biology Guide* (Ed. by D.D. Dindale), pp. 583–665. John Wiley, New York.

Krantz, G.W. & Ehrensing, D.T. (1990) Deuterogyny in the skeleton weed mite, *Aceria chondrillae* (G. Can.) (Acari: Eriophyidae). *Int. J. Acarol.*, **16**, 129–33.

Krantz, G.W. & Khot, N.S. (1962) A review of the family Otopheidomenidae Treat 1955 (Acarina: Mesostigmata). *Acarologia*, **4**, 532–42.

Kreiter, S., Cotton, D., Baleste, N. & Le Scolan, N. (1995) First insecticide resistance in *Typhlodromus pyri* Scheuten (Acari, Phytoseiidae) strains from French vineyards. In: *The Acari, Physiological and Ecological Aspects of Acari–Host Relationships* (Ed. by D. Kropczynska, J. Boczek & A. Tomczyk), pp. 621–8. DABOR, Warsaw.

Kreiter, S., Sentenac, G., Barthes, D. & Auger, P. (1998a) Toxicity of four fungicides to the predaceous mite *Typhlodromus pyri* (Acari: Phytoseiidae). *J. Econ. Entomol.*, **91**, 802–11.

Kreiter, S., Sentenac, G., Weber, M. *et al.* (1998b) Effets non intentionnels de quelques produits phytopharmaceutiques sur *Typhlodromus pyri, Kampimodromus aberrans* et *Phytoseius plumifer. Phytoma*, **505**, 1–6.

Krips, O.E., Willems, P.E.L. & Dicke, M. (1999) Compatibility of host plant and biological control of the two-spotted spider mite *Tetranychus urticae* in the ornamental crop gerbera. *Biol. Cont.*, **16**, 155–63.

Krogh, P.H. (1995) Effects of pesticides on the reproduction of *Hypoaspis aculeifer* (Gamasida: Laelapidae) in the laboratory. *Acta Zool. Fenn.*, **196**, 333–7.

Kuwahara, Y. (1991) Pheromone studies on astigmatid mites – alarm, aggregation and sex. In: *Modern Acarology* (Ed. by F Dusbábek & V. Bukva), Vol. 1, pp. 43–52. SPB, The Hague.

Kuznetzov, N.N. (1979) New species of Tydeidae (Acariformes) from soil and forest litter. *Entomol. Rev.*, **58**, 126–8.

Kuznetzov, N.N. (1984) Two new genera of the family Stigmaeidae (Acariformes). *Zool. Zhur.*, **63**, 1105–7 (in Russian).

Laing, J.E. (1973) Evaluating the effectiveness of *Paracheyletia bakeri* (Acarina: Cheyletidae) as a predator of the two-spotted spider mite *Tetranychus urticae. Ann. Entomol. Soc. Am.*, **66**, 641–6.

Laing, J.E. & Huffaker, C.B. (1969) Comparative studies of predation by *Phytoseiulus persimilis* Athias-Henriot and *Metaseiulus occidentalis* (Nesbitt) (Acarina: Phytoseiidae) on populations of *Tetranychus urticae* Koch (Acari: Tetyranychidae). *Res. Popul. Ecol.*, **11**, 105–26.

Laing, J.E. & Knop, N.F. (1983) Potential use of predaceous mites other than Phytoseiidae for biological control of orchard pests. In: *Biological Control of Pests by Mites* (Ed. by M.A. Hoy, G.L. Cunningham & L. Knutson), pp. 28–35. University of California, Special Publ., No. 3304, Berkeley, CA.

Laird, M. (1947) Some natural enemies of mosquitoes in the vicinity of Palmalmal, New Britain. *Trans. R. Soc. N.Z.*, **76**, 453–76.

Lan, W., Xin, J.-L. & Aoki, J.-I. (1986) Two new species of oribatid mites of economic importance from China (Acari: Oribatida). *Proc. Jpn. Soc. Syst. Zool.*, **34**, 27–31.

Lanciani, C.A. (1978) Parasitism of Ceratopogonidae (Diptera) by the water mite *Tyrrellia circularis. Mosq. News*, **38**, 282–4.

Lanciani, C.A. (1979a) The influence of parasitic water mites on the instantaneous death rate of their hosts. *Oecologia*, **44**, 60–2.

Lanciani, C.A. (1979b) Water mite-induced mortality in a natural population of the mosquito *Anopheles crucians* (Diptera: Culicidae). *J. Med. Entomol.*, **15**, 529–32.

Lanciani, C.A. (1983) Overview of the effects of water mite parasitism on aquatic insects. In: *Biological Control of Pests by Mites* (Ed. by M.A. Hoy, G.L. Cunningham & L. Knutson), pp. 86–90. University of California, Special Publ., No. 3304, Berkeley, CA.

Lanciani, C.A. (1986) Reduced survivorship in *Dasyhelea mutabilis* (Diptera: Ceratopogonidae) parasitized by the water mite *Tyrrellia circularis* (Acariformis: Limnesiidae). *J. Parasitol.*, **72**, 613–14.

Lanciani, C.A. (1987) Mortality in mite-infested, male *Anopheles crucians*. *J. Am. Mosquito Cont. Assoc.*, **3**, 107–88.

Lanciani, C.A. (1988) Defensive consumption of parasitic mites by *Anopheles crucians* larvae. *J. Am. Mosq. Cont. Assoc.*, **4**, 195.

Lanciani, C.A. (1995) Effect of a parasitic water mite on the per capita rate of increase of its host *Hydrometra australis* (Hemiptera: Hydrometridae). *Fl. Entomol.*, **78**, 357–9.

Lanciani, C.A. & Boyt, A.D. (1977) The effect of a parasitic water mite, *Arrenurus pseudotenuicollis* (Acari: Hydrachnellae), on the survival and reproduction of the mosquito, *Anopheles crucians* (Diptera: Culicidae). *J. Med. Entomol.*, **14**, 10–15.

Lanciani, C.A. & Smith, B.P. (1989) Constancy of stylostome form in two water mite species. *Can. Entomol.*, **121**, 439–43.

Lange, A.B., Drozdovskii, E.M. & Bushkovskaya, L.M. (1974a) Collecting and releasing anystis. *Zaschita Rast.*, **1974**, 26–8 (in Russian).

Lange, A.B., Drozdovskii, E.M. & Bushkovskaya, L.M. (1974b) The anystis mite – an effective predator of small polyphages. *Zaschita Rast.*, **1974**, 33–4 (in Russian).

Lawson, A.B. & Walde, S.J. (1993) Comparison of the responses of two predaceous mites, *Typhlodromus pyri* and *Zetzellia mali*, to variation in prey density. *Exp. Appl. Acarol.*, **17**, 811–21.

Leal, W.S., Kuwahara, Y., Suzuki, T. & Nakao, H. (1989) Chemical taxonomy of economically important *Tyrohagus* mites (Acariformes, Acaridae). *Agric. Biol. Chem.*, **53**, 3279–84.

Leal, W.S., Kuwahara, Y. & Suzuki, T. (1990) Hexyl 2-formyl-3-hydroxybenzoate, a fungitoxic cuticular constituent of the bulb mite *Rhizoglyphus robini*. *Agric. Biol. Chem.*, **54**, 2593–7.

Leary, R.F. & Allendorf, F.W. (1989) Fluctuating asymmetry as an indicator of stress: implications for conservation biology. *Trends Ecol. Evol. (TREE)*, **4**, 214–16.

Lee, W.-T. & Lo, K.C. (1990) Integrated control of two-spotted spider mite on strawberry in Taiwan. *Chin. J. Entomol.*, Special Publ., No. **3**, 125–37.

Leetham, J.W. & Jorgensen, D.C. (1969) Overwintering phytoseiid mites in central Utah apple orchards. *Great Basin Natur.*, **29**, 96–104.

van Lenteren, J.C. (1993) Quality control for natural enemies used in greenhouses. *IOBC/WPRS Bull.*, **16**, 89–92.

van Lenteren, J.C. & Steinberg, S. (1991) A preliminary list of criteria for quality control of beneficial arthropods used commercially in greenhouse crops. *Proc. 5th IOBC Working Group 'Quality Control of Mass Reared Organisms'*, Wageningen, pp. 195–9.

van Lenteren, J.C. & Woets, J. (1988) Biological and integrated pest control in greenhouses. *Annu. Rev. Entomol.*, **33**, 239–69.

van Lenteren, J.C., Roskam, M.M. & Timmer, R. (1997) Commercial mass production and pricing of organisms for biological control of pests in Europe. *Biol. Cont.*, **10**, 143–9.

Leppla, N.C. & Fisher, W.R. (1989) Total quality control in insect mass production for insect pest management. *J. Appl. Entomol.*, **108**, 452–61.

Lesna, I., Sabelis, M.W., Bolland, H.R. & Conijn, C.G.M. (1995) Candidate natural enemies for control of *Rhizoglyphus robini* Claparède (Acari: Astigmata) in lily bulbs: exploration in the field and pre-selection in the laboratory. *Exp. Appl. Acarol.*, **19**, 655–69.

Lesna, I., Sabelis, M. & Conijn, C. (1996) Biological control of the bulb mite, *Rhizoglyphus robini*, by the predatory mite, *Hypoaspis aculeifer*, on lilies: predator–prey interactions at various spatial scales. *J. Appl. Ecol.*, **33**, 369–76.

Lester, P.J., Pree, D.J., Thistlewood, H.M.A. et al. (1999) Pyrethroid encapsulation for conservation of acarine predators and reduced spider mite (Acari: Tetranychidae) outbreaks in apple orchards. *Environ. Entomol.*, **28**, 72–80.

Liang, W. & Huang, M. (1994) Influence of citrus orchard ground cover plants on arthropod communities in China: a review. *Agric. Ecosyst. Environ.*, **50**, 29–37.

Lighthart, B., Sewall, D. & Thomas, D.R. (1988) Effect of several stress factors on the susceptibility of the predatory mite, *Metaseiulus occidentalis* (Acari: Phytoseiidae), to the weak bacterial pathogen *Serratia marcescens. J. Invert. Pathol.*, **52**, 33–42.

Lin, J.-Z. & Zhang, Z.-Q. (1999) Tarsonemidae of China (Acari: Prostigmata): an annotated
and illustrated catalogue and bibliography. *Syst. Appl. Acarol., Special Publ.*, No. 3,
1–120.

Lindquist, E.E. (1969) Review of Holarctic tarsonemid mites (Acarina: Prostigmata) parasitizing eggs of pine bark beetles. *Mem. Entomol. Soc. Canada*, **60**, 1–111.

Lindquist, E.E. (1975) Associations between mites and other arthropods in forest floor habitats. *Can. Entomol.*, **107**, 435–7.

Lindquist, E.E. (1983) Some thoughts on the potential for use of mites in biological control,
including a modified concept of 'parasitoids'. In: *Biological Control of Pests by Mites* (Ed.
by M.A. Hoy, G.L. Cunningham & L. Knutson), pp. 12–20. University of California,
Special Publ., No. 3304, Berkeley, CA.

Lindquist, E.E. (1986) The world genera of Tarsonemidae (Acari: Heterostigmata): a morphological, phylogenetic, and systematic revision, with a reclassification of family-group
taxa in the Heterostigmata. *Mem. Entomol. Soc. Can.*, **136**, 1–517.

Lindquist, E.E. (1996) Antiquity of some symbiotic associations of heterostigmatic mite clades
with beetle clades: phylogenetic and biogeographic patterns. In: *Proc. Acarology IX* (Ed.
by R. Mitchell, D.J. Horn, G.R. Needham & W.C. Wellbourn), pp. 301–3. Ohio Biological
Survey, Columbus, OH.

Lindquist, E.E. & Evans, G.O. (1965) Taxonomic concepts in the Ascidae, with a modified
setal nomenclature for the idiosoma of the Gamasina (Acarina: Mesostigmata). *Mem.
Entomol. Soc. Can.*, **47**, 1–64.

Lindquist, E.E. & Smiley, R.L. (1978) *Acaronemus*, a new genus proposed for tarsonemid
mites (Acari: Prosrigmata) predaceous on tetranychoid mite eggs. *Can. Entomol.*, **110**,
655–62.

Lindquist, E.E., Sabelis, M.W. & Bruin, J. (Eds) (1996) *Eriophyoid Mites, Their Biology, Natural
Enemies and Control*. Elsevier, Amsterdam.

Lindquist, R.K. & Wolgamott, M.L. (1980) Toxicity of acephate to *Phytoseiulus persimilis* and
Tetranychus urticae. Environ. Entomol., **9**, 389–92.

Lindquist, R., Buxton, J. & Piatkowski, J. (1994) Biological control of sciarid flies and shore
flies in glasshouses. *1994 Brighton Crop Prot. Conf., – Pests and Diseases*, pp. 1067–72.
Farnham.

Lindqvist, I. & Tiitanen, K. (1989) Biological control of *Thrips tabaci* (Thysanoptera,
Thripidae) on greenhouse cucumber. *Acta Entomol. Fenn.*, **53**, 37–42.

Linford, M.B. & Oliveira, J.M. (1938) Potential agents of biological control of plant-parasitic
nematodes. *Phytopathology*, **28**, 14.

Linnamäki, M., Hulshof, J. & Vänninen, I. (1998) Biology and prospects for enhancing
biocontrol of western flower thrips *Frankliniella occidentalis* in cut roses. *1998 Brighton
Crop Prot. Conf., Pests & Diseases*, Vol. 1, pp. 187–92. Farnham.

Llewellyn, R. (1991) The commercial use of predators in biological control. *P. Protect. Q.*, **6**,
186–7.

Lloyd, R.W. & Krieg, D.R. (1987) Cotton development and yield as affected by insecticides. *J.
Econ. Entomol.*, **80**, 854–8.

Lo, P.K.-C. (1986) Present status of biological control of mite pests in Taiwan. *P. Protect. Bull.
(Taiwan)*, **28**, 31–9 (in Chinese with English summary).

Lo, K.C. (1996) Potential use of mites as biological control agents of insect and mite pests. *P.
Protect. Bull. (Taiwan)*, **38**, 99–110 (in Chinese with English summary).

Lo, K.-c., Tseng, H.-k. & Ho, C.-c. (1984) Biological control of spider mites on strawberry in
Taiwan. *J. Agric. Res. China*, **33**, 406–17 (in Chinese with English summary).

Lo, K.-C., Lee, W.T., Wu, T.-K. & Ho, C.-C. (1990) Use of predators to control spider mites (Acarina: Tetranychidae) in the Republic of China on Taiwan. In: *The Use of Natural Enemies to Control Agricultural Pests* (Ed. by J. Bay-Petersen), pp. 166–77. FFTC Asian & Pacific Region, Ser. 40, Taipei, Taiwan.

Lobbes, P. & Schotten, C. (1980) Capacities for increase of the soil mite *Hypoaspis aculeifer* Canestrini (Mesostigmata: Laelapidae). *Z. Angew. Entomol.*, **90**, 9–22.

Lockwood, J.A. (1987) Entomological warfare: history of the use of insects as weapons of war. *Bull. Entomol. Soc. Am.*, **33**, 76–82.

Lockwood, J.A. (1993) Environmental issues involved in biological control of rangeland grasshoppers (Orthoptera: Acrididae) with exotic organisms. *Environ. Entomol.*, **22**, 503–18.

Lomer, C.J. & Prior, C. (Eds) (1992) *Biological Control of Locusts and Grasshoppers*. CABI, Wallingford.

Londsdale, W.M., Briese, D.T. & Cullen, J.M. (2001) Risk analysis and weed biological control. In: *Evaluating Indirect Ecological Effects of Biological Control* (Ed. by E. Wajnberg, J.K. Scott & P.C. Quimby), pp. 185–210. CABI, Wallingford.

Longworth, J.F. (1987) Biological control of gorse: a case study. *N.Z. Entomol.*, **10**, 8–13.

Lord, F.T. (1947) The influence of spray programs on the fauna of apple orchards in Nova Scotia. II. Oystershell scale. *Can. Entomol.*, **79**, 196–209.

Lord, F.T. (1949) The influence of spray programs on the fauna of apple orchards in Nova Scotia. III. Mites and their predators. *Can. Entomol.*, **81**, 202–30.

Lord, F.T. & MacPhee, A.W. (1953) The influence of spray programs on the fauna of apple orchards in Nova Scotia. VI. Low temperatures and the natural control of the oystershell scale, *Lepidosaphes ulmi* (L.) (Homoptera: Coccidae). *Can. Entomol.*, **85**, 282–91.

Louda, S.M., Kendall, D., Connor, J. & Simberloff, D. (1997) Ecological effects of an insect introduced for the biological control of weeds. *Science*, **277**, 1088–90.

Lozano, A.P. (1979) Environmental control in asthmatic homes. The role of cheylatus mites. Preliminary report. *Allerg. Immunol.*, **7**, 303–6.

Luck, R.F. (1990) Evaluation of natural enemies for biological control: a behavioral evaluation. *Trends Ecol. Evol. (TREE)*, **5**, 196–9.

Luck, R.F., Shepard, B.M. & Kenmore, P.E. (1988) Experimental methods for evaluating arthropod natural enemies. *Annu. Rev. Entomol.*, **33**, 367–91.

Luck, R.F., Jiang, G. & Houck, M.A. (1999a) A laboratory evaluation of the astigmatid mite *Hemisarcoptes cooremani* Thomas (Acari: Hemisarcoptidae) as a potential biological control agent for an armored scale, *Aonidiella aurantii* (Maskell) (Homoptera: Diaspididae). *Biol. Cont.*, **15**, 173–83.

Luck, R.F., Nunney, L. & Stouthamer, R. (1999b) Sex ratio and quality in the culturing of parasitic Hymenoptera. A genetic and evolutionary perspective. In: *Handbook of Biological Control, Principles and Applications of Biological Control* (Ed. by T.S. Bellows & T.W. Fisher), pp. 653–72. Academic Press, San Diego, CA.

Luh, H.-K. & Croft, B.A. (1999) Classification of generalist or specialist life styles of predaceous phytoseiid mites using a computer genetic algorithm, information theory, and life history traits. *Environ. Entomol.*, **28**, 915–23.

Luxton, M. (1985) *Cryptostigmata (Arachnida: Acari) – A Concise Review*. Fauna of New Zealand, DSIR, Wellington.

Lydon, J. & Duke, S.O. (1989) Pesticide effects on secondary metabolism of higher plants. *Pesticide Sci.*, **25**, 361–73.

Lynch, L.D. & Thomas, M.B. (2000) Nontarget effects in the biocontrol of insects with insects, nematodes and microbial agents: the evidence. *Biocont. News Inform.*, **21**, 117–30.

Lysek, H. (1963) Effect of certain soil organisms on the eggs of parasitic roundworms. *Nature*, **199**, 925.

McClanahan, R.J. (1967) Food-chain toxicity of systemic acaricides to predaceous mites. *Nature, Lond.*, **215**, 1001.

McClay, A.S. (1996) Host range, specificity and recruitment: synthesis of session 2. In: *Proc. IX Int. Symp. Biological Control of Weeds* (Ed. by V.C. Moran & J.H. Hoffmann), pp. 105–12. University of Cape Town.

McClay, A.S., Littlefield, J.L. & Kashefi, J. (1999) Establishment of *Aceria malherbae* (Acari: Eriophyidae) as a biological control agent for field bindweed (Convolvulaceae) in the northern Great Plains. *Can. Entomol.*, **131**, 541–7.

McClure, M.S. (1995) *Diapterobates humeralis* (Oribatida: Ceratozetidae): an effective biological control agent of hemlock wooly adelgid (Homoptera: Adelgidae) in *Japan. Environ. Entomol.*, **24**: 1207–15.

McClure, M.S. & Cheah, C.A.S.-J. (1999) Reshaping the ecology of invading populations of hemlock woolly adelgid, *Adelges tsugae* (Homoptera: Adelgidae), in eastern North America. *Biol. Invasions*, **1**, 247–54.

McDermott, G.J. & Hoy, M.A. (1997) Persistence and containment of *Metaseiulus occidentalis* (Acari: Phytoseiidae) in Florida: risk assessment for possible releases of transgenic strains. *Fl. Entomol.*, **80**, 42–53.

McEvoy, P.B. (1996) Evaluation and economics: synthesis of session 7. In: *Proc. IX Int. Symp. Biological Control of Weeds* (Ed. by V.C. Moran & J.H. Hoffmann), pp. 511–15. University of Cape Town.

McFadyen, R.E.C. (1998) Biological control of weeds. *Annu. Rev. Entomol.*, **43**, 369–93.

McGarry, J.W. & Baker, A.S. (1997) Observations on the mite fauna associated with adult *Stomoxys calcitrans* in the U.K. *Med. Vet. Entomol.*, **11**, 159–64.

MacGill, E.I. (1939) A gamasid mite (*Thyphlodromus thripsi* n.sp.), a predator of *Thrips tabaci* Lind. *Ann. Appl. Biol.* **26**, 309–17.

McGregor, E.A. (1950) Mites of the genus *Neophyllobius. Bull. South. Calif. Acad. Sci.*, **49**, 55–70.

McGregor, E.A. & McDonough, F.L. (1917) The red spider on cotton. *U.S. Dept Agr. Bull.* 416.

Machado, M.P. & Thomé, J.W. (1994) Influência do ácaro *Riccardoella limacum* (Schranck, 1776) (Acarina) na Çriacão de caracóis *Helix aspersa* Müller, 1774 (Gastropoda). *Biocienc. Porto Alegre*, **2**, 3–14.

Mackauer, M. (1976) Genetic problems in the production of biological control agents. *Annu. Rev. Entomol.*, **21**, 369–85.

McLean, M.A., Kaneko, N. & Parkinson, D. (1996) Does selective grazing by mites and collembola affect litter fungal community structure? *Pedobiologia*, **40**, 97–105.

McMurtry, J.A. (1963) Diaspidine scale insects as prey for certain phytoseiid mites. In: *Advances in Acarology* (Ed. by J.A. Naegele), Vol. 1, pp. 151–4. Comstock, Ithaca, New York.

McMurtry, J.A. (1969) Biological control of citrus red mite in California. In: *Proc. 1st Int. Citrus Symp.* (Ed. by H.D. Chapman), Vol. 2, pp. 855–62. University of Calfornia, Riverside, CA.

McMurtry, J.A. (1977) Some predaceous mites (Phytoseiidae) on citrus in the Mediterraneam region. *Entomophaga*, **22**, 19–30.

McMurtry, J.A. (1980) Biosysytematics of three taxa in the *Amblyseius finlandicus* group from South Africa, with comparative life history studies (Acari: Phytoseiidae). *Int. J. Acarol.*, **6**, 147–56.

McMurtry, J.A. (1982) The use of phytoseiids for biological control: progress and future prospects. In: *Recent Advances in Knowledge of the Phytoseiidae* (Ed. by M.A. Hoy), pp. 23–48. Div. Agric. Sci., University of California, Special Publ. No. 3284, Berkeley, CA.

McMurtry, J.A. (1992) Dynamics and potential impact of 'generalist' phytoseiids in agroecosystems and possibilities for establishment of exotic species. *Exp. Appl. Acarol.*, **14**, 371–82.

McMurtry, J.A. & Croft, B.A. (1997) Life styles of phytoseiid mites and their roles in biological control. *Annu. Rev. Entomol.*, **42**, 291–321.

McMurtry, J.A. & Flaherty, D.L. (1977) An ecological study of phytoseiid and tetranychid mites on walnut in Tulare County, California. *Environ. Entomol.*, **6**, 287–92.

McMurtry, J.A. & Rodriguez, J.G. (1987) Nutritional ecology of phytoseiid mites. In: *Nutritional Ecology of Insects, Mites and Spiders* (Ed. by F. Slansky. & J.G. Rodriguez), pp. 609–44. John Wiley, New York.

McMurtry, J.A. & Scriven, G.T. (1962) The use of agar media in transporting and rearing phytoseiid mites. *J. Econ. Entomol.*, **55**, 412–14.

McMurtry, J.A. and Scriven, G.T. (1964) The influence of pollen and prey density on the number of prey consumed by *Amblyseius hibisci* (Acarina: Phytoseiidae). *Ann. Entomol. Soc. Am.*, **59**, 147–9.

McMurtry, J.A. & Scriven, G. T. (1965a) Life-history studies of *Amblyseius limonicus*, with comparative observations on *Amblyseius hibisci* (Acarina: Phytoseiidae). *Ann. Entomol. Soc. Am.*, **58**, 106–11.

McMurtry, J.A. & Scriven, G.T. (1965b) Insectary production of phytoseiid mites. *J. Econ. Entomol.*, **58**, 282–4.

McMurtry, J.A. & Scriven, G.T. (1966a) The influence of pollen and prey density on the number of prey consumed by *Amblyseius hibisci* (Acarina: Phytoseiidae). *Ann. Entomol. Soc. Am.*, **59**, 147–9.

McMurtry, J.A. & Scriven, G.T. (1966b) Studies on the feeding, reproduction, and development of *Amblyseius hibisci* (Acarina: Phytoseiidae) on various food substances. *Ann. Entomol. Soc. Am.*, **59**, 649–55.

McMurtry, J.A. & Scriven, G. T. (1971) Predation by *Amblyseius limonicus* on *Oligonychus punicae* (Acarina): effect of initial predator–prey ratios and prey distribution. *Ann. Entomol. Soc. Am.*, **64**, 219–24.

McMurtry, J.A. & Scriven, G.T. (1975) Population increase of *Phytoseiulus persimilis* on different insectary feeding programs. *J. Econ. Entomol.*, **68**, 319–21.

McMurtry, J.A., Huffaker, C.B. & van de Vrie, M. (1970) Ecology of teranychid mites and their natural enemies: a review. 1. Tetranychid enemies: their biological characters and the impact of spray practices. *Hilgardia*, **40**, 331–90.

McMurtry, J.A., Shaw, J.G. & Johnson, H.G. (1979) Citrus red mite populations in relation to virus disease and predaceous mites in southern California. *Environ. Entomol.*, **8**, 160–4.

MacNulty, B.J. (1971) An introduction to the study of Acari–Insecta associations. *Proc. Trans. Br. Entomol. Soc.*, **4**, 46–70.

MacPhee, A.W. & Sanford, K.H. (1954) The influence of spray programs on the fauna of apple orchards in Nova Scotia. VIII. Effects on some beneficial arthropods. *Can. Entomol.*, **86**, 128–35.

MacPhee, A.W. & Sanford, K.H. (1961) The influence of spray programs on the fauna of apple orchards in Nova Scotia. XII. Second supplement to VII. Effects on beneficial arthropods. *Can. Entomol.*, **93**, 671–3.

McRae, C.F. (1988) Classical and inundative approaches to biological weed control compared. *P. Protect. Q.*, **3**, 124–7.

MacRae, I.V. & Croft, B.A. (1993) Influence of temperature on interspecific predation and cannibalism by *Metaseiulus occidentalis* and *Typhlodromus pyri* (Acarina: Phytoseiidae). *Environ. Entomol.*, **22**, 770–5.

MacRae, I.V. & Croft, B.A. (1996) Differential impact of egg predation by *Zetzellia mali* (Acari: Stigmaeidae) on *Metaseiulus occidentalis* and *Typhlodromus pyri* (Acari: Phytoseiidae). *Exp. Appl. Acarol.*, **20**, 143–54.

Maeda, T., Takabayashi, J., Yano, S. & Takafuji, A. (1999) Response of the predatory mite, *Amblyseius womersleyi* (Acari: Phytoseiidae), toward herbivore-induced plant volatiles: variation in response between two local populations. *Appl. Entomol. Zool.*, **34**, 449–54.

Maeda, T., Takabayashi, J., Yano, S. & Takafuji, A. (2000) Effects of light on the tritrophic interaction between kidney bean plants, two-spotted spider mites and predatory mites, *Amblyseius womersleyi* (Acari: Phytoseiidae). *Exp. Appl. Acarol.*, **24**, 415–25.

Magowski, W.L. (1994) Discovery of the first representative of the mite subcohort Heterostigmata (Arachnida: Acari) in the Mesozoic Siberian amber. *Acarologia*, **35**, 229–41.

Mahr, F.A., Kwong, R.M., McLaren, D.A. & Jupp, P.W. (1997) Redistribution and present status of the mite *Aculus hyperici* for the control of St. John's wort, *Hypericum perforatum*, in Australia. *P. Protect. Q.*, **12**, 84–8.

Mahunka, S. (1970) Considerations on the systematics of the Tarsonemina and the description of new European taxa (Acari: Trombidiformes). *Acta Acad. Sci. Hung.*, **16**, 137–74.

Makol, J. (2000) Catalogue of the world Trombidiidae (Acari: Actinotrichida: Trombidioidea). *Ann. Zool.*, **50**, 599–625.

Malezioux, S., Lapchin, L., Pralavorio, M. *et al.* (1992) Toxicity of pesticide residues to a beneficial arthropod, *Phytoseiulus persimilis*. *J. Econ. Entomol.*, **85**, 2077–81.

Malhotra, P.R. & Mahanta, H.C. (1994) An annotated checklist of the mosquitoes parasitised by water mites (*Arrenurus* sp.) in India. *Ind. J. Parasitol.*, **18**, 107–13.

Mangini, A.C., Jr & Hain, F.P. (1991) Vapor pressure deficit differentially affects laboratory populations of *Metaseiulus occidentalis* and *Neoseiulus fallacis* (Acarina: Phytoseiidae) reared together. *Environ. Entomol.*, **20**, 823–31.

Mann, J. (1970) *Cacti Naturalised in Australia and Their Control*. Government Printer, Brisbane.

Manning, M. & Halliday, R.B. (1994) Biology and reproduction of some Australian species of Macrochelidae (Acarina). *Aust. Entomol.*, **21**, 89–94.

Manson, D.C.M. (1972) A contribution to the study of the genus *Rhizoglyphus* Claparede, 1869 (Acarina: Acaridae). *Acarologia*, **13**, 621–50.

Manson, D.C.M. & Gerson, U. (1996) Web spinning, wax secretion and liquid secretion by eriophyoid mites. In: *Eriophyoid Mites, Their Biology, Natural Enemies and Control* (Ed. by E.E. Lindquist, M.W. Sabelis & J. Bruin), pp. 251–8. Elsevier, Amsterdam.

Mansour, F., Ascher, K.R.S. & Omari, N. (1987) Effects of neem (*Azadirachta indica*) seed kernel extracts from different solvents on the predacious mite *Phytoseiulus persimilis* and the phytophagous mite *Tetranychus cinnabarinus*. *Phytoparasitica*, **15**, 125–30.

Mansour, F., Cohen, H. & Shain, Z. (1993) Integrated mite management in apples in Israel: augmentation of a beneficial mite and sensitivity of tetranychid and phytoseiid mites to pesticides. *Phytoparasitica*, **21**, 39–51.

Marei, S.S. (1992) First record of *Pyemotes tritici* (Acari, Prostigmata, Pyemotidae) as ectoparasite of *Chrysopa carnea* (Neuroptera, Chrysopidae). *Acta Entomol. Bohem.*, **89**, 393–4.

Margolies, L., Esch, G.W., Holmes, J.C. *et al.* (1982) The use of ecological terms in parasitology (report of an ad hoc committee of the American Society of Parasitologists). *J. Parasitol.*, **68**, 131–3.

Markin, G.P., Yoshioka, E.R. & Conant, P. (1996) Biological control of gorse in Hawaii. In: *Proc. IX Int. Symp. Biological Control of Weeds* (Ed. by V.C. Moran & J.H. Hoffmann), pp. 371–5. University of Cape Town.

Markkula, M. & Tiittanen, K. (1976) 'Pest in first' and 'natural infestation' methods in the control of *Tetranychus urticae* Koch with *Phytoseiulus persimilis* Athias-Henriot on glasshouse cucumbers. *Acta Agric. Fenn.*, **15**, 81–5.

Markwick, N.P., Wearing, C.H. & Shaw, P.W. (1990) Pyrethroid insecticides for apple pest control. 1. Development of pyrethroid-resistant predatory mites. *Proc. 43th N.Z. Plant Protection Conf.* (Ed. by A.J. Popay), pp. 296–300. Swiftprint Centre, Palmerston North.

Marohasy, J. (1996) Host shifts in biological weed control: real problems, semantic difficulties or poor science? *Int. J. Pest Manage.*, **42**, 71–5.

Marquis, R.J. & Whelan, C. (1996) Plant morphology and recruitment of the third trophic level: subtle and little-recognized defenses? *Oikos*, **75**, 330–3.

Marris, J.W.M. & Chapman, R.B. (1987) Effect of leaf type on the ovicidal activity of hexythiazox. *Proc. 40th N.Z. Weed Pest Control Conf.* (Ed. by A.J. Popay), pp. 112–15. Swiftprint Centre, Palmerston North.

Marshall, J.F. & Staley, J. (1929) A newly observed reaction of certain species of mosquitoes to the bites of larval hydrachnids. *Parasitology*, **21**, 158–60.

Matokot, L., Mapangou-Divassa, S. & Delobel, A. (1987) Évolution des populations de *Caryedon serratus* (Ol.) (Coleoptera: Bruchidae) dans les stocks d'arachnide au Congo. *L'Agron. Trop.*, **42**, 69–74.

Matveev, V.F., Martinez, C.C. & Frutos, S.M. (1989) Predator–prey relationships in subtropical zooplankton: water mites against cladocerans in an argentine lake. *Oecologia*, **79**, 489–95.

May, R.M. & Hassell, M.P. (1988) Population dynamics and biological control. *Phil. Trans. R. Soc. B.*, **318**, 129–69.

Meagher, R.L., Jr & Meyer, J.R. (1990) Influence of ground cover and herbicide treatments on *Tetranychus urticae* populations in peach orchards. *Exp. Appl. Acarol.*, **9**, 149–58.

Mégevand, B., Klay, A., Gnanvossou, D. & Paraiso, G. (1993) Maintenance and mass rearing of phytoseiid predators of the cassava green mite. *Exp. Appl. Acarol.*, **17**, 115–28.

Mehl, R. (1977) *Haemogamasus pontiger* Berlese (Acari, Mesostigmata) in Norway. *Rhizocrinus*, **9**, 1–4.

Mendel, Z. & Gerson, U. (1982) Is the mite *Lorryia formosa* Cooreman (Prostigmata: Tydeidae) a sanitizing agent in citrus groves? *Acta Oecol.*, **3**, 47–51.

Menn, J.J. & Hall, F.R. (1999) Biopesticides: present status and future prospects. In: *Biopesticides, Use and Delivery* (Ed. by F.R. Hall & J.J. Menn), pp. 1–10. Humana Press, Totowa, NJ.

van der Merwe, G.G. (1968) A taxonomic study of the family Phytoseiidae (Acari) in South Africa with contributions to the biology of two species. *South Afr. Dept Agric. Tech. Serv., Entomol. Mem.* **18**, 1–198.

Metz, J.A.J., Sabelis, M.W. & Kuchelin, J.H. (1988) Sources of variation in predation rates at high prey densities: an analytic model and mite example. *Exp. Appl. Acarol.*, **5**, 187–205.

Metz, M.A. & Irwin, M.E. (2001) Microtrombidiid mite parasitization frequencies and attachment site preferences on brachyceran Diptera with specific reference to Therevidae (Asiloidea) and Tachinidae (Oestroidea). *Environ. Entomol.*, **30**, 903–8.

Meyer, M.K.P. (1962) Two new mite predators of the red scale (*Aonidiella aurantii*) in South Africa. *South Afr. J. Agric. Sci.*, **5**, 411–17.

Meyer, M.K.P. (Smith) (1987) African Tetranychidae (Acari: Prostigmata) – with reference to the world genera. *Entomol. Mem. Dept Agric. Wat. Supp. Repub. South Africa*, **69**, 1–175.

Meyer, M.K.P. (Smith) (1996) *Mite Pests and Their Predators on Cultivated Plants in South Africa; Vegetables and Berries*. ARC – Plant Protection Research Institute, Handbook No. 6.

Meyer, M.K.P. (Smith) & Ueckermann, E.A (1984) The family Eupalopsellidae (Acari: Prostigmata) with descriptions of new species from South Africa. *Phytophylactica*, **16**, 121–42.

Meyer, M.K.P. (Smith) & Ueckermann, E.A. (1987) A taxonomic study of some Anystidae (Acari: Prostigmata). *Entomol. Mem. Dept Agric. Wat. Supp. Repub. South Africa*, **68**, 1–37.

Meyer, M.K.P. (Smith) & Ueckermann, E.A. (1989) African Raphignathoidea (Acari: Prostigmata). *Entomol. Mem. Dept Agric. Wat. Supp. Repub. South Africa*, **74**, 1–58.

Meyer, R.H. (1975) Release of carbaryl-resistant predatory mites in apple orchards. *Environ. Entomol.*, **4**, 49–51.

Meyerdirk, D.E. & Coudriet, D.L. (1986) Evaluation of two biotypes of *Euseius scutalis* (Acari: Phytoseiidae) as predators of *Bemisia tabaci* (Homoptera: Aleyrodidae). *J. Econ. Entomol.*, **79**, 659–63.

Michael, P. (1995) Biological control of redlegged earth mite and lucerne flea by the predators *Anystis wallacei and Neomolgus capillatus. P. Protect. Q.*, **10**, 55–7.

Michael, P.J., Dutch, M.E. & Pekin, C.J. (1991) Successful spreading of *Anystis salicinus* (L.), (Acari: Anystidae) and *Neomolgus capillatus* (Kramer), (Acari: Bdellidae) and their effects on redlegged earth mite and lucerne flea. In: *Proc. Nat. Workshop Redlegged Earth Mite, Lucerne Flea and Blue Oat Mite* (Ed. by J. Ridsdill-Smith), pp. 121–6. Dept of Agriculture, WA, Perth.

Micherdzinski, W. (1969) Die Familie Parasitidae Oudemans 1901 (Acarina, Mesostigmata). Panstwowe Wydawnictwo Naukowe, Cracow.

Mihm, J.A. & Chiang, H.C. (1976) Laboratory studies on the life cycle and reproduction of some soil- and manure-inhabiting predatory mites (Acarina: Laelapidae). *Pedobiologia*, **16**, 353–63.

Mikulecky, M. & Zemek, R. (1992) Does the moon influence the predatory activity of mites? *Experientia*, **48**, 530–2.

Milne, W.M. & Milne, A.L. (1987) The role of predators and parasites in the natural regulation of lucerne aphids in eastern Australia. *J. Appl. Ecol.*, **24**, 893–905.

Mitchell, M.J. & Parkinson, D. (1976) Fungal feeding of oribatid mites (Acari: Cryptostigmata) in an aspen woodland soil. *Ecology*, **57**, 302–12.

Mitchell, R. (1968) Site selection by larval water mites parasitic on the damselfly *Cercion hieroglyphicum* Brauer. *Ecology*, **49**, 40–7.

Mitchell, R. (1998) The behavior of *Arrenurus* larvae (Acari: Hydrachnidea) parasitizing Diptera. *Acarologia*, **39**, 49–55.

Mochizuki, M. (1996) Development, fecundity, diapause attribute and reproductive compatibility of a pesticide-resistant strain of the predatory mite, *Amblyseius womersleyi* Schicha (Acari: Phytoseiidae). *Jpn. J. Appl. Entomol. Zool.*, **40**, 121–6.

Mochizuki, M. (1997) Permethrin resistance and stability in the predatory mite, *Amblyseius womersleyi* Schicha (Acari: Phytoseiidae). *Jpn. J. Appl. Entomol. Zool.*, **41**, 1–5 (in Japanese with English abstract).

Mohamed, M.I., Zaher, M.A. & Hassan, M.F. (1982) Observations on *Cheyletus cacahuamilpensis*, a predator of the tenuipalpid mite *Dolichotetranychus floridanus*. *Entomophaga*, **27**, 343–8.

Momen, F. & Lundqvist, L. (1996) A new genus, *Quadrotydeus*, and three new species of the family Tydeidae (Acari: Prostigmata) from southern Sweden. *Int. J. Acarol.*, **22**, 3–10.

Momen, F.M. (1993) Effects of single and multiple copulation on fecundity, longevity and the sex ratio of the predacious mite *Amblyseius barkeri* (Hugh.) (Acari, Phytoseiidae). *Anz. Schädl. Pflanzen. Umwelt.*, **66**, 148–50.

Momen, F.M. (1994) Fertilization and starvation affecting reproduction in *Amblyseius barkeri* (Hughes) (Acari, Phytoseiidae). *Anz. Schädl. Pflanzen. Umwelt.*, **67**, 130–2.

Momen, F.M. (1995) Feeding, development and reproduction of *Amblyseius barkeri* on various kinds of food substances. *Acarologia*, **36**, 101–5.

de Moraes, G.J. & McMurtry, J.A. (1987) Physiological effect of the host plant on the suitability of *Tetranychus urticae* as prey for *Phytoseiulus persimilis* (Acari: Tetranychidae: Phytoseiidae). *Entomophaga*, **32**, 35–8.

de Moraes, G.J, McMurtry, J.A. & Denmark, H.A. (1986) *A Catalog of the Mite Family Phytoseiidae. References to Taxonomy, Synonymy, Distribution and Habitat.* EMBRAPA, Dept Difusao de Tecnologia, Brazil.

de Moraes, G.J., Neto, R.S. & Pinto, H.C.S. (1989) Morphology, biology and pesticide tolerance of *Cheletogenes ornatus* (Acari: Cheyletidae). *Entomophaga*, **34**, 477–88.

de Moraes, G.J., Mesa, N.C., Braun, A. & Melo, E.L. (1994a) Definition of the *Amblyseius limonicus* species group (Acari: Phytoseiidae), with descriptions of two new species and new records. *Int. J. Acarol.*, **20**, 209–17.

de Moraes, G.J., da Silva, C.A.D. & Moreira, A.N. (1994b) Biology of a strain of *Neoseiulus idaeus* (Acari: Phytoseiidae) from northwest Brazil. *Exp. Appl. Acarol.*, **18**, 213–20.

Morewood, W.D. (1992) Cold storage of *Phytoseiulus persimilis* (Phytoseiidae). *Exp. Appl. Acarol.*, **13**, 231–6.

Morgan, C.V.G., Chant, D.A., Anderson, N.H. & Ayre, G.L. (1955) Methods for estimating orchard mite populations, especially with the mite brushing machine. *Can. Entomol.*, **77**, 189–200.

Mori, H. & Chant, D.A. (1966) The influence of humidity on the activity of *Phytoseiulus persimilis* Athias-Henriot and its prey, *Tetranychus urticae* (C.L. Koch) (Acarina: Phytoseiidae, Tetranychidae). *Can. J. Zool.*, **44**, 863–71.

Mori, H. & Saito, Y. (1979) Biological control of *Tetranychus urticae* Koch (Acarina: Tetranychidae) populations by three species of phytoseiid mites (Acarina: Phytoseiidae). *J. Fac. Agric. Hokkaido Univ.*, **59**, 303–11.

Mori, H., Saito, Y. & Nakao, H. (1990) Use of predatory mites to control spider mites (Acarina: Tetranychidae) in Japan. In: *The Use of Natural Enemies to Control Agricultural Pests* (Ed. by J. Ben-Petersen), pp. 142–55. Food and Fertilizer Technology Center, ASPAC, Kuo Thai Color Printing Co., Bangkok.

Mori, H., Saito, Y. & Tho, Y.P. (1999) Co-operative group predation in a sit-and-wait cheyletid mite. *Exp. Appl. Acarol.*, **23**, 643–51.

Mori, N., Kuwahara, Y. & Kurosa, K. (1998) Rosefuran: the sex pheromone of an acarid mite. *J. Chem. Ecol.*, **24**, 1771–9.

Morris, M.A., Croft, B.A. & Berry, R.E. (1996) Overwintering and effects of autumn habitat manipulation and carbofuran on *Neoseiulus fallacis* and *Tetranychus urticae* in peppermint. *Exp. Appl. Acarol.*, **20**, 249–58.

Morris, M.A., Berry, R.E. & Croft, B.A. (1999) Phytoseiid mites on peppermint and effectiveness of *Neoseiulus fallacis* to control *Tetranychus urticae* (Acari: Phytoseiidae, Tetranychidae) in arid growing regions. *J. Econ. Entomol.*, **95**, 1072–8.

Moser, J.C. (1975) Biosystematics of the straw itch mite with special reference to nomenclature and dermatology. *Trans. R. Entomol. Soc. Lond.*, **127**, 185–91.

Moser, J.C. & Cross, E.A. (1975) Phoretomorph: a new phoretic phase unique to the Pyemotidae (Acarina: Tarsonemoidea). *Ann. Entomol Soc. Am.*, **68**, 820–2.

Moser, J.C. & Roton, L.M. (1970) Tagging mites with aerosol paint. *Ann. Entomol. Soc. Am.*, **63**, 1784.

Moser, J.C., Cross, E.A. & Roton, L.M. (1971) Biology of *Pyemotes parviscolyti* (Acarina: Pyemotidae). *Entomophaga*, **16**, 367–79.

Moser, J.C., Keilczewski, B., Wisniewski, J. & Balazy, S. (1978) Evaluating *Pyemotes dryas* (Vitzthum 1923) (Acari: Pyemotidae) as a parasite of the southern pine beetle. *Int. J. Acarol.*, **4**, 67–70.

Moser, J.C., Smiley, R.L. & Otvos, I.S. (1987) A new *Pyemotes* (Acari: Pyemotidae) reared from the Douglas-fir cone moth. *Int. J. Acarol.*, **13**, 141–7.

Mostafa, A.R., DeBach, P. & Fisher, T.W. (1975) Anystid mite: citrus thrips predator. *Calif. Agric.*, **29**(3), 5.

Motoyama, N., Rock, G.C. & Dauterman, W.C. (1971) Studies on the mechanism of azinphosmethyl resistance in the predaceous mite *Neoseiulus (T.) fallacis. Pest. Biochem. Physiol.*, **1**, 205–15.

Moutia, L.A. (1958) Contribution to the study of some phytophagous Acarina and their predators in Mauritius. *Bull. Entomol. Res.*, **49**, 59–75.

Moya Borja, G.E. (1981) Effects of *Macrocheles muscaedomesticae* (Scopoli) on the sexual behaviour and longevity of *Dermatobia hominis* (L.). *Rev. Brasil. Biol.*, **41**, 237–41.

Mozes-Koch, R. & Gerson, U. (1998) The separation of acarine species by enzymatic, immunological and molecular methods. *Syst. Appl. Acarol.*, **3**, 9–18.

Mullen, G.R. (1974) Acarine parasites of mosquitoes. II. Illustrated larval keys to the families and genera of mites reportedly parasitic on mosquitoes. *Mosq. News*, **34**, 183–94.

Mullen, G.R. (1975) Predation by water mites (Acarina: Hydrachnellae) on the immature stages of mosquitoes. *Mosq. News*, **35**, 168–71.

Mullen, G.R. (1977) Acarine parasites of mosquitoes. IV. Taxonomy, life history and behavior of *Thyas barbigera* and *Thyasides sphagnorum* (Hydrachnellae: Thyasidae). *J. Med. Entomol.*, **13**, 475–85.

Mullens, B.A., Hinkle, N.C. & Szijj, C.E. (1996) Impact of alternating manure removal schedules on pest flies (Diptera: Muscidae) and associated predators (Coleoptera: Histeridae, Staphylinidae; Acarina: Macrochelidae) in caged-layer poultry manure in southern California. *J. Econ. Entomol.*, **89**, 1406–17.

Müller-Schorer, H., Lewinsohn, T.M. & Lawton, J.H. (1991) Searching for weed biocontrol agents – when to move on? *Biocont. Sci. Tech.*, **1**, 271–80.

Muma, M.H. (1968) Phytoseiidae of sand-pine litter. *Fl. Entomol.*, **51**, 37–44.

Muma, M.H. (1969) *Typhlodromalus peregrinus* (Muma) (Acarina: Phytoseiidae) on Florida citrus. In: *Proc. 2nd Int. Congr. Acarology* (Ed. by G.O. Evans), pp. 135–48. Akadémiai Kiadó, Budapest.

Muma, M.H. (1975) Mites associated with citrus in Florida. *Univ. Florida Agric. Exp. Stn Bull.*, No. 640A.

Muma, M.H. & Selhime, A.G. (1971) *Agistemus floridanus* (Acarina: Stigmaeidae), a predatory mite, on Florida citrus. *Fl. Entomol.*, **54**, 249–58.

Muraoka, M. & Ishibashi, N. (1976) Nematode-feeding mites and their feeding behaviour. *Appl. Entomol. Zool.*, **11**, 1–7.

Murdoch, W.W., Chesson, J. & Chesson, P.L. (1985) Biological control in theory and practice. *Am. Natur.*, **125**, 344–66.

Murray, N.D. (1985) Rates of change in introduced organisms. In: *Proc. VI Int. Symp. Biological Control of Weeds* (Ed. by E.S. Delfosse), pp. 191–9. Agriculture Canada, Ottawa.

Murray, R.A. & Solomon, M.G. (1978) A rapid technique for analysing diets of invertebrate predators by electrophoresis. *Ann. Appl. Biol.*, **90**, 7–10.

Muttrie, M.P. & Anderson, I.B. (1984) *Pyemotes tritici* (?) – an uncomfortable, puzzling and expensive case in Scotland. In: *Acarology VI* (Ed. by D.A. Griffiths & C.E. Bowman), Vol. 2, pp. 1143–8. Ellis Horwood, Chichester.

Mwango, J., Williams, T. & Wiles, R. (1995) A preliminary study of the predator–prey relationships of watermites (Acari: Hydrachidia) and blackfly larvae (Diptera: Simuliidae). *Entomologist*, **114**, 107–17.

Nagelkerke, C.J. & Sabelis, M.W. (1991) Precise sex-ratio control in the pseudo-arrhenotok-ous phytoseiid mite *Typhlodromus occidentalis* Nesbitt. In: *The Acari: Reproduction, Development and Life History Strategies* (Ed. by P.W. Murphy & R. Schuster), pp. 193–207. Ellis Horwood, Chichester.

Nagelkerke, C.J. & Sabelis, M.W. (1996) Hierarchical levels of spatial structure and their consequences for the evolution of sex allocation in mites and other arthropods. *Am. Nat.*, **148**, 16–39.

Nakao, H., Saito, Y. & Mori, H. (1990) Mass production of predatory mites. In: *The Use of Natural Enemies to Control Agricultural Pests*. FFTC Series No. 40, pp. 184–9.

Naresh, J.S. & Balan, J.S. (1985) *Pyemotes ventricosus* (Newport), an ectoparasitic mite on cotton pink bollworm *Pectinophora gossypiella* (Saunders) in Haryana. *Indian J. Entomol.*, **47**, 239–40.

Navajas, M., Gutierrez, J. & Gotoh, T. (1997) Convergence of molecular and morphological data reveals phylogenetic information on *Tetranychus* species and allows the restoration of the genus *Amphitetranychus* (Acari: Tetranychidae). *Bull. Entomol. Res.*, **87**, 283–8.

Navajas, M., Thistlewood, H., Lagnel, J. *et al.* (2001) Field releases of the predatory mite *Neoseiulus fallacis* (Acari: Phytoseiidae) in Canada, monitored by pyrethroid resistance and allozyme markers. *Biol. Cont.*, **20**, 191–8.

Nawar, M.S., Shereef, G.M. & Ahmed, M.A. (1993) Effect of food on development, reproduction and survival of *Chiropturopoda bakeri* (Acarina: Uropodidae). *Exp. Appl. Acarol.*, **17**, 277–81.

Neale, M. (2000) The regulation of natural products as crop-protection agents. *Pest Manage. Sci.*, **56**, 677–80.

Neilson, M.M. (1963) The analysis of egg survival in the unsprayed area. *Mem. Entomol. Soc. Can.*, **31**, 37–41.

Nelson, E.E., Croft, B.A., Howitt, A.J. & Jones, A.L. (1973) Toxicity of apple orchard pesticides to *Agistemus fleschneri*. *Environ. Entomol.*, **2**, 212–22.

Nentwig, W., Frank, T. & Lethmayer, C. (1998) Sown weed strips: artificial ecological compensation areas as an important tool in conservation biological control. In: *Conservation Biological Control* (Ed. by P. Barbosa), pp. 133–53. Academic Press, San Diego, CA.

Neser, S. & Moran, V.C. (1985) Tactics for evading conflicts in the biological control of South African weeds. In: *Proc. VI Int. Symp. Biological Control of Weeds* (Ed. by E.S. Delfosse), pp. 359–63. Agriculture Canada, Ottawa.

Neuenschwander, P. (1996) Evaluating the efficacy of biological control of three exotic homopteran pests in tropical Africa. *Entomophaga*, **41**, 405–24.

Neuenschwander, P. & Markham, R. (2001) Biological control in Africa and its possible effects on biodiversity. In: *Evaluating Indirect Ecological Effects of Biological Control* (Ed. by E. Wajnberg, J.K. Scott & P.C. Quimby), pp. 127–46. CABI, Wallingford.

Newell, I.M. (1963) Feeding habits in the genus *Balaustium* (Acarina, Erythraeidae), with special reference to attacks on man. *J. Parasitol.*, **49**, 498–502.

Newell, I.M. & Ryckman, R.E. (1966) Species of *Pimeliaphilus* (Acari: Pterygosomidae) attacking insects, with particular reference to the species parasitizing Triatominae (Hemiptera: Reduviidae). *Hilgardia*, **37**, 403–36.

Newman, L.J. & Womersley, H. (1932) Clover springtail (lucerne flea) (*Smynthurus viridis*) investigations. *J. Agric. West. Aust.*, **9**, 289–90.

Newman, R.M., Thompson, D.C. & Richman, D.B. (1998) Conservation strategies for the biological control of weeds. In: *Conservation Biological Control* (Ed. by P. Barbosa), pp. 371–96. Academic Press, San Diego, CA.

Newstead, R. & Duvall, H.M. (1918) Bionomic, morphological and economic report on the acarids of stored grain and flour – Part I. *Reports of the Grain Pests (War) Committee of the Royal Society*, No. 2, pp. 1–48 (Abstract in *Rev. Appl. Entomol., Ser. A*, 91–4).

Newton, P.J., Neale, M.C, Arslan-Bir, M. *et al.* (1996) Full-range pest management with IPM systems – an industry view of the options for non-indigenous biopesticides. In: *Biological Control Introductions: Opportunities for Improved Crop Production* (Ed. by J.K. Waage), pp. 79–97. British Crop Protection Council, Farnham.

Nihoul, P. (1993a) Asynchronous populations of *Phytoseiulus persimilis* Athias-Henriot and effective control of *Tetranychus urticae* Koch on tomatoes under glass. *J. Hort. Sci.*, **68**, 581–8.

Nihoul, P. (1993b) Do light intensity, temperature and photoperiod affect the entrapment of mites on glandular hairs of cultivated tomato? *Exp. Appl. Acarol.*, **17**, 709–18.

Nihoul, P. & Hance, T. (1993) Use of a damage index to evaluate the biological control of the two-spotted spider mite *Tetranychus urticae* Koch (Acari; Tetranychidae) on tomato crops. *J. Hort. Sci.*, **68**, 575–80.

Nihoul, P., Van Impe, G. & Hance, T. (1991) Characterizing indices of damage to tomato by the two-spotted spider mite, *Tetranychus urticae* Koch (Acari; Tetranychidae) to achieve biological control. *J. Hort. Sci.*, **66**, 643–8.

Nomikou, M., Janssen, A., Schraag, R. & Sabelis, M.W. (2001) Phytoseiid predators as potential biological control agents for *Bemisia tabaci*. *Exp. Appl. Acarol.*, **25**, 271–91.

Noronha, A. C. da S., de Moraes, G.J. & Ciociola, A.I. (1995) Biologia de *Amblyseius manihoti* Moraes n. sp. sobre *Mononychellus tanajoa* (Bondar) (Acari: Phytoseiidae: Tetranychidae) em variedades de mandioca. *An. Soc. Entomol. Brasil*, **24**, 305–13.

Norris, J.D. (1958) Observations on the control of mite infestations in stored wheat by *Cheyletus* spp. (Acarina: Cheyletidae). *Ann. Appl. Biol.*, **46**, 411–22.

Norton, A.P., English-Loeb, G., Gadoury, D. & Seem, R.C. (2000) Mycophagous mites and foliar pathogens: leaf domatia mediate tritrophic interactions in grapes. *Ecology*, **81**, 490–9.

Norton, A.P., English-Loeb, G. & Belden, E. (2001) Host plant manipulation of natural enemies: leaf domatia protect beneficial mites from insect predators. *Oecologia*, **126**, 535–42.

Norton, R.A., Kethley, J.B., Johnston, D.E. & OConnor, B.M. (1993) Phylogenetic perspectives on genetic systems and reproductive modes of mites. In: *Evolution and Diversity of Sex Ratio in Insects and Mites* (Ed. by D. Wrensch & M. Ebbert), pp. 8–99. Chapman & Hall, London.

Nuzzaci, G., Mimmocchi, T. & Clement, S.L. (1985) A new species of *Aceria (Acari: Eriophyidae)* from *Convolvulus arvensis* L. (*Convolvulaceae*) with notes on other eriophyid associates of convolvulaceous plants. *Entomologica*, **20**, 81–9.

Nyiira, Z.M. (1970) A note on the natural enemies of lepidopterous larvae in cotton bolls in Uganda. *Ann. Entomol. Soc. Am.*, **63**, 1461–2.

Nyrop, J.P. (1988a) Sequential classification of prey/predator ratios with application to European red mite (Acari: Phytoseiidae) and *Typhlodromus pyri* (Acari: Phytoseiidae) in New York apple orchards. *J. Econ. Entomol.*, **81**, 14–21.

Nyrop, J.P. (1988b) Spatial dynamics of an acarine predator-prey system: *Typhlodromus pyri* (Acari: Phytoseiidae) preying on *Panonychus ulmi* (Acari, Tetranychidae). *Environ. Entomol.*, **17**, 1019–31.

Nyrop, J.P., Minns, J.C. & Herring, C.P. (1994) Influence of ground cover on dynamics of *Amblyseius fallacis* Garman (Acarina; Phytoseiidae) in New York apple orchards. *Agric. Ecosyst. Environ.*, **50**, 61–72.

Oatman, E.R., Gilstrap, F.E. & Voth, V. (1976) Effect of different release rates of *Phytoseiulus persimilis* (Acarina: Phytoseiidae) on the twospotted spider mite on strawberry in southern California. *Entomophaga*, **21**, 269–73.

Ochieng, R.S., Oloo, G.W. & Amboga, E.O. (1987) An artificial diet for rearing the phytoseiid mite *Amblyseius teke* Pritchard and Baker. *Exp. Appl. Acarol.*, **3**, 169–73.

Ochoa, R., Smiley, R.L. & Saunders, J.L. (1991) The family Tarsonemidae in Costa Rica (Acari: Heterostigmata). *Int. J. Acarol.*, **17**, 41–86.

OConnor, B.M. (1982) Astigmata. In: *Synopsis and Classification of Living Organisms* (Ed. by S.P. Parker), pp. 146–69. McGraw-Hill, New York.

OConnor, B.M. & Houck, M.A. (1989) Two new genera of Hemisarcoptidae (Acari: Astigmata) from the Huron Mountains of northern Michigan. *Great Lakes Entomol.*, **22**, 1–10.

O'Donnell, A.E. & Axtell, R.C. (1965) Predation by *Fuscuropoda vegetans* (Acarina: Uropodidae) on the house fly, (*Musca domestica*). *Ann. Entomol. Soc. Am.*, **58**, 403–4.

O'Donnell, A.E. & Nelson, E.L. (1967) Predation by *Fuscuropoda vegetans* (Acarina: Uropodidae) and *Macrocheles muscaedomesticae* (Acarina: Macrochelidae) on the eggs of the little house fly, *Fannia canicularis*. *J. Kansas Entomol. Soc.*, **40**, 441–3.

O'Dowd, D.J. & Willson, M.F. (1989) Leaf domatia and mites on Australian plants: ecological and evolutionary implications. *Biol. J. Linn. Soc.*, **37**, 191–236.

O'Dowd, D.J. & Willson, M.F. (1997) Leaf domatia and the distribution and abundance of foliar mites in broadleaf deciduous forest in Wisconsin. *Am. Midl. Nat.*, **137**, 337–48.

O'Dowd, D.J., Brew, C.R., Christophel, D.C. & Norton, R.A. (1991) Mite-plant associations from the Eocene of Southern Australia, *Science*, **252**, 99–101.

Okabe, K. & Amano, H. (1990) Attractancy of alcohols isolated from culture filtrates of *Fusarium* fungi for the robine bulb mite, *Rhizoglyphus robini* Claparède (Acari: Acaridae) in sand. *Appl. Entomol. Zool.*, **25**, 397–404.

Okamoto, M., Matsumoto, K., Wada, Y. & Nakano, H. (1978) Studies on antifungal effect of mite alarm pheromone citral. I. Evaluation of antifungal effect of citral. *Jpn. J. Sanit. Zool.*, **29**, 255–60.

Oldfield, G.N. & Proeseler, G. (1996) Eriophyoid mites as vectors of plant patogens. In: *Eriophyoid Mites, Their Biology, Natural Enemies and Control* (Ed. by E.E. Lindquist, M.W. Sabelis & J. Bruin), pp. 259–75. Elsevier, Amsterdam.

Oomen, P.A. (1982) Studies on population dynamics of the scarlet mite, *Brevipalpus phoenicis*, a pest of tea in Indonesia. *Med. Landbouww. Wageningen*, No. 82–1.

Oomen, P.A., Romeijn, G. & Wiegers, G.L. (1991) Side-effects of 100 pesticides on the predatory mite *Phytoseiulus persimilis*, collected and evaluated according to the EPPO guidelines. *EPPO Bull.*, **21**, 701–12.

Osman, A.A. & Zaki, A.M. (1986) Studies on the predation efficiency of *Agistemus exsertus* Gonzalez (Acarina, Stigmaeidae) on the eriophyid mite *Aculops lycopersici* (Massee). *Anz. Schädling. Pflazenschutz. Umwelt*, **59**, 135–6.

Osman, A.A. & Zohdi, G. (1976) Suppression of the spider mites on cotton with mass releases of *Amblyseius gossipi* (El-Badry). *Z. Angew. Entomol.*, **81**, 245–8.

Osman, A.A., Abo-Taka S.M. & Zaki, A.M. (1991) *Agistemus exsertus* Gonzalez (Acarina, Stigmaeidae) as a predator of the grapevine mite *Colomerus vitis* (Pgst.) (Acarina: Actinedida). In: *Modern Acarology* (Ed. by F. Dusbábek & V. Bukva), Vol. 2, pp. 689–90. SPB, The Hague.

Otto, J.C. (1992) A new species of *Anystis* von Heyden compared with *Anystis salicinus* (Linnaeus) (Acarina: Anystidae). *Int. J. Acarol.* **18**, 25–35.

Otto, J.C. (2000) A cladistic analysis of Erythracarinae (Acarina: Prostigmata: Anystidae), with the description of a new genus. *Syst. Entomol.*, **25**, 447–84.

Otto, J.C. & Halliday, R.B. (1991) Systematic and biology of a predatory mite (*Anystis* sp.) introduced into Australia for biological control of redlegged mite. *P. Protect. Q.*, **6**, 181–5.

Ouyang, Y., Grafton-Caldwell, E.E. & Bugg, R.L. (1992) Effects of various pollens on development, survivorship, and reproduction of *Euseius tularensis* (Acari: Phytoseiidae). *Environ. Entomol.*, **21**, 1371–6.

Overmeer, W.P.J. (1981) Notes on breeding phytoseiid mites from orchards (Acarina: Phytosiidae) in the laboratory. *Med. Fac. Landbouww. Rijkuniv. Gent*, **46**, 503–9.

Overmeer, W.P.J. (1985) Rearing and handling. In: *Spider Mites, Their Biology, Natural Enemies and Control* (Ed. by W. Helle & M.W. Sabelis), Vol. 1B, pp. 161–70. Elsevier, Amsterdam.

Overmeer, W.P.J. & van Zon, A.Q. (1983) The effect of different kinds of food on the induction of diapause in the predacious *Amblyseius potentillae*. *Entomol. Exp. Appl.*, **33**, 27–30.

Overmeer, W.P.J. & van Zon, A.Q. (1984) The preference of *Amblyseius potentillae* (Garman) (Acarina: Phytoseiidae) for certain plant substrates. In: *Acarology VI* (Ed. by D.A. Griffiths & C.E. Bowman), Vol. 1, pp. 391–6. Ellis Horwood, Chichester.

Painter, R.H. (1951) *Insect Resistance in Crop Plants*. Macmillan, New York.

Palevsky, E., Oppenheim, D., Reuveny, H. & Gerson, U. (1996) Impact of European red mite on Golden Delicious and Oregon Spur apples in Israel. *Exp. Appl. Acarol.*, **20**, 343–54.

Palevsky, E., Reuveny, H., Okonis, O. & Gerson, U. (1999) Comparative behavioural studies of larval and adult stages of the phytoseiids (Acari: Mesostigmata) *Typhlodromus athiasae* and *Neoseiulus californicus*. *Exp. Appl. Acarol.*, **23**, 467–85.

Palmer, A.R. (1994) Fluctuating asymmetry analyses: a primer. In: *Develoment Instability: Its Origins and Evolutionary Implications* (Ed. by T.A. Markow), pp. 335–64. Kluwer, Dordrecht.

Parent, B. (1967) Population studies of phytophagous mites and predators on apple in southwestern Quebec. *Can. Entomol.*, **99**, 771–8.

Parrella, M.P., Heinz, K.M. & Nunney, L. (1992) Biological control through augmentative releases of natural enemies: a strategy whose time has come. *Am. Entomol.*, **38**, 172–9.

Parrott, P.J., Hodgkiss, H.E. & Schoene, W.J. (1906) The apple and pear mites. *New York Agr. Exp. St. Bull.*, No. 283, 281–318.

Parsons, P.A. (1990) Fluctuating asymmetry: an epigenic measure of stress. *Biol. Rev.*, **65**, 131–45.

Paterson, C.G. (1970) Water mites (Hydracarina) as predators of chironomid larvae (Insecta: Diptera). *Can. J. Zool.*, **48**, 610–4.

Pavlitshenko, P.G. (1994) *A Guide to the Ceratozetoid mites (Oribatei, Ceratozetoidea) of Ukraine. Kiev* (in Russian with bilingual keys). Institute of Zoology , Kiev.

Peck, J.H. & Anderson, J.R. (1969) Arthropod predators of immature Diptera developing in poultry droppings in northern California. Part 1. Determination, seasonal abundance and natural cohabitation with prey. *J. Med. Entomol.*, **6**, 163–7.

Peck, J.H. & Anderson, J.R. (1970) Influence of poultry-manure-removal schedules on various Diptera larvae and selected arthropod predators. *J. Econ. Entomol.*, **63**, 82–90.

Pemberton, R.W. (1985) Native weeds as candidates for biological control research. In: *Proc. VI Int. Symp. Biological Control of Weeds* (Ed. by E.S. Delfosse), pp. 869–77. Agriculture Canada, Ottawa.

Pemberton, R.W. (1993) Observations on extrafloral nectar feeding by predaceous and fungivorous mites. *Proc. Entomol. Soc. Wash.*, **95**, 642–3.

Pemberton, R.W. (2000) Predictable risks to native plants in weed biological control. *Oecologia*, **125**, 489–94.

Peña, J.E. & Osborne, L. (1996) Biological control of *Polyphagotarsonemus latus* (Acarina: Tarsonemidae) in greenhouses and field trials using introductions of predacious mites (Acarina: Phytoseiidae). *Entomophaga*, **41**, 279–85.

Penman, D.R. & Chapman, R.B. (1980) Integrated control of apple pests in New Zealand. 17. Relationships of *Amblyseius fallacis* to phytophagous mites in an apple orchard. *N.Z. J. Zool.*, **7**, 281–7.

Penman, D.R., Wearing, C.H., Collyer, E. & Thomas, W.P. (1979) The role of insecticide-resistant phytoseiids in integrated mite control in New Zealand. In: *Recent Advances in Acarology* (Ed. by J.G. Rodriguez), Vol. 1, pp. 59–69. Academic Press, New York.

Pereira, C. & de Castro, M.P. (1945) Contribuição para o conhecimento de espécie tipo di *Macrocheles* Latr. (Acarina): *M. muscaedomesticae* (Scopoli, 1772) emend. *Arq. Inst. Biol. (São Paulo)*, **16**, 153–89.

Pérez-Iñigo, C. (1993) *Fauna Iberica*, Vol. 3. *Acari, Oribatei, Poronota*. Museo Nacional de Ciencias Naturales, CSIC, Madrid.

Pérez-Moreno, I. & Moraza, M.L. (1997) Étude sur le *Typhlodromus pyri* Scheuten en relation avec le *Calepitrimerus vitis* (Nalepa) dans les vignobles de la Rioja. *Bull. l'OIV*, **70**, 832–45.

Perkins, B.D. (1973) Preliminary studies of a strain of the waterhyacinth mite from Argentina. In: *Proc. 2nd. Int. Symp. Biological Control of Weeds* (Ed. by R.H. Dunn), pp. 180–4. Commonwealth Agricultural Bureaux, Slough.

Perring, T.M. & Lackey, L.J. (1989) Temperature and humidity effects on mortality and pre-adult development of two *Phytoseiulus persimilis* strains (Acari: Phytoseiidae). *Int. J. Acarol.*, **15**, 47–52.

Perroti, A. (2001) Prey location and predation rates of predatory mites (Acari: Macrochelidae) on immature stages of pest flies (Diptera: Muscidae). *Syst. Appl. Acarol.*, **6**, 27–33.

Petanović, R. (1993) Mites (Acari) biological control agents. I. Use in orchards and vineyards. *Zaštita bilja*, **44**, 5–20 (in Serbian).

Petrushov, A.Z. (1987) Results of introduction into the USSR of a population of a predaceous mite, *Metaseiulus occidentalis* (Acarina, Phytoseiidae), resistant to pesticides. *Zool. Zhur.*, **66**, 674–80 (in Russian with English summary).

Petrushov, A.Z. (1991) Selection and mechanisms of Ambush resistance in *Metaseiulus occidentalis* (Acarina, Phytoseiidae). In: *Modern Acarology* (Ed. by F. Dusbábek & V. Bukva), Vol. 2, pp. 741–8. SPB, The Hague.

Petrushov, A.Z. (1992) Pyrethroid resistance in the predacious mite *Amblyseius barkeri*. *EPPO Bull.*, **22**, 471–3.

Pfeiffer, D.G. (1986) Effects of field applications of paraquat on densities of *Panonychus ulmi* (Koch) and *Neoseiulus fallacis* (Garman). *J. Agric. Entomol.*, **3**, 322–5.

Piatkowski, J. (1997) Biological control of sciarid flies (*Bradysia* spp.) with predatory mites *Hypoaspis aculeifer* on poinsetia crops in greenhouses. *OILB/WPRS Bull.*, **20**(4), 221–4.

Pickett, A.D. (1965) The influence of spray programs on the fauna of apple orchards in Nova Scotia. XIV. Supplement to II. Oystershell scale, *Lepidosaphes ulmi* (L.). *Can. Entomol.*, **97**, 816–21.

Pickett, C.H. & Gilstrap, F.E. (1986a) Predation of *Oligonychus pratensis* (Aca.: Tetranychidae) by *Phytoseiulus persimilis* and *Amblyseius californicus* (Aca.: Phytoseiidae) under controlled laboratory conditions. *Entomophaga*, **31**, 205–12.

Pickett, C.H. & Gilstrap, F.E. (1986b) Inoculative releases of phytoseiids (Acari) for the biological control of spider mites (Acari: Tetranychidae) in corn. *Environ. Entomol.*, **15**, 790–4.

Pickett, C.H., Gilstrap, F.E., Morrison, R.K. & Bouse, L.F. (1987) Release of predatory mites (Acari: Phytoseiidae) by aircraft for the biological control of spider mites (Acari: Tetranychidae) infesting corn. *J. Econ. Entomol.*, **80**, 906–10.

Pickett, C.H., Morrison, R.K., Gilstrap, F.E. & Bouse, L.F. (1990) Storage of *Phytoseiulus persimilis* for aerial release (Acari: Phytoseiidae). *Int. J. Acarol.*, **16**, 37–40.

Pieterse, A.H. (1972) A preliminary investigation on control of water hyacinth by spider mites. *Proc. 11th. British Weed Control Conf.*, Vol. 1, 1–3. British Crop Protection Council, Nottingham.

Pimm, S.L. & Lawton, J.H. (1978) On feeding on more than one trophic level. *Nature, Lond.*, **275**, 542–4.

Poe, S.L. & Enns, W.R. (1970) Effects of inbreeding on closed populations of predaceous mites (Acarina: Phytoseiidae). *Can. Entomol.*, **102**, 1222–9.

Poinar, G.O., Treat, A.E. & Southcott, R.V. (1991) Mite parasitism of moths: examples of paleosymbiosis in Dominican amber. *Experientia*, **47**, 210–2.

Poinar, G.O., Jr, Pike, E.M. & Krantz, G.W. (1993) Animal–animal parasitism. *Nature*, **361**, 307–8.

Poinar, G.O., Jr, Krantz, G.W., Boucot, A.J. & Pike, E.M. (1997) A unique Mesozoic parasitic association. *Naturwissenschaften*, **84**, 321–2.

Polak, M. & Trivers, R. (1994) The science of symmetry in biology. *Trends Ecol. Evol. (TREE)*, **9**, 122–4.

Popov, N.A. & Khudyakova, O.A. (1989) Development of *Phytoseiulus persimilis* (Acarina, Phytoseiidae) fed on *Tetranychus urticae* (Acarina, Tetranychidae) on various food plants. *Acta Entomol. Fenn.*, **53**, 43–6.

Porres, M.A., McMurtry, J.A. & March, R.B. (1975) Investigations of leaf sap feeding by three species of phytoseiid mites by labelling with radioactive phosphoric acid ($H_3{}^{32}PO_4$). *Ann. Entomol. Soc. Am.*, **68**, 871–2.

Prasad, V. (1973) The role of *Phytoseiulus macropilis* (Banks) (Acarina: Phytoseiidae) in control of spider mites in Hawaii. *Acarologia*, **15**, 400–5.

Prasad, V. (1975) Biology of *Treatia indica* Krantz and Khot (Acarina: Phytoseiidae), a mite parasite of the red-cotton bug in India. *Acarologia*, **17**, 30–55.

Prather, T.S., Callihan, R.H., McCaffrey, J.P. & Shafii, B. (1989) Effects of a gall midge and a gall mite on rush skeletonweed. *Proc. West. Soc. Weed Sci.*, **42**, 284–5.

Pratt, P.D., Schausberger, P. & Croft, B.A. (1999) Prey-food types of *Neoseiulus fallacis* (Acari: Phytoseiidae) and literature versus experimentally derived prey-food estimates for five phytoseiid species. *Exp. Appl. Acarol.*, **23**, 551–65.

Presnail, J.K. & Hoy, M.A. (1994a) Transient expression of a *Drosophila melanogaster hsp70 promoter/lacZ* construct injected into larvae of two species of predatory mites (Acari: Phytoseiidae). *Exp. Appl. Acarol.*, **18**, 301–8.

Presnail, J.K. & Hoy, M.A. (1994b) Transmission of injected DNA sequences to multiple eggs of *Metaseiulus occidentalis* and *Amblyseius finlandicus* (Acari: Phytoseiidae) following maternal microinjection. *Exp. Appl. Acarol.*, **18**, 319–30.

Presnail, J.K., Hoy, M.A. & Jeyaprakash, A. (2000) Developing transgenic phytoseiids for biological control programs. In: *Acarology IX, Symposia* (Ed. by G.R. Needham, R. Mitchell, D.J. Horn & W.C. Wellbourn), Vol. 2, pp. 15–19. Ohio Biological Survey, Columbus, OH.

Price, J.F. (1981) Postharvest control of the twospotted spider mite on chrysanthemum and carnation flowers. *Proc. Fl. State Hort. Soc.*, **94**, 74–6.

Price, J.F. & Nguyen, F.Q. (1997) Fitness of commercially produced, shipped and stored *Phytoseiulus persimilis* Athias-Henriot (Acari: Phytoseiidae) predators for dispersing in Florida strawberry. *Acta Hort.*, **439**, 913–16.

Price, P.W. (1997) *Insect Ecology*, 3rd Edn. Wiley, New York.

Price, P.W., Bouton, C.E., Gross, P. *et al.* (1980) Interactions among three trophic levels: influence of plants on interactions between insect herbivores and natural enemies. *Annu. Rev. Ecol. Syst.*, **11**, 41–65.

Proctor, H.C. (1992) Discord between field and laboratory sex ratios of the water mite *Neumania papillator* Marshall (Acari: Unionicolidae). *Can. J. Zool.*, **70**, 2483–6.

Proctor, H.C. (1996) Sex-ratios and chromosomes in water mites (Hydracarina). In: *Proc. Acarology IX* (Ed. by R. Mitchell, D.J. Horn, G.R. Needham & W.C. Welbourn), pp. 441–5. Ohio Biological Survey, Columbus, OH.

Proctor, H. & Pritchard, G. (1989) Neglected predators: water mites (Acari: Parasitengona: Hydrachnellae) in freshwater communities. *J. North Amer. Benthol. Soc.*, **8**, 100–11.

Pruszynski, S. & Cone, W.W. (1972) Relationships between *Phytoseiulus persimilis* and other enemies of the twospotted spider mite on hops. *Environ. Entomol.*, **1**, 431–3.

Pulpán, J. & Verner, P.H. (1965) Control of tyroglyphoid mites in stored grain by the predatory mite *Cheyletus eruditus* (Schrank). *Can. J. Zool.*, **43**, 417–32.

Pultar, O., Pliva, J. & Muška, J. (1992) *Typhlodromus pyri* Scheut. as a biologcal control agent of spider mites in Czechoslovakian large scale fruit production. *Acta Phytol. Entomol. Hung.*, **27**, 513–15.

Purrington, C.B. (2000) Costs of resistance. *Curr. Opin. P. Biol.*, **3**, 305–8.

Putman, W.L. (1962) Life-history and behaviour of the predacious mite *Typhlodromus (T.) caudiglans* Schuster (Acarina: Phytoseiidae) in Ontario, with notes on the prey of related species. *Can. Entomol.*, **94**, 163–77.

Putman, W.L. (1970) Life history and behavior of *Balaustium putmani* (Acarina: Erythraeidae). *Ann. Entomol. Soc. Am.*, **63**, 76–81.

Putman, W.L. & Herne, H.C. (1964) Relationships between *Typhlodromus caudiglans* Schuster (Acarina: Phytoseiidae) and phytophagous mites in Ontario peach orchards. *Can. Entomol.*, **96**, 925–43.

Rack, G. (1959) *Acarophenax dermestidarum* sp. n. (Acarina, Pyemotidae), ein Eiparasit von *Dermestes-Arten. Z. Parasiten.*, **19**, 411–31.

Rack G. & Rilling, G. (1978) Über das Vorkommen der Modermilbe, *Tyrophagus putrescentiae* (Schrank) in Blatgallen der Reblaus, *Dactylosphaera vitifolii* Shimer. *Vitis*, **17**, 54–66.

Radovsky, F.J. & Gettinger, D. (1999) Acanthochelinae, new subfamily (Acari: Parasitiformes: Laelapidae), with redescription of *Acanthochela chilensis* Ewing and descriptions of a new genus and species from Argentina. *Int. J. Acarol.*, **25**, 77–90.

Ragusa, S. & Ciulla, A.M. (1988) Effects of a development inhibitor on phytoseiid mites inhabiting citrus trees. In: *Progress in Acarology* (Ed. by G.P. Channabasavanna & C.A. Viraktamath), Vol. 1, pp. 411–15. Oxford & IBH, New Delhi.

Ragusa, S. & Swirski, E. (1975) Feeding habits, development and oviposition of the predacious mite *Amblyseius swirskii* Athias-Henriot (Acarina: Phytoseiidae) on pollen of various weeds. *Israel J. Entomol.*, **10**, 93–103.

Ragusa, S. & Swirski, E. (1977) Feeding habits, post-embryonic and adult survival, mating, virility and fecundity of the predacious mite *Amblyseius swirskii* (Acarina: Phytoseiidae) on some coccids and mealybugs. *Entomophaga*, **22**, 383–92.

Ragusa, S. & Zedan, M.A. (1988) Biology and predation of *Hypoaspis aculeifer* (Canestrini) (Parasitiformes, Dermanyssidae) on *Rhizoglyphus echinopus* (Fum. & Rob.) (Acariformes, Acaridae). *Redia*, **71**, 213–25.

Ragusa di Chiara, S. & Tsolakis, H. (1995) The effect of *Phenacoccus madeirensis* Green (Coccoidea: Pseudococcidae) on some biological parameters of four species of phytoseiid mites (Parasitiformes: Phytoseiidae). *Israel J. Entomol.*, **29**, 301–9.

Ragusa Di Chiara, S. & Tsolakis, H. (2001) Phytoseiid faunas of natural and agricultural ecosystems in Sicily. In: Acarology: *Proc. 10th Int. Congr.* (Ed. by R. B. Halliday, D.E. Walter, H.C. Proctor *et al.*), pp. 522–9. CSIRO, Melbourne.

Rajendran, R. & Prasad, R.S. (1989) *Encentridophorus similis* (Acarina: Unionicolidae) an active predator of mosquito larvae. *Curr. Sci.*, **58**, 466–7.

Rajendran, R. & Prasad, R.S. (1992) Influence of mite infestation on the longevity and fecundity of the mosquito *Mansonia uniformis* (Diptera: Insecta) under laboratory conditions. *J. Biosci.*, **17**, 35–40.

Rajendran, R. & Prasad, R.S. (1994) A laboratory study of the life cycle and feeding behaviour of *Arrenurus madaraszi* (Acari: Arrenuridae) parasitizing *Anopheles* mosquitoes. *Trop. Med. Parasitol.*, **88**, 169–74.

Ramakers, P.M.J. (1978) Possibilities for biological control of *Thrips tabaci* Lind. (Thysanoptera: Thripidae) in glasshouses. *Med. Fac. Landbouww. Rijkuniv. Gent*, **43**, 463–9.

Ramakers, P.M.J. (1980) Biological control of *Thrips tabaci* (Thysanoptera: Thripidae) with *Amblyseius* spp. (Acari: Phytoseiidae). *SROP/WPRS Bull.*, **3**, 203–8.

Ramakers, P.M.J. (1988) Population dynamics of the thrips predators *Amblyseius mckenziei* and *Amblyseius cucumeris* (Acarina: Phytoseiidae) on sweet pepper. *Neth. J. Agric. Sci.*, **36**, 247–52.

Ramakers, P.M.J. (1993) Coexistence of two thrips predators, the anthocorid *Orius insidiosus* and the phytoseiid *Amblyseius cucumeris* on sweet pepper. *WPRS/OILB Bull.*, **16**(2), 133–6.

Ramakers, P.M.J. & van Lieburg, M.J. (1982) Start of commercial production and introduction of *Amblyseius mckenziei* Sch. & Pr. (Acarina: Phytoseiidae) for the control of *Thrips tabaci* (Thysanoptera: Thripidae) in glasshouses. *Med. Fac. Landbouww. Rijkuniv. Gent*, **47**, 541–5.

Ramakers, P.M.J. & Voet, S.J.P. (1996) Introduction of *Amblyseius degenerans* for thrips control in sweet peppers with potted castor beans as banker plants. *IOBC/WPRS Bull.*, **19**(1), 127–30.

Ramakers, P.M.J., Dissevelt, M. & Peeters, K. (1989) Large scale introductions of phytoseiid predators to control thrips on cucumber. *Med. Fac. Landbouww. Rijkuniv. Gent*, **54**, 923–9.

Ramos, M. & Rodriguez, H. (1995) Eficiencia en el control de *Tetranychus tumidus* (Banks) por *Phytoseiulus macropilis* (Banks) en platano. 1. Prueba en casa de malla. *Rev. Prot. Veg.*, **10**, 207–11.

Rangel, A.F. (1901) Cuarto informe acerca del picudo del algodon (*Insanthonomus grandis* I. C. Cu.). *Bol. Comis. Parasitol. Agric.*, **1**, 245–61.

Rao, V.P. & Sankaran, T. (1969) The scale insects of sugar cane. In: *Pests of Sugar Cane* (Ed. by J.R. Williams, J.R. Metcalfe, R.W. Mungomery & R. Mathes), pp. 325–42. Elsevier, Amsterdam.

Rasmy, A.H., Abdel Rahman, H.A., Abdel-Kader, M.M. & Hussein, H.E. (1996) The predatory response and prey consumption of the predatory mite *Agistemus exsertus* Gonz. (Stigmaeidae) as affected by rearing on certain diets. In: *Proc. Acarology IX* (Ed. by R. Mitchell, D.J. Horn, G.R. Needham & W.C. Wellbourn), pp. 219–20. Ohio Biological Survey, Columbus, OH.

Rasmy, A.H., Elbagoury, M.E. & Reda, A.S. (1987) A new diet for reproduction of two predaceous mites *Amblyseius gossipi* and *Agistemus exsertus* (Acari: Phytoseiidae: Stigmaeidae). *Entomophaga*, **32**, 277–80.

Rasmy, A.S. (1970) Management of orchard mite populations with particular reference to citrus mites. *Z. Angew. Entomol.*, **66**, 174–7.

Rat-Morris, E. (1999) Biological control of *Thrips tabaci* on protected leek seed crops. *IOLB/WPRS Bull.*, **22**(1), 201–4.

Raut, S.K. (1996) Factors determining the effectiveness of the mites *Fuscuropoda marginata* in the control of the slug pests *Laevicaulis alte*. In: *Slug and Snail Pests in Agriculture* (Ed by I.F. Henderson), pp. 247–54. British Crop Protection Council, Farnham.

Raut, S.K. & Panigrahi, A. (1991) The mite *Fuscuropoda marginata* (C.L. Koch) for the control of pest slugs *Laevicaulis alte* (Férussac). In: *Modern Acarology* (Ed. by F. Dusbábek & V. Bukva), Vol. 2, pp. 683–7. SPB, The Hague.

Raworth, D.A., Fauvel, G. & Auger, P. (1994) Location, reproduction and movement of *Neoseiulus californicus* (Acari: Phytoseiidae) during the autumn, winter and spring in orchards in the south of France. *Exp. Appl. Acarol.*, **18**, 593–602.

Readshaw, J.L. (1975) Biological control of orchard mites in Australia with an insecticide-resistant predator. *J. Aust. Inst. Agric. Sci.*, **41**, 213–14.

Readshaw, L. (1998) The importance of pollen to integrated pest management (IPM) in Australian apple orchards and other horticultural crops. Unpublished talk at the 10th International Congress of Acarology, Canberra, Australia.

Reda, A.S. (1990) The use of artificial diets and natural diets in rearing *Agistemus exsertus* (Acari: Stigmaeidae). *Ann. Agric. Sci., Moshtohor*, **28**, 2633–42.

Rees, D.P. (1994) Biological control – session summary. In: *Proc. 6th Int. Work. Conf. Stored Product Protection* (Ed. by E. Highley, E.J. Wright, H.J. Banks & B.R. Champ), p. 1165. CABI, Wallingford.

Regenfuss, H. (1968) Untersuchungen zur Morphologie, Systematik und Ökologie der Podapolipidae (Acarina, Tarsonemini). *Zeit. Wissen. Zool.*, **177**, 183–282.

Reinhardt, K. (1996) Negative effects of *Arrenurus* water mites on the flight distances of the damselfly *Nehalennia speciosa* (Odonata: Coenagrionidae). *Aquat. Insects*, **18**, 233–40.

Reisen, W.K. & Mullen, G.R. (1978) Ecological observations on acarine associates (Acari) of Pakistan mosquitoes (Diptera: Culicidae). *Environ. Entomol.*, **7**, 769–76.

Reuveny, H., Palevsky, E. & Gerson. U. (1996) Laboratory life history studies of the predaceous mite *Typhlodromus athiasae* (Acari: Phytoseiidae). *Syst. Appl. Acarol.*, **1**, 45–53.

Rezk, H.A. & Gadelhak, G.G. (1996) Relationship between phytophagaous and predatory mites in citrus orchards and the effect of acaricides on their populations. *Alex. J. Agric. Res.*, **41**, 217–24.

Richards, A.M. (1962) The oyster-shell scale, *Quadraspidiotus ostreaeformis* (Curtis), in the Christchurch district of New Zealand. *N.Z. J. Agric. Res.*, **5**, 95–100.

Richardson, R.G. & Hill, R.L. (1998) The biology of Australian weeds. 34. *Ulex europaeus* L. *P. Protect. Q.*, **13**, 46–58.

Riessen, H.P. (1982) Predatory behavior and prey selectivity of the pelagic water mite *Piona constricta*. *Can. J. Fish. Aquat. Sci.*, **39**, 1569–79.

van Rijn, P.C.J. & van Houten, Y.M. (1991) Life history of *Amblyseius cucumeris* and *A. barkeri* (Acarina: Phytoseiidae) on a diet of pollen. In: *Modern Acarology* (Ed. by F. Dusbábek, & V. Bukva), pp. 647–54. SPB, The Hague.

van Rijn, P.C.J. & Tanigoshi, L.K. (1999a) The contribution of extrafloral nectar to survival and reproduction of the predatory mite *Iphiseius degenerans* on *Ricinus communis*. *Exp. Appl. Acarol.*, **23**, 281–96.

van Rijn, P.C.J. & Tanigoshi, L.K. (1999b) Pollen as food for the predatory mites *Iphiseius degenerans* and *Neoseiulus cucumeris* (Acari: Phytoseiidae): dietary range and life history. *Exp. Appl. Acarol.*, **23**, 785–802.

Riley, C.V. (1873) *Fifth Annual Report on the Noxious, Beneficial and Other Insects of the State of Missouri*. Regan & Carter, Missouri.

Riley, C.V. (1874a) Descriptions of two new subterranean mites. *Trans. Acad. Sci. St Louis*, **25**, 215–16.

Riley, C.V. (1874b) *Sixth Annual Report on the Noxious, Beneficial and Other Insects of the State of Missouri*. Regan & Carter, Missouri.

Ripper, W.E. (1956) Effect of pesticides on balance of arthropod populations. *Annu. Rev. Entomol.*, **1**, 403–38.

Ristich, S.S. (1956) Toxicity of pesticides to *Typhlodromus fallacis* (Gar.). *J. Econ. Entomol.*, **49**, 511–15.

Riudavets, J. (1995) Predators of *Frankliniella occidentalis* (Perg.) and *Thrips tabaci* (Lind.): a review. In: *Biological Control of Thrips Pests* (Ed. by A.J.M. Loomans, J.C. van Lenteren, M.G. Tommasini *et al.*), Vol. 95.1, pp. 43–87. Wageningen Agricultural University Papers.

Rivard, I. (1961) The influence of temperature and humidity on longevity, fecundity and rate of increase of the mite *Tyrophagus putrescentiae* (Schrank) (Acarina: Acaridae) reared on mold cultures. *Can. J. Zool.*, **39**, 869–76.

Rizk, G.N., El Badry, E. & Hafez, S.M. (1979) The effectiveness of predacious and parasitic mites in controlling *Tribolium confusum* Duv. *Mesopotamia J. Agric.*, **14**, 167–82.

Robaux, P. (1974) Recherches sur le développement et la biologie des acariens Thrombidiidae. *Mém. Mus. Hist. Nat., Sér. A, Zool.*, **85**, 1–186.

Rock, G.C. & Yeargan, D.R. (1973) Toxicity of apple orchard herbicides and growth-regulating chemicals to *Neoseiulus fallacis* and twospotted spider mite. *J. Econ. Entomol.*, **66**, 1342–3.

Rock, G.C., Monroe, R.J. & Yeargan, D.R. (1976) Demonstration of a sex pheromone in the predaceous mite *Neoseiulus fallacis*. *Environ. Entomol.*, **5**, 264–6.

Rockett, C.L. (1980) Nematode predation by oribatid mites (Acari: Oribatida). *Int. J. Acarol.*, **6**, 219–24.

Roda, A., Nyrop, J., Dicke, M. & English-Loeb, G. (2000) Trichomes and spider-mite webbing protect predatory mite eggs from intraguild predation. *Oecologia*, **125**, 428–35.

Roderick, G.K. (1990) Postcolonization evolution of natural enemies. In: *Selection Criteria and Ecological Consequences of Importing Natural Enemies* (Ed. by W.C. Kauffman & J.R. Nechols), pp. 71–86. Entomology Society of America, Lanham, MD.

Rodriguez, J.G. & Rodriguez, L.D. (1987) Nutritional ecology of phytophagous mites. In: *Nutritional Ecology of Insects, Mites and Spiders* (Ed. by F. Slansky Jr & J.G. Rodriguez), pp. 177–208. John Wiley, New York.

Rodriguez, J.G., Maynard, D.E. & Smith, W.T., Jr (1960) Effects of soil insecticides and absorbents on plant sugars and resulting effect on mite nutrition. *J. Econ. Entomol.*, **53**, 491–5.

Rodriguez, J.G., Singh, P. & Taylor, B. (1970) Manure mites and their role in fly control. *J. Med. Entomol.*, **7**, 335–41.

Rolff, J. & Martens, A. (1997) Completing the life cycle: detachment of an aquatic parasite (*Arrenurus cuspidator*, Hydrachnellae) from an aerial host (*Coenagrion puella*, Odonata). *Can. J. Zool.*, **75**, 655–9.

Rosen, D. (1986) The role of taxonomy in effective biological control programs. *Agric. Ecosyst. Environ.*, **15**, 121–9.

Rosen, D. & Huffaker, C.B. (1983) An overview of desired attributes of effective biological control agents, with particular emphasis on mites. In: *Biological Control of Pests by Mites*, (Ed. by M.A. Hoy, G.L. Cunningham & L. Knutson), pp. 2–11. University California, Special Publ., No. 3304, Berkeley, CA.

Rosenheim, J.A., Kaya, H.K., Ehler, L.E. *et al.* (1995) Intraguild predation among biological-control agents: theory and evidence. *Biol. Cont.*, **5**, 303–35.

Rosenthal, S.S. (1983) Current status and potential for biological control of field bindweed, *Convolvulus arvensis*, with *Aceria convolvuli*. In: *Biological Control of Pests by Mites* (Ed. by M.A. Hoy, G.L. Cunningham & L. Knutson), pp. 57–60. University of California, Special Publ., No. 3304, Berkeley, CA.

Rosenthal, S.S. (1996) *Aceria, Epitrimerus* and *Aculus* species and biological control of weeds. In: *Eriophyoid Mites, Their Biology, Natural Enemies and Control* (Ed. by E.E. Lindquist, M.W. Sabelis & J. Bruin), pp. 729–39. Elsevier, Amsterdam.

Rosenthal, S.S. & Buckingham, G. (1982) Natural enemies of *Convolvulus arvensis* in western Mediterranean Europe. *Hilgardia*, **50**(2), 1–19.

Rosenthal, S.S. & Platts, B.E. (1990) Host specificity of *Aceria (Eriophyes) malherbae* (Acari: Eriophyidae), a biological control agent for the weed, *Convolvulus arvensis* (Convolvulaceae). *Entomophaga*, **35**, 459–63.

Roth, J.P., Macqueen, A. & Bay, D.E. (1988) Predation by the introduced phoretic mite, *Macrocheles peregrinus* (Acari: Macrochelidae), on the buffalo fly, *Haematobia irritans exigua* (Diptera: Muscidae), in Australia. *Environ. Entomol.*, **17**, 603–7.

Rothman, L.D. (1988) Habitat associations, patterns of abundance, and species richness of phytoseiid mites (Acari: Phytoseiidae) on a recent landfill site in lake Ontario. *Proc. Entomol. Soc. Ontario*, **119**, 1–7.

Roush, R.T. (1990) Genetic variation in natural enemies: critical issues for colonization in biological control. In: *Critical Issues in Biological Control* (Ed. by M. Mackauer, L.E. Ehler & J. Roland), pp. 263–88. Intercept, Andover.

Roush, R.T. & Hopper, K.R. (1995) Use of single family lines to preserve genetic variation in laboratory colonies. *Ann. Entomol. Soc. Am.*, **88**, 713–17.

Roush, R.T. & Hoy, M.A. (1981a) Genetic improvement of *Metaseiulus occidentalis*: selection with methomyl, dimethoate, and carbaryl and genetic analysis of carbaryl resistance. *J. Econ. Entomol.*, **74**, 138–41.

Roush, R.T. & Hoy, M.A. (1981b) Laboratory, glasshouse, and field studies of artificially selected carbaryl resistance in *Metaseiulus occidentalis. J. Econ. Entomol.*, **74**, 142–7.

Royalty, R.N. & Perring, T.M. (1987) Comparative toxicity of acaricides to *Aculops lycopersici* and *Homeopronematus anconai* (Acari: Eriophyidae, Tydeidae). *J. Econ. Entomol.*, **80**, 348–51.

Royce, L.A. & Krantz, G.W. (1991) A new rearing method for nematophagous mites. In: *Modern Acarology* (Ed. by F. Dusbábek & V. Bukva), Vol. 2, pp. 619–22. SPB, The Hague.

Ruberson, J.R., Nemoto, H. & Hirose, Y. (1998) Pesticides and conservation of natural enemies in pest management. In: *Conservation Biological Control* (Ed. by P. Barbosa), pp. 207–33. Academic Press, San Diego, CA.

Rudzińska, M. (1998) Life history of the phoretic predatory mite *Arctoseius semicissus* (Acari: Ascidae) on a diet of sciarid fly eggs. *Exp. Appl. Acarol.* **22**, 643–8.

Ruf, A. (1991) Do females eat males? Laboratory studies on the population development of *Hypoaspis aculeifer* (Acari: Parasitiformes). In: *Modern Acarology* (Ed. by F. Dusbábek & V. Bukva), Vol. 2, pp. 487–92. SPB, The Hague.

Rust, H.J. (1933) Many bark beetles destroyed by predaceous mite. *J. Econ. Entomol.*, **26**, 733–4.

Ruther, J. & Steidle, J.L.M. (2000) Mites as matchmakers: semiochemicals from host-associated mites attract both sexes of the parasitoid *Lariophagus distinguendus. J. Chem. Ecol.*, **26**, 1205–16.

Ryba, J., Rödl, P., Bartoš, L., Daniel, M. & Černy, V. (1987) Some features of the ecology of fleas inhabiting the nests of the European suslik (*Citellus citellus* (L.)). II. The influence of mesostigmatid mites on fleas. *Folia Parasitol.*, **34**, 61–8.

Ryu, M.-O. (1993) A review of the Phytoseiidae (Mesostigmata: Acarina) from Korea. *Insecta Koreana*, **10**, 92–137.

Sabelis, M.W. (1985) Sex allocation. In: *Spider Mites, Their Biology, Natural Enemies and Control* (Ed. by W. Helle & M.W. Sabelis), Vol. 1B, pp. 83–94. Elsevier, Amsterdam.

Sabelis, M.W. (1991) Life-history evolution of spider mites. In: *The Acari: Reproduction, Development and Life-history Strategies* (Ed. by R. Schuster & P.W. Murphy), pp. 23–49. Chapman & Hall, London.

Sabelis, M.W. (1996) Phytoseiidae. In: *Eriophyoid Mites, Their Biology, Natural Enemies and Control* (Ed. by E.E. Lindquist, M.W. Sabelis & J. Bruin), pp. 427–56. Elsevier, Amsterdam.

Sabelis, M.W., van Baalen, M., Bakker, F.M. *et al.* (1999) The evolution of direct and indirect plant defence against herbivorous arthropods. In: *Herbivores: Between Plant and Predators* (Ed. by H. Olff, V.K. Brown & R.H. Drent), pp. 109–66. Blackwell, Oxford.

Sabelis, M.W. & van de Baan, H.E. (1983) Location of distant spider mite colonies by phyto-seiid predators: demonstration of specific kairomones emitted by *Tetranychus urticae* and *Panonychus ulmi. Entomol. Exp. Appl.*, **33**, 303–14.

Sabelis, M.W. & Bakker, F. (1992) How predatory mites cope with the web of their tetranychid prey: a functional view on dorsal chaetotaxy in the Phytoseiidae. *Exp. Appl. Acarol.*, **16**, 203–25.

Sabelis, M.W. & Bruin, J. (1996) Evolutionary ecology: life history patterns, food plants choice and dispersal. In: *Eriophyoid Mites, Their Biology, Natural Enemies and Control* (Ed. by E.E. Lindquist, M.W. Sabelis & J. Bruin), pp. 329–66. Elsevier, Amsterdam.

Sabelis, M.W. & De Jong, M.C.M. (1988) Should all plants recruit bodyguards? Conditions for a polymorphic ESS of synomone production in plants. *Oikos*, **53**, 247–52.

Sabelis, M.W. & Dicke, M. (1985) Long-range dispersal and searching behaviour. In: *Spider Mites, Their Biology, Natural Enemies and Control* (Ed. by W. Helle & M.W. Sabelis), Vol. 1B, pp. 141–60. Elsevier, Amsterdam.

Sabelis, M.W. & Janssen, A. (1994) Evolution of life-history patterns in the Phytoseiidae. In: *Mites, Ecological and Evolutionary Analyses of Life-history Patterns* (Ed. by M.A. Houck), pp. 70–98. Chapman & Hall, New York.

Sabelis, M.W. & van Rijn, P.C.J. (1996) Eriophyoid mites as alternative prey. In: *Eriophyoid Mites, Their Biology, Natural Enemies and Control* (Ed. by E.E. Lindquist, M.W. Sabelis & J. Bruin), pp. 757–64. Elsevier, Amsterdam.

Sabelis, M.W. & van Rijn, P.C.J. (1997) Predation by insects and mites. In: *Thrips as Crop Pests* (Ed. by T. Lewis), pp. 259–354. CABI, Wallingford.

Sabelis, M.W. & van der Weel, J.J. (1993) Anemotactic responses of the predatory mite, *Phytoseiulus persimilis* Athias-Henriot, and their role in prey finding. *Exp. Appl. Acarol.*, **17**, 521–9.

Saboori, A. (2000) Two new larval erythraeine mites (Acari: Erythraeidae) from Iran. *Syst. Appl. Acarol.*, **5**, 125–30.

Sah, K., Somchoudhury, A.K., Sarkar, P.K. & Gupta, S.K. (1999) Evaluation of some pesticides for efficacy against tea red spider mite, *Oligonychus coffeae* (Nietner) and their adverse effects on natural enemies belonging to *Amblyseius* spp., *Agistemus* spp. and *Cunaxa* spp. in Dooars areas of West Bengal, India. *Tea*, **20**, 12–20.

Saleh, S.M., El-Helaly, M.S. & El-Gayar, F.H. (1986) Life history of the predatory mite *Cheyletus malaccensis* Oudemans. *Acarologia*, **27**, 37–40.

Salman, A.G.A. & Manna, S.H. (1991) Biology and predatory capacity of *Agistemus exsertus* Gonzalez-Rodriguez feeding on eggs of mites infesting fruit trees in upper Egypt. In: *Modern Acarology* (Ed. by F. Dusbábek & V. Bukva), Vol. 2, pp. 537–42. SPB, The Hague.

Samarasinghe, S. & LeRoux, E.J. (1966) The biology and dynamics of the oystershell scale, *Lepidosaphes ulmi* (L.) (Homoptera: Coccidae) on apple in Quebec. *Ann. Entomol. Soc. Quebec*, **11**, 206–92.

Sampson, C. (1998) The commercial development of an *Amblyseius cucumeris* controlled release method for the control of *Frankliniella occidentalis* in protected crops. *1998 Brighton Crop Protection Conf., Pests and Diseases*, Vol. 2, pp. 409–16. Farnham.

Samšiňáková, A. & Samšiňák, K. (1970) Milben (Acari) als Verbreiter des Pilzes *Beauveria bassiana* (Bals.). Vuill. *Z. parasitenk.*, **34**, 351–5.

Sanderson, J.P. & Zhang, Z.-Q. (1995) Dispersion, sampling and potential for integrated control of twospotted spider mite (Acari: Tetranychidae) on greenhouse roses. *J. Econ. Entomol.*, **88**, 343–51.

Santos, M.A. (1976a) Evaluation of *Zetzellia mali* as a predator of *Panonychus ulmi* and *Aculus schlechtendali*. *Environ. Entomol.*, **5**, 187–91.

Santos, M.A. (1976b) Prey selectivity and switching response of *Zetzellia mali*. *Ecology*, **57**, 390–4.

Santos, M.A. (1982) Effects of low prey densities on the predation and oviposition of *Zetzellia mali*. *Environ. Entomol.*, **11**, 972–4.

Santos, M.A. (1991) Searching behavior and associational response of *Zetzellia mali* (Acarina: Stigmaeidae). *Exp. Appl. Acarol.*, **11**, 81–7.

Santos, P.F., Philips, J. & Whitford, W.G. (1981) The role of mites and nematodes in early stages of buried litter decomposition in a desert. *Ecology*, **62**, 664–9.

Sathiamma, B. (1995) Biological suppression of the white spider mite *Oligonychus iseilemae* (Hirst) on coconut foliage. *Entomon*, **20**, 237–43.

Sayre, R.M. & Walter, D.E. (1991) Factors affecting the efficacy of natural enemies of nematodes. *Annu. Rev. Phytopathol.*, **29**, 149–66.

Schabel, H.G. (1982) Phoretic mites as carriers of entomopathogenic fungi. *J. Invert. Pathol.*, **39**, 410–12.

Schausberger, P. (1998) Survival, development and fecundity in *Euseius finlandicus, Typhlodromus pyri* and *Kampimodromus aberrans* (Acari, Phytoseiidae) feeding on the San José scale, *Quadraspidiotus perniciosus* (Coccina, Diaspididae). *J. Appl. Entomol.*, **122**, 53–6.

Schausberger, P. & Croft, B.A. (2000) Nutritional benefits of intraguild predation and cannibalism among generalists and specialist phytoseiid mites. *Ecol. Entomol.*, **25**, 473–80.

Schausberger, P. & Walzer, A. (2001) Combined versus single species release of predaceous mites: predator–predator interactions and pest suppression. *Biol. Cont.*, **20**, 269–78.

Schicha, E. (1987) *Phytoseiidae of Australia and Neighboring Areas.* Indira, Oak Park, MI.

Schicha, E. & Corpuz-Raros, L.A. (1992) *Phytoseiidae of the Philippines.* Indira, West Bloomfield, MI.

Schöller, M., Prozell, S., Al-Kirshi, A.-G. & Reichmuth, Ch. (1997) Towards biological control as a major component of integrated pest management in stored product protection. *J. Stored Prod. Res.*, **33**, 81–97.

Schroder, R.F.W. (1979) Host specificity tests of *Coccipolipus epilachnae*, a mite parasitic on the Mexican bean beetle. *Environ. Entomol.*, **8**, 46–7.

Schroder, R.F.W. (1981) Biological control of the Mexican bean beetle, *Epilachna varivestis* Mulsant, in the United States. In: *Biological Control in Crop Production* (Ed. by G.C. Papavizas), pp. 351–60. Allanhead, Osmun.

Schroder, R.F.W. (1982) Effect of infestation with *Coccipolipus epilachnae* Smiley (Acarina: Podapolipidae) on fecundity and longevity of the Mexican bean beetle. *Int. J. Acarol.*, **8**, 81–4.

Schruft, G. (1971) *Haleupalus oliveri* nov. spec., eine Dornpalpenmilbe an Reben (*Vitis* spec.) (Acari: Cunaxidae). *Dtsch Entomol. Z.*, **18**, 377–82.

Schruft, G. (1972) Les tydéidés (Acari) sur vigne. *EPPO Bull.*, **3**, 51–5.

Schruft, G.A. (1985) Grape. In: *Spider Mites, Their Biology, Natural Enemies and Control* (Ed. by W. Helle & M.W. Sabelis), Vol. 1B, pp. 359–66. Elsevier, Amsterdam.

Schuler, T.H., Poppy, G.M., Potting, R.P.J. *et al.* (1999a) Interactions between insect tolerant genetically modified plants and natural enemies. In: *Gene Flow and Agriculture: Relevance for Transgenic Crops* (Ed. by P.J.W. Lutman), pp. 197–202. Proc. 1999 British Crop Protection Council Symp., No. 72.

Schuler, T.H., Poppy, G.M., Kerry, B.R. & Denholm, I. (1999b) Potential side effects of insect-resistant transgenic plants on arthropod natural enemies. *Trends BioTech.*, **17**, 210–16.

Schulten, G.G.M. (1985) Pseudo-arrhenotoky. In: *Spider Mites, Their Biology, Natural Enemies and Control* (Ed. by W. Helle & M.W. Sabelis), Vol. 1B, pp. 67–71. Elsevier, Amsterdam.

Schultz, F.W. (1974) Some aspects of the ecology of *Amblyseius* (*Amblyseius*) *teke* Pritchard & Baker and A. (*Mesoseiulus*) *longipes* (Evans) (Acarina: Phytoseiidae). *Phytophylactica*, **6**, 1–9.

Schwartz, A. (1990) Pesticide effect on populations of *Tetranychus urticae* Koch (Acari: Tetranychidae) and a predaceous mite on table grapes in the Hex River Valley. *South Afr. J. Enol. Vitic.*, **11**, 33–7.

Searle, G.G., Penman, D.R. & Chapman, R.B. (1990) The toxicity of herbicides to the gorse spider mite *Tetranychus lintearius*. *Proc. 43th N.Z. Weed Pest Control Conf.* (Ed. by A.J. Popay), pp. 178–81. Swiftprint Centre, Palmerston North.

Secord, D. & Kareiva, P. (1996) Perils and pitfalls in the host specificity paradigm. *BioScience*, **46**, 448–53.

Sell, P. (1988) *Caloglyphus* sp. (Acarina: Acaridae), an effective nematophagous mite on root-knot nematodes (*Meloidogyne* spp.). *Nematologia*, **34**, 246–8.

Sellers, W.F. & Robinson, G.G. (1950) The effect of the miticide Neotran® upon the laboratory production of *Aspidiotus lataniae* Signoret as a coccinellid food. *Can. Entomol.*, **82**, 170–3.

Shahi, K.P. & Krishna, S.S. (1981) A new host of *Hemipteroseius indicus*. *Experientia*, **37**, 1072.

Shaldybina, E.S. (1975) Ceratozetoidea. In: *A Key to Soil-inhabiting Mites. Sarcoptiformes* (Ed. by M.S. Gilyarov), pp. 275–319. Nauka, Moscow (in Russian).

Sharma, G.D., Farrier, M.H. & Drooz, A.T. (1983) Food, life-history, and sexual differences of *Callidosoma metzi* Sharma, Drooz and Treat (Acarina: Erythraeidae). *Int. J. Acarol.*, **9**, 149–55.

Sharma, R.D. (1971) Studies on the plant parasitic nematode *Tylenchorhynchus dubius*. *Meded. Landbouww. Wageningen*, **71**, 1–154.

Shehata, K.K. & Weismann, L. (1972) Rearing the predaceous mites *Phytoseilus persimilis* Athias-Henriot on artificial diet (Acarina: Phytoseiidae). *Biol. Bratislava*, **27**, 607–15.

Shehata, M. & Baker, A. (1996) Mites infesting phlebotomine sandflies in southern Sinai, Egypt. *Med. Vet. Entomol.*, **10**, 193–6.

Shih, C.I., Poe, S.L. & Cromroy, H.L. (1979) Biology and predation of *Phytoseiulus macropilis* on *Tetranychus urticae*. *Fl. Entomol.*, **62**, 48–53.

Shimer, H. (1868a) Notes on the 'apple bark-louse' (*Lepidosaphes conchiformis*, Gmelin sp.) with a description of a supposed new Acarus. *Trans. Am. Entomol. Soc.*, **1**, 361–74.

Shimer, H. (1868b) The apple bark-louse in 1866. Birds vindicated from the charge preferred against them by the State Entomologist. *Trans. Illinois St. Hort. Soc.*, **2**, 227–33.

Shipp, J.L. & Whitfield, G.H. (1991) Functional response of the predatory mite *Amblyseius cucumeris* (Acari: Phytoseiidae), on western flower thrips, *Frankliniella occidentalis* (Thysanoptera: Thripidae). *Environ. Entomol.*, **20**, 694–9.

Simberloff, D. & Stiling, P. (1996) How risky is biological control? *Ecology*, **77**, 1965–74.

Simmonds, F.J. (1958) The oleander scale, *Pseudalacaspis pentagona* (Targ.) (Homoptera, Diaspididae) in Bermuda. *Bull. Bermuda Dept Agric.*, No. 31.

Singh, P., King, W.E. & Rodriguez, J.G. (1966) Biological control of muscids as influenced by host preference of *Macrocheles muscaedomesticae* (Acarina: Macrochelidae). *J. Med. Entomol.*, **3**, 78–81.

Sinha, R.N. (1988) Population dynamics of Psocoptera in farm-stored grain and oilseed. *Can. J. Zool.*, **66**, 2618–27.

Sircom, J. (2000) Photographic sampling: a photographic sampling method for mites on plants. *Exp. Appl. Acarol.*, **24**, 55–61.

Skovgård, H., Tomkiewicz, J., Nachman, G. & Münster-Swendsen, M. (1993) The dynamics of the cassava green mite *Mononychellus tanajoa* in a seasonally dry area in Kenya. *Exp. Appl. Acarol.*, **17**, 59–76.

Slansky, F. Jr. & Rodriguez, J.G. (Eds) (1987) *Nutritional Ecology of Insects, Mites and Spiders*. Wiley, New York.

Slansky, F., Jr & Scriber, J.M. (1985) Food consumption and utilization. In: *Comprehensive Insect Physiology Biochemistry and Pharmacology* (Ed. by G. A. Kerkut & L.I. Gilbert), Vol. 4, pp. 87–163. Pergamon, Oxford.

Slone, D.H. & Croft, B.A. (2000) Changes in intraspecific aggregation and the coexistence of predaceous apple mites. *Oikos*, **91**, 153–61.

Smiley, R.L. (1968) A new genus and three new species of Erythraeoidea (Acarina: Erythraeidae and Smarididae). *Proc. Entomol. Soc. Wash.*, **70**, 13–21.

Smiley, R.L. (1974) A new species of *Coccipolipus* parasitic on the Mexican bean beetle. *J. Wash. Acad. Sci.*, **64**, 298–302.

Smiley, R.L. (1991) Mites (Acari). In: *Insect and Mite Pests in Food. An Illustrated Key* (Ed. by J.R. Gorham), Vol. 1, pp. 3–43. USDA Agricultural Handbook, No. 655.

Smiley, R.L. (1992) *The Predatory Mite Family Cunaxidae (Acari) of the World with a New Classification*. Indira, West Bloomfield, MI.

Smiley, R.L. & Knutson, L. (1983) Aspects of taxonomic research and services relative to mites as biological control agents. In: *Biological Control of Pests by Mites* (Ed. by M.A. Hoy, G.L. Cunningham & L. Knutson), pp. 148–64. University of California, Special Publ., No. 3304, Berkeley, CA.

Smiley, R.L. & Landwehr, V.R. (1976) A new species of *Tarsonemus* (Acarina: Tarsonemidae), predaceous on tetranychid mite eggs. *Ann. Entomol. Soc. Am.*, **69**, 1065–72.

Smirnoff, W.A. (1957) La cochenille du palmier dattier (*Parlatoria blanchardi Targ.*) en Afrique du Nord, comportement, importance économique prédateurs et lutte biologique. *Entomophaga*, **2**, 1–98.

Smith, B.P. (1983) The potential of mites as biological control agents of mosquitoes. In: *Biological Control of Pests by Mites* (Ed. by M.A. Hoy, G.L. Cunningham & L. Knutson), pp. 79–85. University of California, Special Publ., No. 3304, Berkeley, CA.

Smith, B.P. (1988) Host–parasite interaction and impact of larval water mites on insects. *Annu. Rev. Entomol.*, **33**, 487–505.

Smith, B.P. (1998) Loss of larval parasitism in parasitengonine mites. *Exp. Appl. Acarol.*, **22**, 187–99.

Smith, B.P. & Laughland, L.A. (1990) Stimuli inducing detachment of larval *Arrenurus danbyensis* (Hydrachnidia: Arrenuridae) from adult *Coquillettidia perturbans* (Diptera: Culicidae). *Exp. Appl. Acarol.*, **9**, 51–62.

Smith, B.P. & McIver, S.B. (1984a) Factors influencing host selection and successful parasitism of *Aedes* spp. mosquitoes by *Arrenurus* spp. mites. *Can. J. Zool.*, **62**, 1114–20.

Smith, B.P. & McIver, S.B. (1984b) The impact of *Arrenurus danbyensis* Mullen (Acari: Prostigmata: Arrenuridae) on a population of *Coquillettidia perturbans* (Walker) (Diptera: Culicidae). *Can. J. Zool.*, **62**, 1121–34.

Smith, D. & Papacek, D.F. (1985) Integrated pest management in Queensland citrus. *Qld Agric. J.*, **111**, 249–59.

Smith, D. & Papacek, D.F. (1991) Studies of the predatory mite *Amblyseius victoriensis* (Acarina: Phytoseiidae) in citrus orchards in south-east Queensland: control of *Tegolophus australis* and *Phyllocoptruta oleivora* (Acarina: Eriophyidae), effect of pesticides, alternative host plants and augmentative release. *Exp. Appl. Acarol.*, **12**, 195–217.

Smith, F.F., Henneberry, T.J. & Boswell, A.L. (1963) The pesticide tolerance of *Typhlodromus fallacis* (Garman) and *Phytoseiulus persimilis* A.H. with some observations on the predator efficiency of *P. persimilis*. *J. Econ. Entomol.*, **56**, 274–8.

Smith, I.M. (1976) A study of the systematics of the water mite family Pionidae (Prostigmata: Parasitengona). *Mem. Entomol. Soc. Can.*, **98**, 1–249.

Smith, I.M. & Cook, D.R. (1991) Water mites. In: *Ecology and Classification of North American Freshwater Invertebrates* (Ed. by J. Thorp & A. Covich), pp. 523–92. Academic Press, San Diego, CA.

Smith, I.M. & Oliver, D.R. (1986) Review of the parasitic associations of larval water mites (Acari: Parasitengona: Hydrachnida) with insect hosts. *Can. Entomol.*, **118**, 407–72.

Smith, J.C. & Newsom, L.D. (1970) Laboratory evaluation of *Amblyseius fallacis* as a predator of tetranychid mites. *J. Econ. Entomol.*, **63**, 1876–8.

Smith, J.W. & Furr, R.E. (1975) Spider mites and some natural control agents found on cotton in the Delta area of Mississippi. *Environ. Entomol.*, **4**, 559–60.

Smith, L., Cuellar, M.E. & Melo, E.L. (1996) Evaluating prey preference by several phytoseiid predators for *Mononychellus tanajoa* (Bondar) and *M. caribbeanae* McGregor (Acari: Tetranychidae) in cassava. *Biol. Cont.*, **7**, 179–84.

Smith, L.M. & Goldsmith, E.V. (1936) The cyclamen mite, *Tarsonemus pallidus*, and its control on field strawberries. *Hilgardia*, **10**, 53–94.

Smrž, J. & Čatská, V. (1987) Food selection of the field population of *Tyrophagus putrescentiae* (Schrank) (Acari, Acarida). *J. Appl. Entomol.*, **104**, 329–35.

Smrž, J., Svobodová, J. & Čatská, V. (1987) Synergistic participation of *Tyrophagus putrescentiae* (Schrank) (Acari; Acaridida) and its associated bacteria on the destruction of some soil micromycetes. *J. Appl. Entomol.*, **111**, 206–10.

Snetsinger, R. (1956) Biology of *Bdella depressa*, a predaceous mite. *J. Econ. Entomol.*, **49**, 745–6.

Snyder, J.C. & Carter, C.D. (1984) Leaf trichomes and resistance of *Lycopersicon hirsutum* and *L. esculentum* to spider mites. *J. Am. Soc. Hort. Sci.*, **109**, 837–43.

Sobhian, R. & Andres, L.A. (1978) The response of the skeletonweed gall midge, *Cystiphora schmidti* (Diptera: Cecidomyiidae), and gall mite, *Aceria chondrillae* (Eriophyidae) to North American strains of rush skeletonweed (*Chondrilla juncea*). *Environ. Entomol.*, **7**, 506–8.

Sobhian, R., Tunç, I. & Erler, F. (1999) Preliminary studies on the biology and host specificity of *Aceria salsolae* DeLillo and Sobhian (Acari, Eriophyidae) and *Lixus salsolae* Becker (Col., Curculionidae), two candidates for biological control of *Salsola kali*. *J. Appl. Entomol.*, **123**, 205–9.

Soliman, Z.R. & Mohamed, M.I. (1972) Biological studies on the soil-inhabiting bdellid mite, *Spinibdella bifurcata* (Acarina: Bdellidae) in the U.A.R. *Z. Angew. Entomol.*, **70**, 15–23.

Solomon, M.F. (1946) Tyroglyphid mites in stored products. Ecological studies. *Ann. Appl. Biol.*, **33**, 82–97.

Solomon, M.F. (1969) Experiments on predator–prey interactions of storage mites. *Acarologia*, **11**, 484–503.

Solomon, M.G. (1981) Windbreaks as a source of orchard pests and predators. In: *Pests, Pathogens and Vegetation* (Ed. by J.M. Thresh), pp. 273–83. Pitman, London.

Solomon, M.G. (1986) Natural control of red spider mite in English apple orchards. *SROP/WPRS Bull.*, **9**(4), 43–7.

Solomon, M.G. & Fitzgerald, J.D. (1993) Orchard selection for resistance to a synthetic pyrethroid in organophosphate-resistant *Typhlodromus pyri* in the UK. *Biocont. Sci. Tech.*, **3**, 127–32.

Solomon, M.G., Fitzgerald, J. D. & Murray, R.A. (1996) Electrophoretic approaches to predator–prey interactions. In: *The Ecology of Agricultural Pests, Biochemical Approaches* (Ed. by W.O.C. Symondson & J.E. Liddell), pp. 457–68. Chapman & Hall, London.

Sorensen, J.T., Kinn, D.N., Doutt, R.L. & Cate, J.R. (1976) Biology of the mite *Anystis agilis* (Acari: Anystidae): a California vineyard predator. *Ann. Entomol. Soc. Am.*, **69**, 905–10.

Sorensen, J.T., Kinn, D.N. & Doutt, R.L. (1983) Biological observations on *Bdella longicornis*: a predatory mite in California vineyards (Acari: Bdellidae). *Entomography*, **2**, 297–305.

Sörensson, A. & Nedstam, B. (1993) Effect of *Amblyseius cucumeris* and *Orius insidiosus* on *Frankliniella occidentalis* in ornamentals. *IOBC/WPRS Bull.*, **16**(8), 129–32.

Southcott, R.V. (1961) Studies on the systematics and biology of the Erythraeoidea (Acarina), with a critical revision of the genera and subfamilies. *Aust. J. Zool.*, **9**, 367–610.

Southcott, R.V. (1987) A new larval mite (Acarina: Trombidioidea) ectoparasitic on an Australian centipede, and the Trombidiidae reclassified. *Trans. R. Soc. South. Aust.*, **111**, 43–52.

Southcott, R.V. (1988) Two new larval Erythraeinae (Acarina: Erythraeidae) from New Zealand, and the larval Erythraeinae revised. *N.Z. J. Zool.*, **15**, 223–33.

Spolen, K.M. & Isman, M.B. (1996) Acute and sublethal effects of a neem insecticide on the commercial biological control agents *Phytoseiulus persimilis* and *Amblyseius cucumeris* (Acari: Phytoseiidae) and *Aphidoletes aphidimyza* (Diptera: Cecidomyiidae). *J. Econ. Entomol.*, **89**, 1379–86.

Stafford, K.C. & Bay, D.E. (1994) Dispersion statistics and sample size estimates for house fly (Diptera: Muscidae) larvae and *Macrocheles muscaedomesticae* (Acari: Macrochelidae) in poultry manure. *J. Med. Entomol.*, **31**, 732–7.

Stamou, G.P. & Asikidis, M.D. (1992) The effect of certain biotic factors on the demographic parameters of *Scheloribates cf. latipes* (Acari: Oribatida). *Pedobiologia*, **36**, 351–8.

Stechmann, D.-H. (1978) Eiablage, Parasitismus und postparasitische Entwicklung von *Arrenurus*-Arten (Hydrachnellae, Acari). *Z. Parasitenkund.*, **57**, 169–88.

Steinberg, S. & Cohen, M. (1992) Biological control of the two-spotted spider mite (*Tetranychus urticae*) in apple orchards by inundative releases of the predatory mite *Phytoseiulus persimilis* – a feasibility study. *Phytoparasitica*, **20** (supp.), 37–44.

Steinberg, S., Cain, H. & Kaminski, J. (1999) Product control of *Phytoseiulus persimilis* (Athias-Henriot): current practice of a commercial producer. *IOBC/WPRS Bull.*, **22**(1), 237–40.

Steiner, M.Y. (1993) Some observations on the quality of biological control organisms used in greenhouses. *IOBC/WPRS Bull.*, **16**, 165–8.

Steiner, M.Y. & Goodwin, S. (2001) Phytoseiids with potential for commercial exploitation in Australia. In: *Acarology: Proc. 10th Int. Congr.* (Ed. by R.B. Halliday, D.E. Walter, H.C. Proctor *et al.*), pp. 476–83. CSIRO, Melbourne.

Steiner, M., Goodwin, S. & Wellham, T. (1999) A simplified rearing method for *Stratiolaelaps (Hypoaspis) miles* (Acari: Laelapidae). *IOBC/WPRS Bull.*, **22**(1), 241–2.

Steinhaus, E.A. (1958) Crowding as a possible stress factor in insect disease. *Ecology*, **39**, 503–14.

Steinkraus, D.C. & Cross, E.A. (1993) Description and life history of *Acarophenax mahunkai*, n. sp. (Acari, Tarsonemina: Acarophenacidae), an egg parasite of the lesser mealworm (Coleoptera: Tenebrionidae). *Ann. Entomol. Soc. Am.*, **86**, 239–49.

Sterk, G., Hassan, S.A., Baillod, M. *et al.* (1999) Results of the seventh joint pesticide testing programme carried out by the IOBC/WPRS-Working Group 'Pesticides and Beneficial Organisms'. *BioControl*, **44**, 99–117.

Steven, D. & Lofroth, B. (1987) Screening chemicals against an oribatid mite. *Proc. 40th N.Z. Weed Pest Control Conf.*, Nelson (Ed. by A.J. Popay), pp. 116–19. Swiftprint Centre, Palmerston North.

Steyskal, G.C., Murphy, W.L. & Hoover, E.M. (1986) *Insects and Mites: Techniques for Collection and Preservation.* USDA Misc. Publ. No. 1443.

Stirling, G.R. (1991) *Biological Control of Plant Parasitic Nematodes, Progress, Problems and Prospects.* CABI, Wallingford.

Stone, C. (1986) An investigation into the morphology and biology of *Tetranychus lintearius* Dufour (Acari: Tetranychidae). *Exp. Appl. Acarol.*, **2**, 173–86.

Strapazzon, A. & Dalla Monta, L. (1988) Ruolo e distribuzione di *Amblyseius andersoni* (Chant) e *Zetzellia mali* (Ewing) in meleti infestati da *Aculus schlechtendali* (Nalepa). *Redia*, **71**, 39–54.

Strickler, K. & Croft, B.A. (1985) Comparative rotenone toxicity in the predator, *Amblyseius fallacis* (Acari: Phytoseiidae), and the herbivore, *Tetranychus urticae* (Acari: Tetranychidae), grown on lima beans and cucumbers. *Environ. Entomol.*, **14**, 243–6.

Strong, D.R. & Pemberton, R.W. (2001) Food webs, risks of alien enemies and reform of biological control. In: *Evaluating Indirect Ecological Effects of Biological Control* (Ed. by E. Wajnberg, J.K. Scott & P.C. Quimby), pp. 57–79. CABI, Wallingford.

Strong, W.B. & Croft, B.A. (1996) Release strategies and cultural modifications for biological control of twospotted spider mite by *Neoseiulus fallacis* (Acari: Tetranychidae, Phytoseiidae) on hops. *Environ. Entomol.*, **25**, 529–35.

Strong, W.B., Croft, B.A. & Slone, D.H. (1997) Spatial aggregation and refugia of the mites *Tetranychus urticae* and *Neoseiulus fallacis* (Acari: Tetranychidae, Phytoseiidae) on hop. *Environ. Entomol.*, **26**, 859–65.

Sturhan, D. & Hampel, G. (1977) Pflanzenparasitische Nematoden als Beute der Wurzelmilbe *Rhizoglyphus echinopus* (Acarina: Tyroglyphidae). *Anz. Schädl. Pflazensch. Umwelt.*, **50**, 115–18.

Suckling D.M., Walker, J.T.S., Shaw, P.W. *et al.* (1988) Management of resistance in horticultural pests and beneficial species in New Zealand. *Pest. Sci.*, **23**, 157–64.

Sumangala, K. & Haq, M.A. (1995) Nutritional diversity of Acari infesting *Eichhornia crassipes. J. Ecobiol.*, **7**, 289–97.

Summers, F.M. (1960) Several stigmaeid mites formerly included in *Mediolata* redescribed in *Zetzellia* Oudms. and *Agistemus*, new genus. *Proc. Entomol. Soc. Wash.*, **62**, 233–47.

Summers, F.M. (1962) The genus *Stigmaeus* (Acarina: Stigmaeidae). *Hilgardia*, **33**, 491–537.

Summers, F.M. & Schlinger, E.I. (1955) Mites of the family Caligonellidae (Acarina). *Hilgardia*, **23**, 539–61.

Summers, F.M. & Price, D.W. (1970) Review of the mite family Cheyletidae. *Univ. Calif. Publ. Entomol.*, **61**, 1–153.

Sunderland, K.D. (1996) Progress in quantifying predation using antibody techniques. In: *The Ecology of Agricultural Pests, Biochemical Approaches* (Ed. by W.O.C. Symondson & J.E. Liddell), pp. 419–55. Chapman & Hall, London.

Supkoff, D.M., Joley, D.B. & Marois, J.J. (1988) Effect of introduced biological control organisms on the density of *Chondrilla juncea* in California. *J. Appl. Ecol.*, **25**, 1089–95.

Šut'akova, G. & Rüttgen, F. (1978) *Rickettsiella phytoseiuli* and virus-like particles in *Phytoseiulus persimilis* (Gamasoidea: Phytoseiidae) mites. *Acta Virol.*, **22**, 333–6.

Sutherst, R.W. & Maywald, G.F. (1985) A computerised system for matching climates in ecology. *Agric. Ecosyst. Environ.*, **13**, 281–99.

Sutherst, R.W. & Maywald, G.F. (1991) Climate modelling and pest establishment. *P. Protect. Q.*, **6**, 3–7.

Swaddle, J.P., Witter, M.S. & Cuthill, I.C. (1994) The analysis of fluctuating asymmetry. *Anim. Behav.*, **48**, 986–9.

Swan, D.I. (1974) A review of the work on predators, parasites and pathogens for the control of *Oryctes rhinoceros* (L.) (Coleoptera: Scarabaeidae) in the pacific area. *Misc. Publ. Commonwealth Inst. Biol. Cont.*, No. 7.

Swift, S.F. (1987) A new species of *Stigmaeus* (Acari: Prostigmata: Stigmaeidae) parasitic on phlebotomine flies (Diptera: Psychodidae). *Int. J. Acarol.*, **13**, 239–44.

Swift, S.F. & Goff, M.L. (1987) The family Bdellidae (Acari: Prostigmata) in the Hawaiian Islands. *Int. J. Acarol.*, **13**, 29–49.

Swirski, E. & Dorzia, N. (1968) Studies on the feeding, development and oviposition of the predaceous mite *Amblyseius limonicus* Garman and McGregor (Acarina: Phytoseiidae) on various kinds of food substances. *Israel J. Agric. Res.*, **18**, 71–5.

Swirski, E., Amitai, S. & Dorzia, N. (1967) Laboratory studies on the feeding, development and reproduction of the predaceous mites *Amblyseius rubini* Swirski & Amitai and *Amblyseius swirskii* Athias (Acarina: Phytoseiidae) on various kinds of food substances. *Israel J. Agric. Res.*, **17**, 101–19.

Swirski, E., Ragusa de Chiara, S. & Tsolakis, H. (1998) Keys to the phytoseiid mites (Parasitiformes, Phytoseiidae) of Israel. *Phytophaga*, **8**, 85–154.

Szwejda, J. & Nawrocka, B. (1996) Influence of soil-less cultivation on development and control of spider mites on vegetable crops in glasshouses. *Proc. Brighton Crop Protection Conf., 1966 – Pests and Diseases*, Vol. 3, pp. 855–60. Farnham.

Tabashnik, B.E. & Johnson, M.W. (1999) Evolution of pesticide resistance in natural enemies. In: *Handbook of Biological Control, Principles and Applications of Biological Control* (Ed. by T.S. Bellows & T.W. Fisher), pp. 673–89. Academic Press, San Diego, CA.

Taha, H.A., El-Naggar, M.E.E., Abou-El-Ngaga, M.M. & Soliman, S.M, (1988a) Effect of different prey species on the development and fecundity of the predacious mite, *Neocunaxoides andrei* Baker and Hoff. (Acari: Cunaxidae). *Agric. Res. Rev.*, **66**, 129–35.

Taha, H.A., El-Naggar, M.E.E., Ibrahim, G.A. & Soliman, S.M. (1988b) Biological studies on two acarine predators of the grain mite, *Caloglyphus rhizoglyphoides* Zachvatkin. *Agric. Res. Rev.*, **66**, 137–44.

Takabayashi, J., Dicke, M. & Posthumus, M.A. (1991) Variation in composition of predator-attracting allelochemicals emitted by herbivore-plants: relative influence of plant and herbivore. *Chemoecology*, **2**, 1–6.

Takabayashi, J., Dicke, M., Takahashi, S. *et al.* (1994a) Leaf age affects composition of herbivore-induced synomones and attraction of predatory mites. *J. Chem. Ecol.*, **20**, 373–86.

Takabayashi, J., Dicke, M. & Posthumus, M.A. (1994b) Volatile herbivore-induced terpenoids in plant–mite interactions: variation caused by biotic and abiotic factors. *J. Chem. Ecol.*, **20**, 1329–54.

Takafuji, A. (1977) The effect of the rate of successful dispersal of a phytoseiid mite, *Phytoseiulus persimilis* Athias-Henriot on the persistence in the interactive system between the prey and its predator. *Res. Popul. Ecol.*, **18**, 210–22.

Takafuji, A. & Chant, D.A. (1976) Comparative studies of two species of predacious phytoseiid mites (Acari: Phytoseiidae), with special reference to their responses to the density of their prey. *Res. Popul. Ecol.*, **17**, 255–310.

Takahashi, F. & Chant, D.A. (1992) Adaptive strategies in the genus *Phytoseiulus* Evans (Acari: Phytoseiidae). I. Developmental times. *Int. J. Acarol.*, **18**, 171–6.

Takahashi, F. & Chant, D.A. (1993) Phylogenetic relationships in the genus *Phytoseiulus* Evans (Acari: Phytoseiidae). I. Geographical distribution. *Int. J. Acarol.*, **19**, 15–22.

Takahashi, F. & Chant, D.A. (1994) Adaptive strategies in the genus *Phytoseiulus* Evans (Acari: Phytoseiidae). II. Survivorship and reproduction. *Int. J. Acarol.*, **20**, 87–97.

Takahashi, M. & Wada, Y. (1988) Insecticide susceptibility of *Cheyletus malaccensis* (Acari: Cheyletidae). *Jpn. J. Sanit. Zool.*, **39**, 369–70.

Takahashi, F., Inoue, M. & Takafuji, A. (1998) Management of the spider-mite population in a vinylhouse vinery by releasing *Phytoseiulus persimilis* Athias-Henriot onto the ground cover. *Jpn. J. Appl. Entomol. Zool.*, **42**, 71–6.

Takano-Lee, M. & Hoddle, M.S. (2001) Biological control of *Oligonychus perseae* (Acari: Tetranychidae) on avocado: IV. Evaluating the efficacy of a modified mistblower to mechanically dispense *Neoseiulus californicus* (Acari: Phytoseiidae). *Int. J. Acarol.*, **27**, 157–69.

Tandon, P.L. & Lal, B. (1976) New records of predatory mites on mango mealy bug, *Drosicha mangiferae* Green (Margarodidae: Hemiptera). *Curr. Sci.*, **45**, 566–7.

Tanigoshi, L.K. (1982) Advances in knowledge of the biology of the Phytoseiidae. In: *Recent Advances in Knowledge of the Phytoseiidae* (Ed. by M.A. Hoy), pp. 1–22. Div. Agric. Sci., University of California, Special Publ., No. 3284, Berkeley, CA.

Tanigoshi, L.K., Hoyt, S.C., Browne, R.W. & Logan, J.A. (1975) Influence of temperature on population increase of *Metaseiulus occidentalis* (Acarina: Phytoseiidae). *Ann. Entomol. Soc. Am.*, **68**, 979–86.

Tanigoshi, L.K., Fargerlund, J., Nishio-Wong, J.Y. & Griffiths, H.J. (1985) Biological control of citrus thrips, *Scirtothrips citri* (Thysanoptera: Thripidae), in southern California citrus groves. *Environ. Entomol.*, **14**, 733–41.

Tawfik, M.F.S. & Awadallah, K.T. (1970) The biology of *Pyemotes herfsi* Oudemans and its efficiency in the control of the resting larvae of the pink bollworm, *Pectinophora gossypiella* Saunders, in U.A.R. *Bull. Soc. Entomol. Egypte*, **54**, 49–71.

Taylor, A. & James, D.G. (1993) Effect of temperature on development and survival of *Typhlodromus doreenae* Schicha (Acari: Phytoseiidae). *Int. J. Acarol.*, **19**, 185–8.

Teerling, C.R., Gillespie, D.R. & Borden, J.H. (1993) Utilization of western flower thrips alarm pheromone as a prey-finding kairomone by predators. *Can. Entomol.*, **125**, 431–7.

Teich, Y. (1966) Mites of the family Phytoseiidae as predators of the tobacco whitefly, *Bemisia tabaci* Gennadius. *Israel J. Agric. Res.*, **16**, 141–2.

Ten Winkel, E.H., Davids, C. & De Nobel, J.G. (1989) Food and feeding strategies of water mites of the genus *Hygrobates* and the impact of their predation on the larval population of the chironomid *Cladotanytarsus mancus* (Walker) in Lake Maarsseveen. *Neth. J. Zool.*, **39**, 246–63.

Theaker, T.L. & Tonks, N.V. (1977) A method for rearing the predaceous mite *Phytoseiulus persimilis* (Acarina: Phytoseiidae). *J. Entomol. Soc. Br. Columbia*, **74**, 8–9.

Thistlewood, H.M.A. (1991) A survey of predatory mites in Ontario apple orchards with diverse pesticide programs. *Can. Entomol.* **123**, 1163–74.

Thistlewood, H.M.A. & Elfving, D.C. (1992) Laboratory and field effects of chemical fruit thinners on tetranychid and predatory mites (Acari) of apple. *J. Econ. Entomol.*, **85**, 477–85.

Thistlewood, H.M.A., Clements, D.R. & Harmsen, R. (1996) Stigmaeidae. In: *Eriophyoid Mites, Their Biology, Natural Enemies and Control* (Ed. by E.E. Lindquist, M.W. Sabelis & J. Bruin), pp. 457–70. Elsevier, Amsterdam.

Thomas, M.B. & Willis, A.J. (1998) Biocontrol – risky but necessary? *Trends Ecol. Evol. (TREE)*, **13**, 325–9.

Thomas, W.P. & Chapman, L.M. (1978) Integrated control of apple pests in New Zealand. 15. Introduction of two predacious phytoseiid mites. *Proc. 31st N.Z. Weed and Pest Control Conf.*, (Ed. by A.J. Popay), pp. 236–43. Swiftprint Centre, Palmerston North.

Thomas, W.P. & Walker, J.T.S. (1989) *Tetranychus urticae* Koch, two-spotted mite (Acari: Tetranychidae). In: *A Review of Biological Control of Invertebrate Pests and Weeds in New Zealand 1874 to 1987* (Ed. by P.J. Cameron, R.L. Hill, J. Bain & W.P. Thomas), pp. 245–50. CABI, Wallingford.

Thompson, J.N. (1998) Rapid evolution as an ecological process. *Trends. Ecol. Evol. (TREE)*, **13**, 329–32.

Thompson, W.R. & Simmonds, F.J. (1965) *A Catalogue of the Parasites and Predators of Insect Pests*, Section 4, *Host Predator Catalogue*. Commonwealth Agricultural Bureaux, Farnham.

Thorvilson, H.G., Phillips, S.A., Jr, Sorensen, A.A. & Trostle, M.R. (1987) The straw itch mite, *Pyemotes tritici* (Acari: Pyemotidae), as a biological control agent of red imported fire ants, *Solenopsis invicta* (Hymenoptera: Formicidae). *Fl. Entomol.*, **70**, 439–44.

Timmermans, K.R., Spijkerman, E., Tonkes, M. & Govers, H. (1992) Cadmium and zinc uptake by two species of aquatic invertebrate predators from dietary and aqueous sources. *Can. J. Fish. Aquat. Sci.*, **49**, 655–62.

Titayavan, M. & Davis, D.W. (1988) Studies of a uniparental form of *Aphytis vandenboschi* (Hymenoptera: Aphelinidae), a parasite of the San Jose scale in northern Utah. *Great Basin Natur.*, **48**, 388–93.

Tixier, S.-T., Kreiter, S., Auger, P. & Weber, M. (1998) Colonization of Languedoc vineyards by phytoseiid mites (Acari: Phytoseiidae): influence of wind and crop environment. *Exp. Appl. Acarol.*, **22**, 523–42.

du Toit, B.J., Theron, P.D. & Ueckermann, E.A. (1998) A new genus and four new species of the family Camerobiidae (Acari: Raphignathoidea) from South Africa. *Int. J. Acarol.*, **24**, 3–19.

Toko, M., Yaninek, J.S. & O'Neil, R.J. (1996) Response of *Mononychellus tanajoa* (Acari: Tetranychidae) to cropping systems, cultivars and pest interventions. *Environ. Entomol.*, **25**, 237–49.

Tomalski, M.D. & Miller, L.K. (1991) Insect paralysis by baculoviruse-mediated expression of a mite neurotoxin gene. *Nature (Lond.)*, **352**, 82–5.

Tomalski, M.D. & Miller, L.K. (1992) Expression of a paralytic neurotoxin gene to improve insect baculoviruses as biopesticides. *Bio/Technology*, **10**, 545–9.

Tomkins, A.R., Lupton, T., Brown, N. *et al.* (1997) Tydeid mite control on persimmons. *Proc. 50th N.Z. Plant Protection Conf.*, Lincoln (Ed. by M. O'Callaghan), pp. 414–19. Swiftprint Centre, Palmerston North.

Tomlin, C.D.S. (Ed.) (1997) *The Pesticide Manual*, 11th Edn. British Crop Protection Council, Farnham.

Tothill, J.D. (1918a) Natural control investigations in British Columbia. *Proc. Entomol. Soc. Br. Columbia*, **12**, 37–9.

Tothill, J.D. (1918b) Some notes on the natural control of the oyster-shell scale (*Lepidosaphes ulmi*, L.). *Bull. Entomol. Res.*, **9**, 183–96.

Toyama, G.M. & Ikeda, J.K. (1976) An evaluation of fly predators at animal farms on leeward and central Oahu. *Proc. Hawaiian Entomol. Soc.*, **22**, 369–79.

Toyoshima, S. & Amano, H. (1998) Effect of prey density on sex ratio of two predacious mites, *Phytoseiulus persimilis* and *Amblyseius womersleyi* (Acari: Phytoseiidae). *Exp. Appl. Acarol.*, **22**, 709–23.

Traboulsi, R. (1968) Prédateurs et parasites d'*Aphytis* (Hym., Aphelinidae). *Entomophaga*, **13**, 345–55.

Treat, A.E. (1975) *Mites of Moths and Butterflies*. Cornell University Press, London.

Trumble, J.T. (1998) IPM: overcoming conflicts in adoption. *Integr. Pest Manage. Rev.*, **3**, 195–207.

Tseng, I.-H. (1978) Mites of the family Bdellidae from Taiwan (Acarina: Prostigmata). New Taxa. *J. Agric. Assoc. China*, **104**, 25–51.

Tseng, Y.-H. (1977) A contribution to the knowledge of Formosan cheyletid mites (Acarina: Prostigmata). *Proc. Natn. Sci. Counc.*, **10**, 213–64.

Tseng, Y.-H. (1979) Studies on the mites infesting stored food products on Taiwan. In: *Recent Advances in Acarology* (Ed. by J.G. Rodriguez), Vol. 1, pp. 311–16. Academic Press, New York.

Tseng, Y.-H. (1982) Mites of the family Stigmaeidae of Taiwan, with key to genera of the world (Acarina: Prostigmata). *Phytol. Entomol. Nat. Taiwan Univ.*, **9**, 1–52.

Tseng, Y.-H. (1984) Mites associated with weeds, paddy rice and upland rice fields in Taiwan. In: *Acarology VI* (Ed. by D.A. Griffiths & C.E. Bowman), Vol. 2, pp. 770–80. Ellis Horwood, Chichester.

Tseng, Y.-H. (1995) A taxonomic study of free-living gamasine mite family Parasitidae Oudemans (Acari: Mesostigmata) from Taiwan. *J. Taiwan Mus.*, **48**, 11–81.

Tseng, Y.-H. & Lo, P.K.C. (1980) Tarsonemid mites (Acarina: Prostigmata) from Taiwan. *P. Protect. Bull. Taiwan*, **22**, 113–40.

Tseng, Y.-S. (1990) Taxonomical studies on mites associated with weeds, paddy rice and upland rice fields in Taiwan (I). *Chin. J. Entomol.*, Spec. Publ. No. 3, 1–35.

Tuovinen, T. (1994) Influence of surrounding trees and bushes on the phytoseiid mite fauna on apple orchard trees in Finland. *Agric. Ecosyst. Environ.*, **50**, 39–47.

Tuovinen, T. (2000) Integrated control of the strawberry mite (*Phytonemus pallidus*) in the Nordic multi-year growing system. *Acta Hort.*, **525**, 389–91.

Turk, E. & Turk, F. (1957) *Systematik und Ökologie der Tyroglyphiden Mitteleuropas*. Akademische Verlag, Leipzig.

Turk, F.A. & Phillips, S.-M. (1946) A monograph of the slug mite – *Riccardoella limacum* (Schrank). *Proc. Zool. Soc. Lond.*, **115**, 448–72.

Turnbull, A.L. & Chant, D.A. (1961) The practice and theory of biological control in Canada. *Can. J. Zool.*, **39**, 697–753.

Tyndale-Biscoe, M., Wallace, M.M.H. & Morton, R. (1981) Arthropod-induced mortality in immature stages of the bush fly, *Musca vetustissima* Walker (Diptera: Muscidae). *Bull. Entomol. Res.*, **71**, 681–90.

Uchida, T. & Miyazaki, I. (1935) Life-history of a water-mite parasitic on *Anopheles*. *Proc. Imp. Acad. Jpn*, **11**, 73–6.

Ueckermann, E.A. & Loots, G.C. (1988) The African species of the subgenera *Anthoseius* De Leon and *Amblyseius* Berlese (Acari: Phytoseiidae). *Entomol. Mem. Dept Agric. Wat. Supp. Repub. South Africa*, **73**, 1–168.

Ueckermann, E.A. & Smith Meyer, M.K.P. (1987) Afrotropical Stigmaeidae (Acari: Prostigmata). *Phytophylactica*, **19**, 371–97.

Unruh, T.R., White, W., Gonzalez, D. *et al.* (1983) Heterozygosity and effective population size in laboratory populations of *Aphidius ervi* (Hym.: Aphidiidae). *Entomophaga*, **28**, 245–58.

Usher, M.B. & Edwards, M. (1986) A biometrical study of the family Tydeidae (Acari, Prostigmata) in the maritime Antarctic, with descriptions of three new taxa. *J. Zool., Lond.*, **209**, 355–83.

Vacante, V. & Gerson, U. (1989) Three species of *Eryngiopus* (Acari: Stigmaeidae) from Italy, with key to species and summary of habitats. *Redia*, **70**, 385–401.

Vaivanijkul, P. & Haramoto, F.H. (1969) The biology of *Pyemotes boylei* Krczal (Acarina: Pyemotidae). *Proc. Hawaiian Entomol. Soc.*, **20**, 443–54.

Van den Bosch, R. & Hagen, K.S. (1966) Predacious and parasitic arthropods in California cotton fields. *Calif. Agric. Exp. Stn Bull.*, No. 820.

Veerman, A. (1992) Diapause in phytoseiid mites: a review. *Exp. Appl. Acarol.*, **14**, 1–60.

Villanueva, R. & Harmsen, R. (1996) Biological interactions of tarsonemid mites in apple orchards: predation of apple rust mite and use of *Phyllonorycter blancardella* mines. *Proc. Entomol. Soc. Ontario*, **127**, 99–106.

Villanueva, R.T. & Harmsen, R. (1998) Studies on the role of the stigmaeid predator *Zetzellia mali* in the acarine system of apple foliage. *Proc. Entomol. Soc. Ontario*, **129**, 149–55.

Vincenot, D. & Quilici, S. (1995) Lutte integree en vergers d'agrumes a l'Ile de la Reunion: experimentation et developpment. *IOBC/WPRS Bull.*, **18**(5), 140–59.

Volgin, V.I. (1969) *Acarina of the Family Cheyletidae of the World*. Akademia Nauk, Leningrad, (in Russian).

Waage, J.K. (1982) Sex ratio and population dynamics of natural enemies – some possible interactions. *Ann. Appl. Biol.*, **101**, 159–64.

Waage, J. (1990) Ecological theory and the selection of biological control agents. In: *Critical Issues in Biological Control* (Ed. by M. Mackauer, L.E. Ehler & J. Roland), pp. 135–57. Intercept, Andover.

Waage, J.K. (1996) 'Yes, but does it work in the field?' The challenge of technology transfer in biological control. *Entomophaga*, **41**, 315–32.

Waage, J.K. (1997) Biopesticides at the crossroads: IPM products or chemical clones? In: *Microbial Insecticides: Novelty or Necessity?* (Ed. by H.F. Evans), pp. 11–19. British Crop Protection Council, Farnham.

Waage, J.K. & Greathead, D.J. (1988) Biological control: challenges and opportunities. *Phil. Trans. R. Soc. Lond.*, **318**, 111–28.

Wada, T. (1998) Development of Cucumeris® and its future prospect. *Agro. Jpn*, **73**, 17–19.

Wade, C.F. & Rodriguez, J.G. (1961) Life history of *Macrocheles muscaedomesticae* (Acarina: Macrochelidae), a predator of the house fly. *Ann. Entomol. Soc. Am.*, **54**, 776–81.

Waite, G.K. (1988a) Integrated control of *Tetranychus urticae* in strawberries in south-east Queensland. *Exp. Appl. Acarol.*, **5**, 23–32.

Waite, G.K. (1988b) Control of *Tetranychus urticae* Koch by *Phytoseiulus persimilis* Athias Henriot in low chill stonefruit. *Queens. J. Agric. Anim. Sci.*, **45**, 185–8.

Waite, G. (2001) Managing spider mites in field-grown strawberries using *Phytoseiulus persimilis* and the 'pest-in-first' technique. In: *Acarology: Proc. 10th Int. Congr.* (Ed. by R.B. Halliday, D.E. Walter, H.C. Proctor *et al.*), pp. 381–6. CSIRO., Melbourne.

Wajnberg, E., Scott J.K. & Quimby P.C. (Eds) (2001) *Evaluating Indirect Ecological Effects of Biological Control.* CABI, Wallingford.

Walde, S.J. (1995) How quality of host plant affects a predator–prey interaction in biological control. *Ecology*, **76**, 1206–19.

Walde, S.J., Nyrop, J.P. & Hardman, J. M. (1992) Dynamics of *Panonychus ulmi* and *Typhlodromus pyri*: factors contributing to persistence. *Exp. Appl. Acarol.*, **14**, 261–91.

Walde, S.J., Magagula, C.N. & Norton, M.L. (1995) Feeding preferences of *Zetzellia mali*: does absolute or relative abundance of prey matter more? *Exp. Appl. Acarol.*, **19**, 307–17.

Walde, S.J., Hardman, J.M. & Magagula, C.N. (1997) Direct and indirect species interactions influencing within-season dynamics of apple rust mite, *Aculus schlechtendali* (Acari: Eriophyidae). *Exp. Appl. Acarol.*, **21**, 587–614.

Walia, K.K. & Mathur, S. (1994a) Acarine fauna of arable soils and their screening for nematophagy. *Indian J. Nematol.*, **24**, 69–79.

Walia, K.K. & Mathur, S. (1994b) Predatory potential of two nematophagous mites on fungivorous nematode, *Aphelenchoides composticola*, in vitro. *Indian J. Nematol.*, **24**, 243–5.

Walia, K.K. & Mathur, S. (1995) Predatory behaviour of two nematophagous mites, *Tyrophagus putrescentiae* and *Hypoaspis calcuttaensis*, on root-knot nematodes, *Meloidogyne javanica. Nematol. Medit.*, **23**, 255–61.

Walia, K.K. & Mathur, S. (1996) Potential of two nematophagous mites in suppressing root-knot nematode, *Meloidogyne javanica*, on ochra. *Indian J. Nematol.*, **26**, 172–5.

Wallace, M.M.H. (1954) The effect of DDT and BHC on the population of the lucerne flea, *Sminthurus viridis* (L.) (Collembola), and its control by predatory mites, *Biscirus* spp. (Bdellidae). *Aust. J. Agric. Res.*, **5**, 148–55.

Wallace, M.M.H. (1967) The ecology of *Sminthurus viridis* (L.) (Collembola) I. Processes influencing numbers in pastures in Western Australia. *Aust. J. Zool.*, **15**, 1173–206.

Wallace, M.M.H. (1971) The influence of temperature and moisture on diapause development in the eggs of *Bdellodes lapidaria* (Acari: Bdellidae). *J. Aust. Entomol. Soc.*, **10**, 276–80.

Wallace, M.M.H. (1972) A portable power-operated apparatus for collecting epigaeic Collembola and Acari. *J. Aust. Entomol. Soc.*, **11**, 261–3.

Wallace, M.M.H. (1973) The taxonomy and distribution of *Sminthurus viridis* and related species (Collembola: Sminthuridae) in Western Europe and Morocco. *Rev. Ecol. Biol. Sol.*, **10**, 211–24.

Wallace, M.M.H. (1974) An attempt to extend the biological control of *Sminthurus viridis* (Collembola) to new areas in Australia by introducing a predatory mite, *Neomolgus capillatus* (Bdellidae). *Aust. J. Zool.*, **22**, 519–29.

Wallace, M.M.H. (1981) Tackling the lucerne flea and red-legged earth mite. *J. Agric. West. Aust.*, **22**, 72–4.

Wallace, M.M.H. (1986) Some macrochelid mites (Acari: Macrochelidae) associated with Australian dung beetles (Coleoptera: Scarabaeidae). *Acarologia*, **27**, 3–15.

Wallace, M.M.H. & Holm, E. (1983) Establishment and dispersal of the introduced predatory mite, *Macrocheles peregrinus* Krantz, in Australia. *J. Aust. Entomol. Soc.*, **22**, 345–8.

Wallace, M.M.H. & Holm, E. (1984) Competition between species of predatory macrochelid mites in the cow-dung pads. In: *Acarology VI* (Ed. by D.A. Griffiths & C.E. Bowman), Vol. 1, pp. 581–4. Ellis Horwood, Chichester.

Wallace, M.M.H. & Mahon, J. A. (1971) The ecology of *Sminthurus viridis* (L.) (Collembola) III. The influence of climate and land use on its distribution and that of an important predator, *Bdellodes lapidaria* (Acari: Bdellidae). *Aust. J. Zool.*, **19**, 177–88.

Wallace, M.M.H. & Mahon, J.A. (1972) The taxonomy and biology of Australian Bdellidae (Acari). I. Subfamilies Bdellinae, Spinibdellinae and Cytinae. *Acarologia*, **14**, 544–80.

Wallace, M.M.H. & Mahon, J.A. (1976) The taxonomy and biology of Australian Bdellidae (Acari). II. Subfamily Odontoscirinae. *Acarologia*, **18**, 65–123.

Wallace, M.M.H. & Walters, M. C. (1974) The introduction of *Bdellodes lapidaria* (Acari: Bdellidae) from Australia into South Africa for the biological control of *Sminthurus viridis* (Collembola). *Aust. J. Zool.*, **22**, 505–17.

Wallace, M.M.H., Tyndale-Briscoe, M. & Holm, E. (1979) The influence of *Macrocheles glaber* on the breeding of the Australian bushfly, *Musca vetustissima* in cow dungs. In: *Recent Advances in Acarology* (Ed. by J.G. Rodriguez), Vol. 2, pp. 217–22. Academic Press, New York.

Wallwork, J.H. & Rodriguez, J.G. (1963) The effect of ammonia on the predation rate of *Macrocheles muscaedomesticae* (Acarina: Macrochelidae) on house fly eggs. In: *Advances Acarology* (Ed. by J.A. Naegele), Vol. 1, pp. 60–9. Comstock, Ithaca, NY.

Walter, D.E. (1987) Life history, trophic behavior and description of *Gamasellodes vermivorax* n. sp. (Mesostigmata: Ascidae), a predator of nematodes and arthropods in semiarid grassland soils. *Can. J. Zool.*, **65**, 1689–95.

Walter, D.E. (1988a) Nematophagy by soil arthropods from the shortgrass steppe, Chihuahuan desert and Rocky Mountains of the central United States. *Agric. Ecosyst. Environ.*, **24**, 307–16.

Walter, D.E. (1988b) Predation and mycophagy by endeostigmatid mites (Acari: Prostigmata). *Exp. Appl. Acarol.*, **4**, 159–66.

Walter, D.E. (1992) Leaf surface structure and the distribution of *Phytoseius* mites (Acarina: Phytoseiidae) in south-eastern Australian forests. *Aust. J. Zool.*, **40**, 593–603.

Walter, D.E. (1996) Living on leaves: mites, tomentia and leaf domatia. *Annu. Rev. Entomol.* **41**, 101–14.

Walter, D.E. (1999) Review of Australian *Asperoseius* Chant, *Euseius* Wainstein, *Okiseius* Ehara and *Phytoscutus* Muma (Acari: Mesostigmata: Phytoseiidae) with a key to the genera of Australian Amblyseiinae and descriptions of two new species. *Aust. J. Entomol.*, **38**, 85–95.

Walter, D.E. & Beard, J.J. (1997) A review of the Australian Phytoseiinae (Acari: Mesostigmata: Phytoseiidae). *Invert. Taxon.*, **11**, 823–860.

Walter, D.E. & Ikonen, E.K. (1989) Species, guilds and functional groups: taxonomy and behavior in nematophagous arthropods. *J. Nematol.*, **21**, 315–27.

Walter, D.E. & Kaplan, D.T. (1990) A guild of thelytokous mites (Acari: Mesostigmata) associated with citrus roots in Florida. *Environ. Entomol.*, **19**, 1338–43.

Walter, D.E. & Kaplan, D.T. (1991) Observations on *Coleoscirus simplex* (Acarina: Prostigmata), a predatory mite that colonizes greenhouse cultures of rootknot nematode (*Meloidogyne* spp.), and a review of feeding behavior in the Cunaxidae. *Exp. Appl. Acarol.*, **12**, 47–59.

Walter, D.E. & Lindquist, E.E. (1995) The distributions of parthenogenetic ascid mites (Acari: Parasitiformes) do not support the biotic uncertainty hypothesis. *Exp. Appl. Acarol.* **19**, 423–42.

Walter, D.E. & Lindquist, E.E. (1997) Australian species of *Lasioseius* (Acari: Mesostigmata: Ascidae): the *porulosus* group and other species from rainforest canopies. *Invert. Taxon.*, **11**, 525–47.

Walter, D.E. & O'Dowd, D.J. (1992) Leaf morphology and predators: effect of leaf domatia on the abundance of predatory mites (Acari: Phytoseiidae). *Environ. Entomol.*, **21**, 478–84.

Walter, D.E. & O'Dowd, D.J. (1995) Beneath biodiversity: factors influencing the diversity and abundance of canopy mites. *Selbyana*, **16**, 12–20.

Walter, D.E. & Proctor, H.C. (1999) *Mites Ecology, Evolution and Behaviour*. CABI, Sydney.

Walter, D.E., Hudgens, R.A. & Freckman, D.W. (1986) Consumption of nematodes by fungivorous mites *Tyrophagus* spp. (Acarina: Astigmata: Acaridae). *Oecologia*, **70**, 357–61.

Walter, D.E., Hunt, H.W. & Elliott, E.T. (1988) Guilds or functional groups? An analysis of predatory arthropods from a shortgrass steppe soil. *Pedobiologia*, **31**, 247–60.

Walter, D.E., Kaplan, D.T. & Davis, E.L. (1993) Colonization of greenhouse nematode cultures by nematophagous mites and fungi. *J. Nematol.*, **25**, 789–94.

Walter, D.E., Azam, G.N., Waite, G. & Hargreaves, J. (1998) Risk assessment of an exotic biocontrol agent: *Phytoseiulus persimilis* (Acari: Phytoseiidae) does not establish in rainforest in south-east Queensland. *Aust. J. Ecol.*, **23**, 587–92.

Wapshere, A.J., Delfosse, E.S. & Cullen, J.M. (1989) Recent developments in biological control of weeds. *Crop. Protect.*, **8**, 227–50.

Wassenaar, D.P.J. (1988) Effectiveness of vacuum cleaning and wet cleaning in reducing housedust mites, fungi and mite allergen in a cotton carpet: a case study. *Exp. Appl. Acarol.*, **4**, 53–62.

Waterhouse, D.F. (1994) *Biological Control of Weeds: Southeast Asian Prospects*. Australian Centre for International Agricultural Research, Canberra.

Waterhouse, D.F. & Norris, K.R. (1987) *Biological Control Pacific Prospects*. Inkata Press, Melbourne.

Way, M.J. (1986) The role of biological control in integrated plant protection. In: *Biological Plant and Health Protection* (Ed. by J.M. Franz), pp. 289–303. Fischer, Stuttgart.

Wearing, C.H. (1988) Evaluating the IPM implementation process. *Annu. Rev. Entomol.*, **33**, 17–38.

Webster, F.M. (1910) A predaceous and supposedly beneficial mite, *Pediculoides*, becomes noxious to man. *Ann. Entomol. Soc. Am.*, **3**, 15–39.

Wei, Q. & Walde, S.J. (1997) The functional response of *Typhlodromus pyri* to its prey, *Panonychus ulmi*: the effect of pollen. *Exp. Appl. Acarol.*, **21**, 677–84.

Weigmann, G. & Miko, L. (1998) Taxonomy of European Scheloribatidae, 3. Remarks on *Scheloribates* Berlese 1908 with description of two new species of the subgenus *Topobates* Grandjean 1958 (n. stat.) (Arachnida: Acari: Oribatida). *Sencken. Biol.*, **7**, 247–55.

Weiser, J. (1963) Über Massenzuchten von *Pyemotes*-Milben. *Beit. Z. Entomol.*, **13**, 547–51.

Weiser, J. & Sláma, K. (1964) Effects of the toxin of *Pyemotes* (Acarina: Pyemotidae) on the insect prey, with special reference to respiration. *Ann. Entomol. Soc. Am.*, **57**, 479–82.

Welbourn, W.C. (1983) Potential use of trombidioid and erythraeoid mites as biological control agents of insect pests. In: *Biological Control of Pests by Mites* (Ed. by M.A. Hoy, G.L. Cunningham & L. Knutson), pp. 103–40. University of California, Special Publ., No. 3304, Berkeley, CA.

Welbourn, W.C. & Young, O.P. (1987) New genus and species of Erythraeinae (Acari: Erythraeidae) from Mississippi with a key to the genera of North American Erythraeidae. *Ann. Entomol. Soc. Am.*, **80**, 230–42.

Welty, C. (1995) Survey of predators associated with European red mite (*Panonychus ulmi*; Acari: Tetranychidae) in Ohio apple orchards. *Great Lake Entomol.*, **28**, 171–84.

Wergin, W.P., Ochoa, R., Erbe, E.F. *et al.* (2000) Use of low-temperature field emission scanning electron microscopy to examine mites. *Scanning*, **22**, 145–55.

Werren, J.H. & Beukeboom, L.W. (1998) Sex determination, sex ratios and genetic conflict. *Annu. Rev. Ecol. Syst.*, **29**, 233–61.

Westcott, A.E. 1993. Biological control quarantine submissions involving mites in Australia. Unpublished talk, Acarology Symposium, BCRI, Rydalmere, NSW.

Whalon, M.E. & Norris, D.L. (1999) Field management: delivery of new technologies to growers. In: *Biopesticides, Use and Delivery* (Ed. by F.R. Hall & J.J. Menn), pp. 595–608. Humana Press, Totowa, NJ.

Wharton, G.W. & Arlian, L.G. (1972) Predatory behaviour of the mite *Cheyletus aversor. Anim. Behav.*, **20**, 719–23.

White, E.G. & Huffaker, C.B. (1969) Regulatory processes and population cyclicity in laboratory populations of *Anagasta kuhniella* (Zeller) (Lepidoptera: Phyticidae). I. Competition for food and predation. *Res. Popul. Ecol.*, **11**, 57–83.

White, N.D.G. & Laing, J.E. (1977a) Field observations of *Zetzellia mali* (Ewing) (Acarina: Stigmaeidae) in southern Ontario apple orchards. *Proc. Entomol. Soc. Ontario*, **108**, 23–30.

White, N.D.G. & Laing, J.E. (1977b) Some aspects of the biology and a laboratory life table of the acarine predator *Zetzellia mali*. *Can. Entomol.*, **109**, 1275–81.

White, N.D.G. & Sinha, R.N. (1990) Effect of chlorpyrifos-methyl on oat ecosystems in farm granaries. *J. Econ. Entomol.*, **83**, 1128–34.

Wicht, M.C., Jr & Rodriguez, J.G. (1970) Integrated control of muscid flies in poultry houses using predator mites, selected pesticides and microbial agents. *J. Med. Entomol.*, **7**, 687–92.

Wieres, R.W. & Smith, G.L. (1978) Apple mite control, Hudson valley. *Insecticide Acaricide Tests*, **3**, 42–3.

Wiggins, G.J., Grant, J.F. & Welbourn, W.C. (2001) *Allothrombium mitchelli* (Acari: Trombidiidae) in the Great Smoky Mountains National Park: incidence, seasonality and predation on beech scale (Homoptera: Eriococcidae). *Ann. Entomol. Soc. Entomol.*, **94**, 896–901.

Williams, B. & Hallas, T.E. (1988) *Cheyletus eruditus* (Schrank) (Acari) feeding on eggs of the cat flea, *Ctenocephalides felis* (Bouché) (Siphonaptera). *Entomol. Mon. Mag.*, **124**, 144.

Williams, R.H., Whipps, J.M. & Cooke, R.C. (1998) Role of soil mesofauna in dispersal of *Coniothyrium minitans*: mechanisms of transmission. *Soil Biol. Biochem.*, **30**, 1937–45.

Williamson, M. & Fitter, A. (1996) The varying success of invaders. *Ecology*, **77**, 1661–6.

Willis, A.J., Ash, J.E. & Groves, R.H. (1995) The effects of herbivory by a mite, *Aculus hyperici*, and nutrient deficiency on growth in *Hypericum* species. *Aust. J. Bot.*, **43**, 305–16.

Willis, R.R. & Axtell, R.C. (1968) Mite predators of the house fly: a comparison of *Fuscuropoda vegetans* and *Macrocheles muscaedomesticae. J. Econ. Entomol.*, **61**, 1669–74.

Wills, L.E., Mullins, B.A. & Mandeville, J.D. (1990) Effects of pesticides on filth fly predators (Coleoptera: Histeridae, Staphylinidae; Acarina: Macrochelidae, Uropodidae) in caged layer poultry manure. *J. Econ. Entomol.*, **83**, 451–7.

Wilson, F. (1960) *A Review of the Biological Control of Insects and Weeds in Australia and Australian New Guinea*. Commonwealth Agricultural Bureaux, Farnham.

Wilson, L.T., Hoy, M.A., Zalom, F.G. & Smilanick, J.M. (1984) Sampling mites in almonds: I. Within-tree distribution and clumping pattern of mites with comments on predator–prey interactions. *Hilgardia*, **52**(7), 1–24.

Wise, G.U., Henneberry, M.K. & Axtell, R.C. (1988) A new species of manure- inhabiting mite in the genus *Poecilochirus* (Acari: Mesostigmata: Parasitidae) predacious on house fly eggs and larvae. *Ann. Entomol. Soc. Am.*, **81**, 209–24.

Wisniewski, H.M., Sigudarson, S., Rubenstein, R. *et al.* (1996) Mites as vectors of scrapie. *Lancet*, **347**, 1114.

Wittmann, E.J. & Leather, S.R. (1997) Compatibility of *Orius laevigatus* Fieber (Hemiptera: Anthocoridae) with *Neoseiulus (Amblyseius) cucumeris* Oudemans (Acari: Phytoseiidae) and *Iphiseius (Amblyseius) degenerans* Berlese (Acari: Phytoseiidae) in the biocontrol of *Frankliniella occidentalis* Pergande (Thysanoptera: Thripidae). *Exp. Appl. Acarol.*, **21**, 523–38.

Woets, J. (1973) Integrated control in vegetables under glass in the Netherlands. *IOBC/WPRS Bull.*, **4**, 26–31.

Wohltmann, A., Wendt, F.-E. & Waubke, M. (1996) The life cycle and parasitism of the European grasshopper mite *Eutrombidium trigonum* (Hermann 1804) (Prostigmata: Parasitengonae: Microthrombidiidae), a potential agent for biological control of grasshoppers (Saltatoria). *Exp. Appl. Acarol.*, **20**, 545–61.

Womersley, H. (1933) A possible biological control of the clover springtail or lucerne flea (*Sminthurus viridis* L.) in Western Australia. *J. Aust. Counc. Sci. Ind. Res.*, **6**, 83–91.

Wood, T.G. (1967) New Zealand mites of the family Stigmaeidae (Acari, Prostigmata). *Trans. R. Soc. N.Z.*, **9**, 93–139.

Woodburn, A.T. (1995) The market for agrochemicals present and future. *Proc. Brighton 66th Crop Protection Conf. – Weeds*, Vol. 1, pp. 121–8. British Crop Protection Council, Farnham.

Woodring, J.P. & Cook, E.F. (1962) The biology of *Ceratozetes cisalpinus* Berlese, *Scheloribates laevigatus* Koch, and *Oppia neerlandica* Oudemans (Oribatei), with a description of all stages. *Acarologia*, **4**, 101–37.

Woolhouse, M.E.J. & Harmsen, R. (1984) The mite complex on the foliage of a pesticide-free apple orchard: population dynamics and habitat associations. *Proc. Entomol. Soc. Ontario*, **115**, 1–11.

Woolley, T.A. (1988) Acarology, *Mites and Human Welfare*. Wiley, New York.

Worner, S.P. (1991) Use of models in applied entomology: the need for perspective. *Environ. Entomol.*, **20**, 768–73.

Wrensch, D.L. & Bruce, W.A. (1991) Sex ratio, fitness and capacity for population increase in *Pyemotes tritici* (L.-F. & M.) (Pyemotidae). In: *The Acari, Reproduction, Development and Life History Strategies* (Ed. by R. Schuster & P.W. Murphy), pp. 209–21. Chapman & Hall, London.

Wrensch, D. & Ebbert, M. (Eds) (1993) *Evolution and Diversity of Sex Ratio in Insects and Mites*. Chapman & Hall, London.

Wright, D.H. & Coleman, D.C. (1988) Soil faunal vs. fertilization effects on plant nutrition: results of a biocide experiment. *Biol. Fertil. Soils*, **7**, 46–52.

Wright, E.M. & Chambers, R.J. (1994) The biology of the predatory mite *Hypoaspis miles* (Acari: Laelapidae), a potential biological control agent of *Bradysia paupera* (Dipt.: Sciaridae). *Entomophaga*, **39**, 225–35.

Wu, W.-J. & Shen, S.-P. (1990) Preliminary observations on the control of *Hemiberlesia pityosphila* by *Anystis baccarum*. *Natur. Enem. Insects*, **12**, 100 (in Chinese).

Wu, W.-n. (1984) Notes on the genus *Amblyseius* Berlese with descriptions of two new species from citrus orchards in south China (Acarina: Phytoseiidae). In: *Acarology VI* (Ed. by D.A. Griffiths & C.E. Bowman), Vol. 1, pp. 222–7. Ellis Horwood, Chichester.

Wysoki, M. & Bolland, H.R. (1983) Chromosome studies of phytoseiid mites (Acari: Gamasida). *Int. J. Acarol.*, **9**, 91–4.

Wysoki, M. & Cohen, M. (1983) Mites of the family Phytoseiidae (Acarina, Mesostigmata) as predators of the Japanese bayberry whitefly, *Parabemisia myricae* Kuwana (Hom., Aleyrodidae). *Agronomie*, **3**, 823–5.

Xin, J. (1985) Current status and perspective in mites as biological control agents of insect pests. *Chin. J. Biol. Cont.*, **1**, 40–3 (in Chinese).

Xin, J., Lu, J. & Zhang, Z.-Q. (1998) *Predatory Mites. Their Biology and Roles in Biological Control*. Systematic and Applied Acarology Society, London (in Chinese with English abstract).

Yaninek, J.S. & Hanna, R. (1998) Factors affecting the dispersal and spread of exotic phytoseiids in Africa. Unpublished talk at the 10th International Congress of Acarology, Canberra, Australia.

Yaninek, J.S. & Herren, H.R. (1988) Introduction and spread of the cassava green mite, *Mononychellus tanajoa* (Bondar) (Acari: Tetranychidae), an exotic pest in Africa and the search for appropriate control methods: a review. *Bull. Entomol. Res.*, **78**, 1–13.

Yaninek, J.S., de Moraes, G.J. & Markham, R.H. (1989) *Handbook on the Cassava Green Mite Mononychellus tanajoa in Africa*. International Institute of Tropical Agriculture, Cotonou, Benin.

Yaninek, J.S., Mégevand, B., de Moraes, G.J. *et al.* (1991) Establishment of the Neotropical predator *Amblyseius idaeus* (Acari: Phytoseiidae) in Benin, West Africa. *Biocont. Sci. Tech.*, **1**, 323–30.

Yaninek, J.S., Onzo, A. & Ojo, J.B. (1993) Continent-wide releases of neotropical phytoseiids against the exotic cassava green mite in Africa. *Exp. Appl. Acarol.*, **17**, 145–60.

Yaninek, J.S., Mégevand, B., Ojo, B. *et al.* (1998) Establishment and spread of *Typhlodromalus manihoti* (Acari: Phytoseiidae), an introduced phytoseiid predator of *Mononychellus tanajoa* (Bondar) (Acari: Tetranychidae) in Africa. *Environ. Entomol.*, **27**, 1496–505.

Yao, D.S. & Chant, D.A. (1989) Population growth and predation interference between two species of predatory phytoseiid mites (Acarina: Phytoseiidae) in interactive systems. *Oecologia*, **80**, 443–55.

Yao, D.S. & Chant, D.A. (1990) Changes in body weight of two species of predatory mites (Acarina: Phytoseiidae) as a consequence of feeding in an interactive system. *Exp. Appl. Acarol.*, **8**, 195–200.

Yasui, Y. (1992) The existence of sex pheromone eliciting precopulatory mate guarding behavior in a male predatory mite, *Macrocheles muscaedomesticae* (Scopoli). *J. Ethol.*, **10**, 15–19.

Ydergaard, S., Enkegaard, A. & Brødsgaard, H.F. (1997) The predatory mite *Hypoaspis miles*: temperature dependent life table characteristics on a diet of sciarid larvae, *Bradysia paupera* and *B. tritici*. *Entomol. Exp. Appl.*, **85**, 177–87.

Yoshikawa, M. (1985) Skin lesions of papular urticaria induced experimentally by *Cheyletus malaccensis* and *Chelacaropsis* sp. (Acari: Cheyletidae). *J. Med. Entomol.* **22**, 115–17.

Young, O.P. & Welbourn, W.C. (1987) Biology of *Lasioerythraeus johnstoni* (Acari: Erythraeidae), ectoparasite and predaceous on the tarnished plant bug, *Lygus lineolaris* (Hemiptera: Miridae), and other arthropods. *Ann. Entomol. Soc. Am.*, **80**, 243–50.

Young, O.P. & Welbourn, W.C. (1988) Parasitism of *Trigonotylus doddi* (Heteroptera: Miridae) by *Lasioerythraeus johnstoni* (Acari: Erythraeidae), with notes on additional hosts and distribution. *J. Entomol. Sci.*, **23**, 269–73.

Yousef, A.A., Zaher, M.A. & Kandil, M.M. (1982) Effect of prey and temperature on the development and biology of *Cheyletus malaccensis* Oudemans (Acari: Cheyletidae). *Z. Angew. Entomol.*, **93**, 39–42.

Yue, B. & Childers, C.C. (1994) Effects of temperature on life table parameters of *Agistemus exsertus* Gonzalez (Acari: Stigmaeidae) and its attack rate on *Panonychus citri* eggs. *Int. J. Acarol.*, **20**, 109–13.

Yue, B. & Tsai, J.H. (1995) *Agistemus exsertus* Gonzalez (Acari: Stigmaeidae) as a predator of citrus red mite [(*Panonychus citri* McGregor)]. *J. N.Y. Entomol. Soc.*, **103**, 107–13.

Zacharda, M. & Hluchy, M. (1996) Biological control of two-spotted spider mite *Tetranychus urticae* on strawberries by the predatory phytoseiid mite *Typhlodromus pyri*. *OILB/WPRS Bull.*, **19**(4), 226–9.

Zaher, M.A., Soliman, Z.R. & El-Bishlawy, S.M. (1975) Feeding habits of the predaceous mite, *Cunaxa capreolus* (Acarina: Cunaxidae). *Entomophaga*, **20**, 209–12.

Zaher, M.A., Yousef, A.E-T.A. & Kandil, M.M. (1981) Effect of food on the biology of *Cheletogenes ornatus* (C. & F.) (Acari: Prostigmata: Cheyletidae). *Acarologia*, **22**, 361–6.

Zaher, M.A., Soliman, Z.R. & Rakha, M.A. (1984) Biological studies on the predatory mite *Saniosulus nudus* Summers (Raphignathoidea: Eupalopsellidae). In: *Acarology VI* (Ed. by D.A. Griffiths & C.E. Bowman), Vol. 1, pp. 597–600. Ellis Horwood, Chichester,

Zakhvatkin, A.A. (1959) *Fauna of the U.S.S.R., Arachnoidea*, Vol. VI, No. 1, *Tyroglyphoidea [Acari]*. American Institute of Biological Sciences, Washington, DC.

Zdárková, E. (1986) Mass rearing of the predator *Cheyletus eruditus* (Schrank) (Acarina: Cheyletidae) for biological control of acarid mites infesting stored products. *Crop Protect.*, **5**, 122–4.

Zdárková, E. (1991) Application of the bio-preparation 'Cheyletin' in empty stores. In: *Modern Acarology* (Ed. by F. Dusbábek & V. Bukva), Vol. 1, pp. 607–10. SPB, The Hague.

Zdárková, E. (1994) The effectiveness of organophosphate acaricides on stored product mites interacting in biological conrol. *Exp. Appl. Acarol.*, **18**, 747–51.

Zdárková, E. (1997) The susceptibility of different strains of *Cheyletus eruditus* (Schrank) (Acarina: Cheyletidae) to organophosphate acaricides. *Exp. Appl. Acarol.*, **21**, 259–64.

Zdárková, E. & Horák, P. (1990) Preventive biological control of stored food mites in empty stores using *Cheyletus eruditus* (Schrank). *Crop Protect.*, **9**, 378–82.

Zdárková, E. & Pulpán, J. (1973) Low temperature storage of the predatory mite *Cheyletus eruditus* (Schrank) for future use in biological control. *J. Stored Prod. Res.*, **9**, 217–20.

Zemek, R. (1993) Characteristics of development and reproduction in *Typhlodromus pyri* on *Tetranychus urticae* and *Cecidophyopsis ribis*. I. Overwintered females. *Exp. Appl. Acarol.*, **17**, 405–21.

Zemek, R. & Nachman, G. (1998) Interactions in a tritrophic acarine predator–prey meta-population system: effects of *Tetranychus urticae* on the dispersal rates of *Phytoseiulus per-similis* (Acarina: Tetranychidae, Phytsoeiidae). *Exp. Appl. Acarol.*, **22**, 259–78.

Zemek, R. & Prenerová, E. (1997) Powdery mildew (Ascomycotina: Erysiphales) – an alternative food for the predatory mite *Typhlodromus pyri* Scheuten (Acari: Phytsoeiidae). *Exp. Appl. Acarol.*, **21**, 405–14.

Zhang, H. & Li, J. (1996) Sources and dispersal of *Allothrombium ovatum* larvae (Acari: Trombidiidae) in cotton fields and effects of larval mites on *Aphis gossypii* (Homoptera: Aphididae). *Syst. Appl. Acarol.*, **1**, 65–71.

Zhang, H., Li, J. & Liang, Y. (1999) Studies of aphid host selection by parasitic larvae of *Allothrombium ovatum* (Acari: Trombidiidae). *Syst. Appl. Acarol.*, **4**, 91–5.

Zhang, Y. & Lin, J. (1991) Studies of *Gnorimus chaudhrii* (Acari: Phytoseiidae), an important predator of tarsonemid mites in rice fields. *Chin. J. Biol. Cont.*, **7**, 163–5 (in Chinese).

Zhang, Y., Zhang, Z.-Q., Ji, J. & Lin, J. (1999) Predation of *Amblyseius longispinosus* (Acari: Phytoseiidae) on *Schizotetranychus nanjingensis* (Acari: Tetranychidae), a spider mite injurious to bamboo in Fujian, China. *Syst. Appl. Acarol.*, **4**, 63–8.

Zhang, Z. (1988) Progresses and future prospects of Trombidioidea and Erythraeoidea mites as candidates for biological control of insect pests. *Chin. J. Biol. Cont.*, **4**, 79–82 (in Chinese).

Zhang, Z.-Q. (1991a) Phototactic and geotactic response in *Allothrombium pulvinum* larvae (Acari: Trombidiidae). *Exp. Appl. Acarol.*, **15**, 41–7.

Zhang, Z.-Q. (1991b) Parasitism of *Acyrthosiphon pisum* by *Allothrombium pulvinum* (Acariformes: Trombidiidae): host attachment site, host size selection, superparasitism and effect on host. *Exp. Appl. Acarol.*, **11**, 137–47.

Zhang, Z.-Q. (1992) Functional response of *Allothrombium pulvinum* deutonymphs (Acari: Trombidiidae) on two-spotted mites (Acari: Tetranychidae). *Exp. Appl. Acarol.*, **15**, 249–57.

Zhang, Z.-Q. (1995a) Review of the systematics and biology of the Otopheidomenidae (Acari: Mesostigmata) with a description of *Eickwortius* gen. n. from a termite (Isoptera: Termitidae). *Syst. Entomol.*, **20**, 239–46.

Zhang, Z.-Q. (1995b) Variance and covariance of ovipositional rates and developmental rates in the Phytoseiidae (Acari: Mesostigmata): a phylogenetic consideration. *Exp. Appl. Acarol.*, **19**, 139–46.

Zhang, Z.-Q. (1995c) A cladistic analysis of Trombidiidae (Acari: Parasitengona): congruence of larval and adult morphology. *Can. J. Zool.*, **77**, 96–103.

Zhang, Z.-Q. (1996) Parasitism of aphids (Homoptera: Aphidiidae) by larvae of *Allothrombium pulvinum* (Acari: Trombidiidae): host species selection, host size selection, and superparasitism. *Syst. Appl. Acarol.*, **1**, 55–63.

Zhang, Z.-Q. (1998) Biology and ecology of trombidiid mites (Acari: Trombidioidea). *Exp. Appl. Acarol.*, **22**, 139–55.

Zhang, Z.-Q. & Croft, B.A. (1994) A comparative life history study of immature *Amblyseius fallacis, Amblyseius andersoni, Typhlodromus occidentalis and Typhlodromus pyri* (Acari: Phytoseiidae) with a review of larval feeding patterns in the family. *Exp. Appl. Acarol.*, **18**, 631–57.

Zhang, Z.-Q. & Norbakhsh, H. (1995) A new genus and three new species of mites (Acari: Trombidiidae) described from larvae ectoparasitic on aphids from Iran. *Europ. J. Entomol.*, **92**, 705–18.

Zhang, Z.-Q. & Sanderson, J.P. (1992) Effects of host plant experience on foraging behavior of the predatory mite *Phytoseiulus persimilis* (Acari: Phytoseiidae). *Ann. Entomol. Soc. Amer.*, **85**, 775–83.

Zhang, Z.-Q. & Sanderson, J.P. (1993) Association of *Ereynetes* tritonymphs (Acari: Ereynetidae) with the fungus gnat, *Bradysia impatiens* (Diptera: Sciaridae). *Int. J. Acarol.*, **19**, 179–83.

Zhang, Z.-Q. & Xin, J.-L. (1989) Biology of *Allothrombium pulvinum* (Acariformes: Trombidiidae), a potential biological control agent of aphids in China. *Exp. Appl. Acarol.*, **6**, 101–9.

Zhimo, Z. & McMurtry, J.A. (1990) Development and reproduction of three *Euseius* (Acari: Phytoseiidae) species in the presence and absence of supplementary foods. *Exp. Appl. Acarol.*, **8**, 233–42.

van Zon, A. Q. & Wysoki, M. (1978) The effect of some fungicides on *Phytoseiulus persimilis* (Acarina: Phytoseiidae). *Entomophaga*, **23**, 371–8.

van Zon, A.Q., Overmeer, W.P.J. & Veerman, A. (1981) Carotenoids function in photoperiodic induction of diapause in a predacious mite. *Science*, **213**, 1131–3.

Author index

Boczek, J. 117, 279
Boldt, P.E. 122–3, 320
Bolland, H.R. 100, 174, 250
Bollen, G.J. 267
Boller, E.F. 199, 359
Bonde, J. 211
Booth, J.P. 88
Borden, E.E.R. 167
Bornemissza, G.E. 167
Borthakur, M. 239, 246, 259
Boshko, G.V. 34
Bostanian, N.J. 190, 374
Botha, J.H. 180
Bottrell, D.G. 353
Bounfour, M. 174, 216
Bourchier, R.S. 400
Bower, C.C. 186
Bowie, M.H. 25
Bowman, C.E. 171
Boyne, J.V. 174, 189
Boyt, A.D. 86, 392
Braun, A.R. 195
Bravenboer, L. 178
Breeuwer, J.A.J. 311
Brennan, P. 152
Brickhill, C.D. 24, 260
Bridges, J.R. 249, 281
Briese, D.T. 118, 121–2, 279
Broadley, R. 407
Broce, A.B. xiv
Brodeur, J. 260, 300
Brødsgaard, H.F. 207, 212, 294, 361
Bronstein, J.L. 342, 348
Brower, J.H. 284
Brown, A.W.A. 186
Browning, M.W. 129
Bruce, W.A. 24, 174, 228, 230–1, 284, 298, 302, 304, 329, 331
Bruin, J. 323
Brust, G.E. 70
Buchanan, G.A. 201
Buckingham, G. 122
Budge, S.P. 268
Bugg, R.L. 356
Burden, J.P. 231
Burnett, T. 93, 106, 364
Burns, C.W. 220
Burrell, R.W. 209
Butler, K.P. 240

Butler, M.I. 220
Buxton, J.H. 209, 365

Cadogan, B.L. 128–9, 371
Calderon-Arguedas, O. 225
Caldwell, P.M. 310
Calis, J.N.M. 260
Callaway, R.M. 277
Caltagirone, L.E. 187
Camporese, P. 198, 260, 335, 361
Cantwell, G.E. 223
Capua, S. 380
Carèsche, L.A. 118–20, 278, 282, 336
Carey, J.R. 240
Carlton, J.T. 275, 290
Carter, C.D. 346
Casagrande, R.A. 229
Casanueva, M.E. 151
Cassagne-Méjean, F. 84
Castagnoli, M. 124, 203, 205, 208, 210–11, 326, 328
Castineiras, A. 209
Catská, V. 72, 267–8
Cave, F.E. 184
Chambers, R.J. 152, 154, 317
Chant, D.A 1, 10, 145, 173–4, 178–9, 181–3, 213, 235, 245, 325, 340, 343, 361, 385
Chapman, L.M. 317
Chapman, M.H. 373
Chapman, R.B. 190, 351
Charles, J.G. 140–1, 144–5, 175, 317, 320, 400, 404
Charudatan, R. 134
Cheah, C.A.S.-J. 102
Chen, H. 260
Chen, P.-R. 256, 374
Chen, W.-H. 307
Cheng, H.-Y. 78, 80
Chiang, H.C. 151, 157, 274, 394
Childers, C.C. 24, 92, 128, 201, 235–7, 139, 245, 371, 374, 397
Chow, Y.S. 130
Cicolani, B. 162, 168
Ciulla, A.M. 373
Clancy, D.W. 367

Clarke, A.R. 295
Clarke, G.M. 311–13
Clements, D.R. 239–40, 244, 362, 396
Cloutier, C. 209, 300
Cohen, A.C. 139, 301
Cohen, E. 176, 351, 378, 383, 388
Cohen, M. 180, 217
Coleman, D.C. 371
Coli, W.M. 338, 357
Colkesen, T. 192
Coll, M. 341
Collyer, E. 196, 198, 239, 374, 389, 394
Colwell, R.K. 89
Cone, W.W. 181, 361
Congdon, B.D. 22, 173, 214–16, 350
Conijn, C.G.M. 156, 206–7
Cook, D.R. 33, 60–2, 64, 84
Cook, E.F. 101, 232,
Cook, R.J. 266
Cook, W.J. 85, 149, 159
Coons, L.B. 11–12
Copping, L.G. 199, 406–7, 411
Cordo, H.A. 134
Corpuz-Raros, L.A. 103, 112, 173
Cortesero, A.M. 332
Costa, M. xiv, 157, 171
Coudriet, D.L. 217
Coulombe, L.J. 190
Coulson, J.R. 287, 405
Cox, P.D. 285
Craemer, C. 123
Cranham, J.E. 198
Cranshaw, W. 409–10
Croft, B.A. 9, 126, 174–6, 184–6, 188–90, 193–4, 197, 202–3, 218, 241–4, 246, 314, 320, 351, 356, 361–2, 365–7, 373–4, 376–8, 380–1, 389, 395–6
Croft, M.B. 188
Cromroy, H.L. 117, 275, 279–82
Cross, E.A. 51, 75, 227–8
Cross, J.V. 180, 182

Subject index

Tyrophagus putrescentiae 1, 43–4, 69–73,
    108, 152, 157, 267–9, 271, 300, 310, 327
Tyrophagus similes 71
Tyrophagus zachvatkini 270
Tyrrellia circularis 160

UK 208–9, 229, 285, 365
USA 1, 4, 117–18, 120, 122–5, 143, 150, 157,
    187, 191, 222–3, 225, 230, 241, 245, 250,
    252, 257, 276, 284, 315–16, 325, 334,
    343, 357, 364, 410
USSR 124, 186
Uganda 231
Uncinula necator 261
understory vegetation (ground cover)
    189–90, 334–5, 338, 356–8
Unionicolidae 413
Uroobovella marginata (=*Fuscuropoda
    vegetans*) 37, 263–5, 298, 302, 375
Uropodidae 24, 34, 37, 166, 172, 263–5,
    375, 397, 422
Uropodoidea 263
Utah 357

'vaccination' 252–4, 266, 268, 398
Varroa destructor 8, 151, 372
Varroidae 151
Veigaiidae 270
vertebrates 34, 69, 103, 114–15, 133, 136,
    421, 426
Verticillium 266
    Verticillium dahliae 253, 268
Viburnum tinus 347
Vicia faba 157, 197
Victoria (Australia) 95, 316
video 25, 392
viruses 216, 231, 266, 282, 310, 406, 414, 427
vitamin A 418
Vitis riparia 261
voucher specimens 294, 305, 320

walnut 334
Walzia australica 83
Washington (state) 185, 187
Waterhyacinth (*Eichhornia crassipes*) 133–5,
    252, 279, 401

water melons 180, 334
webbing 80, 129, 175, 177–8, 184–5, 191, 193,
    205–6, 214, 217, 228, 244, 250–1, 256,
    288, 346, 354–6, 361, 391, 396
weeds 2–4, 6, 117–25, 133–4, 180, 189, 193,
    248–52, 254, 275–82, 289, 292–3, 295,
    297, 316–17, 319–21, 336–7, 339, 356–8,
    371, 385, 387–8, 391, 395, 398–401,
    405–6, 414
weevils (Curculionidae) 134, 168
Western Australia (WA) 81, 94–9, 291, 315,
    385, 402
western flower thrips (*Frankliniella
    occidentalis*) (WFT) 154, 206–11, 213,
    217, 289, 291, 363–4, 395, 414
wheat 104–8, 228, 230, 256, 338
whiteflies (Aleyrodidae) 9, 179, 194, 214,
    216–17, 307, 336, 342–3, 408
Willamette mite (*Eotetranychus willamettei*)
    252–4
willow (*Salix*) 137, 146, 245, 337, 359
windbreak trees 5, 215, 338, 359
Wolbachia 184, 311, 329, 331, 427
woolly pine needle aphid (*Schizolachnus
    pini-radiatae*) 80

yeast 90, 297, 301–2, 426
yellow pecan aphid (*Monelliopsis
    pecanis*) 80
Yugoslavia 117

Zagloba ornata 140
Zambia 134
Zeiraphera diniana 128
Zerconidae 270
Zetzellia 234, 243
    Zetzellia graeciana 243
    Zetzellia javanica 243
    Zetzellia mali (=*Mediolata novae-scotiae*)
        188, 200, 234–5, 237, 239–45, 335,
        340, 343–4, 352, 362, 366, 374, 376,
        389, 396
Zhenjiang province (China) 260
zineb 369, 371
Zygogrammini 222

Printed and bound by CPI Group (UK) Ltd, Croydon, CR0 4YY

19/12/2022

03174720-0001